ABOUT ISLAND PRESS

Island Press is the only nonprofit organization in the United States whose principal purpose is the publication of books on environmental issues and natural resource management. We provide solutions-oriented information to professionals, public officials, business and community leaders, and concerned citizens who are shaping responses to environmental problems.

In 1994, Island Press celebrated its tenth anniversary as the leading provider of timely and practical books that take a multidisciplinary approach to critical environmental concerns. Our growing list of titles reflects our commitment to bringing the best of an expanding body of literature to the environmental community throughout North America and the world.

Support for Island Press is provided by The Geraldine R. Dodge Foundation, The Energy Foundation, The Ford Foundation, The George Gund Foundation, William and Flora Hewlett Foundation, The James Irvine Foundation, The John D. and Catherine T. MacArthur Foundation, The Andrew W. Mellon Foundation, The Joyce Mertz-Gilmore Foundation, The New-Land Foundation, The Pew Charitable Trusts, The Rockefeller Brothers Fund, The Tides Foundation, Turner Foundation, Inc., The Rockefeller Philanthropic Collaborative, Inc., and individual donors.

STRANGERS IN PARADISE

STRANGERS IN PARADISE

Impact and Management of
Nonindigenous Species in Florida

Edited by   Daniel Simberloff
〜   Don C. Schmitz
〜   Tom C. Brown

Foreword by   Edward O. Wilson

Island Press   Washington, D.C. 〜 Covelo, California

Library of Congress Cataloging-in-Publication Data

Strangers in paradise : impact and management of nonindigenous
    species in Florida / edited by Daniel Simberloff, Don C. Schmitz,
    and Tom C. Brown; foreword by Edward O. Wilson.
        p.   cm.
    Includes bibliographical references and index.
    ISBN 1-55963-429-4 (cloth). —ISBN 1-55963-430-8 (paper)
    1. Non-indigenous pests—Florida.   2. Non-indigenous pests—
Control—Florida.   3. Exotic animals—Florida.   4. Exotic plants
—Florida.   I. Simberloff, Daniel.   II. Schmitz, Don C.   III. Brown,
Tom C.
SB990.5.U6S88   1997
36.7′8—dc21   96-29829
                                                                    CIP

Printed on recycled, acid-free paper ∞

Manufactured in the United States of America
10 9 8 7 6 5 4 3 2 1

CONTENTS

Foreword ᕰ *Edward O. Wilson*   ix
Preface ᕰ *Daniel Simberloff, Don C. Schmitz, and Tom C. Brown*   xi

PART I   Introduction

1. The Biology of Invasions ᕰ *Daniel Simberloff*   3

PART II   How Nonindigenous Species Get to Florida—and What They Do When They Get There

2. Florida's Invasion by Nonindigenous Plants: History, Screening, and Regulation ᕰ *Doria R. Gordon and Kevin P. Thomas*   21
3. The Ecological Impact of Nonindigenous Plants ᕰ *Don C. Schmitz, Daniel Simberloff, Ronald H. Hofstetter, William Haller, and David Sutton*   39
4. The Impact of Natural Disturbances ᕰ *Carol C. Horvitz*   63
5. Immigration and Introduction of Insects ᕰ *J. Howard Frank, Earl D. McCoy, H. Glenn Hall, George F. O'Meara, and Walter R. Tschinkel*   75
6. Nonindigenous Freshwater Invertebrates ᕰ *Gary L. Warren*   101
7. Nonindigenous Fishes ᕰ *Walter R. Courtenay, Jr.*   109
8. Nonindigenous Amphibians and Reptiles ᕰ *Brian P. Butterfield, Walter E. Meshaka, Jr., and Craig Guyer*   123
9. Nonindigenous Birds ᕰ *Frances C. James*   139
10. Nonindigenous Mammals ᕰ *James N. Layne*   157
11. Nonindigenous Marine Invertebrates and Algae ᕰ *James T. Carlton and Mary H. Ruckelshaus*   187

PART III   Managing Nonindigenous Species: Strategies and Tactics

12. Ecological Restoration ᕰ *John M. Randall, Roy R. Lewis III, and Deborah B. Jensen*   205
13. Eradication ᕰ *Daniel Simberloff*   221
14. Maintenance Control ᕰ *Jeffrey D. Schardt*   229

15. Biological Control ⮵ *Ted D. Center, J. Howard Frank, and F. Allen Dray, Jr.*   245

PART IV Managing Nonindigenous Species: Policy and Implementation

16. Management in National Wildlife Refuges ⮵ *Mark D. Maffei*   267

17. Management in Everglades National Park ⮵ *Robert F. Doren and David T. Jones*   275

18. Management on State Lands ⮵ *Mark W. Glisson*   287

19. Management by Florida's Game and Fresh Water Fish Commission ⮵ *James A. Cox, Lt. Thomas G. Quinn, and H. Hugh Boyter, Jr.*   297

20. Management in Water Management Districts ⮵ *Amy Ferriter, Dan Thayer, Brian Nelson, Tony Richards, and David Girardin*   317

PART V The Regulatory Framework

21. The Federal Government's Role ⮵ *Don C. Schmitz and Randy G. Westbrooks*   329

22. The State's Role ⮵ *Tom C. Brown*   339

PART VI Conclusion

23. Why We Should Care ⮵ *Daniel Simberloff, Don C. Schmitz, and Tom C. Brown*   359

References   369
Contributors   433
Index   437

FOREWORD

 Florida has a near magical resonance for me. In the 1940s, growing up as a young naturalist in Pensacola and mostly nearby towns in Alabama, I envisioned the state as a paradise of subtropical plants and animals awaiting discovery. Nowadays I spend several weeks each year on Sanibel Island, in a writer's retreat, surf casting for fish to throw back in and enjoying unfocused sightseeing. Of course I also explore the natural habitats there, the shell-stacked beaches and their foredunes, the mangrove swamps of the Ding Darling Sanctuary, and the scraggly hardwood groves of the shell-mound hammocks. Native species prevail in the more sequestered spots, but mostly I find myself surrounded by exotic species, the subject of the present book.

Parts of South Florida look good to the uninitiated, I grant, but in the naturalist's eye it is substantially a Potemkin facade of foreign species. On Sanibel Island stately Australian pines tower along the roadways and around the golf course—they make good storm breaks I am told—but the thick acidic beds of needles beneath them are almost lifeless. Only an eradication program keeps the Brazilian peppers under control. Dense populations of Cuban brown anoles perch on the trunks of the introduced fig trees of the mall quadrangles and out into virtually all the other habitats, including the hammocks. I have never seen a native green anole. No room, evidently, has been left for them. There are a few native ants, the objects of my own specialized research, but most are newcomers from the West Indies, Africa, and Asia.

It is illuminating to place this new book, *Strangers in Paradise: Impact and Management of Nonindigenous Species in Florida,* alongside the recently published first four volumes of *Rare and Endangered Biota of Florida* and ask: What is being lost? The answer is easy. A precious and irreplaceable part of Florida's, and the nation's, heritage is disappearing. Plants, animals, and entire ecosystems that took tens of thousands to millions of years to evolve are at risk. What is being gained in their place? A hodgepodge of species found in other parts of the world, in some cases all around the world in tropical and subtropical environments. Species often dominate and destabilize the environment. Florida is being homogenized, and everyone, for all time to come, will be the poorer for it.

Conservation biologists like the authors of *Strangers in Paradise: Impact and Management of Nonindigenous Species in Florida* are devoted to making the most of both worlds. They recognize that Florida has changed forever, that 14 million people now seek a better life on limited space, and more are coming. In the midst of the transformation, the scientists aim at halting the flood of unwanted exotic organisms while saving nuclear populations of the natives. On a global basis they recognize that the two great destroyers of biodiversity are, first, habitat destruction and, second, invasion by exotic species. Extinction by habitat destruction is like death in an automobile accident: easy to see and assess. Extinction by the invasion of exotic species is like death by disease: gradual, insidious, requiring scientific methods to diagnose.

Florida rivals Hawaii, America's undisputed capital of extinction, in the magnitude of the threat from exotic species—and for the same reasons: geographic insularity, widespread habitat disturbance, and bombardment from all sides by nonnative plants and animals. For scientists engaged in basic ecological research, the state offers a vast array of experiments in invasion and adaptation, and thus an opportunity to analyze the multiple factors of rapid change. For those directly concerned with conservation, it poses a challenge of critical humanistic importance at a period in history when time is running out. Both domains of study, each essential to the other, are amply represented in this important new book.

Edward O. Wilson

PREFACE

This book arose from our conviction that Florida is a state in serious trouble from nonindigenous species and that the lessons we have learned about the problems these species pose, and how to deal with them, are valuable far beyond Florida. Florida and Hawaii are the states most affected by nonindigenous species, and in Florida can be found versions of virtually all the problems described in the burgeoning literature on invasion biology. Further, some regions and habitats in Florida are much less affected by nonindigenous species than others. Why? In parts of southern Florida, the seeds of introduced plants pollinated by introduced insects are eaten and dispersed by introduced birds. Entire landscapes there are dominated by introduced plants. Such stories and scenes are uncommon in northern Florida. Moreover, some introduced species, though persistent, have seemed ecologically innocuous, while others are scourges. Some plant species have changed entire major ecosystems, while others are harmless curiosities. Nonindigenous species of whole animal groups, reptiles and amphibians, for example, seem ecologically benign. What causes these differences among taxa? What can we predict about potential invaders from observing the effects of past incursions?

Because of its size and the gravity of the problems, Florida has extensive experience with attempting to eradicate introduced species—or at least trying to manage them so that they do not devastate native ecosystems. This effort has been especially intensive for invasive aquatic plants, but there have been both successes and failures for other groups. Some of the approaches are likely to prove useful well beyond Florida's borders. Several agencies of the federal government collaborate with numerous state and regional agencies, local governments, and private organizations to combat nonindigenous species, so the track record is determined as much by the legal and governmental framework as by scientific knowledge and management technology. Again, the ways in which this framework has affected control efforts should be instructive to workers in other states and nations.

There are classic books on the biology of invasions by nonindigenous species, as well as research papers on specific invasions and, in other journals, technical reports on how to deal with them. In yet other journals one can find papers on the legal and governmental dimensions of the problem. But no comprehensive book has addressed a large, diverse region and the full gamut of nonindigenous species of all taxa, the problems they cause, and the methods and impediments to dealing with them. Our aim is to fill that role.

The project originated when the Florida legislature mandated in 1993 that the Florida Department of Environmental Protection recommend a program of research and control of nonindigenous species on public lands in Florida. When that narrowly focused report (*An Assessment of Invasive Non-indigenous Species in Florida's Public Lands,* 1994) attracted wide attention outside the state, it became clear to us that our findings and experiences could be of general interest if put in a more comprehensive framework. We asked authors of relevant chapters in the previous document to rewrite their contributions with the new focus in mind, and we recruited leading authorities for new chapters on topics untreated or insufficiently treated in the state document. We are deeply grateful for the response of the many authors who have contributed to this book; they have dealt with numerous suggestions and draconian deadlines. We also thank the many referees whose detailed criticisms of all the chapters were crucial to whatever success we have achieved. Finally, our copyeditor, Anne Thistle, assisted us and all the authors enormously in producing clear and concise chapters; we cannot thank her enough.

Daniel Simberloff
Don C. Schmitz
Tom C. Brown

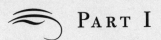

PART I

Introduction

1 The Biology of Invasions

Daniel Simberloff

 Highlighted by a project of the Scientific Committee on Problems of the Environment of the International Council of Scientific Unions (Mooney and Drake 1986; Drake et al. 1989), invasion biology has burgeoned in the last decade into a main focus of conservation biology and ecology. Realization is growing that nonindigenous species are not only a monumental economic threat but an insidious and pervasive conservation problem. To provide a framework for the chapters that follow, this introduction reviews the developing field of invasion biology and points out those features of Florida's geography, topography, and history that render it particularly prone to invasion.

Until the recent burst of interest, conservationists were often complacent about nonindigenous species, assuming that disturbed habitats and communities are those most likely to be affected by these invasions whereas pristine habitats are relatively immune (Simberloff 1995a). This view was strongly propounded by Charles Elton in *The Ecology of Invasions by Plants and Animals* (1958), the early book that introduced generations of biologists to invasion problems. Elton contended that disturbed habitats—by virtue of having fewer or less vigorous species—pose less "biotic resistance" to new arrivals. It is now evident that this view is far too simplistic. Disturbed habitats may, on average, be more invasible, but not simply by virtue of having fewer species or weaker species. Rather, what matters is which particular species are present and their particular interactions with specific invaders (Simberloff 1986). Even species-rich pristine habitats are threatened by nonindigenous species, and both conservation and economic problems generated by these invasions are increasing rapidly—hence the explosion of interest. It also appears that many habitats classified by Elton as "disturbed" could equally be termed "new" and "human-produced," and it is these features rather than the disturbance per se that often renders them vulnerable (Simberloff 1986).

Elton (1958) saw islands as the other sort of site highly susceptible to damage by invasions, because islands, with fewer species than similar-sized tracts of mainland, would pose less biotic resistance to invaders. Although the absence of whole groups of species—terrestrial mammals, for example—may predispose islands to particularly devastating effects by some nonindigenous species, it has now become apparent that, whatever makes islands on average more vulnerable to damage from nonindigenous species, it is not simply a function of lower biotic resistance generated by fewer species. Every sort of havoc wrought by nonindigenous species on islands can be found on the mainland as well (Simberloff 1995b). Rather, island invasions are governed by the particular species present and the biology of the potential invader.

Research on nonindigenous invaders has generally been piecemeal, consisting primarily of reports on specific effects of particular species in particular sites and of reviews of particular kinds of invasions—those by aquatic weeds, by agricultural pests, and so forth. In the United States, Florida and Hawaii suffer the most severe problems caused by nonindigenous species. About 25 percent of many taxa in Florida are nonnative (Ewel 1986; U.S. Congress 1993), and millions of hectares of land and water are dominated by nonindigenous species. This book tabulates these invaders and summarizes what we know about the myriad ways in which these species insinuate themselves into the environment and affect native ecosystems, communities, and species as well as human well-being. In addition to presenting a comprehensive overview of the harm caused by invasive nonindigenous species in a large and ecologically diverse region, various chapters show how such species can interact synergistically to exacerbate matters.

Two problems complicate any study of the impacts of nonindigenous species. First, surviving introduced species are much more likely to be observed than those that quickly disappear, and survivors producing highly visible ecological and economic impacts are much more likely to be detected and studied than are inconspicuous or innocuous ones (Simberloff 1981). Second, a distressing fraction of the invasion literature is anecdotal and sketchy. Thus "stories" abound concerning which nonindigenous species cause what effects. As Dasmann (1971:171) has noted, "the ecology of both exotics and native vegetation and the interrelationships between the two are far more talked about than studied." Though many stories may be correct, they are seldom buttressed by experimental or massive observational evidence. Forest bird species in Guam, for example, have declined dramatically because of the introduction of the large Australian brown tree snake (*Boiga irregularis*) in the late 1940s or early 1950s (Savidge 1984, 1987), but these declines have been confidently assigned to organochlorine pesticide use (Diamond 1984; Diamond and Case 1986) in the absence of any evidence other than the suggestive timing of pesticide use (Grue 1985).

Workers have used many terms for nonindigenous species: invaders, exotics, immigrants, colonists, introduced species, naturalized species. Frank and McCoy (1990) prefer to dichotomize nonindigenous species into introduced species, which are deliberately brought to a site by humans, and immigrants, which arrive on their own (even if inadvertently aided by humans), and to lump these two categories together as "adventive" species. In this volume we retain their dichotomy but use "nonindigenous" rather than "adventive" for all species that have recently arrived in Florida, following the report of the Congressional Office of Technology Assessment (U.S. Congress 1993). For many species, of course, one cannot be certain whether the arrival was deliberately aided by humans. Further, some species were deliberately introduced into sites from which, either on their own or with inadvertent human help, they reached Florida. The European house sparrow (*Passer domesticus*) was brought from Europe to New York deliberately (Chapter 9 in this volume), for example, and the cactus moth *Cactoblastis cactorum* from South America was deliberately released on Nevis in the Leeward Islands (Simmonds and Bennett 1966). From these beachheads, both species were able to reach Florida without further deliberate assistance. In other words, we are concerned here with species that arrived recently, with human help or without, whether directly from their areas of endemism or from some region closer to Florida in which they are also nonindigenous. Finally, just how recent an arrival must be to be termed "nonindigenous" is arbitrary. The term "naturalized" is often used for nonindigenous species that have been ensconced for a very long time. We will not set a cutoff date; the overwhelming majority of nonindigenous species discussed in this book arrived within the last two centuries.

Establishment and Spread: Arrive Alive

Almost all introduced propagules die without establishing ongoing populations, but it is difficult to estimate the fraction that survive and reproduce, because the arrival and disappearance of a few individuals are unlikely to be recorded. Of 154 introduced insect species *recorded* in Florida, approximately 42 have survived (Frank and McCoy 1993), but virtually all of these were deliberate introductions for biological control, with mean founding propagules of over 4000 individuals. How many insect species were deliberately introduced but not recorded and simply disappeared? How many insect species immigrated on their own and disappeared? Surely biological-control introductions would tend to have far greater propagule sizes than those of other nonindigenous insects and to have been screened for traits that would predispose them to survival. The great majority of nonindigenous insects arriving in Florida have probably disappeared, but the exact percentage will never be known.

Several forces conspire against a propagule. First, it must arrive in a suitable habitat or be able to move to one. Most propagules probably land in unsuitable habitats and die there—seeds of terrestrial plants land in water, parasites fail to find hosts, microbes land in environments of unsuitable pH. Even if adults survive at a site, they may be unable to reproduce or their offspring may be unable to live there. The geographic ranges of plant species, for example, may be limited by their inability to produce seed or by failure of seeds to germinate beyond the edges of the range. Biotic as well as physical factors might prevent a propagule from surviving or reproducing: a pathogen or predator could be crucial, for example. In one of the few controlled experimental introductions, the Puerto Rican lizard *Anolis pulchellus* failed to survive on the physically suitable island of Palominitos, almost certainly because of predation by the resident *Anolis cristatellus* (Levins and Heatwole 1973). Absence of another species—for example, a key insect pollinator for a plant—may also doom an introduction.

If the environment at a site is indeed suitable, the propagule's size is probably a major determinant of invasion success. Problems of breeding and increase at low population size—the "Allee effect"—can hinder establishment. Mates may be hard to find. All the forces that act differentially on small populations and increase their likelihood of extinction, collectively viewed as setting a "minimum viable population size" (Shaffer 1981), would threaten small propagules. Demographic stochasticity, inbreeding depression, and genetic depauperation have received most attention (Lande 1988). The expected pattern—that large propagule size contributes to probability of introduction success—is found to hold for several taxa but not all (Smallwood 1990). To some extent, initial survival and reproduction are stochastic processes, so very small initial propagules can lead to huge populations. (See, for example, Simberloff 1989.) Propagules of about 8 and 50 individuals, respectively, of the European house sparrow released in New York City failed to become established, for example, whereas a later propagule of 50 individuals released in the same place grew rapidly and ultimately colonized most of North America (Long 1981). The biological-control literature offers many examples in which propagules of identical size had radically different fates (Simberloff 1989).

Many environmentally significant introductions arose from very small propagules. Several widespread insects used in biological control have spread from releases of fewer than 20 individuals (Simberloff 1989). Five individuals of the North American muskrat (*Ondatra zibethica*) released in Czechoslovakia in 1905 produced a population that spread throughout much of Europe (Elton 1958). Similarly, one cage of European gypsy moth (*Lymantria dispar*) caterpillars that broke in a windstorm in Massachusetts established a population that defoliated large parts of the Northeast (Elton 1958).

Most species have evolved behavior that promotes dispersal, usually during certain parts of their life cycles. Even when movement is passive, some traits tend to facilitate it. For example, spiders spin silk and assume postures that lead to ballooning. Many animals and plants have also evolved behavior that increases the probability that dispersal will end at a site favorable for survival and reproduction—as when planktonic larvae of certain marine invertebrates settle in response to chemical and tactile cues associated with suitable habitat.

Even so, human actions are the key to the increasing numbers and distribution of nonindigenous species. First, transportation of both people and goods is faster today, and its volume has increased enormously. The increased speed increases the likelihood that an organism, transported deliberately or inadvertently, will survive the voyage, whereas the burgeoning volume of traffic will increase the number of propagules concomitantly. In the past, many propagules were transported in ballast water or soil, so it is no surprise that ports were the major sites of plant introductions after Europeans colonized other continents. Now that an airplane can land in, say, Orlando, the coast is not the only possible invasion theater.

Further, a growing taste for exotic pets and ornamental plants has combined with Florida's subtropical climate to exacerbate the influx enormously. On the loose in southern Florida in recent times have been piranhas, walking catfish, blue tilapia, electric eels, Cuban anoles, iguanas, Asian water monitors, caimans, boa constrictors, pythons, mambas, red-whiskered bulbuls, monk parakeets, howler monkeys, gibbons, green African savanna monkeys, crab-eating macaques, and a herd of 300 buffalo, not to mention numerous pets and perhaps 25,000 nonindigenous plant species that have not established populations in Florida (Frank and McCoy 1995a). There seems no economic or aesthetic limit to people's desire to be first on the block with a dramatic new pet. Recently Florida pet stores offered giant Madagascan hissing cockroaches (*Gromphadorhina* sp.), initially at $6.00 each, a price quickly pushed to $19.95 by news stories (Thomas 1995). Many people released these playthings in their yards when they tired of them.

Most dispersal mechanisms can be assigned to one of two categories: diffusion or jump dispersal (Pielou 1979). Diffusion consists not of Brownian movement but rather of gradual and regular spread, such that a species' growing range resembles increasing concentric circles with increasingly distorted edges.

The rate of expansion depends on the biology of the species, whereas the distortion is probably caused by the chance movement and survival of individuals more in some directions than in others, combined with spatial environmental heterogeneity. The initially circular spread of the Japanese beetle (Elton 1958) from its site of introduction in New Jersey, for example, was eventually distorted by the presence of the Atlantic Ocean and Chesapeake Bay. In this instance, the cause of the distortion is obvious, but other important habitat gradients may be far subtler.

Jump dispersal, by contrast, consists of long-distance movement to a point well outside the range of an established nonindigenous species. Often diffusion and jump dispersal are combined. Many nonindigenous species initially spread by diffusion, for example, but eventually long-distance jumps initiate several other foci, each of which becomes a base for both diffusion and other jumps. The gypsy moth spread rather gradually in the Northeast but later was found, discontinuously, hundreds of miles away in the Midwest and West. These jumps probably arose from eggs attached to motor vehicles. Other jump-dispersal incidents may be due to the rare transport of a propagule by a bird or an upper air current. Not only do jump dispersal and diffusion often occur simultaneously, but the categories are somewhat arbitrary. Each move in a gradual diffusion can be seen as a jump, for example, and the length that characterizes jump dispersal is arbitrary. Even so, a jump is usually defined as a propagule landing far outside the existing range of a species.

Because dispersal means are so varied and jump-dispersal events seem so idiosyncratic, one might be pessimistic about predicting the rate and direction of spread. Nevertheless, modeling the spread of nonindigenous species is an active focus of research. These models include post facto attempts to fit curves to observed range changes (as in Hengeveld 1989) and prior prediction based on life-history traits and assumptions about dispersal means (as in van den Bosch et al. 1992). Although the most extensively studied models rest on classical reaction-diffusion models (see the references in Hastings 1996a and 1996b), recent approaches include cellular automata—computer-intensive methods in which each spatial location is explicitly simulated and assigned to one of a small number of states (say, occupied or not) (Czaran and Bartha 1992). At each time step, each location changes state according to fixed rules. Although these models are deterministic, stochastic models are available (Hastings 1996a; 1996b). One of the most recent diffusion models (van den Bosch et al. 1992) and a stochastic model (Durrett 1988) both predict that the square root of the area occupied by an invading species will increase linearly with time, at least for a large fraction of the invasion process, and data for several animal species that disperse by various means conform to this prediction (van den Bosch et al. 1992).

Nevertheless, we still cannot predict the changing shape of an invading species' range very accurately at a fine scale—probably because such range expansions entail several dispersal means simultaneously (flight, transport by humans, storms), including stochastic events. Further, it is odd that, despite their increasing sophistication, models of range expansion have not been used extensively in the management of nonindigenous species. Suppose we were able to predict with some assurance exactly where a newly arrived species will go and when it will get there—how might such knowledge help us to deal with the invasion?

The spread of a nonindigenous species may be caused by habitat change. The Colorado potato beetle (*Leptinotarsa decemlineata*) of the western United States fed originally on wild plants of the genus *Solanum*. As the potato (*Solanum tuberosum*) was introduced and spread widely in the nineteenth century, the beetle spread accordingly—and into Florida (Chapter 5 in this volume).

Effects

The effects of nonindigenous species are often subtle. Indeed, though their impact on the biotic community can be astounding, to the casual observer of nature they may not seem to be a major threat. After all, if one plant species completely replaces another, the landscape may still be green, nutrients are cycled, energy still flows. Thus the ecologically disastrous invasion of *Melaleuca quinquenervia* (Chapter 3 in this volume) can be viewed optimistically in newspaper articles (see Zaneski 1995) because it forms forests, and a plethora of introduced animals may still represent nature to the average city dweller: millions of tourists in Hawaii who stay near Honolulu and see many colorful bird species do not even realize that all of them are nonindigenous and that those native species not already extinct are almost completely restricted to inaccessible upland forests. Similarly Rackleff (1972), a journalist writing about Florida's environmental crisis, mentioned aquatic weeds and fire ants in passing but saw the main threats as chemical pollution and direct habitat destruction.

As noted earlier, the literature on nonindigenous species tends to be anecdotal, and alternative hypotheses for the claimed effects of such species are often not considered. Yet many well-documented cases demonstrate an array of potential effects on native species. These effects can be classified as direct and indirect.

Direct Effects

A nonindigenous parasite or pathogen can cause a disease in another species. For example, the mosquito *Culex quinquefasciatus* was inadvertently introduced into the Hawaiian Islands in 1826. This insect is a vector of the avian malaria parasite, *Plasmodium relictum capistranoae,* also nonindigenous in the Hawaiian Islands. Many of the nonindigenous birds of the Hawaiian Islands (largely from Eurasia) are at least somewhat resistant to the plasmodium, whereas the native birds are highly susceptible. Thus the presence of the non-indigenous birds in the low elevations serves as a reservoir for the parasite, which helps to exclude the native species from lower-elevation sites (Warner 1968; van Riper et al. 1986).

A nonindigenous species can eat native animals or plants. Bird species have been eliminated all over the world by nonindigenous rats, mustelids, cats, dogs, and pigs (Atkinson 1989). In the best-known case, the lighthouse keeper's cat arrived on Stephen Island (New Zealand) and discovered the endemic wren *Xenicus lyalli,* which it extinguished within a year (Greenway 1967). Caterpillars of the European gypsy moth have largely defoliated many native species in the Northeast, greatly changing forest composition in some areas (Elton 1958).

Despite the frequent claim of such an impact, demonstrating resource competition by a nonindigenous species against a native one has not been easy. This problem is just part of the larger difficulty of demonstrating resource competition nonexperimentally even when it is occurring (see, for example, Wiens 1989), but in a number of cases the observational evidence is suggestive. The decline of the native encyrtid wasp *Pseudhomalopoda prima* in Florida, for example, corresponded closely with the introduction in 1960 of another wasp, *Aphytis holoxanthus* for the biological control of Florida red scale (F. D. Bennett 1993). This scale is very common, and the encyrtid had previously been the main parasitoid found on it. A limiting resource need not be used in the same way for a nonindigenous species to harm an indigenous one. In Bermuda, nest boxes for eastern bluebirds (*Sialia sialis*) are used as perches by the nonindigenous great kiskadee (*Pitangus sulphuratus*), which prevents bluebirds from nesting (Samuel 1975).

Interference competition is more easily detected than resource competition—and the effects of nonindigenous species are no exception. Three non-indigenous ants of Florida—the imported fire ant (*Solenopsis invicta*), the Argentine ant (*Linepithema humile*), and the big-headed ant (*Pheidole megacephala*)—have greatly affected native ant communities in many parts of the world (Williams 1994), largely because they are aggressive toward other ants. Imported fire ant workers often attack other ants (Jones and Phillips 1987; Bhatkar 1988), for example, and this aggression, combined with habitat disturbance that favors imported fire ants over other species, is leading to wholesale replacement of native ant communities in parts of Florida (Wojcik 1994).

The plant analog to interference behavior by animals is allelopathy. The African ice plant (*Mesembryanthemum crystallinum*), for example, is an annual introduced into California. It accumulates salt, which is leached by rain and fog into the soil when the plant dies, and the salt suppresses growth and germination of native coastal plant species (Macdonald et al. 1989).

The most important direct effect of nonindigenous species on native communities is habitat modification. In the eighteenth and nineteenth centuries, for example, the northeastern North American coast consisted largely of mudflats and salt marshes. The European periwinkle (*Littorina littorea*), which eats algae on rocks and rhizomes of marsh grasses, has transformed much of that coast into a rocky shore since its introduction into Nova Scotia about 1840. Because the physical structure of the whole intertidal zone changed, most of the biotic community dependent on that structure changed also, even though the periwinkle did not compete with or eat most of these species (Bertness 1984). Introduced feral pigs (*Sus scrofa*) have modified entire ecosystems in the Great Smoky Mountains National Park and the Hawaiian Islands by selectively rooting out and feeding on plant species with starchy belowground parts, modifying soil characteristics by thinning litter and mixing organic and mineral soil layers, and increasing mineral leaching (Singer et al. 1984; Ebenhard 1988). In this way, habitat of numerous native species is destroyed.

Nonindigenous plants can modify an entire biotic community by changing the fire regime, as described for *Melaleuca quinquenervia* by Schmitz et al. (Chapter 3 in this volume), and they can modify the habitat by forming forests where none had existed, as Eurasian salt cedars (*Tamarix* spp.) have done along rivers in the arid Southwest (Vitousek 1986) and red mangrove (*Rhizophora mangle*) has done in sheltered bays and estuaries in the Hawaiian archipelago (Walsh 1967). The consequences of such new forests for the entire community must be enormous. Healthy mangrove swamps drop about 4000 kg of leaves annually per hectare, for example, while the roots trap sediment and serve as habitat for fishes and crustaceans (Holdridge 1940; Carey 1982). The introduction of South American water hyacinth into Florida in the late nineteenth century (Chapter 3 in this volume) similarly affected an entire community by completely changing the habitat: it clogged over 20,000 ha of waterways by the 1950s, lowering dissolved-oxygen levels, increasing temperatures, smothering beds of indigenous plants, and reducing populations of some indigenous animals.

Finally, introduced species can hybridize with native species (Rhymer and Simberloff 1996). Mallard ducks (*Anas platyrhynchos*), which have a holarctic breeding range, have hybridized with closely related endemic species, producing introgression. In New Zealand and Hawaii, this introgression threatens native species with extinction. In Florida, migratory mallards return north to breed, but domesticated mallards that escaped or were released for hunting have bred with the endemic Florida mottled duck (*A. fulvigula fulvigula*) and introgression may threaten the existence of this subspecies (Mazourek and Gray 1994). Plants are similarly threatened. *Lantana depressa,* restricted to native vegetation on stabilized dunes, relictual dunes, coastal prairies, and limestone ridges in peninsular Florida, hybridizes with *L. camara,* a Latin American or West Indian species brought to Europe as an ornamental in the seventeenth century (Sanders 1987). There it was artificially selected by horticulturists, and by the late eighteenth century it had been carried to the New World. It is found almost exclusively in disturbed habitat. Because so much of the habitat of *L. depressa* is disturbed, there is frequent hybridization and introgression between the species, and *L. depressa* is probably doomed by this process. Hybridization between a nonindigenous species and a native one can produce a new pest. North American cord grass (*Spartina alterniflora*), for example, was carried in shipping ballast to southern England. A subsequent hybrid with the noninvasive native *S. maritima* was sterile, but one of these hybrid individuals underwent a doubling of chromosome number to produce a fertile and invasive species, *S. anglica* (Thompson 1991).

Indirect Effects

An indirect effect occurs when one species affects the interaction between others. For nonindigenous species, these effects can be complex. The mite *Pediculoides ventricosus* was accidentally introduced into Fiji, for example, where it attacked larvae and pupae but not eggs and adults of the coconut leafmining beetle during the dry season (DeBach 1974). Adult beetles then oviposited and died, so the mite had effectively forced the beetle population into a new pattern of synchronous, nonoverlapping generations. The subsequent absence of larvae and pupae during some periods caused the mite population to crash along with those of two native parasitoids that had controlled the beetle. And because the mite and parasitoids did not live long enough to survive the periods between occurrences of the host stages they needed for oviposition, the beetle population exploded.

Although this example is probably no more complex than many others in nature, it is apparent that detecting such an effect would be difficult. Often such effects are demonstrated only when a species is introduced or removed. Because we fail to notice many species introductions or disappearances, it is very possible that many features of nature are governed by such interactions without our knowing it. Even if we are aware of a removal or an introduction, are we likely to associate a change in the status of some other species with it?

Because species can interact indirectly through shared prey or hosts, predators or herbivores, parasites, and pathogens, as well as through many forms of habitat modification, the number of possible indirect effects is enormous. In the previous example of avian malaria in Hawaii, the differential impact of the plasmodium on native and introduced birds affects interactions of the bird species with one another, so the plasmodium has both direct and indirect effects. Here is another example: the chestnut blight fungus (*Cryphonectria parasitica*) arrived in New York from Asia on nursery stock in the late nineteenth century; in less than 50 years it had spread over 91 million hectares of the eastern United States. The American chestnut (*Castanea dentata*) had been a dominant in many forests, composing up to 25 percent of canopy trees. It was virtually eliminated, precipitating many effects on these forest communities, both direct and indirect (Krebs 1985). Several insects host-specific to the chestnut are endangered or extinct (Opler 1978), whereas the oak wilt disease (*Ceratocystis fagacearum*) has increased on many native oak species (an indirect effect) because the particularly susceptible red oak (*Quercus rubra*) increased greatly as the chestnut disappeared (Quimby 1982). And because senescent leaves of chestnut contain less lignin than those of the oaks that are replacing them (Cromack and Monk 1975), it is very likely that litter-decomposition rates are now substantially lower (Cromack, pers. comm.), with major but unknown consequences.

Any nonindigenous species that substantially modifies the habitat can produce myriad indirect effects by changing the balance between two native competitors, between a predator and its prey, or between a parasite, pathogen, or phytophage and its host. The habitat modification by feral pigs described earlier, for example, could easily favor one native plant species to the detriment of another by modifying the nutrient regime, thus creating indirect as well as direct effects on the existing community.

Synergism

As nonindigenous species proliferate, they come to interact with one another as well as with the native community—and these interactions can render the damage by a combination of species far greater than what might have been expected from individual consideration of the component species. In Hawaii, lowland forest destruction has been the main threat to native birds in recent years, but, as noted earlier, the combination of Eurasian bird species, a mosquito, and the malaria plasmodium has helped to devastate this avifauna. Mack (1986, 1989) has described how the invasion by Old World plants in the North American Intermountain West was facilitated by the introduction of livestock; the native caespitose or bunch grasses were unadapted to large, congregating mammals. In turn, the livestock benefited from the nonindigenous plants. Crosby (1986) depicts an "ecological imperialism" of Eurasian species spreading over the earth during the past millennium by virtue of a juggernaut comprising plants, animals, and pathogens that erased native ecosystems in their path.

Why Florida?

Dasmann (1971), an ecologist writing about the environmental destruction of Florida, devoted an entire chapter to nonindigenous species and saw them as an overwhelming threat. His predictions are eerily confirmed in many chapters of this book. *Melaleuca, Hydrilla, Casuarina,* Brazilian pepper—the invaders that dominate his pages have indeed become the scourges he foresaw.

Dasmann had recourse to a commonplace of invasion biology, the idea that two sorts of sites seem particularly prone to devastation by nonnative species: islands and habitats created and disturbed by humans (Elton 1958; Simberloff 1986). Florida fulfills both criteria. Much of the state has become a patchwork of habitats created by humans—agriculture, residential tracts, planted forests—and the southern third of the peninsula is a habitat island, bounded on three sides by water and on the fourth by frost and typified, as are oceanic islands, by an impoverished native flora and fauna (Myers and Ewel 1990b). These geographic features alone would predispose Florida to great invasibility: there are many new habitats and not many native species that might be preadapted to them. It is hardly a coincidence that the two states with the most severe nonnative species problems are Florida and Hawaii. Hawaii is an archipelago of oceanic islands and is also dominated, in the lowlands where nonnatives are particularly troublesome, by human-produced habitats (U.S. Congress 1993). These two features—insularity and disturbed or novel habitats—go far toward explaining why the proportion of nonindigenous species in many taxa in Florida is far greater than, for example, in California (Mooney et al. 1986), though less than in Hawaii (U.S. Congress 1993).

Insularity and the amount of new or disturbed habitat are not the only features shared by Florida and Hawaii that have rendered them peculiarly vulnerable to damage from nonnative species. Both have large tropical areas (the south of Florida, the lowlands of Hawaii) that are the locus of most damage by nonnatives (U.S. Congress 1993). The absence of freezes allows many species to survive that would otherwise have been eliminated. This point is demonstrated in Florida by the numerous species, both native and nonnative, whose geographic ranges terminate quite abruptly where frost becomes routine. The fact that both Florida and Hawaii are partly tropical also increases the number of nonnative species introduced as ornamental plants and pets. Many species sold in nurseries and pet stores in these two states would not be available in, for example, New York—or if they were, their escape or deliberate release out of doors would be innocuous because they would not survive. But plants such as banana poka in Hawaii and Brazilian pepper in Florida, as well as animals such as bulbuls in both states, not only survive but spread and achieve pest status. Most of Florida's nonindigenous fish species, for example, escaped from aquarium-culture facilities (Chapter 7 in this volume), as did the plant hydrilla (Chapter 3 in this volume). Furthermore, pets and ornamental plants carry disease organisms that may attack native species.

Both Florida and Hawaii are transportation hubs and centers of tourism from other regions. Miami is the port of entry for most visitors from Latin America and has numerous flights from other regions as well. Of all plant shipments into the United States, 85 percent pass through Miami—333 million plants in 1990 (U.S. Congress 1993). Hawaii is not only an enormously popular vacation destination, primarily for North Americans and Asians, but also a center for military and civilian movement through the Pacific. Small wonder, then, that inadvertent introductions accompany such a mass of traffic.

A feature not shared with Hawaii that predisposes Florida to invasion by nonnative species is the abundance of lakes, streams, and other wetland habitats. Some 7800 lakes comprise about 6 percent of Florida's area, for example, and 1700 rivers dissect the state. Not only are these habitats available to aquatic nonnative species of plants and animals introduced by pet, fishing, and ornamental plant enthusiasts, but the modification of the state's waterways for irrigation, water supplies, flood control, and recreation has facilitated the spread of some of these species and worked to the disadvantage of several key native species (U.S. Congress 1993). Pollution of Florida's water bodies by human activity has also favored growth of several nonnative plant species over that of natives. Florida has also undergone extensive disturbance of soils through rock plowing, diking, strip mining, and bedding, particularly in the southern part of the state. These newly created habitats tend to support novel ecosystems whose function is dominated by nonnative species (Myers and Ewel 1990b).

The terrestrial habitats of northern Florida, though under assault by nonnative species as are those of every other state, have not suffered nearly so much as those of southern Florida (Abrahamson and Hartnett 1990; Myers and Ewel 1990b). The pinelands that dominate the north—at least where they have not been replaced or greatly changed by human activity—do not seem particularly susceptible to invasion. (Where they are intact, the same is true of Florida's dwindling scrub forests.) The imported fire ant (*Solenopsis invicta*), for example, which is spreading throughout the South, is a minor component of undisturbed pine forests in northern Florida. When it is found in these habitats, it is almost exclusively along roads rather than within the forest itself (Tschinkel 1988; McInnes 1994). We do not know the full range of reasons why this habitat is not very invasible. We do know, however, that the native species have all evolved to tolerate the natural regime of frequent fires that maintains these pinelands, whereas most invading nonnatives are maladapted to this milieu. Moreover, upland pine forests tend to be on excessively drained soil that is often poor in nutrients (Myers 1990).

Another reason why nonnative species may be more prominent in southern than in northern Florida is that there are fewer native species of most large taxa in the south than in the north. For example, birds (Wamer 1978), amphibians and reptiles (Means and Simberloff 1987), freshwater fishes (Swift et al. 1986), and woody plants (Schwartz 1988) all have fewer native species in southern Florida than in northern Florida. It has been suggested that this gradient is a historical artifact—that Pleistocene climatic fluctuations eliminated species and that insufficient time has passed for replacement species to have colonized, particularly in the south because of its island-like features. But for particular taxa that have been closely studied (the birds, herpetofauna, and plants just mentioned), the dearth of native species in southern Florida seems to be more parsimoniously explained by the fewer and poorer habitats found there.

Whatever the reason for it, the impoverishment in the south may contribute to the relatively greater impact of nonnatives there. Chances may be lower, for example, that some native species will be able to thrive in a new or disturbed habitat. Several writers (Dasmann 1971; Ewel 1986; Myers and Ewel 1990b) have suggested that, because there are fewer native species in southern Florida, nonnative species are more likely to be able to outcompete the natives—but as noted earlier, there are very few examples demonstrating such competition (Simberloff 1986; U.S. Congress 1993). Further, as Dasmann (1971) points out, the native species in tropical hardwood hammock forests in southern Florida have not in fact been outcompeted by nonindigenous species. Rather, once the hammocks have been greatly disturbed by either humans (canalization, clearing, roads, fire prevention) or nature (hurricanes), the early colonists are often nonindigenous species adapted to the new habitat. Without the disturbance, there would have been no replacement.

In sum, then, there are two key factors predisposing Florida, particularly southern Florida, to invasion by and damage from nonnative species: one is the destruction and disturbance of native habitats and their replacement by novel habitats; the other is the geographic features of tropicality, insularity, and the great expanse of aquatic habitats. The problems are exacerbated by the amount of tourism and transportation into the state. Given these forces promoting the arrival of nonindigenous species in Florida, it is not surprising that synergisms among these species, of the sort described here, have worsened the situation. For example, several fig (*Ficus*) species imported as ornamentals into southern Florida have become invasive because their host-specific fig wasps (Agaonidae) have independently immigrated, while their fruits may be dispersed by introduced parrots (Chapter 9 in this volume). Numerous writers (Cox et al., Ferriter et al., Glisson, and Layne, all in this volume) point to a variety of impacts by feral pigs in Florida similar to those described earlier for other regions. It is worth noting that in the Hawaiian Islands, feral pigs disperse nonindigenous plants (Loope and Scowcroft 1985) and feed particularly on certain native species, while they aid nonindigenous invertebrates by their rooting and defecation (Stone 1985). There is every reason to believe that they also aid nonindigenous invaders in Florida.

Prospects

From both a scientific standpoint and a management perspective, despite a decade of intensive activity, invasion biology leaves many questions. The spread and effects of invasions are known primarily as a collection of anecdotes rather than as parts of a science of invasion biology. Although the literature on spread, as described here, is rapidly developing a theoretical basis that may make this component of invasions more predictive, invasions in their entirety are still not viewed in the context of well-established paradigms that characterize a mature science. Peters (1991) argues that ecology as a whole is not very predictive and is thus not a mature science. Such a criticism would apply a fortiori to invasion biology, as well, and challenges researchers. The response to this challenge must include both theoretical contexts to guide data gathering and more extensive, accurate data. It is entirely possible that theory in ecology will be less general and more bound to particular cases than that in the physical sciences (Shrader-Frechette and McCoy 1993). But the field of invasion ecology is still too young for us to know just what sorts of theories will prove useful.

Management of nonindigenous species is hamstrung partly by the immaturity of the science on which it must be based, but also by the fact that much of what is known is scattered in the gray literature of agency memos and other reports. Further, one finds little formal coordination between scientists and managers. This book is a contribution, we hope, to redressing both of these problems.

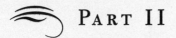

PART II

How Nonindigenous Species Get to Florida— and What They Do When They Get There

2

Florida's Invasion by Nonindigenous Plants: History, Screening, and Regulation

Doria R. Gordon and Kevin P. Thomas

 Nonindigenous plant species have arrived in Florida by a variety of pathways, almost all associated with human activity (Austin 1978a; Williams 1980; Schmitz et al. 1991). The vast majority of these introduced species pose no threat to Florida's natural systems. They become "established"—persist and grow in sites where they are planted, often with the assistance of added water, fertilizers, and sometimes pesticides—but they are not "invasive." This term is reserved for species that spread, vegetatively or by seed, into areas where they are not planted, persist without help, and displace native species. These invasive plant species have well-documented negative impacts and are increasingly expensive to control (U.S. Congress 1993).

This chapter summarizes the primary pathways by which nonindigenous plants have arrived in Florida and focuses on the relatively few invasive species. We then list the common characteristics and patterns of invasive species for potential incorporation into a screening process designed to preclude further introductions of species with invasive potential. We conclude by suggesting additional measures and specific legislative approaches for avoiding introduction of these problem species.

Sources

Perhaps thousands of nonindigenous plant species have, deliberately or accidentally, been introduced into Florida. Deliberate introductions include species brought in for agricultural purposes (including silvicultural and forage species) and for ornamental, industrial, or pharmaceutical reasons. The ornamental category includes both terrestrial landscaping plants and those imported by the aquarium and aquaculture trades. Species like kudzu (*Pueraria montana*) and Japanese honeysuckle (*Lonicera japonica*) have been imported for soil conservation along roads and other disturbed sites.

Southern Florida has been the destination for hundreds of nonindigenous plants, many selected for the local ornamental trade. Moreover, one of the first U.S. Department of Agriculture (USDA) plant introduction gardens was established in Miami in 1910 (Ryerson 1967). The Miami Inspection Station is the port of entry for 85 percent of all shipments of nonindigenous plants into the United States (U.S. Congress 1993). In 1990, some 333 million plants were brought into Florida through Miami International Airport (U.S. Congress 1993).

Accidental introductions occur when plants enter as contaminants in agricultural seed, packing materials, shipping containers, or ballast or on vehicles and other conveyances (U.S. Congress 1993). Several species may have entered Florida following shipwrecks (Austin 1978a). Researchers have also intentionally and unintentionally released nonindigenous plants. Many species have been tested by the USDA and the University of Florida Institute of Food and Agricultural Sciences (IFAS) research laboratories for agricultural and ornamental uses and have probably escaped from those sites. The USDA continues to identify and bring in agricultural and ornamental species and may be responsible for introducing most of the nonindigenous species currently identified as invasive. (See, for example, Fairchild 1938; Ryerson 1967; Williams 1980.)

In Florida, plant exploration activities by public agencies in the nineteenth century have been supplemented by the activities of plant collectors and botanical gardens. David Fairchild, of Fairchild Tropical Garden and the USDA, documented his extensive private and public role in introducing numerous nonindigenous plants (Fairchild 1938; Ryerson 1967). Many species were destined for the ornamental landscape and aquarium plant trades, which still represent a significant source of new nonindigenous species. Private efforts alone have introduced several hundred woody species into the state, supporting a billion-dollar industry (U.S. Congress 1993).

In this chapter we focus on the most invasive nonindigenous plant species identified by the Florida Exotic Pest Plant Council (EPPC), a nonprofit organization founded in 1984. In 1993, the EPPC updated the list of Florida's most invasive species, arranging the species into four categories of decreasing invasiveness in native vegetation (Austin 1993). Examining the history and characteristics of Florida's most invasive species will give us some background for assessing approaches for preventing importation of similarly destructive species.

We focus here on the 94 species listed in the 1993 EPPC Categories I and II as most critical. Category I species are widespread in Florida and have an established potential to invade and disrupt native plant communities. Category II species have local distributions but either have rapidly expanding populations or have shown a potential to invade and disrupt native vegetation in other areas (including areas in other countries with climates similar to Florida's). (Use of the 1995 update of the list (Exotic Plant Pest Council 1995), published after this chapter was written, would not significantly change the patterns reported here. It comprises only two categories and lists 115 species, most of which overlap with Categories I and II of the 1993 list.)

Table 2.1 details the dates and reasons for introduction of species. In several cases, the date listed may not represent the first introduction of a species, only the first record of its presence. Perhaps most striking is our inability to associate a date with 10 percent of the Category I and II species listed by the EPPC. For even more of those species, no reason for introduction was documented. The majority (90 percent of those for which records exist) of species listed were deliberately introduced. Forty-six percent of the EPPC's most invasive plants have been recorded as imported for ornamental reasons. The ornamental trade probably accounts for at least another 15 percent of the species for which no introduction reason was found, as they have been commercially available for this purpose. This contribution by the ornamental trade is not unique to Florida: 42 percent of the established nonindigenous plant species in South Africa (Hughes and Styles 1987) and 31 percent in Australia (Panetta 1993) were imported as ornamentals. In contrast, accidental introductions appear to account for only a few of the most invasive nonindigenous plant species in Florida (Table 2.1).

History

The earliest known plant-collecting expedition occurred in roughly 2500 B.C. when the Sumerians introduced species from Asia Minor (Ryerson 1967). Native American cultures, which inhabited Florida for at least 10,000 years before European settlement, imported and cultivated species like corn, beans, squashes, and peppers from the Caribbean region (Winters 1967). Arrival of Europeans by 1513 accelerated the import of species like pineapple, sugarcane, guava, papaya, and citrus (Austin 1978a), as well as ornamentals like oleander and plumbago (Knott 1976). But human population growth and species imports intensified dramatically after the early 1900s. Europeans brought numerous agricultural species (broccoli and its relatives, lettuce, melons, onions) and ornamental nonindigenous species.

USDA international collection expeditions started in 1897 with exploration

TABLE 2.1.

Date of Introduction (or First Record) of the 1993 Florida EPPC List of Category I and II Most Invasive Plant Species

Latin name	Common name	Introduction to Florida	For sale[a]	Reason for introduction	References
EPPC CATEGORY I					
Abrus precatorius	rosary pea	pre-1932		ornamental	Morton (1976)
Acacia auriculiformis	earleaf acacia	by 1932	X	ornamental	Morton (1985)
Ardisia elliptica (= *A. humilis*)	shoebutton ardisia	by 1900[b]		ornamental	Reasoner (1900)
Casuarina equisetifolia (= *C. litorea*)	Australian pine	1887		agriculture (forage, pulp); windbreak	Mack (1991) Schmitz et al. (1991)
Casuarina glauca	suckering Australian pine	1890		ornamental	Mack (1991) Johnson and Barbour (1990)
Cinnamomum camphora	camphor tree	pre-1933	X	windbreak	Small (1933) Bailey (1924b)
Colubrina asiatica	lather leaf	pre-1933		ornamental	Small (1933) Schmitz et al. (1991)
Cupaniopsis anacardioides	carrotwood	1968	X	escaped cultivation on Asian islands	USDA[c] Oliver (1992)
Dioscorea bulbifera	air potato	1905		ornamental	Schmitz et al. (1991)
Eichhornia crassipes	water hyacinth	1884		agriculture	Crowder (1974)
Ficus microcarpa (= *F. nitida* = *F. retusa* var. *nitida*)	laurel fig	pre-1912 wasp ca. 1975	X	agriculture (forage); ornamental ornamental	Fairchild (1938) McKey and Kaufmann (1991)
Hydrilla verticillata	hydrilla	1950–1951		aquarium ornamental	Schmitz et al. (1991)
Ipomoea aquatica	water spinach	pre-1950		agriculture	Snyder et al. (1981)
Jasminum dichotomum	Gold Coast jasmine	pre-1947		ornamental	Read (1962) Mack (1991)
Lantana camara	lantana	1804	X	ornamental	Habeck (1976)

Species	Common name	For sale[a]	Date introduced	Use	Reference
Lonicera japonica	Japanese honeysuckle	X	1875	ornamental	Mack (1991)
				soil stabilization	Sasek and Strain (1991)
Lygodium microphyllum	Old World climbing fern		pre-1958	not documented[d]	Nauman and Austin (1978)
Melaleuca quinquenervia	melaleuca		1906	ornamental; agriculture	Crowder (1974)
Melia azedarach	chinaberry		ca. 1830 SC & GA	ornamental	Miller (1990)
Mimosa pigra	cat-claw mimosa		1926	ornamental	Schmitz et al. (1991)
Neyraudia reynaudiana	Burma reed; cane grass		1916		USDA[c]
Paederia foetida	skunk vine		1897	agriculture (fiber)	Morton (1976)
Panicum repens	torpedo grass		1920s	agriculture (forage)	Schmitz et al. (1991)
Pistia stratiotes	water lettuce		pre-1765	accidental contaminant	Center (1992a)
Pueraria montana (= P. lobata)	kudzu		1899	ornamental; soil stabilization agriculture (forage)	Mack (1991)
Rhodomyrtus tomentosus	downy myrtle	X[e]		agriculture (forage)	Winberry and Jones (1974)
Sapium sebiferum	popcorn tree; Chinese tallow tree	X	pre-1784 SC 1900s FL	agriculture (fruit)?	Bailey (1924b)
					Hunt (1947)
				agriculture (soap/seed oil)	Jamieson and McKinney (1938)
Scaevola taccada var. sericea (= S. frutescens = S. sericea)	scaevola; half flower	X		not documented[d]	
Schefflera actinophylla (= Brassaia actinophylla)	schefflera	X	1927	not documented[d]	Morton (1976)
Schinus terebinthifolius	Brazilian pepper		1840s	ornamental	Schmitz et al. (1991)
Solanum viarum	tropical soda apple		ca. 1985	accidental contaminant	Mullahey and Colvin (1993)

continues

[a] Species commercially available in 1994 ("For Sale") are found in PlantFinder (1994) except as noted.

[b] The Ardisia polycephala in the Reasoner Bros. catalog is probably A. elliptica (treated similarly by Small 1933).

[c] Earliest plant introduction date in Miami as listed by the USDA; not necessarily the date of first introduction.

[d] Has been used as an ornamental.

[e] D. Schmitz (pers. comm.).

TABLE 2.1. *Continued*

Latin name	Common name	For sale[a]	Introduction to Florida	Reason for introduction	References
EPPC CATEGORY II					
Adenanthera pavonina	red sandalwood		1930		USDA[c]
				ornamental (seeds)	Bailey (1924b)
Agave sisalana	sisal hemp		1836	agriculture	Morton (1976)
Albizia lebbeck	woman's tongue		1883		Knott (1976)
				agriculture (pulp); ornamental	Morton (1983)
Alternanthera philoxeroides	alligator weed		1894	accidental contaminant in ballast	Schmitz et al. (1991)
Antigonon leptopus	coral vine	X	pre-1924	ornamental	Bailey (1924b)
Asparagus densiflorus	asparagus fern	X		not documented[d]	
Asystasia gangetica	Ganges primrose	X	1930		USDA[c]
				agriculture?	Morton (pers. comm.)
Bauhinia variegata	orchid tree	X	1936		USDA[c]
				ornamental	Bailey (1924b)
Bischofia javanica	bischofia	X		not documented[d]	Morton (1976)
Callisia fragrans	inch plant; spironema	X			
Calophyllum calaba (= *C. inophyllum*)	mast wood; Alexandrian laurel	X[e]		not documented[d]	
Casuarina cunninghamiana	Australian pine	X	pre-1924	ornamental; windbreak; soil stabilization	NRC (1984)
Cereus undatus	night-blooming cereus		1775		Austin (1978a)
				ornamental	Bailey (1924b)
Cestrum diurnum	day jasmine		pre-1933		Small (1933)
				ornamental	Bailey (1924b)

continues

Species[a]	Common name	For sale	Date introduced	Use	Reference
Colocasia esculenta	taro; elephant ear		1910	agriculture	Schmitz et al. (1991)
Cryptostegia grandiflora	Palay rubber vine	X	pre-1938	agriculture (rubber)	Fairchild (1938)
Dalbergia sissoo	Indian dalbergia; sissoo	X	1952	ornamental	USDA[c]
Dichrostachys cinerea	"aroma" in Cuba		1934	not documented[d]	Morton (1976)
Enterolobium contortisiliquum	ear-pod tree		1930	not documented[d]	USDA[c]
Epipremnum pinnatum cv. *Aureum*	pothos	X	1974		USDA[c]
Eugenia uniflora	Surinam cherry	X	1931	agriculture (fruit)	USDA[c]
Ficus altissima	banyan tree; lofty fig		pre-1926; wasp ca. 1975	ornamental	Bailey (1924b); Morton (pers. comm.); Nadel et al. (1992); McKey and Kaufmann (1991)
Ficus benjamina	weeping fig	X	1926	ornamental	USDA[c]
Ficus elastica	Indian rubber tree	X	ca. 1897	ornamental	Fairchild (1938)
Flacourtia indica	governor's plum			agriculture (fruit)	Bailey (1924b)
Flueggea virosa	flueggea			not documented[d]	Bailey (1924b)
Hibiscus tiliaceus	mahoe	X[e]	pre-1924 U.S.; 1986	ornamental	USDA[c]
Hygrophila polysperma	green hygro		1950s	aquarium ornamental	Schmitz et al. (1991)
Hyptage benghalensis	hyptage				
Imperata brasiliensis[f]	Brazilian satintail		1905		Westbrooks and Eplee (1989)
Imperata cylindrica[f]	cogon grass		1911	accidental contaminant in packing material; agriculture (forage); soil stabilization	Westbrooks and Eplee (1989)

[a] Species commercially available in 1994 ("For Sale") are found in PlantFinder (1994) except as noted.

[c] Earliest plant introduction date in Miami as listed by the USDA; not necessarily the date of first introduction.

[d] Has been used as an ornamental.

[e] D. Schmitz (pers. comm.).

[f] *Imperata brasiliensis* and *I. cylindrica* are classified as distinct species by Westbrooks and Eplee (1989).

TABLE 2.1. *Continued*

Latin name	Common name	For sale[a]	Introduction to Florida	Reason for introduction	References
Jasminum fluminense	jasmine		1923	not documented[d]	USDA[c]
Jasminum sambac	Arabian jasmine		pre-1933		Small (1933)
				ornamental	Bailey (1924b)
Leucaena leucocephala	lead tree		1898	agriculture (biomass energy crop)	Morton (1976)
					Fitzpatrick and Carter (1984)
Ligustrum sinense	privet	X	pre-1947 SC		Hunt (1947)
				ornamental	Bailey (1924b)
Lygodium japonicum	Japanese climbing fern		1932		Hall (1992)
				ornamental	Bailey 1924b
Macfadyena unguis-cati	cat's claw		pre-1947	not documented[d]	Morton (1976)
Manilkara zapota	sapodilla		1883	agriculture (fruit)	Knott (1976)
Melinis minutiflora	molasses grass		ca. 1914	agriculture (forage)	Campbell et al. (1987)
Merremia tuberosa	wood rose			not documented[d]	Fairchild (1938)
Murraya paniculata	orange jasmine	X	1936	not documented[d]	USDA[c]
Myriophyllum spicatum	Eurasian water milfoil		1940s	aquarium ornamental	Schmitz et al. (1991)
Nephrolepis multiflora	Asian sword fern		pre-1971	not documented[d]	Austin (1978a)
Ochrosia parviflora (= O. elliptica)	kopsia	X	1932	not documented[d]	USDA[c]
Oeceoclades maculata	ground orchid		pre-1974	escaped cultivation or natural expansion	Stern (1988)
Oryza rufipogon	red rice		1930–1959	accidental contaminant in rice seed?	Morton (1976)
					Westbrooks and Eplee (1989)
Paspalum notatum	Bahia grass		1945	agriculture (forage)	Killinger et al. (1951)
					Morton (pers. comm.)

Pennisetum purpureum	napier grass		1915	agriculture (forage); aquarium ornamental	Thompson (1919)
Pittosporum pentandrum	pittosporum		1939	ornamental	USDA[c]
Pouteria campechiana	canistel		pre-1933	agriculture (fruit)	Knott (1976)
Psidium guajava	guava		pre-1765	agriculture	Morton (1976)
Psidium littorale (= *P. cattleianum*)	strawberry guava	X	1887	ornamental; agriculture	Austin (1978a); Mack (1991)
Rhoeo spathacea (= *R. discolor*)	oyster plant	X	pre-1933	ornamental	Small (1933)
Sansevieria hyacinthoides (= *S. trifasciata*)	bowstring hemp	X	ca. 1800	agriculture (fiber); ornamental	Bailey (1924b); Marlatt (1975); Morton (pers. comm.)
Solanum torvum	turkey berry		pre-1899	not documented[d]	Westbrooks and Eplee (1989)
Syngonium podophyllum	arrowhead vine	X	1979	ornamental	USDA[c]
Syzygium cumini	jambolan; Java plum	X	1920	agriculture (fruit)	USDA[c]; Bailey (1924b)
Syzygium jambos	rose apple	X[e]	pre-1980	agriculture (fruit)	USDA[c]; Bailey (1924b)
Tectaria incisa	incised halberd fern		1929	ornamental	USDA[c]
Terminalia catappa	tropical almond		pre-1933	agriculture	Morton (pers. comm.); Small (1933)
Thespesia populnea	seaside mahoe	X	pre-1928	ornamental	USDA[c]; Bailey (1924b)
Triphasia trifoliata	lime berry	X	pre-1933	ornamental	Small (1933); Bailey (1924b)
Wedelia trilobata	wedelia	X	pre-1933	not documented[d]	Small (1933)

[a] Species commercially available in 1994 ("For Sale") are found in PlantFinder (1994) except as noted.

[b] The *Ardisia polycephala* in the Reasoner Bros. catalog is probably *A. elliptica* (treated similarly by Small 1933).

[c] Earliest plant introduction date in Miami as listed by the Reasoner Bros. catalog; not necessarily the date of first introduction.

[d] Has been used as an ornamental.

[e] D. Schmitz (pers. comm.).

[f] *Imperata brasiliensis* and *I. cylindrica* are classified as distinct species by Westbrooks and Eplee (1989).

for forage and grain crops. By the 1930s, the principal goal was to identify ornamentals (Ryerson 1967). The USDA Brooksville Plant Introduction Garden, established in 1910 in Miami, was replaced in 1915 by the Chapman Field Station (Austin 1978a). Overall the USDA was responsible for over 300,000 plant accessions (one or more individuals of a taxon from a particular location) introduced between 1898 and 1967 in the United States (Winters 1967).

Federal legislation restricting importation of plant species identified as threats to natural or agricultural systems has increased since 1900 (see Chapter 21 in this volume), but the 1939 Federal Seed Act and 1974 Federal Noxious Weed Act (amended in 1990) have focused regulation primarily on agricultural weeds from other countries. The 1993 report by the Congressional Office of Technology Assessment identifies several weaknesses in these acts, including the need to list species native to the United States but not to the areas where they are problems. Further, introduction or trade in nonagricultural terrestrial and wetland species that invade natural areas rarely is restricted under current federal laws.

Aquatic plants introduced for outdoor and aquarium ornamental purposes originally were propagated through release into canals and other water bodies and harvested by commercial interests. Only in 1969 were these practices prohibited and propagation of aquatic plants regulated by Florida Statute (403.271) (Schmitz et al. 1991). Invasive aquatic plants have recently received federal attention under the 1990 Nonindigenous Aquatic Nuisance Prevention and Control Act.

Characteristics

It is hard to predict which species will be invasive in new habitats. Although the vast majority appear to be noninvasive (U.S. Congress 1993), in several cases species that appeared to be established for several decades at low abundances then sustained accelerated population growth (Moody and Mack 1988). Brazilian pepper trees (*Schinus terebinthifolius*), for example, were present in Florida for 50 years before becoming pervasive (Ewel 1986). Melaleuca (*Melaleuca quinquenervia*) was introduced into Florida in 1906 but was not identified as a problem species until the 1960s (U.S. Congress 1993). Over 80 years passed (1806–1890) before Japanese honeysuckle escaped cultivation to become a serious weed across much of the northern and southern United States (Hardt 1986). Several species of figs (such as *Ficus microcarpa*) were unable to produce seed in Florida for 45 to 60 years until their native pollinators were also introduced (McKey and Kaufmann 1991).

Ecological theory provides little basis for consistent prediction of invasiveness in species (Newsome and Noble 1986; Crawley 1987, 1989). Because they are generally introduced without their predators and pathogens, nonindigenous plants should have an advantage over native "weeds." Moreover, one might expect invasive nonindigenous species to share many characteristics proposed for "ideal weeds" by Baker (1974). These include:

- Wide environmental tolerance in germination and growing conditions

- Self or wind pollination or nonspecialized pollinators

- Rapid growth to reproductive age or size

- High and continuous seed production during the growing season

- Seed dormancy

- Short- and long-distance seed or vegetative dispersal

- Vegetative as well as sexual reproduction

- Resistance to disturbance

- Growth form or chemistry suited for successful competition for resources

But many of the most invasive species share few of Baker's suggested traits (Newsome and Noble 1986; Perrins et al. 1992a, 1992b). Relative growth rate, for example, was not correlated with weediness in nonindigenous Australian pasture species (Lonsdale 1994). Further, a species with any combination of Baker's traits may be invasive in one location and not in another (Perrins et al. 1992a, 1992b).

Whether or not a nonindigenous species becomes established in a new location depends on the life history of the species, the new habitat, and management practices in that habitat (Perrins et al. 1992a). Chance events (demographic and environmental stochasticity) and timing interact with these factors to determine establishment success (Crawley 1989). The complexity of these relationships explains why the majority of introductions are not successful (Crawley 1987) and has resulted in pessimism about our ability to predict which species are likely to be invasive in a new habitat (Newsome and Noble 1986; Crawley 1987, 1989; Scott and Panetta 1993).

Where life form, breeding system, and dispersal mechanism have been examined for patterns among an invasive flora, all combinations of these traits have been found (Newsome and Noble 1986; Crawley 1987). The single trait common to many nonindigenous invasive species in England was their ability to form dense thickets (Crawley 1987). In Australia, Newsome and Noble (1986) concluded that no single group of ecophysiological traits conclusively predicts invasiveness. Further, they found that environmental factors and site history rather than physiological factors are most likely to influence establishment of invasive species.

Although plants of some families have shown greater tendency to be invasive than others (Heywood 1989; Rejmánek 1994), taxonomic affiliation does not conclusively predict invasiveness. One reason is the nonrandom taxonomic composition of nonindigenous species; some families are disproportionately represented among taxa transported by humans (Crawley 1987). Even within genera some species are invasive and others are not (*Lonicera* spp., Woods 1993; *Echium* spp., Forcella et al. 1986), but in the pine family (Pinaceae), more species in the subgenus *Pinus* have been invasive than have those in *Strobus* (Rejmánek 1994). Among the suite of species native to South Africa and invasive in Australian agricultural areas, the presence of other invasive species in the same genus predicted whether a species would be invasive 60 percent of the time (Scott and Panetta 1993). Conversely, this relationship did not hold for species invading nonagricultural systems. Panetta (1993) concludes that both family and generic relationships can be used as weak predictors of potential invasiveness.

Although three families comprise a disproportionate number of species on the Florida EPPC list (Fabaceae, nine; Poaceae, eight; and Myrtaceae, seven), we cannot conclude a priori that new species within these families pose a greater threat to Florida's natural systems than might species from other families. We do not know how many species in these families have been introduced without becoming invasive, nor do we know whether species in these families have been preferentially introduced relative to species in other families in Florida.

Although many researchers have concluded that prediction is difficult, they have identified a few predictive characteristics of nonindigenous invaders following successful establishment. Perhaps the best predictor is whether the species is invasive elsewhere in a similar climate (Forcella et al. 1986; Crawley 1989; Scott and Panetta 1993). Because many species have no history of prior introduction, this predictor cannot always be used. Weediness (Scott and Panetta 1993) or a wide distribution in the native habitat (Forcella et al. 1986) may also, but not consistently, indicate invasiveness elsewhere. For example, roughly 90 percent of Australia's "noxious weeds" are also invasive in other countries (Panetta 1993).

Some characters may be predictors of invasiveness in specific groups of species. In pines, age at first reproduction, interval between fruiting events, and animal seed dispersal are significant predictors of invasiveness (Rejmánek 1994). Similarly, by focusing on species within *Pinus* and *Banksia*, Richardson et al. (1990) conclude that combined consideration of the structure and dynamics of the recipient habitat and life histories of potential invaders allows some prediction of invasion success.

Rejmánek (1994) has reviewed characteristics of invasive nonindigenous species and notes the influence of characteristics like seed size, vertebrate fruit dispersal, and bark thickness (perhaps conferring fire tolerance). Better ability to acquire resources—nitrogen and light, for example—than the native vegetation also has enabled some invaders to displace native species. The latitudinal range of species in their native habitats has allowed some prediction of the probable range of expansion in their new location (Rejmánek 1994). But lack of information on the biology of many invasive species, and on factors that restrict invasion by others, limits prediction.

Examination of the species in the EPPC's two most invasive categories may suggest some predictors that overlap the characteristics just described. Some of the species (such as *Mimosa pigra* and *Imperata cylindrica*) are invasive elsewhere on other continents; if they were proposed for import now, we would have a basis for predicting their impacts. Moreover, most of the species in Table 2.1 have some mechanism for long-distance seed transport by animals, wind, or water. Many are also capable of vegetative growth from either aboveground or belowground tissue (stems, rhizomes, roots, bulbils), resulting in large clones and new colonizations by vegetative fragments. Almost all of these invasive nonindigenous species disperse rapidly to new sites or form dense interconnected stands that are difficult for other species to penetrate.

Species with Invasive Potential

As documented in Table 2.1, some 39 percent of the 1993 Category I and II invasive species were still commercially available in Florida in 1994. Moreover, several commercial ornamental and silvicultural species, not on the EPPC list, are potentially invasive. For example, species in the genera *Acacia, Azadirachta, Bauhinia, Cassia, Eucalyptus, Ficus, Imperata, Jasminum, Pinus, Prosopis, Sesbania,* and *Solanum* have been invasive both in Florida and elsewhere. Even though taxonomic affiliation is not a strong predictor of invasiveness, certain taxonomic trends have been identified (Panetta 1993; Panetta et al. 1994; Rejmánek 1994). These patterns can be used to suggest species of potential concern, but each taxon must then be individually assessed for characters promoting invasiveness in Florida's habitats.

There are mechanisms, of course, for limiting the importation of potentially invasive plants. Florida may benefit from import regulations already developed elsewhere. A multiagency working group in Hawaii, for example, is currently developing a coordinated approach to control plant imports (Alien Species Action Plan Working Group 1994); the number of plant species imported into Hawaii without permits is thought to average five per year (Nature Conservancy and Natural Resources Defense Council 1992). New Zealand (Nature Conservancy and Natural Resources Defense Council 1992), Australia (Panetta et al. 1994), and South Africa (R. Hobbs, pers. comm.) also have, and are refining, mechanisms to restrict, screen, and control invasive nonindigenous plant species.

Early identification and importation restrictions would reduce the number of invasive species introduced into Florida. One approach would be to establish a screening process for new species proposed for importation (Williams 1980; Mack 1991). Prediction of invasiveness may only be possible if experimental field trials are conducted in several habitats (Scott and Panetta 1993; Lonsdale 1994). Such trials are already required for proposed biological control species and for genetically engineered species. But this approach might be even more useful for annuals or biennials proposed for import because of the time required for woody-species trials as well as the lag time before invasion occurs for several species. Simultaneous testing of different-aged cohorts of longer-lived species would reduce this first difficulty.

Despite the difficulty of identifying invasive species, continued research on a screening process is necessary if we are to reduce imports of new, ecologically destructive species. Under the series of protocols for screening imports developed by the Australian Weeds Committee (Panetta et al. 1994), new species proposed for import would be subjected to a three-tiered review somewhat similar to the process Hawaii is developing (Alien Species Action Plan Working Group 1994). If the species is already listed as permitted or prohibited for import into Australia, the existing decision would be applied. If not, the species would be subjected to a computer-based risk assessment, which would result in a decision to import without restriction, to import and evaluate, or not to import. This assessment includes information about:

- The current status of the species in Australia

- Whether the species is invasive elsewhere

- Whether the native habitat climate and soil types are similar to those in Australia

- Historical patterns of distribution and spread in native and nonnative habitats

- Noxious attributes

- Biological and ecological traits

- Reproductive mechanism and seed dormancy characteristics

- Dispersal mechanisms

- Traits that confer resistance to control methods

- Nonweedy traits of significance (such as history of domestication)

In one proposed screening process, each of these categories is further divided into several questions with associated point ranges. Under the first category, for example, if the species has been repeatedly introduced into Australia without showing evidence of invasion from the introduction site, the points assigned would be low. The scores have been empirically determined and are currently undergoing further testing (Panetta et al. 1994).

If insufficient information is available to address many of the questions in the screening procedure, further evaluation would be required (Panetta et al. 1994). Experimental trials under quarantine conditions would be conducted. The working group has specified that all costs associated with obtaining permission to import a species should be borne by the importer. Although the screening process has not yet been implemented, the costs to Australia incurred during control efforts for nonindigenous species warrant the scale of screening recommended (Panetta et al. 1994).

To reduce the threat posed by invasive nonindigenous plants, the Florida Department of Agriculture and Consumer Services or IFAS should institute a process for screening new species before their widespread availability in Florida. Species already available should be screened, and any species determined to be a threat should be removed from the market. The state agency given responsibility for this screening should review the Florida EPPC list and determine an appropriate list of species that should not be commercially available within Florida or legally imported from other states. Additional research is necessary to refine the screening process and our understanding about the potential invasiveness of certain species.

Potential introduction in seed form requires the same considerations as introduction of plant tissue (Mack 1991). Any commercially available seed—whether for ornamental or agricultural propagation, for animal feed (including bird seed), or for other purposes—should be screened for invasive species.

The recently completed study by the Congressional Office of Technology Assessment (U.S. Congress 1993) suggests a number of approaches to preclude importation of potentially harmful species. These measures range from federal and state legislation to increased public education. Specific recommendations include modification of the Federal Noxious Weed Act and Federal Seed Act, which have now been made more consistent, and effective control of invasive nonindigenous plants on federal lands. In general, the report recommends options both for increasing the rigor of screening before import and release of nonindigenous species and for defining new state roles for increased consistency and effectiveness. Since the report's publication, Congress has enacted many of the recommendations in legislation, regulation, or agency policy; consideration of additional measures is ongoing in several congressional committees as of this writing (Windle, pers. comm.).

Recommendations

While Congress addresses federal responses to the problem of nonindigenous species, Florida should develop an independent approach compatible with proposed model state law frameworks. Approaches compatible with that outlined in the Congressional Office of Technology Assessment report (1993:229) include:

- Development of a procedure to create objective, scientific criteria for identifying species to be included on a single comprehensive Florida Noxious Weed list (including terrestrial and nonagricultural nonindigenous species, which have traditionally received little attention at either the federal or the state level). The list should include all species on the Federal Noxious Weed and Seed lists, plus other species known to be (or with high potential to become) invasive in Florida. The EPPC may provide a useful framework and initial list.

- Strengthening of existing legislation to prohibit intrastate or interstate sale or trade of species on this comprehensive Florida Noxious Weed list (vegetation or seed). Legislation should also prohibit transportation, propagation, or planting of these species. Persons violating this legislation would be subject to fines or other punishment (mitigation might include removal of prohibited species on public lands), and the plant material would be seized by the state.

- Development of legislation requiring quarantine of all livestock or other animals being moved from a location with animal-dispersed invasive nonindigenous plant species to another location until seeds or other propagules (such as burrs) have been cleared from their systems.

- Development of a comprehensive policy addressing the introduction of nonindigenous species on state lands.

- Establishment of a time frame for reduction to zero of the percentage of nonindigenous plant species used by public agencies with no agricultural mandate.

- Required screening of new plant species intended for release in Florida to address their potential for invasiveness. Species identified as likely to be invasive should be subjected to testing in quarantine facilities.

- Development of two educational programs: a public education program to increase awareness of the problems and costs associated with the importation of potentially invasive species and a program developed for businesses involved in commercial trade of ornamental species and for botanical gardens.

Because Florida, like Hawaii, is particularly vulnerable to invasion (Ewel 1986; U.S. Congress 1993), state regulations will probably be more stringent than those imposed at the federal level. Legislation and educational programs developed in Hawaii and Australia may provide useful models for this state. Florida should see that all species on the Florida Noxious Weed list are placed on the Federal Noxious Weed and Seed lists except those demonstrated not to invade in other areas. Florida may need to develop cooperative mechanisms to ensure that federal inspection policies at ports and other entry points are sufficient to prevent most illegal importation of new nonindigenous plant species. Given the unpredictability of invasiveness, Florida should consider enacting a policy similar to that established for fish and wildlife in Hawaii: species should be considered invasive and subject to import restrictions until demonstrated otherwise (U.S. Congress 1993).

ACKNOWLEDGMENTS

We thank Nadja Chamberlain for bringing the letter by Judge Knott (1976) to our attention and Wendy Robertson-Woerner for assistance. Richard Mack, John Randall, Don Schmitz, Daniel Simberloff, and Jora Young provided helpful reviews of this chapter. Jan Bowman of the USDA's ARS National Germplasm Resources Laboratory searched the Germplasm Resources Information Network for plant introduction records for the species in Table 2.1.

3 The Ecological Impact of Nonindigenous Plants

Don C. Schmitz, Daniel Simberloff,
Ronald H. Hofstetter, William Haller,
and David Sutton

The past century has seen a massive restructuring of the biotic components of ecosystems in Florida as the human population has grown from 528,542 in 1900 to 13,698,627 in 1993 (Bergquist et al. 1994). Influences associated with this growth have substantially altered waterways and what remains of the natural landscape (Long 1974b; Blake 1980; Myers and Ewel 1990b; Kautz 1993). Wetland and upland ecosystems, once contiguous throughout the state, have often been reduced to fragments by agriculture and urban expansion, and many of these remnants have been invaded by nonindigenous plant species.

Because plants form the biological foundation of all terrestrial and freshwater communities (Krebs 1994), displacement of a diverse native plant community by an invading plant species seems likely to alter an ecosystem substantially. Vitousek (1986), however, from a summary of the limited research on ecosystem-level consequences of biological invasions, concludes that nonindigenous animals often have an even greater effect on native ecosystems. Ramakrishnan and Vitousek (1989) suggest the reason for this difference is that invading species most often occupy disturbed habitats, where plants have less drastic effects. When plant species do invade undisturbed areas and alter ecosystem characteristics, they often differ in life form from the natives—for example, producing forest growth where none had existed (Vitousek 1986; Simberloff 1991, 1995a). Moreover, plant invaders gaining earlier access to resources, or using those resources more efficiently than natives do, can alter productivity, consumption, decomposition, water fluxes, nutrient cycling and loss, soil fertility, and erosion (Vitousek 1986). Invading plants can also modify entire communities by increasing the frequency of fires (Vitousek 1986; Simberloff 1995a).

Two types of plant invaders are particularly likely to have major impacts on ecosystems: species that constitute new habitats and species that modify the habitat, often by altering ecological processes (Vitousek 1986; Simberloff 1991). Of course, some species do both. Because the physical habitat and ecological processes of an ecosystem are crucial to all its species, such invaders can affect entire suites of plants and animals, as many have done in Florida. Here we describe how the known ecological effects of the most prominent nonindigenous plant species in Florida place them in these two categories. We also discuss the apparent differences in invasibility between different regions of the state and speculate about future ecological impacts of nonindigenous plants.

Prominent Plant Invaders

Surprisingly few studies have addressed the ecosystem impacts of nonindigenous plant invasions in Florida. Although more than 900 nonindigenous plant species have become established in Florida and constitute at least 27 percent of the flora (Ward 1989), most information on this subject is anecdotal. The species whose ecological impacts have been closely examined are generally either the most widespread or those that, because of their explosive growth rates (here or in other countries), threaten to expand their present ranges greatly. In 1993, the Florida Exotic Pest Plant Council identified 31 nonindigenous plants as Category I species—that is, highly invasive or disruptive to Florida's native plant communities (Chapter 2 in this volume). The impacts of 19 species (most in Category I) that form monospecific stands and dense canopies over native vegetation in many of Florida's ecosystems are summarized in the following sections. For the remaining Category I species, we lack even rudimentary knowledge of their effects. Although many Category I species affect the quality of human life in Florida—they may impede flood control, navigation, and water use; reduce property values; invade agricultural land; increase brushfire risk; be unsightly—our summaries are limited to the ways in which they alter Florida's native ecosystems.

No comprehensive survey of nonindigenous plant populations invading all of Florida's natural areas has ever been conducted, but smaller surveys have been performed on the most ubiquitous nonindigenous plant populations in southern Florida and in the state's public waterways. In 1993, the South Florida Water Management District surveyed by air the area from the north rim of Lake Okeechobee south to Florida Bay for Australian pine (*Casuarina* spp.),

Brazilian pepper (*Schinus terebinthifolius*), Old World climbing fern (*Lygodium microphyllum*), and melaleuca (*Melaleuca quinquenervia*) (see Chapter 20 in this volume). This survey indicated that Brazilian pepper had the largest range, followed by melaleuca, Australian pine, and Old World climbing fern (Table 3.1). Additional vegetation surveys conducted by the Florida Natural Areas Inventory along Florida's remaining undeveloped coastal areas indicate that Australian pine trees have invaded 33 percent of the Gulf coast (Pasco to Collier counties) and 46 percent of the Atlantic coast (Indian River to northern Dade County) (Johnson and Muller 1993; Johnson et al. 1993). Surveys conducted by the Florida Department of Environmental Protection during the 1990s show that hydrilla (*Hydrilla verticillata*) (see Chapter 14) and torpedo grass (*Panicum repens*) are the most abundant nonindigenous aquatic plant species found in the state. No regional surveys of wetland and upland nonindigenous plant populations in central and northern Florida have ever been conducted.

Air Potato

Introduced into Florida in 1905 (Nehrling 1944), the air potato vine (*Dioscorea bulbifera*) aggressively invades mesic habitats, disturbed areas, and hardwood hammocks throughout the state. In southern Florida, this native of the Old World tropics is almost always found in tropical hardwood hammocks. Its primary ecological threat is its ability to cover mature native trees and shade out understory vegetation. Disturbance, especially of the canopy, promotes its spread. Increased ground light levels after Hurricane Andrew removed the upper portions of trees, for example, allowed the growth of air potato (and several other

TABLE 3.1.
Approximate Ranges of Nonindigenous Plant Species
in Southern Florida and in Florida Waterways

Species	Year surveyed	Hectares
Casuarina spp. (Australian pine)	1993	151,246[a]
Eichhornia crassipes (water hyacinth)	1993	680[b]
Hydrilla verticillata (hydrilla)	1994	39,458[c]
Lygodium microphyllum (climbing fern)	1993	10,434[a]
Melaleuca quinquenervia (melaleuca)	1993	197,827[a]
Panicum repens (torpedo grass)	1992	7100[d]
Schinus terebinthifolius (Brazilian pepper)	1993	284,708[a]

Note: One infested hectare can represent anything from presence of an individual plant to a densely packed monospecific stand.
[a] 1993 South Florida Water Management District Survey (see Chapter 20 in this volume).
[b] Schardt and Ludlow (1993).
[c] J. D. Schardt (pers. comm.).
[d] Schardt (1992).

native and nonindigenous vines) to disrupt tree canopy closure by native species in Dade County's hardwood hammocks (Exotic Pest Plant Council 1992; Chapter 4 in this volume). Wildlife changes associated with this invasion have not been determined.

Australian Pine

Several species of Australian pine (*Casuarina cunninghamana, C. equisetifolia,* and *C. glauca*) were probably introduced into Florida during the 1890s in the Miami and Palm Beach areas (Martin 1949) and had escaped into natural areas by the early 1900s (Small 1927). Although commonly called pines, these plants are angiosperms, not conifers. Extensive hybridization between female *C. equisetifolia* and male *C. glauca* trees makes identification of these two species difficult (Austin, pers. comm.), but *Casuarina equisetifolia* is believed to be the most widely planted and common Australian pine found in Florida's coastal areas (Morton 1980).

Once commonly planted to form windbreaks around canals, agricultural fields, houses, and roads, Australian pine has spread extensively throughout southern and, to a lesser extent, central Florida. Habitats disturbed by both human activities and natural events seem particularly prone to invasions by Australian pine. In Everglades National Park, Australian pines began to dominate coastal areas after hurricanes destroyed existing vegetation during the 1960s (Klukas and Truesdell 1969). By 1993, they had become established on more than 150,000 ha from just north of Lake Okeechobee to Florida Bay (see Chapter 20 in this volume).

Australian pine alters ecosystems by quickly colonizing disturbed beach plant communities (Austin et al. 1977), thus preempting establishment of native species. Its invasions have displaced diverse plant communities beneath the trees (Mazzotti et al. 1981) by producing a dense litter that smothers most herbaceous vegetation (Figure 3.1).

Comprehensive wildlife studies associated with Australian pine forests are lacking. Anecdotal information suggests that few insect species are associated with Australian pine in Florida (D. Habeck, pers. comm.); even when herbivorous insects were found on the trees, the damage was negligible (Mead and Bennett 1987). Populations of small mammals in freshwater wetlands in southern Florida invaded by Australian pine may be depauperate compared to those in native plant habitats (Mazzotti et al. 1981). The direct impact of Australian pine forests on beach ecology in Florida is poorly understood, but anecdotal information suggests their dense, shallow roots interfere with the ability of threatened and endangered sea turtles and the American crocodile to excavate suitable nesting cavities (Klukas and Truesdell 1969; Jensen and Vosick 1994).

FIGURE 3.1 Leaf litter spreading out from a stand of Australian pine (*Casuarina* sp.) trees on a southwest Florida beach. Photo by Don Schmitz.

Brazilian Pepper

Introduced during the nineteenth century, Brazilian pepper (*Schinus terebinthifolius*), an evergreen shrub, has invaded many habitats in central and southern Florida. Local dispersal of its seeds is primarily by raccoons and opossums; long-distance spread is facilitated by frugivorous birds such as migratory American robins, *Turdus migratorius* (Ewel et al. 1982; Anonymous 1988). This shrub, the most widespread of Florida's invasive nonindigenous plant species, has become established on more than 280,000 ha in southern Florida alone (Table 3.1). Although primarily an invader of landscapes in which the substrate is disturbed and fire is excluded (Loope and Dunevitz 1981a), it has formed large, dense forests in relatively undisturbed areas adjacent to mangroves along the southwestern portion of Everglades National Park (Chapter 17 in this volume). It thrives in better-drained sites (Ewel 1986).

Although Brazilian pepper is better studied than most other nonindigenous plant species in Florida, comprehensive investigations of its ecosystem effects have not been conducted. It is known, however, to form dense monospecific stands in southern Florida that almost completely displace native understory plants (Ewel 1986). Because its leaves decompose rapidly, little leaf litter accumulates beneath these stands (Ewel 1986). Although 46 phytophagous insect species have been collected from Brazilian pepper in Florida, they do not appear to retard plant growth or vigor (Cassani 1986; Cassani et al. 1989; Bennett and

Habeck 1991). A single-season breeding-bird survey in an Everglades Brazilian pepper forest (Curnutt 1989) showed total density to be only 73 pairs/100 ha and species richness to be just 6 species, whereas Robertson (1955) found 28 species and more than 113 pairs/100 ha of pineland in the Everglades and 18 species and more than 255 pairs/100 ha in forest edge. On the one hand, the lower recent bird density and richness may reflect an overall decline in breeding birds found in upland forests of Everglades National Park since 1955 and may be related to factors other than habitat disruption (Snyder et al. 1990). On the other hand, anecdotal information suggests that mangrove bird rookeries surrounded by Brazilian pepper have been abandoned (Beever 1994).

A few native amphibian and reptile species were collected (though rarely) in Brazilian pepper forest habitats in the Long Key–Paradise Key region of Everglades National Park, whereas two nonindigenous species, Cuban tree frogs (*Osteopilus septentrionalis*) and brown anole lizards (*Anolis sagrei*), were most common (Dalrymple 1988). Dalrymple believes that most of the herpetofauna in Brazilian pepper forests in this area was responding to basic microhabitat requirements and not to the species composition of the vegetation. The herpetofauna of Brazilian pepper forests is similar in species numbers and foraging guilds to those of southern Florida's hammock communities, probably because of the closed canopy conditions and soil development found in both (G. Dalrymple, pers. comm.). In Everglades National Park, Brazilian pepper is threatening nesting habitat of the gopher tortoise (*Gopherus polyphemus*), a species threatened in Florida (Chapter 17 in this volume).

Burma Reed

Burma reed (*Neyraudia reynaudiana*), a tall, robust grass with plumelike inflorescences, is found primarily in Dade and Collier counties, although it has been cultivated as far north as Georgia. Introduced from Asia, it usually grows in disturbed areas. It is well established in most remaining pine rockland parcels in Dade County and is viewed as a threat to this rare habitat. Anecdotal information suggests it is highly flammable and promotes frequent fires, enhancing its spread and displacement of native pineland understory vegetation (Maguire 1990). Further, fires linked to Burma reed have led to abnormally high mortality of native fire-adapted slash pine trees (Exotic Pest Plant Council 1993). In 1993, Burma reed was established on nearly three-fourths of the remaining pine rockland areas outside of Everglades National Park (Exotic Pest Plant Council 1993).

Carrotwood

The Australian carrotwood tree (*Cupaniopsis anacardioides*), introduced into Florida around 1980 as an ornamental (Oliver 1992), rapidly escaped cultivation, invading human-made spoil islands of the Intracoastal Waterway in the Indian River lagoon (Oliver 1992), ditch banks and other low-lying areas of Martin County (Jordan 1990), and the Oscar Scherer State Recreation Area in Sarasota County (Oliver 1992). Seedlings can survive in Australian pine leaf litter, an environment that few other plants in Florida tolerate (J. Jordan, pers. comm.). Carrotwood's potential range and impact are poorly understood.

Chinese Tallow

A small to medium-sized tree native to China, Chinese tallow (*Sapium sebiferum*) was introduced into the United States as an ornamental by Benjamin Franklin, who sent seeds to a gentleman farmer in Georgia in 1772 (Bell 1966). It was not introduced in significant numbers on the U.S. Gulf coast until the early twentieth century (Jamieson and McKinney 1937). In northern and central Florida, the tree has escaped cultivation and is invading closed-canopy forests, bottomland hardwood forests, lake shores, and wetlands (Jubinsky 1993). Insect herbivory on Chinese tallow is low in the United States (Jones and McLeod 1989); no comparable data from China are available. Anecdotal information suggests that Chinese tallow can rapidly displace native vegetation in northern Florida wetlands by forming dense monospecific stands (Jubinsky 1993). It may also increase nutrient loadings of aquatic systems through leaf drop and fast decay, which may lead to much higher concentrations of phosphorus, potassium, nitrates, zinc, manganese, and iron (Cameron and Spencer 1989). Wildlife changes associated with Chinese tallow invasion are unknown.

Cogon Grass

Since its accidental introduction near Mobile, Alabama, in 1911, cogon grass (*Imperata cylindrica*), a perennial from Southeast Asia, has become established in at least 27 Florida counties (Tanner and Werner 1986) from the Panhandle to southern Florida (Colvin et al. 1994). There is dispute over whether the southern Florida species is *I. cylindrica* or its Latin American congener *I. brasiliensis*, which may be native there (C. Lippincott, pers. comm.). Cogon grass invades relatively undisturbed sandhill and pine flatwoods (C. Lippincott, pers. comm.), as well as disturbed habitats such as agricultural forests, roadsides, fallow pastures, and disturbed phosphate mining lands (D. Shilling, pers. comm.). It is also invading disturbed pinelands (Myers and Ewel 1990b). Preliminary data indicate that cogon-grass invasion increases fire intensity in sandhill communities and significantly raises mortality of sandhill seedlings, including those of longleaf pine (*Pinus palustris;* C. Lippincott, pers. comm.). Fire is believed to facilitate its spread (Myers and Ewel 1990b).

Hydrilla

Hydrilla (*Hydrilla verticillata*), a submersed caulescent plant from Sri Lanka first introduced in Florida in the early 1950s (Schmitz et al. 1991), has spread to more than 39,000 ha of waterways throughout the state. It forms canopies in which as much of 70 percent of the vegetative biomass is in the top 0.5 m of water (Langeland 1990). This type of growth reduces light penetration, alters oxygen-transfer dynamics, and reduces water circulation (Buscemi 1958; Frodge et al. 1990; Honnell et al. 1993). Beneath this canopy, extremely low levels of dissolved oxygen and significant changes in turbidity, color, chlorophyll, pH, alkalinity, specific conductivity, and phosphorus are common. Changes in the trophic-state classification of lakes, based only on water quality, can be attributed to dense hydrilla infestations (Schmitz et al. 1993). Moreover, native bottom vegetation is shaded out and ecological processes are substantially altered (Schmitz et al. 1993). Shifts in densities and species richness of zooplankton and epiphytic and benthic macroinvertebrates have been linked to hydrilla expansion in Florida lakes. (See Watkins et al. 1983; Schmitz and Osborne 1984; Richard et al. 1985; Schramm et al. 1987; Schramm and Jirka 1989.) When hydrilla canopies cover a substantial portion of a lake, populations of largemouth bass (*Micropterus salmoides*), bluegill (*Lepomis macrochirus*), redear (*L. microlophus*), and black crappie (*Pomoxis nigromaculatus*), all popular game fishes, become skewed to smaller individuals because of insufficient predator cropping (Colle and Shireman 1980). Hydrilla, however, like other submersed plants, may provide food and habitat for waterfowl in Florida (Montegut et al. 1976).

Japanese and Asian Climbing Ferns

First reported in Florida during the late 1950s in Martin County (Beckner 1968), Asian Old World climbing fern (*Lygodium microphyllum*) has spread to 10,117 ha in Collier, Glades, Highlands, Palm Beach, and Polk counties (D. Austin, pers. comm.; A. Ferriter, pers. comm.). Another Asian species, Japanese climbing fern (*L. japonicum,* an Exotic Pest Plant Council [EPPC] Category II species), has become established in northern Florida. Because both species appear to be spreading, Nauman and Austin (1978) have speculated that their ranges may eventually overlap. Recently, scattered populations of *L. japonicum* were found as far south as Broward and Dade counties (R. Wunderlin, pers. comm.). The ecological impact of *L. japonicum* is unknown, although it can become locally abundant (R. Wunderlin, pers. comm.). *Lygodium microphyllum* invades bottomland hardwood plant communities throughout the Southeast. It prefers wet, disturbed sites (Nauman and Austin 1978) and primarily invades cypress swamps disturbed either by logging or by a lowered water table (D. Austin, pers. comm.). In areas managed by ground fires, the climbing fern can serve as a pathway allowing fire to reach the tree crown, which normally would not burn. Moreover, it forms dense blankets over native trees, shading out native understory vegetation (Nauman and Austin 1978). Effects on wildlife communities are unknown.

Kudzu

Kudzu (*Pueraria montana*), a vine native to China, was introduced into Japan, then from Japan into the United States, where it has spread to about 809,400 ha of forestland in Alabama, Georgia, Mississippi, Tennessee, and the Carolinas (Miller and Edwards 1983). It has formed extensive populations on disturbed land throughout the Panhandle region of northern Florida. Kudzu was once widely planted, and populations are now also found from central Florida south to Homestead (Dade County) (Buschman et al. 1977). Kudzu was recently discovered growing on a levee in southern Florida's Everglades Conservation Areas and also in several natural areas in Dade County (Hammer 1995b). Kudzu primarily invades disturbed landscapes ranging from road rights-of-way to old fields (Figure 3.2). At least some native land snails can use kudzu as habitat and cover (Beetle 1972). In northern Florida, a severe kudzu infestation can alter forest-edge communities by shading existing vegetation. Relatively undisturbed natural drainage areas can also be invaded. Moreover, kudzu may interfere with normal plant succession in abandoned agricultural fields. Although the comprehensive ecosystem impacts of kudzu are unknown, it is a nitrogen-fixing legume and hence might alter biogeochemistry.

Figure 3.2 Kudzu (*Pueraria montana*) along a roadside in Tallahassee. Photo by Don Schmitz.

Lather Leaf

Lather leaf (*Colubrina asiatica*), a native of tropical Asia, is a vinelike shrub characterized by its rambling growth over other vegetation. First observed in Florida at Big Pine Key (Monroe County) in the early 1950s (Dickson et al. 1953), lather leaf is now found in frost-free coastal areas of southern Florida and the Florida Keys (Godfrey and Wooten 1981). It invades beach-dune plant communities (Long and Lakela 1976; Wunderlin 1982), spoil sites (M. Renda, pers. comm.), maritime hammocks (Wunderlin 1982), tidal marshes, and mangrove swamps (Olmstead et al. 1981). Its key ecological impact is the overgrowth of native tree crowns and shading of native plants. Its potential for spread throughout Florida's frost-free coastal areas is serious. Effects on wildlife communities are unknown.

Melaleuca

Melaleuca (*Melaleuca quinquenervia*), an Australian tree first introduced in 1906 into southern Florida, has become established on about 197,000 ha from north of Lake Okeechobee south to Florida Bay (Figure 3.3). Additional smaller populations are found throughout central Florida. The largest melaleuca populations in southern Florida are concentrated mostly in the areas of historical introduction: southern Broward and northern Dade counties, Lee County near Estero, Monroe Station (Collier County) at the Big Cypress National Preserve, and along the rim canal of Lake Okeechobee (Laroche 1994). Melaleuca alters ecosystems by forming dense, monospecific stands in poorly drained areas and in marshes previously devoid of forests. Melaleuca invades disturbed areas, although many southern Florida ecosystems are resistant (though not necessarily immune) to colonization (Ewel et al. 1976). Melaleuca is particularly prevalent in Florida wetlands where hydroperiods have been shortened (Hofstetter and Sonenshein 1990) and especially invades muhly (*Muhlenbergia* sp.) prairies and saw-grass (*Cladium jamaicense*) marshes (Wade et al. 1980). Melaleuca has also invaded undisturbed ecotones (transition zones) between wetland and upland forests, especially the ecotone between southern Florida pine and cypress (Myers 1983, 1984).

Published data and anecdotal evidence suggest that melaleuca forests may affect fire regimes (Wade et al. 1980; Wade 1981; Hofstetter and Sonenshein 1990; Flowers 1991), cause higher water loss than native saw grass through evapotranspiration (Hoffstetter, unpublished data), almost totally displace native vegetation (Richardson 1977; Wade et al. 1980; Di Stefano and Fisher 1983; Alexander and Hofstetter, unpublished data), raise soil elevations (which may inhibit normal water flow in the Everglades; Flowers 1991), and provide poor habitat for wildlife (Laroche 1994). Significant insect herbivory of melaleuca has not been observed (T. Center, pers. comm.). Small-mammal pop-

Figure 3.3 A mature melaleuca (*Melaleuca quinquenervia*) tree island in the East Everglades in 1991. Note the seedling spread radiating from the older and taller trees. Photo by Don Schmitz

ulation densities in melaleuca forests were substantially lower than those reported for native hammock and tree island communities (Ostrenko and Mazzotti 1981). Similarly, Sowder and Woodall (1985) consider melaleuca forests to be poor habitats for rodent populations. Melaleuca forests may provide nesting and roosting sites for some species of birds (Schortemeyer et al. 1981) but degrade or reduce habitat for others (Bancroft et al. 1992). A preliminary study reports that melaleuca's relatively low moisture and high crude fiber content may render it poor forage for large herbivores (Schortemeyer et al. 1981).

Skunk Vine

Skunk vine (*Paederia foetida*), imported from Asia about 60 years ago, has recently infested large tracts of land in central Pasco County (Exotic Pest Plant Council 1991). Although it favors disturbed, moist areas, its spread into drier locations is apparent (Exotic Pest Plant Council 1991). It blankets native trees (limbs of large trees have broken off from its weight) and may kill canopy and subcanopy trees as well as understory vegetation (Anonymous 1994). One of its Asian congeners, sewer vine (*Paederia cruddasiana*), has been found only in Dade County and may be a recent introduction. It extensively invaded urban tropical hardwood hammocks (of which only 202 ha remain in Dade County) after Hurricane Andrew (R. Line, pers. comm.).

Torpedo Grass

Old World torpedo grass (*Panicum repens*) was first introduced into North America near Mobile, Alabama, in 1876 (Hodges and Jones 1950). First reported in Florida in the wet prairies of the lower Kissimmee Valley (Osceola County) in the early 1920s (Kretchman 1962), torpedo grass is found most often in central and southern Florida. It had become established on approximately 7100 ha by 1992 (Table 3.1). Extensive monospecific stands of torpedo grass are now found in the marshes of Lake Okeechobee (J. Schardt, pers. comm.). Its key ecosystem impact is displacement of native vegetation along freshwater shorelines. Anecdotal evidence suggests it has little value for fishes, waterfowl, or songbirds (Tarver et al. 1985).

Water Hyacinth

Introduced during the late nineteenth century, the floating South American water hyacinth (*Eichhornia crassipes*) covered more than 48,500 ha of Florida's waterways during the 1950s. Because of intensive management efforts during the past 30 years, its populations have covered less than 1600 ha since 1988 (Schardt and Ludlow 1993; Chapter 14 in this volume). The key environmental effect of water hyacinth, like that of hydrilla, is the formation of a cover, or an aquatic plant canopy, over a body of water. Consequently, alterations to waters infested with it may include low light intensities and dissolved-oxygen levels, higher water temperatures, greater water loss (through evapotranspiration), higher sediment loading as a result of leaf decay, lower fish production and possible kills, and smothered beds of native submersed vegetation (Schmitz et al. 1993). We should note, however, that a diverse, dense macroinvertebrate fauna has been found on water hyacinth roots in Florida, in which the most common species is an amphipod, *Hyalella azteca* (O'Hara 1967; Hansen et al. 1971). Because *H. azteca* consumes living and dead root tissue, it may influence the rate of water hyacinth decomposition (Bartodziej 1992). Large floating water-hyacinth mats, uprooted by heavy winds, have destroyed emergent plants important to waterfowl (Schmitz et al. 1993).

A Closer Look at Species and Habitats

Now that we have outlined what is known about many of Florida's most damaging plant invaders, it is of interest to see what generalizations are suggested by these facts and similar information about other nonindigenous species. Why do some species have such great environmental impacts, while others do not? Do certain traits characterize damaging species? Can these traits be used to predict future invaders? Are invading species likely to cause certain kinds of impacts rather than others? Are certain habitats and regions of the state particularly heavily invaded, and if so, why?

Regional Differences

Many biologists believe that invasive nonindigenous plants flourish only in communities poor in native species—particularly if these communities are subjected to disturbance (Elton 1958; Parsons 1982; Ewel 1986; Chapter 1 in this volume). This traditional belief is based on the view that all available resources in an undisturbed ecosystem, especially a species-rich one, are used by native species that are as competitive as any introduced species. Southern Florida is the part of the state most heavily invaded by nonindigenous species, and Howard (cited by Long 1974a) contends that its native plant community is depauperate because this region is isolated by geological and climatic factors that prevent substantial natural immigration from other tropical floras. The southern portion of the Florida peninsula is island-like (surrounded on three sides by water and on the fourth by frost), and, like other islands, it appears to have an impoverished native biota and thus to be especially vulnerable to invasions, according to the traditional view (Ewel 1986; Myers and Ewel 1990b). But, as is true of other invocations of the "biotic resistance" hypothesis (Chapter 1 in this volume), there is no direct evidence showing that northern Florida ecosystems "resist" invaders better than southern Florida ones do. Among our species accounts, it is difficult to point to a specific invasive nonindigenous plant whose success is due to the absence of a sufficient number of native species to repel it.

Another possible reason for southern Florida's heavier invasion is that it is the locus of more introductions (C. Lippincott, pers. comm.). A disproportionate fraction of the import activity (see Chapter 2 in this volume) is in southern Florida, partly because the climate is conducive to subtropical ornamentals and botanical gardens.

Disturbance of the landscape has played a prominent role in the invasions of many nonindigenous plant species in Florida. Humans have long modified the landscape by farming (thus destroying natural structure and increasing soil aeration and fertility), dredging canals (which help spread invaders), creating spoil banks (new habitats) along coastal areas, constructing roads and fire lanes, interfering with natural fire frequency, and accelerating eutrophication in waterways. Air potato, Australian pines, Brazilian pepper, carrotwood, cogon grass, kudzu, lather leaf, melaleuca, Old World climbing fern, and skunk and sewer vines most frequently invade ecosystems disturbed by nature or humans. Disturbed conditions in waterways, such as nutrient enrichment, can lead to expanding hydrilla populations in Florida lakes (Canfield et al. 1983b). Hofstetter (1991) argues that all remaining freshwater wetlands south of Lake Okeechobee are disturbed by dry-season fires, have altered hydrological regimes because of human activity, and are thus vulnerable to invasions.

Ewel (1986) contends that human modification of southern Florida's ecosystems, rather than low native species richness, has rendered them especially vulnerable to invasions; he suggests that when these ecosystems were more pristine, they were more "invasion resistant" than they are today.

Is southern Florida more disturbed than northern Florida? Superficially, it may seem so. There are large, relatively continuous forests in the north, and fragmentation appears far less. Hydrological regimes are certainly more modified in the south. Fire regimes are highly altered in all parts of the state, but the main change in the north—the season of burn rather than the intensity and frequency of fires—somehow seems less drastic than those in the south. And certainly terrestrially the north seems less affected by nonindigenous plants. For example, pine flatwoods are the most widespread terrestrial ecosystem found from northern to southern Florida and once covered 50 percent of the state (Davis 1967). Although pine flatwoods in southern Florida have been successfully invaded by several nonindigenous plant species, particularly Brazilian pepper and melaleuca, more northern and drier pine flatwoods appear less vulnerable to invasion (Abrahamson and Hartnett 1990). Human alteration of fire regimes is the most common cause of successional changes in pine flatwoods (Richardson 1977; Peroni and Abrahamson 1986). Increased human settlement and land development throughout southern Florida have probably limited the area burned by each lightning-originated fire.

It is difficult to measure degrees of disturbance, however, because it can take myriad forms and have various effects. No one, therefore, has quantified the relative degrees of disturbance in northern and southern Florida. The appearance of homogeneous wilderness in the north is deceptive; the great majority of the primary forest was logged and grazed. In northern Florida, pine flatwoods are highly managed—primarily for increased pine seedling establishment and total wood yields but also to reduce abundances of other plant species in favor of pines (Abrahamson and Hartnett 1990). This goal is accomplished by prescribed burns that differ greatly from pre-European-settlement natural burns in that they occur primarily in the winter; natural fires occur in the spring and summer (Streng et al. 1993). Burns are still frequent—and Abrahamson and Hartnett (1990) argue that frequency alone may preclude potential invaders—but a fire-adapted nonindigenous species can be favored by frequent fires. As noted earlier, cogon-grass invasion is facilitated by fire, and this plant has invaded habitats that are otherwise relatively undisturbed. The greater invasion success of Brazilian pepper and melaleuca in the south than in the north can almost certainly be traced to the fact that these species simply cannot tolerate the climate of the north.

Northern Florida's species-rich temperate hardwood forests have suffered fewer nonindigenous plant invasions than have southern Florida hardwood forests, but there is no evidence that the former are inherently less invasible or less disturbed. Both *Ardisia crenulata* and *Tradescantia fluminensis* have invaded northern Florida hardwood forests, for example, forming dense mono-specific understory layers and apparently inhibiting regeneration of native species (C. Lippincott, pers. comm.). Fewer species may have been introduced into Florida that are adapted to the northern, closed-canopy conditions than to the high-light conditions of southern Florida (C. Lippincott, pers. comm.).

Some nonindigenous species found in northern Florida do not depend on disturbance for spread. They just slowly replace the native flora. A good example of a nonindigenous plant species that outcompetes native plants occurs in the spring-fed rivers of Florida. These clear, shallow rivers (Wacissa River, Jefferson County; Wakulla and St. Marks rivers, Wakulla County) have historically been occupied by native submersed species of *Vallisneria* and *Sagittaria*. Once introduced, hydrilla gradually increases its coverage because of its ability to grow more rapidly to the water surface. Eventually hydrilla shades out the more slowly growing native species (Haller and Sutton 1975). A statewide decrease in *Vallisneria* in 1983 was attributed primarily to competition with hydrilla (van Dijk 1985). Chinese tallow, which has rapidly spread around several northern Florida lakes and may one day equal southern Florida's melaleuca in ecological impact if not in areal coverage (Jubinsky 1993), does not appear to rely on disturbance.

Invasions by nonindigenous plants in Florida have substantially altered stand structures by creating novel habitats, as we shall see, but not all of Florida's plant communities are equally vulnerable to massive invasion. Brothers and Spingarn (1992) suggest that invasions in fragmented Indiana old-growth forests are discouraged by the edge response of forests themselves: development of a dense wall of bordering vegetation (native or nonindigenous) that reduces interior light levels and wind speeds. Similarly, kudzu can rapidly invade a disturbed area along the edge of a temperate hardwood forest and quickly form a dense wall, but the interior portions of the forest may remain uninvaded because of lower light levels. Southern Florida may have its kudzu analogs in other vines, such as sewer vine and Old World climbing fern, that exhibit a similar growth pattern in tropical hardwood hammocks. Although tropical hammocks are usually surrounded by more open vegetation, such as pineland and prairie, the margins are often densely vegetated and nearly impenetrable (Snyder et al. 1990).

Species That Constitute New Habitats

Probably the best-known examples of plant species that constitute new habitats and dramatically alter the landscape are those that establish forests where none had existed (Vitousek 1986; Simberloff 1991). The formation of melaleuca forests in southern Florida marshes is a good example. Large areas of the Everglades Conservation Areas marshes are dominated by only one species: saw grass (Wade et al. 1980; Kushlan 1990). Although saw grass may exceed 3 m in height, melaleuca is typically 20 m tall, allowing it to outcompete marsh plants for sunlight. In the past, invasion of these Everglades marshes by woody vegetation has been precluded by frequent fires and fluctuating water levels (Alexander 1971; Hofstetter 1975; Wade et al. 1980), but melaleuca is well adapted to fire, which can actually facilitate its spread. Since the creation of water-impoundment areas (now called conservation areas), flood control and water regulation have profoundly affected these Everglades marsh-plant communities (Kushlan 1990), probably producing ideal conditions for melaleuca forest formation. Australian pines are another example of a "forest maker" in disturbed habitats. Because *Casuarina equisetifolia* is more resistant to salt spray than is native woody vegetation and can grow taller and closer to the water (Johnson and Barbour 1990), it has invaded naturally disturbed, formerly treeless coastlines throughout southern Florida.

Disturbance in Florida has created new opportunities for other species to form novel habitats. For example, kudzu canopies typically cover abandoned agricultural fields in northern Florida. When gaps occur in adjacent temperate hardwood forests that allow enough light to reach the forest floor, kudzu rapidly invades these gaps, forming a dense vine canopy. The removal of tree canopies in Dade County's tropical hardwood forests by Hurricane Andrew in 1992 is another example. Invasions by nonindigenous vine species, particularly *Paederia cruddasiana,* produced a dense nonindigenous plant cover that threatens to alter hammock ecology permanently by preventing new hardwood canopy regeneration (S. Vardaman, pers. comm.; see Chapter 4 in this volume).

Disturbance is not always necessary (Wade et al. 1980; Ewel 1986). Hydrilla and water hyacinth, for example, both constitute new habitat and can outcompete most native species in existing habitats. The growth of native submersed aquatic plants in many of Florida's naturally eutrophic lakes is limited by light penetration (Canfield et al. 1985). Thus, historically, in lakes such as Orange Lake (Alachua County), Lake Harris (Lake County), and Lake Trafford (Collier County), no native vascular plants have colonized areas much deeper than 1.2 to 1.8 m because of light attenuation by phytoplankton. Hydrilla, which has a lower light compensation point than native species, expands rapidly into deeper waters (Van et al. 1976). The growth habit of hydrilla causes it to form a habitat that some waterfowl accept as an alternative to the shallow marshes that are rapidly being lost in Florida (Johnson 1987). Within a few years of its introduction in the late 1880s, water hyacinth colonized open-water habitats in northern Florida rivers, creating dense, impenetrable stands (Buker 1982) in areas where formerly there were no vascular plants. No free-floating North American plant can withstand open-water wind or waves as well as water hyacinth can. Finally, disturbance may have played a lesser role in melaleuca invasion of pine–cypress ecotones (Myers 1975, 1983, 1984; Ewel et al. 1976; Duever et al. 1979). Myers (1976) and Duever et al. (1979) suggest that either melaleuca is better suited to a transition zone that is too dry for native wetland species and too wet for native upland species or it has no significant competitors. Hofstetter (1991), however, believes these melaleuca invasions in pine–cypress ecotones are more common where hydroperiods have been shortened.

Species That Modify Habitats

Many of the plant species mentioned in the previous section can affect ecological processes in addition to constituting new habitat. Certainly a number of invasive nonindigenous plant species in Florida can modify nutrient supply and evapotranspiration rates. Because of low fixation rates in strongly acid soils and high rates of leaching, for example, nitrogen concentrations are low in many Florida soil types (Brown et al. 1990). Invasions by introduced nitrogen fixers—like Australian pines (symbiotic with an actinomycete; NRC 1984) and kudzu (Tanner et al. 1980)—may ultimately alter soil conditions and lead to changes in nitrogen cycling and increased productivity. Leaf drop and decay associated with Chinese tallow can lead to higher phosphorus, potassium, nitrate, zinc, manganese, and iron soil concentrations and may also increase productivity (Cameron and Spencer 1989). Water hyacinth and hydrilla invasions, however, can lead to dramatically lower dissolved phosphorous concentrations during the growing season (McVea and Boyd 1975; Canfield et al. 1983a; Schmitz and Osborne 1984). Higher evapotranspiration rates have been linked to water hyacinth in container studies (Penfound and Earl 1948; Timmer and Weldon 1967; Rogers and Davis 1972) and possibly to melaleuca. Hofstetter (unpublished data) found that transpiration rates per unit leaf area of saw-grass and melaleuca leaves were similar—but because the amount of leaf area on mature melaleuca trees per unit ground area is about four times that of robust saw-grass communities, it seems logical to conclude that melaleuca forests will have higher evapotranspiration rates than saw-grass communities, leading to lower water tables, especially during times of drought. Data are not yet available for tests of this hypothesis.

Monospecific stands of Australian pines, Brazilian pepper, melaleuca, and water hyacinth have all altered litter accumulation. Brazilian pepper leaves little leaf litter beneath its canopy because of high decomposition rates (Ewel 1986), but melaleuca forests create thick layers of undecomposed leaf litter, leading to higher soil elevations (Flowers 1991). Australian pine forests also produce a litter layer of at least 10 cm (Klukas and Truesdell 1969). In aquatic environments 1 ha of water hyacinths can deposit between 404 and 1886 metric tons (wet weight) of detritus per year (Penfound and Earl 1948; Center and Spencer

1981), which can lead in turn to pronounced modification of water chemistry, such as low dissolved oxygen and a change in available phosphorus concentrations (Reddy and Sacco 1981).

Changes in disturbance frequency have been attributed to several nonindigenous plant species in Florida, including several that do not themselves constitute a substantial new habitat. Nonindigenous grasses can set in motion a grass–fire cycle in which an invader colonizes an area and provides the fuel necessary for more frequent and, perhaps, more intense fires (D'Antonio and Vitousek 1992). Both Burma reed and cogon grass may change fire intensities and frequencies. Because the spongy outer bark and highly flammable foliage of melaleuca are structured to increase fire intensity—thus increasing the probability of top-kill in less fire-resistant competitors (Wade et al. 1980; Wade 1981)—melaleuca probably invades partly by modifying the fire regime (Flowers 1991). Because of the heavier fuels under melaleuca forests, ground fires have become more common there than in adjacent herbaceous communities. Pine–cypress ecotones may be particularly at risk for the higher native tree mortalities associated with melaleuca fires (T. Pernas, pers. comm.). At least one species of vine, Old World climbing fern, appears to act as a pathway for fire spreading into tree canopies that rarely burn in bottomland hardwood communities.

Community Effects of Invasive Plants

Although some introduced plant species can provide a limited "new habitat" for native species—such as structure for birds to perch on and underwater structure for fish to hide in—several of the most widespread nonindigenous plants in Florida (air potato, Australian pine, Old World climbing fern, hydrilla, melaleuca, and torpedo grass) are little used by native insects (Mead and Bennett 1987; T. Center, pers. comm.; D. Habeck, pers. comm.). Moreover, Australian pine, Brazilian pepper, melaleuca, and water hyacinth are rarely grazed by larger herbivores (U.S. Congress 1957; Schortemeyer et al. 1981; Schmitz et al. 1991).

It seems logical to conclude that plant populations which offer little forage value will ultimately lower wildlife abundances or alter species composition: available food, after all, is an essential component of any wildlife habitat (Shaw 1985). Many invading plant species in Florida form dense monospecific stands entirely excluding native plants, but the few preliminary Florida wildlife studies attribute changes in wildlife densities to physical or chemical factors rather than to a lack of suitable forage. For example, lower small-mammal populations were found in melaleuca and Australian pine forests than in native plant habitats (Mazzotti et al. 1981; Ostrenko and Mazzotti 1981; Sowder and Woodall

1985). Sowder and Woodall (1985) attribute the difference to lack of ground cover in melaleuca forests. Similarly, the herpetofauna in Brazilian pepper forests may be related more to physical factors such as closed canopy conditions than to vegetation composition (Dalrymple 1988). In aquatic environments, hydrilla and water-hyacinth populations can dramatically affect zooplankton, benthic macroinvertebrate, and fish populations. (See Colle and Shireman 1980; Watkins et al. 1983; Schmitz and Osborne 1984; Richard et al. 1985; Schramm et al. 1987; Schramm and Jirka 1989; Schmitz et al. 1993.)

Once again, many of these changes are due not to a lack of suitable forage material but rather to pronounced changes in environment, such as low dissolved-oxygen concentrations and changed physical structure. Nevertheless, water hyacinth may benefit native fishes and invertebrates by providing shelter and perhaps senescing plant material as food (W. Bartodziej, pers. comm.). In sum, food shortages imposed by the unpalatability of nonindigenous plants are much more difficult to demonstrate than are their physical or chemical effects, though such shortages surely must arise at times. Competition between species for food may be notoriously hard to prove, but this does not mean it is not important (Wiens 1989).

Why Some Species Become Invasive

Williamson and Fitter (1996) propose the "10-10" rule for introduced plants: they have observed that, for pasture plants in Australia and for angiosperms and Pinaceae in Britain, about 10 percent of introduced species become established (naturalized) and about 10 percent of those become invasive pests. They also suggest that, in both places, many more species could become established (and, presumably, about 10 percent of those would become new pests) if only they could immigrate. The precise reasons why some species become invasive and others do not frequently seem mysterious or idiosyncratic (Chapter 1 in this volume).

It is often argued that a lack of coevolved diseases, parasites, and herbivores attacking nonindigenous plant invaders has given them a competitive advantage over natives (Hansen et al. 1971; Westman 1990; Dray et al. 1993). Many of Florida's nonindigenous plant invaders were deliberately imported (Morton 1976; Austin 1978a) and were cleansed of their phytophages and pathogens on arrival. Melaleuca, for example, is associated with several hundred insect species in its native Australia (Laroche 1994) but attracts few, if any, insects in Florida (T. Center, pers. comm). Melaleuca trees in Florida have greater girth and height and exhibit greater growth rates than do those in Australia (Laroche 1994). In Australia, Balciunas and Burrows (1993) have concluded from short-term studies on saplings that ambient, nonoutbreak levels of insect herbivory suppress the growth of melaleuca. They therefore suggest that freedom from insect attack is the main reason for the invasive character of melaleuca in Florida. Others, however, have argued that habitat change (destruction of wetlands) is what has made melaleuca scarce in parts of Australia (D. Thayer, pers. comm.). It seems at least plausible that it is habitat change which allowed melaleuca to become invasive in Florida.

For an example in which coevolved predators rapidly suppressed a nonindigenous plant, consider the case of alligatorweed (*Alternanthera philoxeroides*). Densities of this plant plummeted in Florida waterways after the release of three South American insect species beginning in the 1960s (Coulson 1977). But introduced, coevolved natural enemies, even when they survive, do not always stem the spread of an invasive nonindigenous plant. In Florida, despite the deliberate release of several insects and plant pathogens (Goyer and Stark 1981; Center and Van 1989; Center et al. 1990), water hyacinth continues to invade waterways in areas where it is not managed (Schardt 1994). Other factors besides biological interactions may be responsible for controlling water-hyacinth expansion in its native range. Seasonal flood events, for example, have played an important role in determining water-hyacinth abundances on rivers in Amazonia (Brazil) (Gopal 1987). Similarly, periodic water-level fluctuations in the St. Marks River naturally reduce water-hyacinth abundances by flushing plants into the Gulf (W. Bartodziej, pers. comm.).

Future Invaders

Several nonindigenous plant species seem poised to expand their range greatly in Florida, especially in southern Florida wetlands and hammocks. Cat-claw mimosa (*Mimosa pigra*), for example, a Central America native, was introduced into Florida probably through an ornamental plant nursery or plant collector in the early 1950s. It has also been spread around the world as a cover crop for erosion control. In Thailand, cat-claw mimosa has formed large thickets, causing flooding along rivers and irrigation systems by obstructing water flow and increasing the sedimentation rate. An intact natural floodplain in Australia has been extensively overrun (Lonsdale et al. 1989). In northern Australia, the species invades sedgeland and grassland communities on floodplains, particularly where fires have removed the vegetation, and paperbark (*Melaleuca* spp.) swamp forests fringing the floodplain (Lonsdale 1992a). Populations of many birds and lizards there are lowered where cat-claw mimosa has formed an almost monospecific shrubland (Braithwaite et al. 1989). It now infests 395 ha in five locations in Florida: southern Martin County, Highlands County near Sebring, along the St. Lucie River, Hobe Sound (Martin County), and on a golf course in Hollywood (Broward County). This species can form dense, impenetrable thickets that exclude all other vegetation and even small animals. The Everglades may be particularly vulnerable because this plant can grow on its low-fertility soils (Sutton and Langeland 1993). The hydrological cycle in southern Florida may provide ideal conditions for germination and subsequent growth of cat-claw mimosa (Sutton and Langeland 1993).

Other species such as laurel fig (*Ficus microcarpa*) and tropical soda apple (*Solanum viarum*) also appear to be rapidly expanding their ranges in Florida. Although fig trees from Southeast Asia have been in Florida since the 1930s, they were introduced without their pollinating wasps and remained sterile until recently. Within the past 15 years, however, pollinators have been introduced for at least three *Ficus* species (McKey and Kaufmann 1991; Nadel et al. 1992). The laurel fig appears to be able to produce the largest number of seedlings in southern Florida and is of great concern as a potential invader in Florida's natural environment (McKey and Kaufmann 1991). A more recent introduction, tropical soda apple, has spread in Florida since the early 1980s, now infests at least 61,000 ha (Mullahey et al. 1993), and can be found in several other states. This native of Argentina and central Brazil has formed dense monospecific stands on agricultural and pasture lands, on ditch banks, and along roadsides. Wildlife, such as raccoons, deer, and feral pigs, apparently help spread its seed by ingestion and defecation (Mullahey et al. 1993), and Coile (1993) suggests that natural areas are at risk for invasion.

Species that have already been introduced but are not especially invasive may also become problematic in the future. Many of Florida's most widespread nonindigenous plant species were introduced long before they became noteworthy. Brazilian pepper, for example, introduced during the nineteenth century (Barkley 1944; Morton 1978), started becoming noticeable on the landscape only in the early 1960s (D. Austin, pers. comm.). Although long time lags between introduction and observed rapid population expansion might be related to other factors such as unnoticed growth, some sites act as staging areas from which nonindigenous species shower the surrounding landscape with seeds, and populations may eventually produce genetic variants adapted to local conditions (Ewel 1986). The specter of a worsened invasion of *Hydrilla verticillata* may seem farfetched, but it is a very real possibility. All hydrilla in Florida is dioecious and female, so reproduction is entirely asexual. In 1982, a conspecific monoecious form was discovered near Washington, D.C., in the Potomac River (Steward et al. 1984), and it has now spread as far as California and Washington state. It will probably reach Florida eventually, and the greater diversity of genotypes that sexual reproduction will confer may well lead to the selection of even more invasive strains, particularly in northern Florida (Schmitz et al. 1993; Steward 1993), while seeds can provide additional means of dispersal and overwintering (Steward 1993).

Prospects

Most of the information on the impact of invading nonindigenous plant species in Florida, as everywhere else, is anecdotal and observational. Nevertheless, numerous species have clearly had ecosystemic impacts—primarily by virtue of modifying the physical habitat or ecological processes. Not all of these species required disturbance for establishment, and it is far from proved that the apparently greater ecological impact of nonindigenous plants in southern than in northern Florida is due to greater disturbance in the south. In fact, some ecosystems (lakes and rivers, for example) are at least as heavily affected by nonindigenous species in the north as in the south. Greater damage from invaders in the south may result largely from a greater influx of introductions there.

The specific reasons why a few nonindigenous species have been ecological scourges, whereas most are relatively innocuous, remain mysterious. The observation that the high-impact species either provide a novel structural habitat or greatly modify physical structure or ecological processes falls far short of giving us a tool with which to forecast the impact of an as-yet unintroduced species or one that is already present but has not become invasive. We should be particularly humble about our predictive ability in light of the observation that several of the major pests remained restricted and unproblematic for extended periods before spreading (Ewel 1986).

Without detailed scientific investigations, particularly experiments, designed to uncover the processes and mechanisms by which invaders affect native species, communities, and ecosystems, it is difficult to predict how these invaders will alter Florida. The statistics on invasive and noninvasive plants amassed by Williamson and Fitter (1996) point to a shortcoming of research to date on invasive nonindigenous plants in Florida. But they also suggest avenues of investigation that might illuminate the trajectories of species that have become pests and, as well, aid in predictions about future introductions.

Three areas of research seem particularly appealing. First, we have not talked about the failures—species that have been introduced but disappeared (or else remained restricted to cultivation)—or about the innocuous established species. It would be interesting to examine the more than 900 established nonindigenous species, especially those closely related to pest species, in a search for patterns: traits that typify pests, geographic or temporal aspects of invasion history in which pests and other species differ, propagule sizes, and so forth. If introduced species that did not survive could also be identified, such comparisons would be all the more interesting. Second, substantially more intensive, largely autecological studies of particular species are needed. It might be especially profitable to examine in great detail the history and effects of a pair or small group of related nonindigenous species with seemingly differing impacts (Ehrlich 1986). Two nonindigenous congeners of air potato are established in

Florida, for example, but neither seems invasive or widespread (Hammer 1995a): *Dioscorea alata* is found in scattered locations in southern and central Florida, while *D. sansibarensis* was recently found in one Dade County hammock. Why is air potato a scourge while these species are so far innocuous? Finally, it might be instructive to compare the autecologies of native and nonindigenous congeners—in *Ardisia* and *Tradescantia,* for example (C. Lippincott, pers. comm.)—in terms of differing invasiveness and impact. Such a comparison for native and introduced fire ants (*Solenopsis invicta*) (Simberloff 1985; McInnes 1994) was enlightening. Such studies would be ideal doctoral dissertations, although their large component of natural history would make them somewhat unfashionable nowadays.

ACKNOWLEDGMENTS

We thank Jack Ewel and Carol Lippincott for numerous suggestions and access to unpublished data and Tom Brown for thoughtful criticism.

4 The Impact of Natural Disturbances

Carol C. Horvitz

 This chapter examines what we can learn from the impact of Hurricane Andrew and the effects of natural disturbances on nonindigenous invaders of natural communities in southern Florida. I will draw on studies of subtropical hardwood hammock forest communities (Koptur et al. 1994; Armentano et al. 1995; Horvitz et al. 1995, in press; Slater et al. 1995), specialist and generalist plant/insect interactions of a native shrub (Pascarella 1995, unpublished data), demography of a native shrub (Pascarella 1995; Pascarella and Horvitz in press), a native fig–fig-wasp mutualism (Bronstein and Hossaert-McKey 1995), fruit-frugivore interactions in a subtropical hardwood hammock and a nearby disturbed bay-head forest in Everglades National Park (C. C. Horvitz, T. H. Fleming, R. Seavey, J. Seavey, and J. Nassar, unpublished data), and observations on ant community changes (T. H. Fleming and the fall 1995 class of BIL 631, University of Miami, unpublished data), as well as speculation about a nonindigenous chrysomelid beetle.

My main conclusion is that, although a natural disturbance may facilitate invasion by some nonindigenous species, not all invasions are disturbance-mediated and some native organisms may be more resilient to disturbance than certain nonindigenous species. The natural regeneration dynamics of natives and nonindigenous species may be illuminating if the invasion of a natural area is the issue at hand. The success of a nonindigenous species in a human-disrupted environment does not always predict the species' ability to displace native species in a natural system.

Hurricanes in Forests

Although hurricane disturbance is to be expected every 8 to 12 years in southern Florida (Simpson and Lawrence 1971; Chen and Gerber 1990; Pimm et al. 1994), it is not always as severe as that caused by Hurricane Andrew. In 1926, the last time a hurricane of similar magnitude traversed southern Florida (Simpson 1932), relatively few nonindigenous organisms were present and natural areas were much less fragmented. Andrew was a small but intense hurricane (class IV on the Saffir/Simpson Hurricane Scale) with estimated maximum sustained wind speeds estimated between 230 km/hr (Mayfield et al. 1994) and 242 km/hr (Pimm et al. 1994) and gusts up to 282 km/hr. It crossed southern Florida early on 24 August 1992 (Mayfield et al. 1994) and traversed the few remaining forest preserves of the Miami Rock Ridge (Dade County), an area characterized by unique communities including a subtropical hardwood forest comprising principally tropical families of trees (Alexander 1967; Tomlinson 1980; Snyder et al. 1990).

The main effect of a hurricane in a natural forest community is to open up the canopy (Brokaw and Walker 1991), usually by causing complete defoliation and major limb loss in the canopy trees. In the most intense hurricane winds, many canopy trees will also be tipped up, snapped off, and even killed. In general, a pulse of light (Fernandez and Fetcher 1991; Horvitz et al. 1995; S. Oberbauer and S. Koptur, unpublished data) and probably nutrients (Brokaw and Walker 1991) is introduced into the forest understory. Species native to hurricane-prone regions are adapted to this kind of disturbance and show vigorous re-sprouting (Brokaw and Walker 1991; Yih et al. 1991; Howard and Schockman 1995; Slater et al. 1995). Whether such disturbance favors invasion of forests by nonindigenous species is an issue of considerable interest.

Some tropical-cyclone-prone forests may be dominated by nonindigenous species. In the Mariana Islands, for example, the nonindigenous papaya is considered an "indicator of storms and canopy gaps" (Craig 1993). In one forest site on Mauritius, Lorence and Sussman (1986) noted that "exotic invasion [was] accelerated by severe cyclone damage." Similarly, Geldenhuys et al. (1986), studying evergreen forests of South Africa, noted that some (though not all) nonindigenous species use cyclone-opened areas to establish themselves. Looking at an effect rather than a cause of domination of forests by nonindigenous species, Macdonald et al. (1991) noted that because nonindigenous species in the Mascarene Islands are less adapted to cyclones than are native species, as forests become dominated by nonindigenous species they become less resilient and more susceptible to severe canopy damage. Moreover, heavy nonindigenous vine cover may cause more trees to tip up during hurricanes because the vine tangles link tree to tree (Lorence and Sussman 1986; Macdonald et al. 1991; Putz 1991).

Dade County Hardwood Hammocks

In fall of 1992, eight months after Hurricane Andrew, my colleagues and I (Horvitz et al. 1995) initiated research in three hammocks in Dade County parks (Matheson, Deering, and Castellow hammocks). A parallel study was initiated in three hammocks (Redd, Pilsbry, and Grimshawe hammocks) inside Everglades National Park (ENP) by Koptur and Oberbauer (Koptur et al. manuscript).

During the first year after the hurricane, the Dade County study plots contained nearly 30 percent nonindigenous plant species; those in ENP contained less than 3 percent. In ENP, the nonindigenous species accounted for less than 1 percent of the stems (Koptur et al. unpublished data), whereas in the Dade County preserves they accounted for 33.7 percent (N = 2666 live stems, all size classes and all hammocks pooled for the eight-month post-hurricane census) (C. C. Horvitz, S. McMann, A. Freedman, and R. H. Hofstetter, unpublished data). The most abundant nonindigenous vines noted in the Dade County hammocks (in observations both on and off the study plots) included skunk vine (*Paederia*), jasmine (*Jasminum*), air potato (*Dioscorea*), wood rose (*Merremia*), and pothos (*Epipremnum*). The most abundant nonindigenous trees and shrubs included papaya (*Carica papaya*), castor bean (*Ricinus communis*), Bishopwood (*Bischofia javanica*), shoebutton ardisia (*Ardisia elliptica*), and Surinam cherry (*Eugenia uniflora*) (Horvitz 1994; Horvitz et al. 1995). None of these species was brought into the hammocks by the hurricane. They were already present when the hurricane occurred, although some were "hidden" in the seed bank.

These hammocks were at different distances from the eye of the hurricane and thus represent a disturbance gradient. Canopy disturbance and regeneration were measured from fish-eye canopy photographs. The hammock located to the north beyond the eye wall of the hurricane (Matheson hammock) suffered only half the canopy disturbance experienced by the hammocks at the outer edge of the northern eye wall (Deering hammock) and at the inner edge of the northern eye wall (Castellow hammock).

The plants in the study plots were measured, classified (species, type of damage, survival, and size), and tagged for long-term study. The first census was completed eight months after the hurricane. Damage differed significantly according to life form, tree size, and hammock. Trees suffered more damage than shrubs or vines. Large trees suffered more damage than smaller trees, and the hammock at the inner edge of the eye wall suffered more damage than the others. The nonindigenous tree species *Carica papaya*, recruiting rapidly and abundantly from a seed bank, varied in its importance among the hammocks; where it was present, it created canopy within four months, perhaps negatively affecting native pioneer tree species. Nonindigenous vine cover also varied significantly among the hammocks (from most in Matheson to least in Castellow).

The most frequently occurring nonindigenous vines were *Jasminum* spp. (Horvitz et al. 1995).

The Dade County subtropical hardwood hammocks differed from those of Everglades National Park not only in the amount of nonindigenous vegetation but also in the degree to which they were affected by Hurricane Andrew. Two of the Dade County hammocks were more directly in the path of the most intense winds of the eye of the hurricane. Tree mortality in these two hammocks (Castellow and Deering) was 67 percent and 48 percent; in the third hammock (Matheson), located north of and outside the eye wall, tree mortality was only 32 percent (Horvitz et al. 1995), closer to that reported for ENP hammocks: 20 to 30 percent, estimated aerially by Loope et al. (1994), 11.5 percent in long-term study plots at some hammocks (Slater et al. 1995), and 20 percent in long-term study plots at others (Koptur et al. unpublished data).

Within Everglades National Park, some subtropical hardwood hammocks are at risk of invasion that is not necessarily hurricane related—especially by shoebutton ardisia (*Ardisia elliptica*), which has formed dense, nearly mono-specific stands in human-disturbed bay-head forests very near Paradise Key Hammock. This species invades forests with closed canopies and was taking over much of the understory of Paradise Key Hammock several years before the hurricane but had been controlled there by a massive and vigilant program of hand-weeding and selective herbicide application (1987–1990), organized by Rick and Jean Seavey, volunteers at ENP. The risk of invasion by this fleshy-fruited animal-dispersed species may have increased a few years after Hurricane Andrew because of an increase in frugivore activity in hammocks. This species, too, often forms a dense carpet of seedlings in the understory of Brazilian pepper (*Schinus terebinthifolius*) dominated forests (C. C. Horvitz, R. Seavey, and J. Seavey, unpublished data).

Three species of nonindigenous trees in Everglades National Park have formed nearly monospecific stands: Australian pine (*Casuarina equisetifolia*), Brazilian pepper, and melaleuca (*Melaleuca quinquenervia*) (Chapter 17 in this volume). Armentano et al. (1995) have reported on the effects of Hurricane Andrew on these forest communities. All three species suffered de-foliation and loss of limbs. No new invasions or spread of invasions were re-ported as a direct result of the hurricane, although there was concern that wind dispersal of the Australian pine and the melaleuca would have occurred, as both species had many wind-adapted seeds available at the time of the hurri-cane. Although the animal-dispersed Brazilian pepper was not in fruit at the time of the hurricane, there was concern that the next fruiting period could be associated with movements of seeds and establishment of seedlings in newly opened areas, including disturbed hammocks but especially mangroves. In contrast to those in Everglades National Park, the Australian pines at Bill

Baggs State Park on Key Biscayne, closer to the strongest winds, were nearly all blown down. Managers there took the opportunity to initiate a restoration plan that included removing the Australian pine stumps and replanting with natives (Westervelt 1995).

Invasive Guilds in Forest Regeneration Niches

To explore how nonindigenous plants may interfere with the process of forest regeneration after a natural disturbance, my colleagues and I (Horvitz et al. in press) developed a conceptual model of nonindigenous invasion in forest communities based on the ways nonindigenous species interfere with sources of natural regeneration. We proposed six classes of nonindigenous species: seed-bank robbers, seedling-layer "oskar" winners, ground-level-resprout stealers, canopy-layer thieves, vine blankets, and the seed rain of terror. The first four refer to species that interfere with certain sources of forest regeneration present at the time of the disturbance: the seed bank; "oskars" (the suppressed seedling layer; Silvertown 1993); resprouts from roots and fallen stems that are horizontal; and resprouts from standing snapped trees or trees that have lost major limbs. The fifth class, the vine blankets, interferes with all of the last three sources of regeneration. The sixth class, the seed-rain-of-terror species, takes over the "seed rain" (Denslow and Gomez Diaz 1980; Janzen 1986) both at the time of the disturbance and for years afterward from both internal and external sources of seeds. These species are not thought to be stored in a dormant seed pool but are believed rather to contribute to seedlings less than one year after they are dispersed.

Using empirical studies of the post-hurricane regeneration of subtropical forests during the first 26 months after Hurricane Andrew as an example, my colleagues and I (in press) reported that the vegetation in the subtropical hardwood hammock study plots in Dade County forest preserves in southern Florida contained 28 percent nonindigenous species (N = 90 species). Among these, several could readily be assigned to the proposed classes: at least one seed-bank robber (*Carica papaya*), three seedling-layer "oskar" winners (*Harpullia arborea, Adenanthera pavonina,* and *Ardisia elliptica*), two ground-level-resprout stealers (*Jasminum fluminense* and *J. dichotomum*), two vine blankets (*Paederia cruddasiana* and *Dioscorea bulbifera*), two canopy-layer thieves (*Ficus microcarpa* and *Bishofia javanica*), and two seed-rain-of-terror species (*Bishofia javanica* and *Schinus terebinthifolius*).

Nonindigenous species as a group did not belong to any particular seed size category. Because the seed sizes of native hammock species were significantly correlated with a successional age index calculated by Ross et al. (1995) (Horvitz et al. in press), seed size seemed to be an indicator of shade tolerance.

Among the nonindigenous species, three trees (*Syzygium cumini, Harpullia arborea,* and *Adenanthera pavonina*), one shrub (*Ardisia elliptica*), and four vines (*Abrus precatorius, Jasminum fluminense, J. dichotomum,* and *Merremia tuberosa*) were shade tolerant. Four trees were shade intolerant (*Carica papaya, Schinus terebinthifolius, Bischofia javanica,* and *Pittosporum pentandrum*). Shade-tolerant nonindigenous species are particularly troublesome because they can invade a forest before disturbance. They may outcompete the native shade-tolerant sources of regeneration (shrubs and suppressed seedlings) such that, when the canopy is opened, they are there in greater abundance than the natives.

During the first twenty-six months after the hurricane, the most dynamic layers of the forest were the seedling layer and middle layer (2–4 m and 4–8 m); these layers experienced dramatic increases in stem density. As early as eight months after the hurricane, the 2–4-m layer at Deering Hammock was quite dense and predominantly composed of *Carica papaya,* which grew in height quite rapidly after appearing in the forest immediately after the hurricane from a dormant seed pool. In contrast, eight months after the hurricane, the 2–4-m layers of Castellow and Matheson hammocks were predominantly occupied by native plants consisting of stem suckers from fallen trees, resprouting branches from standing damaged trees, and plants that had grown up from the previously suppressed seedling state. A native tree that had recruited right after the hurricane from a dormant seed pool (*Solanum erianthum*) did not occupy the 2–4-m layer in great numbers eight months after the hurricane, but by twenty-six months it had become an important element of this layer at Castellow Hammock. The least dynamic layer of the forest was the layer above 8 m, indicating that canopy trees take longer than twenty-six months to regenerate (Horvitz et al. in press).

Both nonindigenous and native vine covers were reduced by a removal program. In control areas, nonindigenous vine cover remained high at one site heavily infested prior to Hurricane Andrew. Generally, nonindigenous vine coverage was denser than that of native vines. Of the various types of vine life forms, nonindigenous lianas and adventitious root climbers of tropical origin may present a greater potential for strong negative interactions with subtropical hardwood forest regeneration than do herbaceous vines or woody hemiepiphytes (Horvitz et al. in press).

Our conceptual model was successfully applied to subtropical forests and helped elucidate the invasiveness of several taxa in the different forest habitats of southern Florida. The model was also successfully applied to other regions invaded by similar taxa, including a review of the behavior of 50 taxa in diverse geographic regions, among them Western Australia, the Mariana Islands (South

Pacific), Hawaii, the Mascarene Islands (Indian Ocean), and South Africa. Ten of the genera in the review were represented by single species, whereas the genus *Rubus* claimed six invasive species; *Solanum* and *Acacia* five each; *Pittosporum, Ligustrum,* and *Albizia* three each; and the seven remaining genera two each. Of these 50 taxa, we could assign 25 to classes: eight were seed-bank robbers, six were seedling-layer "oskar" winners, five were seed-rain-of-terror species, four were vine blankets, one was a canopy-layer thief, and one was a ground-level-resprout stealer (Horvitz et al. in press).

One result of the study is the suggestion that understanding the roles played by nonindigenous species in terms of natural forest dynamics may provide clues to managers about which life-history stages of particular species require the most control and the critical times for action.

Effects of Hurricane Andrew on a Native Understory Shrub

Pascarella (1995, unpublished data) has reported dramatic positive effects of hurricane-caused canopy removal on the demography and reproductive success of a very common native understory shrub, marlberry (*Ardisia escallonioides*). Using the disturbance gradient created by Hurricane Andrew among several hammocks, he studied the effects of differing canopy removal on this species. Marlberry suffered very little mortality as a direct effect of the hurricane and had a dramatic increase in flowering due to the abundant light in the understory and rapid branching and growth in the year following Andrew. Another effect of Hurricane Andrew was the release of plants from normally high levels of infestation by a native insect enemy, a specialist seed-eating and flower-feeding moth whose populations were devastated by the hurricane. Before the hurricane, levels of predation reached 80 to 100 percent of flowering individuals—and 73 to 90 percent of all "apparent fruits" were actually insect galls and not fruits with viable seeds (Pascarella 1995, unpublished data). These moths emerge from pupae in the soil in the late summer and early fall to lay their eggs in flowers.

In the fall of 1992, after the severe defoliation of the plants caused by the hurricane, there were very few flowers. By the following fall, predation levels had dropped and flowering had increased—dramatically increasing the reproductive success of the plants. Moreover, the plant's pollinators (generalist halictid bees) suffered no negative effects from the hurricane. Although natural seedling recruitment had never been observed in the field during the two years before the hurricane, it did occur in the two years after the hurricane as a result of the pulse of successful reproduction. The life history of this native plant fits neither the pioneer mode nor the shade-tolerant mode (Pascarella 1995). (See Denslow 1980 and Whitemore 1989 for descriptions of these modes for tropical forest species.) The post-hurricane reproductive success was a pulse; by 1994, the mean seed production had begun to decline as the canopy began to close and the moth recolonized some areas.

A sensitivity analysis of a megamatrix model of population dynamics in a dynamic environment posed two questions: how rare are large-gap events, and how important are they to population dynamics of this native plant? (See Pascarella 1995 and Pascarella and Horvitz in press.) The answer is that, although large gaps are relatively infrequent, the life-history events that occur in them contribute disproportionately to population growth. Moreover, prereproductive juveniles are more important than large reproductives, even though the latter would have been judged most important if demographic analysis had been restricted to the most common habitat. These results underscore the relationship between the biology of native species and the natural disturbance regime of hurricanes.

In contrast to the increased growth in response to increased light seen in this native species, a nonindigenous congeneric species, shoebutton ardisia, appeared to yellow and exhibit stress in response to the higher light (R. Seavey, pers. comm. and unpublished data), at least during the first year after the hurricane. Unlike the native, this species does not require a gap to flower and fruit.

A Native Fig–Fig-Wasp Mutualism

Bronstein and Hossaert-McKey (1995) had been studying the relationship between the strangler fig (*Ficus aurea*) and its species-specific pollinator, a fig wasp (*Pegoscapus jimenezi*), since 1991. Even though the trees lost all their leaves and figs (and any immature wasps inside the figs) during the hurricane, five months later the plants had reestablished their pre-hurricane flowering phenology and level of fig-wasp pollination. The authors attribute this remarkable result to long-distance dispersal by wasps back into the area from outside the region of hurricane disturbance.

Increased Abundance of Frugivorous Birds

T. H. Fleming, R. Seavey, J. Seavey, and I, along with several field assistants, had been mist netting birds in Paradise Key Hammock and a nearby human-disturbed area in Everglades National Park since 1991 and studying seed dispersal of a nonindigenous shrub by examining seeds in fecal samples of the captured birds. The human-disturbed area was previously agricultural. The principal seed disperser was the gray catbird, *Dumetella carolinensis,* which readily consumed fruits of native plant species (marlberry; passionflower vine, *Passiflora suberosa;* wild coffee, *Psychotria nervosa;* redbay, *Persea borbonia;*

dahoon holly, *Ilex cassine;* strangler fig; snowberry, *Chiococca alba;* white stopper, *Eugenia axillaris;* Spanish stopper, *E. foetida;* wax myrtle, *Myrica cerifera;* greenbriar, *Smilax* sp.; and Florida trema, *Trema micranthum*) as well as nonindigenous plant species (shoebutton ardisia and Brazilian pepper).

Before Hurricane Andrew, catbirds were absent in the hammock sample (0 percent of 11 birds captured in the hammock) but abundant in the human-disturbed bay-head area (74.5 percent of 51 birds captured). Fruit resources were also much scarcer in the hammock than in the disturbed area. About 16 months after the hurricane, fruit resources were abundant in both areas and so were catbirds: in the hammock, catbirds made up 63.4 percent ($N = 41$) of the captures; in the human-disturbed area they made up 92.3 percent ($N = 39$). By the following year, the catbirds were considerably less abundant in the understory of the hammock (12.5 percent of 8 birds captured) but remained abundant in the human-disturbed area (86.2 percent of 29 birds captured). (See C. C. Horvitz, T. H. Fleming, R. Seavey, J. Seavey, and J. Nassar, unpublished data).

The potential for dispersal of nonindigenous plant species into the hammocks seemed greater in the year following the hurricane, although their relative abundance in the seed rain was likely to be lower than that of the abundant natives (especially white stopper, marlberry, and snowberry). None of the birds captured in the hammock produced a fecal sample that contained seeds of the nonindigenous plants—despite the abundance of these species in the fecal samples of the birds captured at the nearby human-disturbed site. These data suggest that most seeds were from very local sources, but our sampling effort was so limited that this hypothesis is only tentative.

The Little Fire Ant

A tuna-bait study of ant communities was carried out in November 1995, some 38 months after Hurricane Andrew, in three subtropical hardwood hammocks and a pineland (Castellow Hammock and Matheson Hammock county parks and a hammock and pineland in Long Pine Key along the road to the Research Center in ENP) (T. H. Fleming and the fall 1995 class of BIL 631, University of Miami, unpublished data). On the basis of tree mortality (Horvitz et al. 1995; Slater et al. 1995; Koptur et al. unpublished data), the least disturbed by the hurricane was the hammock in Long Pine Key, south of the eye wall. Matheson Hammock, slightly north of the eye wall, had greater disturbance. Castellow, at the inner edge of the northern eye wall of the hurricane, was the most severely disturbed. The absolute and relative abundances of the nonindigenous little fire ant (*Wasmannia auropunctata*) were dramatically higher at the hammock where canopy disturbance was greatest despite equivalent sampling effort. The percentage of ants of this species at baits was 97.6 percent at Castellow, 82.5 percent at Matheson, and only 5.7 percent at the hammock in Everglades National Park ($N = 17{,}190$, 1740, and 1540 ants, respectively; T. H. Fleming and the fall 1995 class of BIL 631, University of Miami, unpublished data). The pineland had no *Wasmannia* out of 1733 ants. More abundant well-developed pioneer trees with hollow stems at the more disturbed sites may provide nest cavities for these ants. Ant activity seemed greater near the swollen nodes of the native pioneer tree *Solanum erianthum* and on the hollow stems of papaya (C. C. Horvitz, S. McMann, and G. R. Burgess, pers. obs.).

The ENP sites had also been sampled in March 1992, before the hurricane, and in March 1994, some 19 months after the hurricane (T. H. Fleming and the spring 1992 and spring 1994 classes of BIL 631, University of Miami, unpublished data). The sampling effort at these censuses differed from that of the 1995 data, but I include the data for temporal comparisons. The ENP hammock site, out of 153 ants in 1992 and 640 ants in 1994, had no *Wasmannia* at either of these censuses. The pineland site had more ants before the hurricane (1085 ants) than the hammock site, and 4.7 percent were *Wasmannia*; 19 months after the hurricane it had fewer ants (350), of which about 7.1 percent were *Wasmannia*. These data together with those presented earlier indicate that the relative abundance of this ant is lower in Everglades National Park than in the fragmented nature preserves of Dade County.

Can this result be reasonably attributed to Hurricane Andrew? Because no data are available from the most hurricane-disturbed sites (outside ENP) for the 1994 census, and, as the data from 1995 are from fall and those from 1992 and 1994 are from spring, the data are insufficient to resolve this issue. My best guess is that the superabundance of *Wasmannia* at Castellow in fall 1995 is an effect of the changes in the forest structure and composition that are ultimately attributable to the hurricane. Fall is the wet season and spring the dry season in southern Florida. The wet season may favor *Wasmannia*. A dramatic increase in its abundance was associated with increased rainfall in the Galápagos caused by the 1982–1983 El Niño Southern Oscillation (Meier 1994). Thus, in Florida, we should compare falls to other falls, not to spring. There are no quantitative data from fall of other years, but anecdotal impressions are consistent with the idea that, at Castellow Hammock, *Wasmannia* was far more abundant in fall 1995 than in previous falls (C. C. Horvitz and S. McMann, pers. obs. on number of stings suffered per person-hour in the field).

An Invasive Chrysomelid Tortoise Beetle on *Ipomoea*

A nonindigenous cassidine leaf beetle (*Chelymorpha cribraria*) that feeds on the genus *Ipomoea* was found in southern Florida about a year after Hurricane Andrew (Thomas 1994b). It appeared to be abundant in Broward and Dade counties. *Ipomoea* spp. are vines that became especially abundant in regenerating subtropical hardwood hammocks within the first year after the hurricane (Armentano et al. 1995; Horvitz et al. in press). Apparently the beetles also increased in abundance and distribution at this time. By two years after the hurricane, the vines had decreased considerably, apparently because of herbivory (Horvitz et al. in press). A native chrysomelid, *Cassida* sp., specializes on the same food source (G. R. Burgess, pers. comm.). No quantitative data are available on the relative abundance or effects of these two beetles on the *Ipomoea* vines in southern Florida.

Perspectives

It is impossible to discuss the effects of natural disturbances on nonindigenous species without also examining their effects on native species. There are two opposing views on the success of invasive species: one is that the invader will be most successful in niches not occupied by natives; the other is that the invader will be most successful by displacing native species from their own niches, either stochastically or by superior competitive ability.

The first view emphasizes the success of nonindigenous species in human-altered habitats characterized by altered substrate, drainage, fire regimes, or grazing intensities. The success of nonindigenous species in human-disturbed habitats is not surprising: these are not the habitats to which the native species are expected to be adapted. In such cases, one cannot view the nonindigenous species as causing the problem: the principal cause is the altered habitat. A new niche created by human activity is most susceptible to occupation by species that have a colonizing or "weedy" life history (Chapter 23 in this volume).

In contrast, the invasion of native niches by nonindigenous species is a more serious threat to biodiversity conservation. Such invasions imply that nature preserves where human-caused disturbances are excluded may still not be safe havens for conservation of native species and their interactions. Within the context of natural disturbances, one must ask how nonindigenous species may or may not be better than particular natives at particular responses. Because some disturbances create large forest gaps that favor colonizing life histories, these may be invaded by colonizing nonindigenous species with preestablished seed banks. Papaya, for example, may do better than the native pioneers in post-hurricane gaps. Nevertheless, advance regeneration seems critical to post-hurricane response of many native tropical trees that resprout from fallen logs or use a suppressed seedling bank. For these species, shade-tolerant nonindigenous species that invade before hurricanes, such as jasmine or shoebutton ardisia, may make the biggest difference in who dominates the regeneration process following hurricanes. To the extent that newly arriving seeds find room to germinate and to recruit seedlings, the ratio of nonindigenous species to natives in the seed rain may influence whether natives or nonindigenous species win out at these few recruitment sites. The landscape context and the amount of long-distance dispersal will determine the extent to which seed rain from disturbed areas will alter seedling abundances within a habitat (Janzen 1986; Denslow and Gomez-Diaz 1990).

Whether plant-animal interactions are disrupted by large natural disturbance, and whether nonindigenous organisms become involved in these interactions to a greater extent, depends on the detailed phenology and recolonization ability of the animals. The specialist obligate mutualism between native figs and fig wasps was very resilient to Hurricane Andrew, as was the generalist native halictid bee pollination system of marlberry. In contrast, the native specialist insect moth predator of marlberry was extirpated from several local populations and took several years to rebound to pre-hurricane levels. One plant-animal interaction that may have been taken over by a nonindigenous insect is herbivory by beetles on *Ipomoea,* a vine that became a very abundant resource after the hurricane. Also potentially affected are the fruit-frugivore interactions, as the balance and spatial distribution of nonindigenous and native fruit resources may be quite different after a major disturbance.

Finally, the extent to which other animals, such as ants, are affected will depend on the ways the resource base that most affects their population dynamics changes as a result of the disturbance. For example, abundant nesting cavities in the stems of both native and nonindigenous pioneer plants coupled with high rainfall seem to have facilitated the spread of the nonindigenous little fire ant.

Although natural disturbance may facilitate invasion by certain nonindigenous species, not all invasions are disturbance-mediated—and some native organisms may in fact be more resilient to the disturbance than are some nonindigenous species. Understanding natural regeneration dynamics of both the natives and the nonindigenous species may be illuminating if the invasion of a natural area is the issue at hand. The success of a nonindigenous species in a human-disrupted environment does not necessarily predict the species' ability to displace native species in a natural system.

5 Immigration and Introduction of Insects

J. Howard Frank, Earl D. McCoy,
H. Glenn Hall, George F. O'Meara, and
Walter R. Tschinkel

A thorough account of nonindigenous insect species in Florida would require at least a book, not just a chapter, because such species number almost 1000. Preparation for such a book would require completion of taxonomic studies of the entire Florida insect fauna, estimated at about 12,500 species, which is improbable within the next century because insect taxonomy is so poorly funded. This lack of funding for insect taxonomy—indeed, for almost all studies of nonpest insects—is linked to the public perception that, with the possible exception of some butterflies, ladybird beetles, one subspecies of bee, some fish bait, and a Madagascan hissing cockroach or perhaps Chinese mantis as a pet, the only good insect is a dead one (Frank and McCoy 1991). Nevertheless, in Florida, insect species outnumber those of vertebrate animals and flowering plants combined (Table 5.1). Without insect pollination of plants, insect feeding on weeds, and insect decomposition of dead plant and animal materials, human life could not exist.

Florida's Pest Problem

Losses caused by pest insects have been estimated at well over $1 billion annually in Florida, and many of the worst pests are nonindigenous. Only one of them was deliberately introduced; the others arrived as immigrants, by walking, flying, swimming, rafting, drifting in aerial plankton, or phoresy. More recently, insect immigrants have arrived by hitchhiking or stowing away in cargoes (Sailer 1978; Frank and McCoy 1990, 1995b). These

TABLE 5.1.

Comparison of Florida's Flora and Insect and Bird Faunas

Status	Number of species		
	Plants	Insects	Birds
Native	2523[a]	11,509[b]	461[cd]
Nonindigenous (established in nature)			
Immigrant	0[e]	949[b]	11[fd]
Introduced	925[ae]	44[g]	12[fd]
Cultivated (not established in nature)	25,000[h]	154[i]	hundreds[j]

[a] After Ward (1989).

[b] Estimates by Frank and McCoy (1995a).

[c] Number inflated by inclusion of species that do not breed in Florida.

[d] From Stevenson and Anderson (1994).

[e] Some of the plants reported by Ward (1989) as "introduced" may in fact be immigrants. It is scarcely conceivable that some of the weeds among them were introduced deliberately; their seeds may have arrived on the wind, in sea drift, on or in birds, or as contaminants of shipments of other seeds or materials.

[f] Breeding species only.

[g] Biological control agents; after Frank and McCoy (1993) plus the European honeybee and cottony-cushion scale.

[h] After comments by David Hall and Thomas Sheehan.

[i] House crickets and mealworms as fish bait, silkworms, a mantis, and about 150 butterfly species in a zoo.

[j] Large numbers of species (because the United States has been the world's largest importer of exotic birds; see James 1994).

undocumented invaders continue to arrive at a rate of one major new pest species annually, while minor pests and insects that are not pests arrive and become established at a rate of more than 10 species annually (Frank and McCoy 1992). Insects have been immigrating to Florida ever since Florida existed above sea level and will continue to do so indefinitely.

A major source of immigrant insects is commercial importation of infested plants. Inspection facilities at Florida's ports, airports, and borders with neighboring states are inadequate to stop this flow. Under current policies that favor commerce in plants regardless of consequences, all the major pests of all the major crops and ornamental plants in Florida will arrive and become established. Many have done so already. The arrival rate of these pests overwhelms the ability of researchers to solve the problems they cause, and the importers of the infested plants are not required to supplement the already inadequate research funds. Other kinds of cargo (wooden materials, packing materials, stored products, skins of vertebrate animals, living vertebrate animals, scrap tires) also contribute new pests. The chemical pesticide industry has benefited from the arrival of these immigrant pests but cannot cope with certain major pests because of their cryptic lifestyles or resistance to chemicals. As more species of pests arrive, chemical usage increases.

The two most promising approaches to reversing the trend are slowing the influx and searching for environmentally benign means of control. The U.S. Department of Agriculture's Animal and Plant Health Inspection Service offices, however, charged with stemming the flow of pests at ports and airports, do not have enough personnel to inspect more than about 2 percent of the living plants (more than 350 million annually at Miami airport alone) or cut flowers (over a billion annually at Miami airport alone). Researchers charged with finding environmentally benign controls of pests already have a backlog (of several decades) of incomplete projects.

Importation of plants, vertebrate animals, and other materials, as well as the sale of chemical pesticides, support an industry worth billions of dollars annually, which doubtless generates important revenues for Florida. It is not unreasonable to expect that business and government should work together to fund a much higher level of prevention of arrival of new immigrants and research into ways to control established immigrants.

Some insects that arrive as contaminants of imported plants become pests of native plants. Generally speaking, no attention is paid to these species unless they cause an obvious and serious problem. Examples discussed here are *Metamasius callizona, Chelymorpha cribraria, Cactoblastis cactorum,* and three species of agaonid wasps. That no public agency has yet been willing to provide grant funds for research toward control of these insects points to a general problem.

Other chapters in this book address introduced vertebrates—those brought here as farm animals, for hunting or fishing, and above all as pets. In contrast, the insect species that have been deliberately introduced and established in Florida are (except for the European honeybee and cottony-cushion scale) carefully selected biological control agents of insect pests and weeds (Frank and McCoy 1990, 1993, 1995b). These 44 species form only 0.3 percent of the insect fauna. Introduction of insects into Florida is now illegal under Florida law except under permit (Thomas 1995).

Although no one documented Florida's insect fauna until well into the nineteenth century—and documentation is still incomplete—we must determine which species, of those now documented, were present some hundreds of years ago. One method (Whitehead and Wheeler 1990) is simply to accept as native any species mentioned in the early literature as being so and to treat others as nonindigenous. This method was used by Frank and McCoy (1992), who declared the "early" literature to be anything more than 20 years old and derived a list of 271 immigrant species first reported in the literature since 1970 (misconstrued by U.S. Congress 1993:257), but of course some species present in

Florida in the fifteenth century may still remain unrecorded, and most immigrant insects are known or suspected to have arrived before 1971, so the method has defects whatever cutoff date is selected.

The extreme southern end of Florida presents another dilemma. Many West Indian insects, the major part of whose range is in Cuba or on other islands, also inhabit a small part of Florida, typically the Florida Keys and adjacent mainland. Lack of baseline data for some from 20 years ago, much less 200 years ago, makes it impossible to determine how long they have been in Florida. Some doubtless go extinct in Florida from time to time and then recolonize by flight and winds from the south. Others are so poorly studied that, when they are reported for the first time from Monroe County or Dade County, they are recorded as immigrants simply because no earlier information exists. Six of the former are butterflies (*Chlorostrymon maesites, Eunica tatila, Strymon acis, Eumaeus atala, Heraclides aristodemus,* and *Anaea troglodyta,* the Florida populations of which are called, respectively, the maesites hairstreak, the Florida purplewing, Bartram's hairsteak, the Florida atala, Schaus' swallowtail, and the Florida leafwing) and are listed among Florida's rare and endangered invertebrate animals. Butterflies receive unequal treatment under the law because they have popular appeal, so more information about them is available. Inadequate knowledge of the insect fauna of Cuba and the Bahamas compounds the problem. Florida-based entomologists have for years been discouraged from working in Cuba for political reasons.

In a renewed attempt to enumerate immigrant species, 28 specialist taxonomists were questioned about the published information on insect taxa with which they were most familiar. They were asked to use all available information to arrive at their best estimates of immigrant species numbers (Frank and McCoy 1995b). They were told that species initially described from Georgia, Alabama, the Bahamas, or the Greater Antilles and subsequently detected in Florida should be considered native to Florida unless good evidence indicates otherwise. They were not asked to name the immigrant species. Table 5.2 summarizes their judgment for about a third of the insect fauna. Expansion of the table to deal with the entire recorded fauna cannot be accomplished so readily because of lack of taxonomic expertise. Even if it were somehow accomplished, it would exclude all those species that are not yet reported in the literature.

Taxonomic research on nonpest insects is progressing at a snail's pace because it has little popular appeal. Consider, for a moment, the job of the conservationist required to make decisions about faunal conservation in Florida. For bird identification, one book with colored illustrations and a pair of binoculars are all that is required. For insect identification, a set of at least 27 books of similar size and a microscope would be needed—and the books have not

TABLE 5.2.
Variation Among Taxa in Proportion of Immigrant Insect Species

Taxon	Indigenous	Immigrant	Common name
Ephemeroptera	23	48	mayflies
Odonata	144	12	dragonflies, damselflies
Blattodea	25	15	cockroaches
Isoptera	14	4	termites
Orthoptera	232	10	grasshoppers, crickets
Hemiptera			
Lygaeidae	105	10	seed bugs
Miridae	175	10	plant bugs
Homoptera			
Fulgoroidea	214	6	plant hoppers
Coccidae	14	30	soft scales
Neuroptera	85	0	lacewings, ant lions
Coleoptera			
Carabidae	365	3	ground beetles
Staphylinidae	328	15	rove beetles
Scarabaeidae	275	17	dung beetles
Lampyridae	49	1	fireflies
Nitidulidae	51	6	sap beetles
Flat bark beetles[a]	38	18	
Bruchidae	30	14	seed beetles
Curculionidae[b]	526	45	weevils
Lepidoptera			
Sesiidae	41	0	clearwing moths
Tortricidae	239	9	leaf-roller moths
Butterflies[c]	199	1	
Geometridae	244	5	measuring worms
Diptera			
Culicidae	74	4	mosquitoes
Tephritidae	52	2	fruit flies
Tabanidae	99	0	horse flies
Hymenoptera			
Ichneumonidae	340	5	ichneumon wasps
Aphelinidae	30	1	aphelinid wasps
Formicidae	149	52	ants
Total (this sample)	4160	343	
Total species:	12,500[d]		
Indigenous species:	11,507[e]		
Immigrant species:	949[e]		
Introduced species:	44		

Note: The taxa were selected by availability of taxonomic expertise. Addition of the numbers in the two columns (156 for Odonata, for example) gives the number of species recorded in the literature by the end of 1994 as being present in Florida. (The number underrepresents lesser-known taxa, as is explained in the text, and does not include the forty-four introduced species.)

Source: Modified from Table 1 of Frank and McCoy (1995b).

[a] Silvanidae, Passandridae, Laemophloeidae, and Cucujidae.

[b] Excluding Brentidae, Anthribidae, Scolytidae, and Platypodidae.

[c] Papilionoidea and Hesperioidea.

[d] Estimate.

[e] Estimates based on proportions shown in sample: 12,500 − 44.

been written. Armies of amateur bird-watchers help to document bird populations, but nobody in Florida documents populations of all insect species continuously in even one locality. As a result, the conservationist usually ignores
the vast majority of animal species (insects and other invertebrates) and deals
only with the easily identified species, almost all of which are birds and other
vertebrates. Sometimes an insect species makes its presence known by its
unusually large numbers, but unusually small numbers are unlikely to be
noticed.

Although some of the 44 introduced species are mentioned, this chapter
highlights immigrant pests because much more information is available about
them than about the much larger numbers of innocuous immigrants or about
beneficial introduced species. The examples selected are intended to reflect diversity among immigrants. Because of space limitation, hundreds of immigrant
species and many pests are not mentioned at all.

Losses (including costs of control) in Florida to some pest insects were estimated at $1 billion for 1984 by members of a committee of the Entomological
Society of America (Hamer 1985). For most pests, it can safely be assumed that
losses have risen since then—as indicated by the growing demand for control by
Florida's increasing human population. For a few, such as sweet-potato whitefly
(and its sibling species silverleaf whitefly), costs have risen dramatically because
of resistance to chemical pesticides. For a few, costs have decreased (though this
fact is poorly documented) because of the successful introduction of classical biological-control agents. Overall, costs have increased also because of the arrival
of new pests. None of these pests was introduced, but many are immigrants.

Selected Nonindigenous Species

For the convenience of readers not familiar with entomological classifications,
the insect orders are arranged alphabetically here and families and other sub-
groupings are arranged alphabetically within orders. The subheadings Immigrant and Introduced are used only when species of both these categories are
mentioned under the name of one taxon.

We emphasize that the following account does not mention all of Florida's
nonindigenous insect species or even all of the pests among them. Nor does it
mention immigrant arthropods other than insects, which include some very important pests including, but not restricted to, *Phyllocoptruta oleivora* (citrus
rust mite), *Acarapis woodi* (honeybee tracheal mite), *Varroa jacobsoni* (Asian
honeybee mite), broad mite, several species of spider mites, and mite pests of
livestock and pets. Citrus rust mite alone was estimated to cause $95 million in
losses in 1984 (Hamer 1985).

Anoplura (Sucking Lice)

Pediculus humanus (human louse) doubtless arrived in Florida with Amerindians many centuries ago. In contrast, *Haematopinus* spp. (hog lice and short-nose cattle lice), *Linognathus* spp., and *Solenoptes capillatus* (little blue cattle louse) probably arrived as immigrants on infested livestock imported by Europeans.

Blattaria (Cockroaches)

The Florida fauna includes 25 native species and 15 immigrant species that arrived from the Old World. The major pest species of cockroaches are all immigrants (Koehler and Brenner 1995). One recent immigrant, *Blattella asahinai* (Asian cockroach), was given much publicity when it was detected in 1986. Cockroaches in general have a bad public image, though not all are pests. Losses they caused in Florida were estimated as $235 million in 1984 (Hamer 1985). Nevertheless, in the 1990s pet stores imported and sold *Gromphadorhina* sp., a very large cockroach from Madagascar, until ordered to desist by the Florida Department of Agriculture and Consumer Services (FDACS) under a new law (Thomas 1995).

Coleoptera: Bruchidae (Seed Beetles)

There are 30 native and 14 immigrant species in Florida. Of the latter, four are cosmopolitan "tramp" species in stored legume seeds, three are South American, and seven are Central American. No species was introduced (Kingsolver 1995).

Coleoptera: Carabidae (Ground Beetles)

Most adults and larvae of these beetles are predatory, but a few feed on plants. There are 365 native species reported for Florida, and three immigrant species (Choate 1995). Two species of *Calosoma* were released for biological control purposes some decades ago, but they did not become established.

Coleoptera: Chrysomelidae (Leaf Beetles)

Adults and larvae of this family feed on leaves of plants, generally with a fairly high level of specialization to plants within one genus.

IMMIGRANT   *Leptinotarsa decemlineata* (Colorado potato beetle) is native to the western United States and fed originally on wild plants of the genus *Solanum*. When potato (*Solanum tuberosum*) was introduced and cultivated extensively in the nineteenth century, the beetle followed its host through the eastern United States. *Chelymorpha cribraria,* a leaf beetle native to South

America and the West Indies, was detected in Broward County in 1993 and had spread to Dade and Monroe counties by 1994. It feeds on plants of the genus *Ipomoea* and may have arrived as a contaminant of imported *Ipomoea* (sweet potato). It now threatens *Ipomoea microdactyla* and *Ipomoea tenuissima*, Florida's endangered species of morning glories (Thomas 1994a).

INTRODUCED   *Agasicles hygrophila*, an Argentinean specialized feeder on *Alternanthera philoxeroides* (alligatorweed), was released in Florida in 1964. Together with *Vogtia malloi* (a pyralid moth) and *Amynothrips andersoni* (a thrips), it produced spectacular biological control of that aquatic weed (Coulson 1977).

Coleoptera: Coccinellidae (Ladybird Beetles)

Ladybird beetles have a public image as being beneficial because adults and larvae of many species are predators of sap-sucking insects such as aphids, scale insects, and whiteflies. Although some ladybird beetles may suppress sap-sucking insect pests of cultivated plants, these beetles have additional roles. Most ladybird beetles prey on insects, but some prey on mites, some (subfamily Epilachninae) feed on plants and are pests, and others (tribe Psylloborini) feed on fungi. Populations of weeds may be suppressed by sap-sucking insects; in such a system, a predator is not necessarily beneficial.

IMMIGRANT   *Rhyzobius lophanthae* is an Australian natural enemy of scale insects including *Aonidiella aurantii* (California red scale) and *Coccus hesperidum* (brown soft scale), both of which are pests of citrus. Its exact means of arrival in California (where it was detected in 1892) and subsequently in Florida are unknown, but it may have arrived in shipments of plant material. It is viewed as a beneficial species because of its suppression of scale-insect populations.

Epilachna varivestis (Mexican bean beetle) is a pest of legume crops; its larvae destroy leaves of beans, peas, and the like. Native to Central America, Mexico, and neighboring parts of the southwestern United States, it began to expand its range about 1920, until it occupied all of the eastern and southern United States. Satisfactory control has been achieved by annual releases of a eulophid wasp, *Pediobius foveolatus*. This wasp is a parasitoid of another *Epilachna* species, native to India, and attacks *E. varivestis* readily but seems unable to survive winters in the eastern United States (Stevens et al. 1975).

INTRODUCED   At least 32 species of ladybird beetles have been imported into Florida for study and potential release as biological control agents. Of the smaller number of species released, seven eventually became established:

Rodolia cardinalis from Australia, released in 1899 against *Icerya purchasi* (cottony-cushion scale); *Harmonia dimidiata* from China, released in 1925–1926 against *Aphis spiraecola* (spirea aphid); *Cryptolaemus montrouzieri* from Australia, imported in 1930 against *Planococcus citri* (citrus mealybug); *Cryptognatha nodiceps* from Trinidad, released in 1936–1938 against *Aspidiotus destructor* (coconut scale); *Coelophora inaequalis* from Hawaii, released in 1939 against *Sipha flava* (yellow sugarcane aphid); *Hippodamia variegata* from Asia, released in 1957 against *Acyrthosiphon pisum* (pea aphid), *Myzus persicae* (green peach aphid), *S. flava,* and *Therioaphis maculata* (spotted alfalfa aphid); and *Coccinella septempunctata* from Europe, released in 1976–1977 against *M. persicae* and *T. maculata.*

Coleoptera: Cucujidae, Laemophloeidae, Silvanidae, Passandridae (Flat Bark Beetles)

In Florida occur 56 species of Laemophloeidae, Silvanidae, and Passandridae, but no Cucujidae. These 56 include 38 considered to be native and 18 immigrants. Half of the immigrant species are pests of stored products with wide distributions outside Florida. For example, *Oryzaephilus surinamensis* (sawtoothed grain beetle), *O. mercator* (merchant grain beetle), and *O. acuminatus,* none of which is native to Florida, have often been intercepted at ports in shipments of products including cereals, dried fruits, copra, carob, nuts, and neem seeds (Thomas 1993).

Coleoptera: Curculionidae (Weevils)

Just as ladybird beetles have a public image as being beneficial, so weevils have a public image as being harmful. This simplistic image arises solely because weevils feed on plants and plant products. Yet weevils typically have highly specialized diets, and those that fed on weeds must be judged beneficial. Furthermore, native weevils that feed on native plants not only are elements of the native fauna but also may prevent some of those plant species from becoming weeds. About 526 native species have been reported (C. W. O'Brien 1995).

IMMIGRANT   Forty-five immigrant species have been reported (C. W. O'Brien 1995). Among the most notorious are *Cylas formicarius* (sweet-potato weevil) from Asia, *Anthonomus grandis* (boll weevil) from Mexico, *Hypera posticata* (alfalfa weevil) from Europe, *Graphognathus* spp. (white-fringed beetles) from South America, *Metamasius hemipterus* (cane weevil) from the neotropical region, and *Diaprepes abbreviatus* (known in Puerto Rico as *la vaquita* and in Florida as Apopka weevil) from Puerto Rico and Hispaniola, all of which attack crop plants. Sweet-potato weevil and boll weevil were together estimated to cause $3 million in losses in 1984 (Hamer 1985). Specialist biological-control agents have been introduced and established against alfalfa weevil and Apopka weevil.

Metamasius callizona, a Mexican weevil, was detected in Broward County in 1989, and its population spread to Dade and Palm Beach counties. Its spread to Charlotte and Lee counties may have been due to commerce in infested plants. There is good evidence that it arrived initially as a contaminant of shipments of ornamental bromeliads of the genus *Tillandsia.* Although it will damage, and sometimes kill, ornamental bromeliads of at least 12 genera, including *Ananas* (pineapple), its most serious effects are on a few species of *Tillandsia.* Most of Florida's native bromeliads, the majority of which are protected by law, are of the genus *Tillandsia,* and populations of *T. utriculata* have been devastated in southeastern Florida by this weevil (Frank and Thomas 1984).

INTRODUCED   Five weevil species have been introduced and established for the biological control of weeds (C. W. O'Brien 1995). All are specialized natural enemies of hydrilla, water hyacinth, or water lettuce (aquatic weeds).

Coleoptera: Lampyridae (Fireflies)
Firefly adults and larvae are predators. All established species in Florida are believed to be native. One other species appears to be a repeated immigrant from Central America and may occasionally survive a year or so before disappearing (Lloyd 1995).

Coleoptera: Platypodidae and Scolytidae (Ambrosia Beetles and Bark Beetles)
Adults and larvae of these families feed on tree seeds, on phloem or xylem of trees, or on fungi growing on trees, and several of them are considered to be major or minor pests to the extent that they damage trees. In tropical southern Florida (Broward, Collier, Dade, and Monroe counties) occur 86 species in 38 genera, of which 20 are immigrants (Atkinson and Peck 1994).

Coleoptera: Scarabaeidae (Dung Beetles)
Dung feeding (coprophagy) is far from a universal habit in this family, and larvae of many species (together called "white grubs") are notorious feeders on plant roots. There are 275 native species in Florida, and another 17 are immigrants (Woodruff 1995). Two coprophagous African species of the genus *Onthophagus* have been detected in Florida, one of them after a deliberate release by the USDA in Georgia to aid in decomposition of cattle dung and the other questionably as a hitchhiker aboard a military plane.

Coleoptera: Staphylinidae (Rove Beetles)

Many Staphylinidae are obligate or facultative predators or are fungivores. Members of one genus (*Aleochara*) are parasitoids of dipterous pupae, and some are inquilines in nests of ants and termites. The 1994 count from published records is of 328 native species and 15 immigrant species, although the total will amount to at least 450 when taxonomic studies are completed (Frank 1995). No species have been deliberately introduced. Arrival of the immigrant species is unexplained.

Diptera: Bibionidae (March Flies)

Because flying and walking insects can freely cross Florida's northern and western borders, many of its insect species (including pests) are shared with neighboring states. A familiar example is *Plecia nearctica* (love bug), a Mexican and Central American species that has extended its range to include the Gulf coast of the United States. After moving into Florida from Alabama in 1949, its population spread to reach southern Florida in 1975 (Buschman 1976). It is an immigrant to Florida. Thoughtless use by some entomologists and newspapers of the term "introduced" may have led to the widespread erroneous notion among the Florida public that some entomologist introduced it deliberately. Its adults swarm along highways in spring and autumn, are splattered against windshields and paint of cars, and are difficult to remove.

Diptera: Braulidae (Bee Lice)

The only known host of *Braula caeca* (bee louse), an Old World species first reported in Florida in 1983, is *Apis mellifera* (honeybee). It probably arrived as an immigrant contaminating imported honeybees. If imported shipments of honeybees had been subjected to confinement and examination in quarantine, as are biological control agents, their three major pests, the Asian honeybee mite, honeybee tracheal mite, and bee louse, would not now occur in Florida. These pests have severely affected the beekeeping industry.

Diptera: Calliphoridae (Blowflies)

Three Old World species of *Chrysomyia* have been detected as immigrants since 1990. The spined larvae develop in wounds in vertebrate animals and in cadavers. Thus far, populations in Florida have not built up, for example on poultry farms, where they might be expected to cause a problem. They play a part in forensic entomology (the stage reached by larvae helps to pinpoint time of death).

Diptera: Chironomidae (Midges or "Blind Mosquitoes")
Members of this family cannot bite or sting humans and harm neither agriculture nor horticulture. They cause complaints from the public only when adults emerge in enormous numbers from eutrophic lakes such as Lake Apopka (Orange and Lake counties) and Lake Monroe (Volusia and Seminole counties) in Florida. *Goeldichironomus amazonicus,* a neotropical species, was detected in Florida in 1977. It probably immigrated to Florida as eggs or larvae on aquarium plants or other aquatic plants (Wirth 1979).

Diptera: Culicidae (Mosquitoes)
All but four of Florida's mosquito species are native (Table 5.2). Historical accounts emphasize the hordes of native, biting mosquitoes encountered by early explorers and settlers. It has been suggested that the comparatively slow rate of human immigration into Florida was in large part due to biting and disease-carrying mosquitoes. The human population surged in the latter part of the twentieth century after mosquito control was organized to drain or impound marshes and apply chemical pesticides. The standard use of window screens for buildings helped to separate people from mosquitoes. Losses to mosquitoes were estimated at $61 million in 1985 (Hamer 1985).

IMMIGRANT   The most notable immigrants are *Aedes aegypti* (yellow-fever mosquito), which arrived before 1850, and *Aedes albopictus* (Asian tiger mosquito or forest day mosquito), which was found in 1986. *Aedes aegypti,* of African origin, was the vector of yellow fever in sporadic outbreaks in Florida cities until the twentieth century and remained a vector of dengue until after World War II. The diseases, but not the mosquito, were eliminated from Florida by mosquito-control endeavors. *Aedes albopictus* is believed to have arrived in Texas from Asia in a shipment of scrap tires and then spread through the southern United States. It is a competent vector of those two diseases and can transmit viral encephalitis.

INTRODUCED   Two tropical species of the genus *Toxorhynchites* have been imported and released in Florida. Their adult females, unlike those of most mosquito genera, do not take blood, and their larvae prey on larvae of other mosquitoes in water in junk containers. There is little evidence of their effectiveness in controlling other mosquito populations, however, or of their ability to persist during Florida winters.

Diptera: Hippoboscidae (Louse Flies)
Melophagus ovinus (sheep ked), a European species parasitic on sheep, and *Hippobosca longipennis,* an African species parasitic on cats, presumably arrived in Florida as immigrants infesting imported hosts.

Diptera: Muscidae (Muscid Flies)

All nonindigenous muscid flies in Florida are immigrant species. *Musca domestica* (housefly) and *Stomoxys calcitrans* (stable fly) may be native to tropical Africa and *Haematobia irritans* (horn fly) to southern Europe. Horn fly arrived in the northeastern United States in 1887, probably with a shipment of cattle, and spread rapidly. Housefly and stable fly arrived much earlier. Horn-fly adults feed on the blood of cattle. The housefly is mainly a pest in dairies and poultry and pig farms, from which burgeoning populations of adults invade human habitations and cause annoyance; excretions of the flies contaminate poultry and dairy products. Stable-fly adults feed on the blood of horses, cattle, and people. This trio of flies caused estimated losses of $94 million in 1984 in Florida (Hamer 1985).

Diptera: Oestridae (Bot and Warble Flies)

Larvae of *Gasterophilus haemorrhoidalis* (nose botfly), *G. intestinalis* (horse botfly), and *G. nasalis* (throat botfly) are internal parasites of horses, mules, and donkeys; those of *Hypoderma bovis* (northern cattle grub) and *H. lineatum* (common cattle grub) are parasites of cattle; and those of *Oestrus ovis* (sheep botfly) are parasites of sheep. Presumably all arrived as immigrants with infected hosts.

Diptera: Tachinidae (Tachinid Flies)

Several species of tachinid flies have been released in Florida as biological control agents of pest insects. The longest campaign was to establish a population of *Lixophaga diatraeae* against *Diatraea saccharalis* (sugarcane moth borer). Beginning in 1926 and ending in 1974, over 200,000 of the flies, from Cuba and Trinidad, were released. Populations did not persist more than a few months. The only successful introduction was of *Ormia depleta,* from Brazil, a specialist natural enemy of *Scapteriscus vicinus* (tawny mole cricket) and *S. borellii* (southern mole cricket). The first releases of this nocturnal, phonotactic (attracted to the song of the tawny mole cricket and southern mole cricket, not to the native northern mole cricket), larviparous (hatching inside the mother, who then lays active parasitoid larvae) fly were made in 1988 in Alachua and Manatee counties. By the end of 1992, an apparently continuous population extended from Alachua County to Dade County (Frank 1994).

Diptera: Tephritidae (Fruit Flies)

Six immigrant fruit-fly species have been recorded from Florida, but only two (Caribbean fruit fly and papaya fruit fly) have colonized successfully. The other four immigrant species either have been eradicated (such as *Ceratitis capitata,* Mediterranean fruit fly) or never became successfully established (such as *Anastrepha ludens,* Mexican fruit fly) (Steck 1995). The Caribbean fruit fly, first detected in Florida in the 1920s, is mainly a pest of soft tropical fruits such as guava and does little damage to citrus. Nevertheless, countries to which Florida citrus is exported run a risk that Caribbean fruit fly will hitchhike in shipments of citrus fruits, and they require stringent safeguards against this possibility.

Hemiptera: Pentatomidae (Stink Bugs)

Nezara viridula (southern green stink bug) is of African origin and immigrated into Florida before 1850. An omnivorous pest of many vegetable, fruit, and field crops, it caused a large share of the estimated losses of $4 million in Florida in 1984 to damage by stink bugs (Hamer 1985).

Hemiptera: Tingidae (Lace Bugs)

Leptodictya tabida (sugarcane lace bug), an immigrant, was detected in Florida in 1990. Its arrival, from either Central or South America, was viewed with concern because of its potential as a pest of sugarcane, a threat which so far it seems not to have fulfilled (Hall 1991).

Homoptera: Aleyrodidae (Whiteflies)

All nonindigenous whitefly species in Florida are immigrants. *Aleurocanthus woglumi* (citrus blackfly), a serious pest of citrus, was twice an immigrant from Asia. In the 1930s it was detected soon after its arrival and eradicated. In the 1970s it had colonized a wider area before it was detected, and chemicals failed to eradicate it. It was then brought under control, very successfully, by importation and release of two minute wasps, *Amitus hesperidum* (a platygastrid wasp) and *Encarsia opulenta* (an aphelinid wasp), which specialize on this host.

Dialeurodes citri (citrus whitefly) and *D. citrifolii* (cloudy-winged whitefly) are immigrants from Asia detected in Florida before 1900. The former has been largely controlled by introduction in 1977 of *Encarsia lahorensis* (an aphelinid wasp), which specializes on this host.

Bemisia tabaci (sweet-potato whitefly) and *B. argentifolii* (silverleaf whitefly) are immigrants from Asia; the former arrived before 1900 and the latter apparently in the 1980s. Their extremely similar appearance (some specialists do not believe they are distinct species) led to much confusion about why sweet-potato whitefly, which had for decades been a controllable pest, was suddenly very difficult to control. Losses caused by this species (or these species, if there really are two) are greatly increased by their transmission of viruses to crop plants.

Trialeurodes vaporariorum (greenhouse whitefly) now occurs widely in the Western Hemisphere, Europe, Australia, and New Zealand and attacks greenhouse-grown vegetable crops and ornamental plants. Its geographic origin is obscure. It can be controlled successfully by releasing *Encarsia formosa* (an aphelinid wasp) in greenhouses, and these wasps are marketed commercially in Florida (Frank and McCoy 1994).

Homoptera: Aphididae (Aphids)

Immigrant aphids include *Acyrthosiphon pisum* (pea aphid) from Eurasia, detected about 1900; *Aphis spiraecola* (spirea aphid) from Asia, detected about 1920; *Therioaphis maculata* (spotted alfalfa aphid) from Eurasia, detected about 1960; and *Trichosiphonaphis polygoni* from Asia, detected in 1974. The origin of *Melanaphis sacchari,* detected in 1977, is hard to pinpoint because it now occurs in the tropics worldwide. *Toxoptera citricida* (brown citrus aphid) transmits tristeza virus of citrus in some tropical countries. The discovery of *T. citricida* for the first time in Florida in November 1995 alarmed the citrus industry because the virus causes severe disease.

Homoptera: Coccidae (Soft Scales)

Thirty immigrant species of soft scales are reported from Florida (Hamon 1995). *Coccus capparidis* and *Philephedra tuberculosa,* both from the neotropical region, arrived within the last 20 years.

Homoptera: Diaspididae (Armored Scales)

All the nonindigenous armored scales are immigrants. *Aonidiella aurantii* (California red scale), *Chrysomphalus aonidum* (Florida red scale), *Lepidosaphes beckii* (purple scale), and *Unaspis citri* (citrus snow scale) are Asian pests of citrus detected in Florida before 1900. *Aspidiotus destructor* (coconut scale), *Fiorinia theae* (tea scale, a pest of camellias), *Pseudaulacaspis cockerelli* (false oleander scale), *Pseudaulacaspis pentagona* (white peach scale), and *Unaspis euonymi* (euonymus scale) are likewise of Asian origin and were first detected at various dates from 1910 to 1960. Most of these are or were considered consequential pests. Florida red scale and coconut scale, however, have been brought under substantial control by introduction of biological control agents (*Aphytis holoxanthus* and *Cryptognatha nodiceps*, respectively). Species detected since 1971 include *Parlatoria ziziphi* (black parlatoria scale) and *Lepidosaphes laterochitinosa,* both from Asia, and half a dozen others.

Purple scale is controlled by an immigrant species, *Aphytis lepidosaphes* (an aphelinid wasp). This minute wasp had been imported from China and released in California and Texas. It was detected in Florida in 1958 before releases had been made there.

Homoptera: Fulgoroidea (Plant Hoppers)

There are six immigrants among Fulgoroidea, including three pantropical pests of corn and sugarcane that arrived in this century (L. B. O'Brien 1995). Among them is *Perkinsiella saccharicida* (sugarcane delphacid), which was first detected in Florida in 1982. Its arrival from the Pacific was viewed with concern because of its potential as a devastating pest of sugarcane, which so far seems not to have been fulfilled (Sosa 1985).

Homoptera: Margarodidae (Margarodid Scales)

Icerya purchasi (cottony-cushion scale) seems to be the one pest insect that was introduced into Florida and became established. Cottony-cushion scale was detected in California as an immigrant about 1868 and, when no physical or chemical treatment was found to be effective in its control, threatened to ruin the citrus industry. Introduction of *Cryptochetum iceryae* (an agromyzid fly) and *Rodolia cardinalis* (a ladybird beetle) from Australia as biological control agents brought it under control.

A Florida citrus grower, hearing of events in California, failed to understand that *Rodolia cardinalis* is a specialized biological-control agent of cottony-cushion scale, which was not present in Florida. Perhaps he thought the beetle would be of some use against other citrus pests. In 1893 he obtained a shipment of *R. cardinalis* and its prey, cottony-cushion scale, from California, long before permits were required for such shipments. When the container was opened in a grove, the beetles emerged rapidly, dispersed, and presumably perished for lack of food. The scales emerged slowly and infested citrus trees. Despite the history of control efforts in California, physical and chemical control methods were tried in Florida and again failed. In 1899, beetles were imported again from California. This time, when they were released, they found prey and brought the infestations under control (Frank and McCoy 1992). This account illustrates the specialized feeding habits desirable in biological control agents, the public's lack of understanding of the biological control process, and the need to regulate importation of bioagents.

Homoptera: Pseudococcidae (Mealybugs)

Chloris gayana (Rhodes grass) was brought from Africa to the southern United States early in the twentieth century to improve pastures. In 1942, *Antonina graminis* (Rhodes-grass mealybug), an Old World species, was detected in Texas. Populations of this immigrant mealybug spread through the South, reached Florida and Mexico, and infested numerous other grasses. Introduction of two species of encyrtid wasps (*Anagyrus antoninae* from Japan and *Neodusmetia sangwani* from India) as biological control agents was estimated to save the Texas economy $17 million annually (Dean et al. 1979). Florida benefited from research performed in Texas, when wasps of the two species and a third (*Pseudectroma europaea*) were obtained in the 1950s, and *N. sangwani* and *P. europaea* became established and solved the problem at little cost (Bennett 1994).

Hymenoptera: Agaonidae (Fig Wasps)

More than 60 nonnative *Ficus* species (fig trees) have been introduced into southern Florida as landscape ornamentals. Years ago, it was thought that none of them would set viable seed because each is pollinated only by its own species of agaonid wasp, and these were not present in Florida. Now, however, *Ficus altissima* (lofty fig), *F. benghalensis* (banyan), and *F. microcarpa* (laurel fig) have become invasive because their species of agaonid wasps (*Eupristina altissima, E. masoni,* and *Parapristina verticillata*) have immigrated into Florida within the past three decades, allowing the trees to produce viable seed by routine pollination. Fertile seeds of these enormous trees germinate in southern Florida. Fig-tree seedlings sprout on public and private lands and on structures such as highway bridges (where they pose a maintenance problem because they can destroy the structures if allowed to grow). There is evidence that the pollinating wasps of laurel fig arrived in seeds brought from Hawaii, and fruits (and thus seeds) of the other two fig species may be spread by introduced parrots. The three wasp species are from Asia (Nadel et al. 1992).

Hymenoptera: Aphelinidae (Aphelinid Wasps)

Several species of these minute parasitoid wasps have been introduced into Florida and established. They include *Aphytis holoxanthus* from Hong Kong, *A. lingnanensis* from Hong Kong, *Encarsia lahorensis* from Pakistan, *E. opulenta* from India, *E. sankarani* from India, and *E. smithi* from India. For specifics about these releases and those of parasitoid wasps of the families Braconidae, Encyrtidae, Eucoilidae, Eulophidae, Platygastridae, and Pteromalidae see Frank and McCoy (1993). Such specialist wasps have had great success in the classical biological control of pest insects, especially Homoptera.

Hymenoptera: Apidae (Bumble, Carpenter, and Honeybees)

Apis mellifera mellifera (European honeybee) was probably brought to Florida late in the sixteenth century by Spanish colonists. The emergent beekeeping industry provided honey and pollination of crop plants and continued to do so for hundreds of years. Inevitably some bees became feral and nested in cavities in trees, but these probably had trivial effects on the native fauna and flora. In the twentieth century, through human carelessness, further importation of bees allowed the arrival of bees infested with Asian honeybee mite, honeybee tracheal mite, and bee lice. These pests demand constant attention from the beekeeping industry in managed hives in Florida and have all but eliminated feral honeybees (M. T. Sanford, pers. comm.).

Importation of *Apis mellifera scutellata* (African honeybee) into southern Brazil in 1956 for experimentation, and its subsequent escape, had enormous effects on the Brazilian beekeeping industry because African honeybees are much more defensive of their nests and require different management techniques. There is little documentation of the effect of feral African bees on native vertebrates in Brazil. Populations of African honeybees spread through South and Central America and Mexico and have arrived in Texas (Hengeveld 1992). They have not yet immigrated successfully into Florida, but their arrival is predicted. Newspaper accounts of African honeybees have concentrated on incidents in which humans were stung by them, leading the public to infer an exaggerated risk to human life.

Hymenoptera: Formicidae (Ants)

There are 201 species reported from Florida, of which 52 (25.8 percent) are immigrants (Deyrup 1995). Several of these immigrants are pests, including *Iridomyrmex humilis* (Argentine ant), *Monomorium pharaonis* (pharaoh ant), *Paratrechina longicornis* (crazy ant), and *Solenopsis invicta* (red imported fire ant).

The red imported fire ant, native to South America, is the most notorious of the pests, probably because it bites and stings humans most frequently. It arrived in the port city of Mobile, Alabama, in the late 1930s, spread through the southern United States, and now infests all of Florida. It has displaced some native ants. It feeds on seeds, seedlings, fruits, and vegetables; it invades buildings to feed on processed foods; it bites people and farm animals. Losses attributed to it in Florida in 1984 were $1 million (Hamer 1985). Not accounted for in this total was the benefit it provides as a predator of pest insects (boll weevil, flea larvae, horn-fly larvae, sugarcane borers, and others), the undocumented damage it may do as a predator of native nonpest insects, or any mortality it may inflict on nestling wild birds. From the overall viewpoint of agriculture, the dollar value of the control it exerts on pest insects may far exceed the $1 million in losses that it causes. From the viewpoint of human health, it is one of the most disliked insect species in Florida, even though it rarely causes serious problems except to people hypersensitive to its venom. From the environmental viewpoint, it may cause considerable undocumented damage, though mainly in disturbed habitats (Tschinkel 1993).

Hymenoptera: Sphecidae (Digger Wasps)

All species of the genus *Larra* are parasitoids of mole crickets. The native species *Larra analis* is a parasitoid of *Neocurtilla hexadactyla* (northern mole cricket), the only native mole cricket in Florida, but several neotropical species are parasitoids of *Scapteriscus* mole crickets. Introduction of *Larra bicolor* from Puerto Rico in 1981 and from Bolivia in 1989, for control of *Scapteriscus* mole crickets, succeeded in establishing small populations of this wasp near Fort Lauderdale (Broward County) and in Gainesville (Alachua County), respectively (Frank et al. 1995).

Hymenoptera: Torymidae (Torymid Wasps)

Two immigrant species were detected in the 1980s: *Megastigmus transvaalensis* and *Philotrypesis emeryi* (Habeck et al. 1989; Nadel et al. 1992). Larvae of these species destroy seeds of *Schinus terebinthifolius* (Brazilian pepper) and *Ficus microcarpa* (laurel fig), respectively. Years ago, when these trees were viewed as valuable ornamentals, the wasps would have been looked upon as pests. Now that the trees have been declared weeds, the wasps must be viewed as beneficial. It can only be supposed that the wasps were contaminants of imported fruits (containing seeds) of these trees.

Hymenoptera: Trichogrammatidae (Trichogrammatid Wasps)

Larvae of these minute wasps develop as parasitoids in the eggs of other insects. Mass production and release of various species of the genus *Trichogramma* were used, and may still be used, to control populations of pest moths over millions of hectares in Russia and China and to a much lesser extent in other countries. Research accompanying these programs revealed existence of far more species (and subspecies and strains) of *Trichogramma* and far more specialization to climate and habitat than had previously been recognized. That the method gained little acceptance in the United States may have been due to inadequate knowledge of taxonomy, behavior, and host specialization, as well as to higher labor costs, competition in the marketplace with chemical pesticides, and variable effectiveness. Two species, *T. minutum* and *T. pretiosum*, have been marketed in Florida (Frank and McCoy 1994).

Isoptera (Termites)

The Florida fauna includes 14 native species and 4 immigrant species (Scheffrahn 1995 and pers. comm.). The most notable of the immigrants, generally considered to be the most damaging termite species in Florida, is *Coptotermes formosanus* (Asian subterranean termite), which was detected in Florida in 1980. Presumably it arrived in cargo shipped from Asia. Losses caused by termites in Florida in 1984 were estimated as $425 million (Hamer 1985). Almost all losses are due to damage to buildings, but damage to living trees may also occur.

Lepidoptera: Gelechiidae (Gelechiid Moths)

The latest major insect threat to Florida's citrus industry is *Phyllocnistis citrella* (citrus leaf miner), a native of Asia. It was detected in Florida early in 1993 and soon spread through the citrus-growing area. Citrus industry funds sponsored importation of two parasitoid wasps that, imported into Australia, had shown good suppression of the leaf miner there. Permanent establishment of the wasps is not yet demonstrated. Detection of this leaf miner and several other new pests within a few months after Hurricane Andrew led some to speculate that the insects had been blown to Florida on high winds. But because importation of plants after the hurricane increased to replace those destroyed by the storm, infested planting stock is a more likely source (Chapter 4 in this volume).

Lepidoptera: Lymantriidae (Tussock Moths)

Lymantria dispar (gypsy moth) was brought from Europe about 1869 (long before the days of regulation) to the laboratory of an astronomer in Bedford, Massachusetts, who wished to study silk production by its larvae. Unfortunately, moths escaped to unleash a new pest on forest trees in New England. Introduction of insect parasitoids from Europe as biological control agents has not solved the problem. In recent years the southern border of the area occupied by the moth has moved southward. When fully grown larvae spin their silken cocoons on travel trailers in the northeast and the owners of the vehicles then holiday in Florida, the moths accompany them to begin local pockets of infestation. Thus far, it seems, the moth is not well adapted to the climate of Florida, and the infestations fail to persist, but adaptation may yet occur (Allen et al. 1993).

Lepidoptera: Papilionoidea and Hesperioidea (Butterflies)

The Florida fauna contains 199 native species and one immigrant species (Emmel 1995): *Pieris rapae* (imported cabbageworm), a pest of cole crops. Most of the species of insects that are imported into Florida but not intended for release are butterflies kept as living displays in at least two butterfly zoos (Boender 1995).

Lepidoptera: Pyralidae (Pyralid Moths)

Caterpillars of most pyralid moths feed cryptically within plants. Many species of this large family are pests of crop plants, and some are pests of stored products.

IMMIGRANT Caterpillars of *Cactoblastis cactorum,* native to South America, feed inside cladophylls of some *Opuntia* species (prickly pear). Introduced into Australia in 1925, it saved 12 million acres of pastureland that had been rendered useless by an infestation of two species of *Opuntia* which had unwisely been imported from the Gulf of Mexico coast as ornamental plants. Between 1957 and 1970, the moth was introduced into Nevis, Montserrat, Antigua, and Grand Cayman, where *Opuntia* species were considered to be weeds. Later, the moth appeared in Puerto Rico, Haiti, the Dominican Republic, and the Bahamas, where its introduction had not been sanctioned by government. In 1989, it was found in the Florida Keys, where it threatened the cacti *O. corallicola* (which is rare) and *O. stricta.* The unsanctioned arrivals might have come with infested cacti shipped as ornamental plants, or when adult moths flew or hitchhiked aboard boats, or if moths were smuggled by private agricultural interests to combat *Opuntia* species viewed as troublesome weeds. Although none of these possibilities can now be ruled out, the most probable origin is that ornamental *Opuntia* species imported over a period of several years from the Dominican Republic were infested with *C. cactorum* (Pemberton 1995). See also Center et al. (Chapter 15 in this volume).

INTRODUCED   *Vogtia malloi,* from Argentina, was one of the three introduced species with specialized feeding habits that succeeded in controlling alligatorweed (Coulson 1977).

Lepidoptera: Tineidae (Clothes Moths)

Opogona sacchari (banana moth), an immigrant species from the Old World tropics, was detected in Florida in 1963. Its larvae damage plants, unlike others of the genus, which are feeders on plant detritus. In various parts of the world it has been a serious pest of banana, maize, potato, sweet potato, sugarcane, and a long list of ornamental plants (Davis and Peña 1990). Its effects on native vegetation in Florida seem unknown.

Lepidoptera: Yponomeutidae (Ermine Moths)

Caterpillars of *Plutella xylostella* (diamondback moth), an immigrant species of Mediterranean origin, eat cruciferous plants. This moth was first reported in the United States in 1854 and has spread to all states. It is a major pest of cole crops (broccoli, cabbage, cauliflower, and the like) in Florida and is now resistant to many chemical pesticides. Development of biological control methods for it is hindered by use of chemicals against lesser pests of the same crops, because most biological control agents are highly susceptible to chemicals.

Odonata (Dragonflies and Damselflies)

Twelve species of immigrant Odonata are now established in Florida: one (*Crocothemis servilia*) from Asia, three (*Celithemis elisa, Enallagma basidens,* and *E. civile*) from North America, and eight from the neotropics. Moreover, seven species have been found as vagrants without breeding populations (Dunkle 1995). All are predatory as adults and nymphs.

Orthoptera (Crickets, Mole Crickets, and Grasshoppers)

Ten species of Orthoptera are post-Columbian immigrants, and no species have been introduced (Walker 1995). The most consequential of the immigrants are three species of mole crickets that arrived in ships' ballast from southern South America about 1900: *Scapteriscus abbreviatus* (short-winged mole cricket), *S. borellii* (southern mole cricket), and *S. vicinus* (tawny mole cricket). They were estimated to cause $41 million in losses in turf in Florida in 1984 (Hamer 1985). They also damage vegetable seedlings. If the tawny mole cricket can be said to have a preferred food, it is *Paspalum notatum* (Bahia grass), which is perhaps the most widespread pasture grass in Florida and is also used extensively in road margins and to some extent in lawns. It is ironic that this South American grass should have been introduced (in 1945) and provided as food to a South American mole cricket. No estimate of damage to pastures has been published, but tawny mole cricket is by far the worst insect pest of pasture grasses in Florida, sometimes causing total destruction. A biological control program aimed at the three species has succeeded in establishing some specialist natural enemies (Frank 1994).

Siphonaptera (Fleas)

All nonindigenous species of fleas in Florida are immigrants. *Pulex irritans* (human flea) may well have arrived with Amerindians in pre-Columbian times. Pests of pets and livestock such as *Ctenocephalides felis* (cat flea) and *Echidnophaga gallinacea* (sticktight flea) arrived with European immigrants. *Xenopsylla cheopis* (oriental rat flea) probably arrived with house rodents on ships. Fleas were estimated to cause losses of $51 million in 1984 in Florida (Hamer 1985).

Thysanoptera (Thrips)

Although thrips are best known as herbivores, some are predatory.

IMMIGRANT   Dozens of immigrant species of thrips, many of Asian origin, have been detected in Florida. Among the most important are *Selenothrips rubrocinctus* (redbanded thrips), from Asia; *Frankliniella occidentalis* (western flower thrips), from the western United States; and *Thrips palmi* (palm thrips), from Asia, detected in 1990. *Frankliniella occidentalis* and some others are important not just because they feed on crop and ornamental plants, but also because of their ability to transmit plant viruses such as tomato spotted wilt virus. A further 12 immigrant species of thrips were detected for the first time in 1993–1994 alone. Pest species of thrips were estimated to cause $2 million in losses in 1984 (Hamer 1985), but the current total caused by additional thrips species and by thrips-transmitted viruses probably is much greater.

INTRODUCED   *Amynothrips andersoni,* an Argentinean species, was one of the specialist natural enemies released and established for biological control of alligatorweed in 1967–1972 (Coulson 1977).

Perspectives

It is convenient here to consider two subcategories of immigrant species and two of introduced species.

Immigrant and Introduced Species

Among the immigrant species we separate fly-ins from hitchhikers and stow-aways. Among the introduced species we separate those imported for commercial purposes from those imported for classical biological control.

FLY-INS   Immigration was the means by which the ancestors of all native insects colonized Florida. Immigration, for that reason, is the means of arrival of the vast majority of insect species, and it continues today as it always has. An example in the foregoing account is of the arrival of *Plecia nearctica* (love bug). Major sources of immigrants are Alabama and Georgia and the neotropical region, especially the Greater Antilles, the Bahamas, and the Yucatan peninsula of Mexico (Frank and McCoy 1992). The fact that the Hawaiian Islands have a highly evolved native insect fauna, yet are greatly isolated, demonstrates that some insects are able to migrate long distances.

HITCHHIKERS AND STOWAWAYS   The remaining immigrants accompanied or followed the introduction of plants and vertebrate animals from Europe, Asia, Africa, and the neotropics. A prime example is the importation of citrus. Citrus is native to Asia, and many of its Asian pest insects had been detected in Florida by the end of the nineteenth century. More arrived in the twentieth century: citrus blackfly was detected in 1976, black parlatoria scale in 1985, and citrus leaf miner in 1993. We can assume these pests were not introduced deliberately. They probably arrived as contaminants of imported citrus, but their arrival as aerial plankton cannot be absolutely discounted. They would not survive if citrus were not grown in Florida.

Insects feeding on noxious imported plants have sometimes arrived as contaminants and proved useful as biological control agents. Examples are *Parapoynx diminutalis* (an Asian pyralid moth), whose larvae feed on hydrilla, and *Megastigmus transvaalensis* (an African torymid wasp), whose larvae feed on Brazilian pepper fruits. Parasitoids have also sometimes accompanied pests contaminating shipments of useful plants and have regulated the pest population somewhat. An example is *Trichospilus diatraeae*, an Asian eulophid wasp discovered as a parasitoid of pupae of *Epimecis detexta* (a geometrid moth pest of avocado). Another is *Arrhenophagus albitibiae*, an Asian encyrtid wasp discovered as a parasitoid of false oleander scale and white peach scale. Finally, some insects arrive in cargoes or in ballast that merely provide shelter. For example, ballast from South American riverbanks sheltered *Scapteriscus* mole crickets immigrating to the United States.

Introduced Species

COMMERCE IN INSECTS   Commerce in introduced plants and vertebrate animals has been enormous. Commerce in insects has been limited largely to the

European honeybee and, to a trivial extent, *Bombyx mori* (oriental silkworm). Much more recently, other insects, including *Tenodera aridifolia* (a Chinese mantis), *Gromphadorhina* sp. (a Madagascan cockroach), *Acheta domesticus* (a European cricket), and *Zophoba* sp. (a giant mealworm of unknown origin), have been imported and sold to the public as pets, for educational purposes, and as fishing bait. Their owners sometimes release them into the wild, or they escape (Frank and McCoy 1993, 1994; Thomas 1995). About 150 butterfly species are imported annually for living displays in escape-proof facilities (Boender 1995). To reduce future risk from these avenues, importations are now allowed only after review and under permit (Thomas 1995). Continued importation of *Acheta domesticus* is permitted because this species, from temperate Europe, has been unable to establish outdoor populations in Florida.

Twenty-one nonnative insect species have been imported commercially as biological control agents since 1980. At least four of these already had established populations in Florida, and some of the others are native to the United States (Frank and McCoy 1994). Importations from abroad are allowed only after federal review and under federal permit. Florida, virtually alone among the states, now requires its own review and an additional permit from the Division of Plant Industry (FDACS) even for importations from other parts of the United States. It is in the interests of the companies selling these biological control agents that they not establish populations in Florida, or at least not sustain large populations, for these would eliminate or reduce repeated sales.

CLASSICAL BIOLOGICAL CONTROL   Classical biological-control introductions are noncommercial and initially are brought into secure quarantine laboratories. If, after testing, they prove to be natural enemies specific to pest species, then a second round of permits is required before their progeny may be released into the environment. The intent is establishment of feral populations that will suppress target pests permanently. Targets are mostly immigrant pest insects and weed species (Frank and McCoy 1993). Such releases are made by state and federal biological-control specialists. This is the most tightly regulated of all forms of animal introduction: insects imported into Florida from abroad require federal (USDA) and state (Division of Plant Industry, FDACS) permits for importation into quarantine and federal and state permits for release into the wild. They may need documentation of importation as wildlife from the U.S. Fish and Wildlife Service and various collection and export permits from their countries of origin (depending on the laws of the countries in question).

Despite all the effort, most of these species do not establish populations. Records show that 151 nonnative insect species have been released in Florida as biological control agents, 139 of them against pest insects and 12 against weeds (Frank and McCoy 1993). Among those that became established (34 against insects, 8 against weeds), some proved highly beneficial. Examples are the minute wasps (*Amitus hesperidum* and *Encarsia opulenta*) that attack citrus blackfly as well as the flea beetle (*Agasicles hygrophila*) that feeds on alligator-weed. Although regulations governing introduction of biological control agents were more lax 50 years ago, none of the 42 established insect species has been shown to have detrimental effects. When biological control agents succeed, they give enormous economic and environmental advantages over other means of control.

ACKNOWLEDGMENTS

We thank Jerry Butler, Tom Sanford, and Rudy Scheffrahn (University of Florida) for current information about insects of veterinary importance, feral honeybees, and termites; Fran James (Florida State University) for information about birds; and David Hall and Thomas Sheehan (both formerly of the University of Florida) for information about imported plants.

6 Nonindigenous Freshwater Invertebrates

Gary L. Warren

Numerous international ports of entry, coupled with a subtropical climate, make Florida a gateway and haven for nonindigenous biota. The state is especially suited for the proliferation of nonindigenous aquatic species because of the abundance and immense diversity of freshwater habitats. Because much of this habitat is remote and unmonitored, and most freshwater invertebrate species are small (less than 10 mm in length) and inconspicuous to casual observers, nonindigenous species may be discovered only after large populations have become established.

The native freshwater invertebrate fauna of Florida is depauperate (but numerically abundant) relative to those of South and Central America, most Caribbean islands, and neighboring states (Kushlan 1990). The lack of species richness is attributable to the peninsula's geographic isolation and geologically young age compared to surrounding landmasses. Approximately 1200 aquatic invertebrate species currently inhabit Florida. Nearly 2 percent of these are nonindigenous, reproductively viable, and potentially disruptive of native ecosystems.

Typical indigenous dominants of Florida freshwater invertebrate communities are segmented worms (Oligochaeta), larval midges (Chironomidae, Ceratopogonidae, Chaoboridae), mollusks (Gastropoda and Pelecypoda), crustaceans (Copepoda, Cladocera, Ostracoda, Amphipoda, Decapoda), and mayflies (Ephemeroptera). Nonindigenous invertebrates often become successful colonists because of behavioral or reproductive competitive advantages that help them displace native dominants. Most mussel species native to North America, for example, produce nearly microscopic veligers (larvae) that require attachment to the gills of a fish host in order to develop to the juvenile stage, but veligers of the introduced Asiatic clam *Corbicula fluminea* require no such host. As the physical habitat requirements for many native mussels and *Corbicula* are similar, large populations of *Corbicula* develop rapidly and often physically displace less prolific native species whose veligers must, by chance, contact the appropriate host. The result of such displacements is usually degradation of natural aquatic systems.

Not all nonindigenous species are ecologically disruptive. By burrowing into lake bottoms, the immigrant segmented worm *Branchiura sowerbyi* (Oligochaeta: Tubificidae) promotes oxygenation of sediments. Larvae of two USDA introductions, the moth *Parapoynx diminutalis* (Lepidoptera: Pyralidae) and the fly *Hydrellia pakistanae* (Diptera: Ephydridae), are considered ecologically desirable because they aid in the control of infestations of the introduced aquatic plant *Hydrilla verticillata*.

Avenues of Entry

Nonindigenous aquatic invertebrates can enter Florida by a variety of pathways. A number have entered by more than one pathway nearly simultaneously. The two most common avenues have been release (either inadvertent or deliberate) of species imported for sale by commercial and individual aquarists or aquaculturists and accidental release of species incidentally imported with shipments of nonindigenous ornamental plants for the aquarium industry. Some nonindigenous invertebrates have entered Florida naturally, with streams flowing from northern states, after being accidentally or purposefully introduced into other regions of North America. The potentially damaging zebra mussel *Dreissena polymorpha* could soon enter Florida by this natural pathway or by another, more contemporary, avenue: transport from temperate latitudes in the bilge and live-well water of boats trailered into the state.

Other natural dispersion mechanisms cannot be discounted. Winged adult insects of species with aquatic immature stages have been collected from jet streams flowing from South and Central America at altitudes of 1800 m (Berner and Pescador 1988). The ecologically nondisruptive midge *Djalmabatista pulcher* may have entered the United States from South America by this pathway. The introduction of such obscure insect species may long go undetected but can have serious ecological implications. A final avenue of entry is state permits for the importation of species to be grown commercially, for human consumption, by the aquaculture industry. The state enforces strict regulations upon growers in order to prevent accidental releases into the wild. The Australian red-claw crayfish (*Cherax quadricarinatus*), for example, was imported into

Florida and cultured for food and aquarium potential. *Cherax* are large and aggressive crayfish that, if accidentally released and able to reproduce, could seriously affect native crayfish populations and destroy fish-spawning habitat in Florida. Aquaculturists who imported the crayfish were required to recycle makeup water and to have no effluents to natural water sources. As of this writing, commercial production of the Australian crayfish has been discontinued in Florida because it could not produce a marketable product. To date, there are no documented sightings of this nonindigenous crayfish in natural habitats.

Ecological Consequences

Nonindigenous species can substantially alter native biotic communities by preying on indigenous species, competing for food and space, or altering physical habitat. Because nonindigenous species evolved with habitat and biotic constraints that are frequently absent where they are introduced, they are often able to outcompete native species for resources. Displacement of a single native species can have profound ecosystem-wide ramifications.

In Florida, the most ecologically significant and visibly prominent introduced aquatic invertebrates have been clams and snails (phylum Mollusca) (Table 6.1). These nonnatives may outcompete native invertebrates for space and food because of a reproductive advantage or lack of a significant predator, or their lifestyles and routine activities may change the habitat to the point that native species are excluded. Sometimes the results are not as serious as they were in the cases of the Asiatic clam (*Corbicula fluminea*), spike-topped apple snail (*Pomacea bridgesi*), and zebra mussel (*Dreissena polymorpha*), but they may have severe long-range effects on intricate food webs or endangered species. The following species accounts detail the current status and ecological consequences of the most disruptive nonindigenous invertebrates inhabiting natural freshwater ecosystems of Florida.

Asiatic Clam (*Corbicula fluminea*)

Corbicula is native to the Asian continent and Pacific Asian islands, where it is often harvested for human consumption (Villadolid and Del Rosario 1930; Miller and McClure 1931). *Corbicula* is easily identified by its small size (relative to native unionids), raised umbones, and the washboard-like concentric ridges on its thick brown valves. The clam is believed to have been purposefully introduced into North America (British Columbia) during the mid-1920s by Asian laborers. *Corbicula* entered the United States via the Columbia River, Washington, where it was first collected in 1938 (Sinclair and Isom 1963; Sinclair 1971). Since its introduction, the clam has spread rapidly across the

TABLE 6.1.
Disruptive Nonindigenous Freshwater Invertebrates Presently
or Potentially Occurring in Florida

Common name	Species name	Ecological significance
Asiatic clam	*Corbicula fluminea*	displaces native mussels; disrupts native habitat; clogs hydro installations
Zebra mussel[a]	*Dreissena polymorpha*	disrupts native habitat; attaches to other organisms; clogs hydro installations
Spike-topped apple snail	*Pomacea bridgesi*	displaces native apple snails; not consumed by Everglade kites
Golden-horn marisa	*Marisa cornuaurietus*	consumes native plants
Quilted melania	*Tarebia granifera*	displaces native snails; vectors human lung fluke and sheep liver fluke
Red-rimmed melania	*Melanoides tuberculata*	displaces native snails; vectors human lung fluke, avian lung fluke, and sheep lung fluke
Faune melania	*Melanoides turricula*	displaces native snails; vectors human lung fluke, avian lung fluke, and sheep liver fluke

a Not observed in Florida but probably soon to enter the state.

United States, using a variety of pathways, including natural range extension (stream transport of veligers), boat bilgewater releases, and releases by the bait and aquarium industries (Heinsohn 1955; McMahon 1983).

Corbicula is believed to have entered Florida naturally by veliger transport from northern streams. The Florida observations were in the Escambia (1960) and Apalachicola rivers, both in the western Panhandle (1961) (Schneider 1967). Since these initial observations, the clam has spread rapidly throughout the state and now occurs in virtually all watersheds (Heard 1964, 1966, 1979; Clench 1970; Bass and Hitt 1974).

Corbicula has a reproductive advantage over most native unionid mussels because a fish host is not required to support the veliger stage. As a result, Asiatic clam populations can become extremely abundant quickly; reproductively constrained native unionid mussel populations require many years to develop substantial numbers. This advantage, in combination with tolerance of extreme environmental conditions (temperatures to 40°C, dissolved oxygen below 4.0 ppm), has prompted researchers to regard the Asiatic clam as a biological pollutant (Fuller 1974). Dense *Corbicula* populations affect native invertebrate and fish communities by altering bottom habitat and reducing availability of planktonic and suspended organic detrital food sources (Cohen et al. 1984; Lauritsen 1986). Lake Okeechobee researchers have expressed concern that postmortality accumulation of Asiatic clam shells in critical areas may change the nature of the soft mud–sand bottom, restricting habitat available to burrowing insect populations (Chironomidae) critical to the support of the high-quality, economically valuable, black crappie sport fishery (Warren and Vogel 1991). "Reefs" of dead shells can blanket the bottom in littoral areas of lakes and impede fish spawning. Moreover, infestations of *Corbicula* can cause localized extirpation of more slowly reproducing native unionid mussel species (Clarke 1986).

The Asiatic clam has also affected human industrial activities. Large infestations have clogged irrigation canals, intake pipes of municipal water treatment plants, and intake screens, pipes, and condensers at electric power generation facilities. These industries have spent large sums studying control methods, but in many cases infestations must be removed manually (at substantial cost) because chemical agents that prevent biofouling also harm native aquatic species.

Although aquatic biologists agree that introduction of the Asiatic clam has had mostly negative ecological consequences, positive effects have also been documented. High metabolic and filtering rates may make *Corbicula* useful for increasing water clarity in eutrophic lakes (T. Crisman, pers. comm.).

Zebra Mussel (*Dreissena polymorpha*)

Although not yet observed in Florida, the zebra mussel is discussed here because it will probably soon inhabit at least a portion of the state and because its incursion could have severe ecological and economic impacts.

The zebra mussel is a small (up to 5 cm in length) European species that immigrated to the United States in 1985 or 1986 in ballast water released from European cargo ships visiting the Laurentian Great Lakes (Hebert et al. 1989). In 10 years it spread throughout the Great Lakes, the Mississippi River system, and into the Southeast via the Tennessee River. It has most recently been reported in the Tombigbee system in Alabama and will probably enter at least the northernmost tier of Florida counties via stream transport of veligers within the next several years.

Zebra mussels, like the Asiatic clam, do not require a fish host to ensure survival of the veliger and hence are extremely prolific. Densities of 50,000 per square meter have been observed in southern Lake Michigan (K. Cummings, pers. comm.). Although infestations have been reported on a variety of substrates ranging from steel pilings to soft mud, zebra mussels most often attach to hard substrates in dense colonies up to 20 cm thick. Like the Asiatic clam, zebra mussels develop nuisance infestations at hydro installations. Because chemical agents affect nontarget species, their colonies must often be removed manually.

Zebra mussels spend up to five weeks in the nearly microscopic, planktonic, veliger stage, so Florida waters may be inoculated with veligers in bilge-water and live wells of boats brought to the state from the Great Lakes area by vacationers and winter residents. The higher temperatures of southern Florida waters may preclude establishment of reproducing populations of the mussel, but the species has demonstrated enough genetic plasticity for researchers to conclude that it may rapidly adapt to thermal and other environmental extremes.

Zebra mussels can dramatically affect freshwater ecosystems. Adults have extremely high metabolic rates and can filter-feed at a rate of 36 million algal cells per hour. Entire colonies filtering at this rate can substantially reduce the amount of primary production (planktonic algae) available to other aquatic species. Reduction of food quantity at the level of primary production affects all other food chain levels. Increased water clarity resulting from plankton removal may facilitate ecological reorganization at the community level by providing increased opportunity for predator-prey interaction.

Zebra mussels colonize any surface on which they can establish a firm attachment, including other animals. By attaching to native mussels (including endangered species), zebra mussels prevent feeding and reproductive activities, ultimately killing the host. By attaching to crayfish, zebra mussels hinder natural movements associated with feeding and burrow maintenance. Blanketing of lake and stream littoral zones by zebra mussels prevents fish spawning activities and feeding by wading birds.

Spike-Topped Apple Snail (*Pomacea bridgesi*)

The spike-topped apple snail is similar in size and appearance to the native Florida apple snail (*Pomacea paludosa*): the spike-topped apple snail has a raised and pointed spire; the spire of the Florida apple snail is rounded and not elevated. The spike-topped apple snail, native to Brazil, was introduced by aquarists and aquaculturists into southern Florida and isolated water bodies as far north as Alachua County (Thompson 1984). The snail is a large and appealing aquarium specimen but is extremely prolific and is probably discarded occasionally into canals or natural habitats. Two color variants are present in the state. The banded brownish-green form inhabits southern Florida canals and the Everglades; a lighter yellow variant is grown commercially and sold as the "albino mystery snail" (Thompson 1984). Specimens of the dark form are collected from canals south of Lake Okeechobee and sold to aquarists.

Researchers working in the Everglades have determined that the spike-topped apple snail can displace the native Florida apple snail from natural habitats. Florida apple snails are the primary food source for the North American population of the endangered Everglade kite (*Rostrhamus sociabilis*), which is apparently unable to feed on the spike-topped apple snail. Widespread competition-induced reduction of native apple snail numbers could therefore create a serious food shortage for this endangered bird.

Golden-Horn Marisa (*Marisa cornuaurietus*)

The golden-horn marisa is a planorbiform (spireless) snail introduced from South America into canals, ponds, and marshes in the southern Florida counties of Monroe, Dade, Broward, and Palm Beach (Thompson 1984). The snail was imported by the aquarium industry because of its large size (up to 5.7 cm in diameter) and showy, banded appearance.

The golden-horn marisa is ecologically important because it feeds heavily on aquatic plants, severely damaging both the target food plants and the habitat of many macrophyte-associated species. Although the golden-horn marisa has been used as a biocontrol agent against invasive nonindigenous plant species such as hydrilla (*Hydrilla verticillata*) and water hyacinth (*Eichhornia crassipes*), the snail is nonselective and also feeds indiscriminantly on many desirable native plant species.

Melanias

Melania snails are native to tropical Asia, Africa, and the Pacific Asian islands. They are characterized by small size (about 1.25 cm), sharply elevated spires, and ribbed or nodal sculpturing. Three melania species—the quilted melania (*Tarebia granifera*), the red-rimmed melania (*Melanoides tuberculata*), and the faune melania (*Melanoides turricula*)—have been introduced into Florida (Thompson 1984), probably in association with shipments of nonindigenous plants for the aquarium industry and by releases by individual aquarists. The quilted and faune melanias are restricted to springs and small streams, while the red-rimmed melania is distributed in all types of water bodies throughout the state.

Melanias are extremely prolific and can rapidly become very abundant. The red-rimmed melania has been collected in densities near 10,000 per square meter from Lake Okeechobee (Beck et al. 1970) and the St. John's River (Thompson 1984).

Melanias are ecologically significant because they displace trophically similar native snail populations of the genus *Elimia* (*Goniobasis*). They may be medically important, as well, because they can serve as intermediate hosts for the human lung fluke, but the chance of their spreading paragonimiasis, the parasitic disease caused by the lung fluke, is remote because the disease has been eradicated from the United States.

Prospects

Accidental release of aquatic invertebrates into Florida's fresh waters will undoubtedly continue as the aquarium industry increases in popularity. The result will probably be degradation of the state's native freshwater communities. Past invasions by other exotic plant and animal species have demonstrated that once a foothold is established, control and extirpation are nearly impossible without severe perturbation to native communities. Attempts to limit the number of introductions by increased governmental inspection at ports of entry would probably be prohibitively costly.

The most noticeable, and documented, introductions will probably continue to be molluscan species. Entry of nonnative insects will continue, but will be less well documented because of the small size, difficulty of identification, and temperature restriction of tropical species having aquatic immature stages.

7 Nonindigenous Fishes

WALTER R. COURTENAY, JR.

Because Florida is a coastal state, native fishes inhabiting fresh waters include species restricted to fresh or very low-salinity water, brackish-water forms, and marine taxa with wide salinity tolerances (euryhaline fishes). Thus a more appropriate term than "freshwater" to describe this fauna is "inland" fishes. A preliminary list prepared in 1991 by Carter R. Gilbert (Florida State Museum of Natural History, Gainesville) and James D. Williams (National Biological Service, Gainesville) included 184 taxa of inland fishes native to the state. Approximately 55 of these are primarily marine, and several others occupy brackish waters; both groups are periodically present in freshwater habitats. Warren and Burr (1994), however, list 119 freshwater fish species as native to Florida. In comparison with those of other southeastern states, some with over 250 species of freshwater fishes and no brackish-water or marine species, Florida's native inland fish fauna is somewhat impoverished. There are several reasons why. Florida, particularly peninsular Florida, is younger, geologically speaking, than most other parts of the Southeast, and the age and far greater geographic and geological stability of other areas has allowed more extensive speciation of fishes. Moreover, few drainages from those species-rich areas enter Florida (the southernmost is the Suwannee) to allow natural range expansion of fishes from north to south, except by periodic flooding events (Swift et al. 1986).

This pattern is reflected in the distribution of Florida's 119 native freshwater fishes. The most species-rich fauna occurs from the Suwannee River northwestward; the greatest richness is found in the Panhandle, where several river systems from the north drain into the Gulf of Mexico. South of the Suwannee, the number of native fishes drops dramatically; south of Lake Okeechobee, fewer than 40 species occur. There are no native freshwater fishes in the Florida Keys. The few inland waters of the Keys are inhabited by brackish-water, euryhaline, or nonindigenous species.

Thirty-five nonindigenous fish species have established either permanent or intermittent—that is, periodically established, often replenished by escapes—reproducing populations (Table 7.1), mostly in fresh water. If one subtracts the 55 primarily marine fishes from Gilbert and Williams' list of Florida's 184 inland species, nonindigenous taxa represent an addition equivalent to 27 percent of the state's inland fish fauna and, south of Lake Okeechobee, well over 50 percent.

At least 44 nonestablished nonindigenous fish species have been collected from Florida's open waters but are not known to have established populations (Table 7.2). Another 8 were formerly established but were eradicated or died from unusually cold weather during the winter of 1977 or perhaps earlier (Table 7.3). Many of the nonestablished species could become established if introduced in sufficient numbers, but many may be excluded by their reproductive behavior or other requirements for successful reproduction and range expansion.

The earliest established nonindigenous fish species in Florida was the common carp (*Cyprinus carpio*). It is present in the Apalachicola River and is reported from the Choctawhatchee, Escambia (Bass 1993), and Ochlockonee, all in the western Panhandle (Shafland 1996a). No records indicate that this fish was purposefully introduced, but it doubtless entered from intentional releases in adjoining states late in the nineteenth century. In contrast, all other fishes in Tables 7.1 and 7.3 have become established within the past 50 years—and all the fishes listed in Table 7.2 were released in that same period.

Consequences of Introductions

None of the nonindigenous fishes in Florida has caused extinction of a native species. In other parts of the United States, however, nonindigenous species are listed as a major reason for the threatened or endangered status of many native fishes (Williams et al. 1989). Lassuy (1994, 1995) cites nonindigenous fishes as one of the causes in 68 percent of native fish extinctions in the nineteenth century and in 70 percent of the cases in which fishes are now listed as threatened or endangered. Most of these cases have occurred in western states, especially where native fish diversity is often low, endemism is high, and native species have been exposed to little or no predation or competition.

TABLE 7.1.

Nonindigenous Fishes Permanently (P), Probably (P?), or
Intermittently (I) Established in Open Waters of Florida

FAMILY CYPRINIDAE (MINNOWS AND CARPS)
  Cyprinus carpio (common carp) (P; F)[a]
  Luxilus chrysocephalus isolepis (southern striped shiner) (P; F)
  Nocomis leptocephalus bellicus (southern bluehead chub) (P; F)
  Notropis baileyi (rough shiner) (P; F)
FAMILY COBITIDAE (LOACHES)
  Misgurnus anguillicaudatus (oriental weatherfish) (P; F)
FAMILY ICTALURIDAE (FRESHWATER CATFISHES)
  Pylodictus olivaris (flathead catfish) (P; F)
FAMILY CLARIIDAE (AIR-BREATHING CATFISHES)
  Clarias batrachus (walking catfish) (P; F)
FAMILY LORICARIIDAE (SUCKERMOUTH CATFISHES)
  Hypostomus sp. (armored catfish) (P; F)
  Liposarcus disjunctivus (vermiculated sailfin catfish) (P; F)
  Liposarcus multiradiatus (sailfin catfish) (P; F)
FAMILY POECILIIDAE (LIVEBEARERS)
  Belonesox belizanus (pike killifish) (P; B, F)
  Poecilia reticulata (guppy) (I; F)
  Xiphophorus helleri (green swordtail) (I; F)
  Xiphophorus maculatus (southern platyfish) (I; F)
  Xiphophorus variatus (variable platyfish) (P; F)
FAMILY CENTRARCHIDAE (SUNFISHES)
  Lepomis cyanellus (green sunfish) (P?; F)
  Lepomis humilis (orange-spotted sunfish) (P?; F)
FAMILY CICHLIDAE (CICHLIDS)
  Astronotus ocellatus (oscar) (P; F)
  Cichla ocellaris (peacock cichlid) (P; F)
  Cichlasoma bimaculatum (black acara) (P; F)
  Cichlasoma citrinellum (Midas cichlid) (P; F)
  Cichlasoma cyanoguttatum (Rio Grande cichlid) (P; F)
  Cichlasoma managuense (jaguar guapote) (P; F)
  Cichlasoma meeki (firemouth) (P; F)
  Cichlasoma octofasciatum (Jack Dempsey) (P; F)
  Cichlasoma salvini (yellow-belly cichlid) (P; F)
  Cichlasoma urophthalmus (Mayan cichlid) (P; B, F, M)
  Geophagus surinamensis (red-striped eartheater) (P; F)
  Hemichromis letourneauxi (African jewelfish) (P; F)
  Oreochromis aureus (blue tilapia) (P; B, F)
  Oreochromis mossambicus (Mozambique tilapia) (P; B, F, M)
  Oreochromis niloticus (Nile tilapia) (P; F)
  Sarotherodon melanotheron (blackchin tilapia) (P; B, F, M)
  Tilapia mariae (spotted tilapia) (P; B, F)
FAMILY ANABANTIDAE (GOURAMIES)
  Trichopsis vittatus (croaking gourami) (P; F)

[a] Habitats: B, brackish; F, freshwater; M, marine.

TABLE 7.2.

Nonindigenous Fishes Collected (But Not Known to Be Established) in Open Waters of Florida

FAMILY CYPRINIDAE (CARPS AND MINNOWS)
  Barbodes schwanefeldi (tinfoil barb) (F)[a]
  Carassius auratus (goldfish) (F)
  Ctenopharyngodon idella (grass carp) (F, B)
  Danio malabaricus (Malabar danio) (F)
  Danio rerio (zebra danio) (F)
  Hypophthalmichthys nobilis (bighead carp) (F)
  Labeo chrysophekadion (black sharkminnow) (F)
  Pimephales promelas (fathead minnow) (F)
  Puntius conchonius (rosy barb) (F)
  Puntius gelius (dwarf barb) (F)
  Puntius tetrazona (tiger barb) (F)
FAMILY CHARACIDAE (CHARACINS)
  Colossoma spp. (pacu) (F)
  Colossoma macropomum (tambaqui) (F)
  Gymnocorymbus ternetzi (black tetra) (F)
  Leporinus fasciatus (banded leporinus) (F)
  Metynnis sp. (silver dollar) (F)
  Piaractus brachypomus (pirapatinga) (F)
  Pygocentrus nattereri (red piranha) (F)
  Serrasalmus rhombeus (redeye piranha) (F)
FAMILY DORADIDAE (THORNY CATFISHES)
  Platydoras costatus (Raphael catfish) (F)
  Pseudodoras niger (ripsaw catfish) (F)
  Pterodoras granulosus (granulated catfish) (F)
FAMILY PIMELODIDAE (LONG-WHISKERED CATFISHES)
  Phractocephalus hemioliopterus (red-tail catfish) (F)
FAMILY CALLICHTHYIDAE (PLATED CATFISHES)
  Callichthys (cascarudo) (F)
  Corydoras sp. (corydoras) (F)
FAMILY POECILIIDAE (LIVEBEARERS)
  Poecilia hybrids (B, F)
FAMILY SCORPAENIDAE (SCORPIONFISHES)
  Pterois volitans (lionfish) (M)
FAMILY SERRANIDAE (SEA BASSES)
  Cromileptes altivelis (barramundi cod; panther grouper) (M)
FAMILY MORONIDAE (TEMPERATE BASSES)
  Morone chrysops (white bass) (F)
FAMILY CENTRARCHIDAE (SUNFISHES)
  Pomoxis annularis (white crappie) (F)

TABLE 7.2 (*Continued*)

FAMILY PERCIDAE (PERCHES)
   Stizostedion canadense (sauger) (F)
   Stizostedion vitreum (walleye) (F)
FAMILY CICHLIDAE (CICHLIDS)
   Cichla temensis (speckled pavon) (F)
   Geophagus brasiliensis (pearl eartheater) (F)
   Heros severus (banded cichlid) (F)
   Labeotropheus sp. (mbuna) (F)
   Pterophyllum scalare (freshwater angelfish) (F)
   Tilapia sparmanni (banded tilapia) (F)
FAMILY ANABANTIDAE (GOURAMIES)
   Colisa labiosa (thick-lip gourami) (F)
   Colisa lalia (dwarf gourami) (F)
   Helostoma temmincki (kissing gourami) (F)
   Macropodus opercularis (paradisefish) (F)
   Trichogaster leeri (pearl gourami) (F)
   Trichogaster trichopterus (three-spot gourami) (F)

Source: Modified from Courtenay et al. (1991).
[a] Habitats: B, brackish; F, freshwater; M, marine.

TABLE 7.3.
Extinct Populations of Nonindigenous Fishes
Formerly Established in Open Waters of Florida

FAMILY CHARACIDAE (CHARACINS)
   Hoplias malabaricus (trahira)
   Serrasalmus humeralis (pirambeba)
FAMILY CICHLIDAE (CICHLIDS)
   Aequidens pulcher (blue acara)
   Cichlasoma trimaculatum (three-spot cichlid)
FAMILY ANABANTIDAE (GOURAMIES)
   Anabas testudineus (climbing perch)
   Betta splendens (Siamese fighting fish)
   Ctenopoma nigropannosum (two-spot ctenopoma)
   Macropodus opercularis (paradisefish)

Note: All these fishes are freshwater species.
Source: Modified from Courtenay et al. (1991).

There is reason to believe that nonindigenous fishes in Florida do pose a threat to native taxa. Many are predators, and others compete with native fishes for spawning sites and other resources. Most are tropical and therefore confined to the southern third of the peninsula, and most occupy disturbed or artificially created habitats, although some (such as the walking catfish, *Clarias batrachus;* black acara, *Cichlasoma bimaculatum;* Mayan cichlid, *Cichlasoma uroph-thalmus;* and spotted tilapia, *Tilapia mariae*) have invaded close to pristine waters in Everglades National Park, Big Cypress National Preserve (Loftus 1989), and Corkscrew Swamp Sanctuary (Collier County). Further habitat disturbance can, and probably will, lead to dominance by nonindigenous species in open waters, particularly in metropolitan areas, as has already occurred in much of Dade, Broward, Collier, and Palm Beach counties. Because native predators such as largemouth bass (*Micropterus salmoides floridanus*) and Florida gar (*Lepisosteus platyrhincus*) have failed to stem invasions of almost pristine areas, further ingression by the same and perhaps "new" nonindigenous fishes is likely. Although it has yet to be demonstrated that disturbed habitats are requisite for colonization by nonindigenous species, human-altered habitats do appear to be more easily invasible.

The degree to which nonindigenous fishes have affected native species in Florida has not been investigated, largely because no preintroduction quantitative data on native species populations exist. The many population samples gathered by university and agency personnel for decades—whether by electrofishing or detonation cord (Metzger and Shafland 1986), rotenone, seines, or observation—are qualitative and therefore cannot show changes in species abundance. Moreover, collections, especially those by agencies, have often omitted shallow waters, nongame species, species not affected by electroshock (some cichlids, pers. obs.), and fishes too small to be caught in dipnets. Even where recent quantitative data are available, none exist from the past to permit comparisons that could reveal changes in species composition or abundance. Thus, although little or no evidence exists that shows that nonindigenous fishes have altered native fish diversity or abundance in Florida (Shafland 1986, 1996a, 1996b), this fact gives little comfort. Even if nonindigenous species *had* had a major impact on native ones, the lack of empirical research would probably have obscured this impact.

Introductions of nonindigenous fishes are sometimes considered beneficial, usually by sport or commercial fishermen. The accidental introduction in the late 1950s of the oscar (*Astronotus ocellatus*), an aquarium species that can grow to about 38 cm in length, is said to have provided a new game species. The Florida Game and Fresh Water Fish Commission (FGFC) used the same rationale for the intentional release, beginning in 1986, of the peacock cichlid (*Cichla ocellaris*) and for a more recent release of speckled pavon (*Cichla temensis*), termed "speckled peacock" by its promoters (Huffstodt 1989; Shafland 1993, 1995, 1996a). The Mayan cichlid, apparently first introduced within Everglades National Park (Loftus 1989), is now expanding its distribution well beyond park borders. It has been received enthusiastically by anglers because it readily takes artificial lures (Shafland,1996a; pers. obs.); it also recognizes and avoids monofilament gill nets, devours eyes and body parts of centrarchids entangled in those nets, and will not bleed, even after injection with heparin (pers. obs.; R. Reiners, pers. comm.). The more recently introduced jaguar guapote (*Cichlasoma managuense*), described by R. R. Miller as "highly piscivorous" (Shafland, 1996a), may prove popular with anglers.

In recent years, the flathead catfish (*Pylodictus olivaris*) has become established in the Apalachicola River and possibly the Escambia, owing to upstream introductions. This large, voracious predator, sought after by some anglers, may prove to be the most detrimental introduction to Florida waters to date. C. R. Gilbert (pers. comm.) suggests that it has already seriously affected native catfishes in the Apalachicola basin, particularly the snail bullhead (*Ameiurus brunneus*) and spotted bullhead (*A. serracanthus*) in the Flint River (Georgia) system. Haul-seine fisheries have been created in some central Florida lakes following establishment of blue tilapia (*Oreochromis aureus;* Hale et al. 1995), perhaps only to the benefit of commercial fishermen. In many waters occupied by this species, it has become the most abundant fish (V. Williams, pers. comm.), probably impeding successful spawning by native fishes (Noble and Germany 1986). Both the Mozambique tilapia (*Oreochromis mossambicus*) and blackchin tilapia (*Sarotherodon melanotheron*) are caught commercially in coastal marine fisheries. Nevertheless, none of the other fishes listed in Tables 7.1 to 7.3 can be considered to have improved fishery values in Florida waters. In fact, none of the species listed in these tables has been shown to have enhanced the biological resources of Florida.

There have been numerous attempts to introduce marine species, some by professional collectors and others through aquarium-fish releases. Species native to the Bahamas and Indo-Pacific have been collected, photographed, or observed in inshore marine waters of southeastern Florida and Tampa Bay. One report from an aquarist/diver indicates several venomous Indo-Pacific lionfish (*Pterois volitans*) observed in Biscayne Bay, Dade County, after Hurricane Andrew in 1992, near a damaged home that held large marine aquaria on a waterfront patio before the storm (R. Speiler, pers. comm.). A specimen of this species, doubtless from a separate introduction, was caught from Lake Worth Pier, Palm Beach County (R. E. McAllister, pers. comm.). One specimen of barramundi cod (*Cromileptes altivelis*) has been caught from Tampa Bay (M. M. Leiby, pers. comm.), and another was speared in the Atlantic north of Jupiter Inlet, Palm Beach and Martin counties (T. R. and L. A. Brandt, pers.

comm.). Beyond the nonindigenous fish species cited as collected in Table 7.2, an additional four species (*Gramma loreto,* royal gramma from the Bahamas; *Pomacanthus xanthometopon,* yellow-mask angelfish, and *Rhinecanthus verrucosus,* blackpatch triggerfish, from the western Pacific; and *Pomacanthus asfur,* Arabian angelfish, from the Red Sea and Gulf of Oman) have been observed or photographed in inshore marine waters (S. L. Cummings, C. Lavin, R. E. McAllister, K. S. Norris, and C. R. Robins, pers. comm.). None of the marine introductions, however, has become established. The complex life histories of many nonindigenous marine fishes may preclude colonization, but the possibility remains of concern (C. R. Robins, pers. comm.).

Over the next 10 to 20 years, additional urbanization, drainage of wetlands, pollution, nonindigenous fish introductions, and other habitat disturbances will doubtless beset native aquatic species. Such changes facilitate the establishment of nonindigenous fishes, which are often more tolerant of disrupted habitats. Because habitat disturbance, alteration, nonindigenous species, and pollution have jeopardized native fishes nationwide (many are listed as threatened and endangered), past and future introductions can be expected to increase threats to Florida's native fishes, several of which are endemic. Warren and Burr (1994) have reviewed the history and status of imperiled fishes in the United States and conclude that the Southeast will soon equal, or even exceed, the American West in the number of threatened and endangered fishes.

The introduction of flathead catfish into Panhandle drainages will doubtless harm and perhaps locally eliminate, not only the other catfish species mentioned earlier but other small species, often endemic to these drainages, and larger species such as Gulf sturgeon (*Acipenser oxyrynchus desotoi*) and shoal bass (*Micropterus* n. sp.) (J. D. Williams, pers. comm.). Population increases and adaptation to cooler temperatures, like those of walking catfish (Courtenay and Miley 1975), will allow some tropical nonindigenous fishes now established in southern peninsular Florida to invade northward gradually, a move that would be hastened by global warming. Hybrids between blue and Nile tilapia (*Oreochromis niloticus*) are said to be tolerant of cooler temperatures than is either parent species. Both species are established in Florida waters and are of great interest for aquaculture purposes; at some point, these mouth-brooding tilapias may hybridize as tilapiine fishes have done elsewhere, particularly where they have been introduced (Chervinsky 1967; Pullen and Lowe-McConnell 1982; Trewavas 1983) including Florida (Taylor et al. 1986).

Gaps in Our Knowledge

Although, for the reasons cited earlier, we cannot determine quantitatively the impacts of introduced species, qualitative observations indicate changes in native fish abundance and, to a lesser degree, species composition. During the mid-1950s, for example, canals in southeastern Dade County were occupied by native fishes ranging from small killifishes and livebearers to larger sunfishes and Florida largemouth bass. Those same canals are now dominated by South and Central American and African fishes. The abundance of native fishes has decreased dramatically. Much the same is true of certain canals and some open waters of Everglades National Park and the Big Cypress Swamp (Loftus 1989). In the 1950s and early 1960s, Lake Okeechobee contained no nonindigenous fishes. Now it is home to at least four species—all of which were introduced into other parts of southern and central-western peninsular Florida and later invaded the lake from canals. Since 1991 a number of formerly common native species, although not endangered or threatened, have declined substantially, some to the point of local absence: goldspotted killifish (*Floridichthys carpio*), golden topminnow (*Fundulus chrysotus*), Seminole killifish (*Fundulus seminolis*), bluefin killifish (*Lucania goodei*), flagfish (*Jordanella floridae*), least killifish (*Heterandria formosa*), brook silverside (*Labidesthes sicculus*), and other small fishes in canals of southeastern Palm Beach County. These canals are now being invaded by nonindigenous fishes that are expanding their ranges northward from Dade and Broward counties (pers. obs.; G. K. Reid, pers. comm.).

The question remains whether the population declines of natives are related to factors other than the impacts of range expansion of nonindigenous species; they are probably not just coincidental. Many of these native fishes lay their eggs on, and their larvae develop in, vegetation now being devoured by nonindigenous species. Moreover, postlarval and juvenile stages of nonindigenous fishes, like those of natives, rarely have the same feeding habits or strategies they do as adults, so various life-history stages of nonindigenous fishes may be harming native fishes in ways not expected and not yet investigated (Courtenay 1995).

The first comprehensive survey of nonindigenous fishes in Florida was reported by Courtenay et al. (1974). As part of a national inventory of introduced fishes, updated reports appeared in Courtenay et al. (1984, 1986, 1991). Monitoring of nonindigenous fishes in Florida should be ongoing but has been only sporadic. On a few occasions, the FGFC has learned of new introductions so isolated that they could be eradicated and has taken such action (Courtenay et al. 1986). Eradication was appropriate because once a nonindigenous fish begins to extend its range of distribution beyond the point of introduction eradication becomes virtually impossible and control costs soar.

Florida needs a continuing monitoring program that could detect new introductions before control is out of the question. Such a program should be a joint effort among state natural resource and water management agencies, universities, and county environmental units. To date, efforts have been piecemeal: the FGFC, National Park Service, various Florida water management districts, and university personnel conduct separate, uncoordinated monitoring efforts. A lead agency in monitoring nonindigenous fishes in Florida and nationwide has been the National Biological Service's laboratory in Gainesville. This laboratory maintains a national database on nonindigenous fishes, currently provides a listing of established species by means of its World Wide Web address on the Internet (http://www.nfrcg.gov/noni.fish), and will soon be able to supply distributional data on all known nonindigenous fishes in the United States.

We also need to determine the spectrum of biological interactions between nonindigenous fishes and native fishes in Florida: we cannot draw valid conclusions so long as our only data are qualitative or anecdotal. Where intentional introductions are contemplated, extensive research on potential negative effects—not just temperature tolerances and some plan for future management objectives—must be requisite and the results circulated for review and commentary before final approval or denial. Without such studies, predictions of impacts are speculative and could lead to mistakes costly to biological communities and to taxpayers.

Major Pathways of Fish Introductions

Of the fishes listed in Table 7.1, only the common carp, flathead catfish, green sunfish (*Lepomis cyanellus*), and orange-spotted sunfish (*L. humilis*) have invaded Florida from the north. The FGFC is responsible for introductions of blue tilapia, peacock cichlid, and, more recently, speckled pavon. The recent introduction of the Nile tilapia may have been a deliberate release by aquaculturists attempting to establish the species in Florida in order to ease permitting procedures for its culture as a food fish.

Introduction of the peacock cichlid is particularly interesting. The first attempt was made by FGFC biologist Vernon E. Ogilvie in 1964 but failed (Courtenay and Robins 1989; Courtenay 1993; Shafland 1996a). Peacock cichlids were again brought to Florida in the early 1970s after the FGFC opened its Non-Native Fish Research Laboratory in Boca Raton (Palm Beach County) (P. L. Shafland, pers. comm.). The fish were held in outdoor experimental ponds but brought indoors each winter to prevent exposure to lethal temperatures. In the early 1980s, renewed interest in the species led to an introduction proposal (Shafland 1984), which was circulated to some persons outside the FGFC for review and comment. The proposal indicated the species' lower temperature tolerance to be 15°C and stated that this factor would limit its range to certain clear canals in southeastern Dade County that are confluent with the warm Biscayne Aquifer. These same canals host large populations of introduced cichlids, proposed to become the forage for peacock cichlids. Thus the introduction had two goals: creating a new sport fishery and providing biolog-

ical control of other nonindigenous fishes. On the basis of the information provided, I (and obviously others) felt the introduction would not create more
problems than already existed in the area proposed for release; other reviewers,
however, were more critical.

Qualification as a sport fish requires that the species be a predator—and indeed, peacock cichlids are renowned for their angling qualities, particularly
among North American anglers who have traveled to South America to catch
them in their native waters. They are morphologically similar to and as predatory as the native Florida largemouth bass. What the introduction proposal
omitted is that the fish would be released in sites beyond those described and
that its chances of survival and establishment in those waters were good. Peacock cichlids are now established in Dade, Broward, and parts of Palm Beach
counties.

The frequently offered justification (Shafland 1993, 1995, 1996a) that peacock cichlids were successfully introduced into Hawaii and Puerto Rico
without demonstrable damage to native fish faunas is not very relevant to
Florida—Hawaii and Puerto Rico have no native freshwater fish faunas
(Courtenay 1993). Moreover, papers (Shafland 1993, 1995, 1996a) reporting
that peacock cichlids feed mostly on introduced spotted tilapia—not surprising, as this species often composes a significant or major portion of the fish
biomass—fail to report the proportion that consists of native fishes or whether
peacock cichlids have significantly reduced populations of the tilapia or other
nonindigenous fishes. For reasons that remain largely uninvestigated, largemouth bass and other native predators have been ineffective for several decades
in controlling these invasions.

Finally, although there is no doubt that peacock cichlids have met the goal of
providing a new sport fishery in southeastern Florida (Shafland 1995, 1996a),
the introduction appears to have divided anglers. As a sport angler, I often visit
sporting-goods and bait stores, where I have heard dedicated largemouth bass
anglers and store employees damn the FGFC for introducing "peacock bass,"
adding that where this species exists you cannot catch largemouth bass; others
seek information on where to catch the biggest "peacock bass." The rate of urbanization in southern Florida is such that the peacock cichlid may remain an
"urban" fish, but its range expansion since introduction suggests it may invade
far less disturbed habitats.

In fact, it has yet to be demonstrated that the peacock cichlid introduced is
truly *Cichla ocellaris*. Initial stocks imported to the FGFC laboratory in Boca
Raton came from Guyana (P. L. Shafland, pers. comm.), where *C. ocellaris* occurs only in a few interior drainages. Additional stocks were sent to the Texas
Department of Parks and Wildlife, which was also experimenting with introducing *Cichla* into some Texas heated reservoirs, a project since abandoned.

The fish actually introduced by the FGFC came from stock imported from Guyana, Brazil, and Peru and may have included other species of *Cichla* or hybrids. It is not uncommon for agencies lacking taxonomic expertise to import something under one scientific name and to discover later that they have introduced a different fish (Courtenay and Robins 1989).

The pike killifish (*Belonesox belizanus*) was introduced in November 1957 into a canal along SW 87th Avenue (Galloway Road), Dade County, after funding was terminated for a research project in which this species was being used (Belshe 1961; Miley 1978). It is now widespread in Dade, Collier, and parts of Monroe counties. Among its preferred foods is the native eastern mosquitofish (*Gambusia holbrooki*), a predator of mosquito larvae. It prefers shallow, nearshore, weedy areas (as do eastern mosquitofish) and is preyed on by larger fishes in more open waters. Despite predation, this fish continues to expand its distributional range.

Other than the few exceptions listed here, every nonindigenous fish species now established permanently or intermittently in open waters of Florida escaped from aquarium-fish culture or, more rarely, was released by aquarists. In Table 7.2 only fishes listed within the families Moronidae, possibly some Centrarchidae, and Percidae were intentional introductions, made by the FGFC, by private angler groups, or as unanticipated "passengers" in shipments of other fishes intended for release; speckled pavon was introduced by the FGFC.

Eighty percent or more of the aquarium fishes sold in North America are cultured in Florida. This business is a multimillion-dollar industry, but it has not been leakproof. (See Courtenay and Robins 1973, 1975; Courtenay et al. 1974; Courtenay and Hensley 1979; Shafland 1986; Courtenay 1989, 1990; Courtenay and Stauffer 1990; Courtenay and Williams 1992.) Nonindigenous fishes have escaped through effluent pipes, flooding of culture facilities, and intentional pumping of culture ponds containing species mixed by birds (particularly kingfishers and terns). These "biological pollutants" present a more serious threat than most industrial effluents because they are self-perpetuating and cannot be turned off or cleaned up later (Courtenay 1993).

The aquarium-fish industry is important to Florida's economy, as is the developing aquaculture industry, but they present a constant danger. Two of this nation's leading experts on aquaculture (Shelton and Smitherman 1984) have stated that culture of nonindigenous species almost invariably results in escapes into open waters. Both the aquarium-fish industry and food-fish aquaculture in Florida and other states have proved this statement correct (Courtenay and Stauffer 1990; Courtenay and Williams 1992).

In an effort to stem escapes from aquarium-fish farms, the FGFC trained a series of "inspectors" beginning in the early 1970s. Part of their job is periodic survey of fish farms to verify confinement of culture fishes. In cooperation with the Florida Tropical Fish Farmers Association, this effort has helped to slow the increasing tide of escapes that resulted in established wild populations, but several aquarium-fish farms are still allowing almost daily escapes (pers. obs.).

Food-fish aquaculture, often interested in culturing nonindigenous species, sought to avoid FGFC regulations by declaring itself to be aligned with agriculture—a trend followed in most states with developing aquaculture interests. Florida, however, is unusual in that the FGFC exists under constitutional, rather than legislative, authority, so aquaculture of any kind must abide not only by regulations set by the FGFC but also by those of the Department of Agriculture and Consumer Services. Aquaculturists and some legislators view this procedure as an impediment, but others see it as an important "check and balance" situation. Despite arguments to the contrary, culture of wild fishes is not in fact comparable to agriculture. In agriculture, most species are so far removed from their ancestors genetically and by husbandry that they are incapable of surviving to cause environmental problems if allowed to become feral. Rates of introduction largely parallel the growth of the aquarium-fish-culture industry in Florida.

Further Problems

Some of the problems discussed here result from failure to consider the ecology and historical biogeography of native fishes in southern Florida. Shafland (1986, 1993, in press b) for example, interprets the natural low species diversity there as representing "vacant niches" and blames it on "disturbed habitats." Yet disturbed habitats do not automatically have few species—it depends on the type and degree of disturbance (Connell 1978). Moreover, it is extremely difficult for a human observer to tell what constitutes available niches independently of the species occupying them (Stiling 1996). He also suggests that fishes introduced from more temperate climates pose less of a threat to Florida species than do those from tropical regions. This is true only for native fishes south of the Suwannee River; the invasion of flathead catfish in northern Florida demonstrates the kind of damage that can occur there. Collette (1990) has pointed out the serious problems of reliance on the reams of unpublished "gray literature" generated annually by agencies. If there are in fact data that imply introductions pose little or no threat to native biotic communities, they should be published so they can be subjected to critical review.

Even more serious is the pressure—inherent in the way most state fish and game agencies are funded—to favor utilization and manipulation over conservation or even ecological evaluation of native biotas. The difficulty is that the FGFC derives virtually all of its state funding from the sale of hunting and fishing licenses and the bulk of its federal funding from programs that support hunting and fishing under the guise of conservation (the Pittman-Robertson,

Dingell-Johnson, and Wallop-Breaux acts), usually in the form of matching (often 2:1 or 3:1) of the license-fee income. If license sales fall, so too does overall income. It is not surprising, then, that similarly funded agencies do not consider protecting native faunas a high priority. Some state "conservation" agencies, particularly in the American West, actually view legislation such as the Endangered Species Act as detrimental to their major mission—the provision of a steady supply of game animals, regardless of the effect on native species.

That nonindigenous fishes now outnumber native species in southern Florida is not "symptomatic of major environmental disturbances"; areas with naturally low native fish diversity are simply more easily invasible by species that find the climate suitable. Similar invasions have happened repeatedly in the American West in habitats subject to little or no disturbance and where the fish fauna is often depauperate. In those areas, introductions have often proved cat-astrophic (Minckley and Deacon 1991). I have never suggested that the same will necessarily happen in southern Florida—Florida's native fish fauna is more resilient, and no endangered species are involved—but to ignore the possibility is irresponsible.

Florida now has the national "distinction" of hosting the largest number of nonindigenous fishes in the contiguous United States—not a ranking in which Floridians can take pride. We must recognize that nonindigenous fishes, as they have elsewhere, will at some point create more and perhaps threatening prob-lems for native fish populations. We need much more stringent regulation of what can be imported for culture in Florida (particularly cichlids) and a far better effort to educate the public not to release nonindigenous fishes. Aquarium-fish farms need more frequent inspections to assure compliance with existing regulations, and fines for noncompliance should be increased substan-tially. Agencies with the legal authority to conduct introductions should be re-quired to conduct thorough studies on biological impacts (negative and posi-tive) and, moreover, to subject the results to independent review to be certain they are scientifically sound. Without these minimal steps, the number of non-indigenous fishes inhabiting Florida's waters will continue to escalate—with yet unknown but probably dire consequences for the native biota that makes Florida's inland waters unique.

ACKNOWLEDGMENTS

I gratefully acknowledge Carter R. Gilbert and William F. Smith-Vaniz for pro-viding helpful information. My thanks go also to the two anonymous reviewers for their suggestions on manuscript revision and to Don C. Schmitz and Daniel Simberloff for their invitation to participate in this effort.

8 Nonindigenous Amphibians and Reptiles

BRIAN P. BUTTERFIELD, WALTER E.
MESHAKA, JR., AND CRAIG GUYER

 Since documentation began more than 100 years ago, with Cope (1875, 1889) and Garman (1887), addition of nonindigenous species to Florida's herpetofauna has escalated. (See Barbour 1910; Fowler 1915; Stejneger 1922; Carr 1940; Duellman and Schwartz 1958; King and Krakauer 1966; Wilson and Porras 1983; Dalrymple 1994.) Conservation concerns associated with these taxa were summarized in a seminal paper by Wilson and Porras (1983). These authors described the mechanisms of colonization, discussed the role of the pet trade, and addressed potential threats to native amphibians and reptiles, but they considered environmental modifications associated with uncontrolled human population growth to be the greatest threat to indigenous taxa. Twelve years after their work, we resurvey the nonindigenous herpetofauna of Florida and reassess its impact on indigenous species. Our objectives are to update the list of nonindigenous species, describe where in Florida they have colonized, document where they have come from and how they arrived, discuss their potential to expand their ranges and colonize natural habitats, and speculate upon their impacts on indigenous species and ecosystems.

Diversity of Founding Species

In Florida, 36 species of nonindigenous amphibians and reptiles have become established (Table 8.1). Our list substantially exceeds that of Dalrymple (1994) because we include seven species described in very recent literature not available to Dalrymple: *Litoria caerulea* and

TABLE 8.1.
Nonindigenous Amphibians and Reptiles Established in Florida

Species	Date[a]	Source[b]	Means[c]	Counties[d]	Climate[e]	Location[f]	SVL[g]
Bufo marinus	1958	NT	I	10	TR	S	225
Eleutherodactylus planirostris	1875	B	M	50	TR	S	36
Litoria caerulea	1994	OW	I	2	TE,TR	C,S	100
Osteopilus septentrionalis	1931	B	M	27	TR	S	140
Trachemys scripta elegans	1958	NA	I	3	TE	C,S	289
Cosymbotus platyurus	1984	OW	I	3	TE,TR	C,S	60
Gekko gecko	1983	OW	I	9	TR	N,C,S	180
Gonatodes albogularis	1939	WI	M	2	TR	S	40
Hemidactylus frenatus	1993	OW	I	2	TR	S	60
H. garnotii	1963	OW	I	20	TR	S	64
H. mabouia	1991	OW	M	6	TR	S	68
H. turcicus	1915	OW	M	16	TR	S	60
Sphaerodactylus argus	1954	B	M	1	TR	S	33
S. elegans	1922	WI	M	1	TR	S	39
Calotes versicolor	1993	OW	I	1	TR	C	140
Anolis chlorocyanus	1987	WI	I	2	TR	S	76
A. cristatellus	1975	WI	I	1	TR	S	75
A. cybotes	1973	WI	I	2	TR	S	77
A. distichus	1948	B	M	5	TR	S	58
A. equestris	1952	WI	I	4	TR	S	188
A. garmani	1975	WI	I	2	TR	S	131
A. sagrei	1887	B	M	40	TR	C,S	70
Basiliscus plumifrons	1994	NT	I	1	TR	S	177
B. vittatus	1976	NT	I	2	TR	S	175
Phrynosoma cornutum	1953	NA	I	2	TE	N	130
Ctenosaura pectinata	1972	NT	I	2	TR	S	348
Iguana iguana	1980	WI	I	3	TR	S	500
Leiocephalus carinatus	1945	B	I	3	TR	S	130
L. personatus	1994	WI	I	1	TR	S	86
L. schreibersi	1978	WI	I	1	TR	S	107
Ameiva ameiva	1957	WI	I	1	TR	S	149
Cnemidophorus lemniscatus	1964	NT	I	1	TR	S	113
C. motaguae	1994	NT	I	1	TR	S	145
Ramphotyphlops braminus	1979	OW	M	5	TR	S	173
Boa constrictor	1990	WI	I	1	TR	S	3000
Caiman crocodilus	1960	NT	I	1	TR	S	2640

[a] Date: first known date of establishment.

[b] Source (geographic source of introduction): B = natural populations found in the Bahamas (may include other locations in the West Indies); NA = temperate North America; NT = neotropics exclusive of the West Indies; WI = West Indies exclusive of the Bahamas; OW = Old World.

[c] Means (most probable means of introduction into Florida): I = introduced either accidentally or intentionally by humans; M = at least one population established by immigration with cargo.

[d] Counties: number of counties occupied (Monroe Co. and the Keys are counted separately).

[e] Climate (predominant climate at source area): TE = temperate; TR = tropical.

[f] Location (sites of initial introductions): N = northern Florida; C = central Florida; S = southern Florida.

[g] SVL: maximum snout–vent length in millimeters.

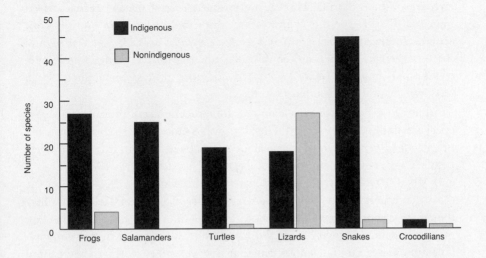

FIGURE 8.1   Comparison of indigenous and nonindigenous species of amphibians and reptiles in Florida.

Leiocephalus personatus (Bartlett 1994); *Cosymbotus platyurus* (Meshaka and Lewis 1994); *Hemidactylus frenatus* (Meshaka et al. 1994); *Hemidactylus mabouia* (Lawson et al. 1991); *Anolis chlorocyanus* (Bartlett 1988; Butterfield et al. 1994); and *Cnemidophorus motaguae* (Bartlett 1995a). We also include two species (*Basiliscus plumifrons* and *Calotes versicolor*) not described in the literature but known to be established in Florida (R. D. Bartlett, pers. comm.) and one species—*Phrynosoma cornutum* (King and Krakauer 1966; Jensen 1994)—that Dalrymple considered not to be reproductive. We do not include *Eleutherodactylus coqui* as established, contrary to Dalrymple (1994), because this frog persists only in greenhouses in the greater Miami area.

Florida's nonindigenous herpetofauna includes four anurans, one turtle, twenty-eight lizards, two snakes, and one crocodilian (Fig. 8.1). The distribution of nonindigenous taxa among taxonomic groups differs significantly from that of indigenous species (maximum likelihood test; $\chi^2 = 35.21$, d.f. = 4, $p <$ 0.0001). Lizards dominate the nonindigenous herpetofauna. In fact, Florida now has more nonindigenous species of lizards than indigenous ones.

Colonization by nonindigenous amphibians and reptiles in Florida is by far most common in southern Florida. We define southern Florida as the area south of the southern rim of Lake Okeechobee, central Florida as the area between the northern border of Alachua County and the northern rim of Lake Okeechobee, and northern Florida as the area between the northern border of Alachua

County and the state line. Our literature survey found that 29 species first arrived in southern Florida, whereas one species (*Calotes versicolor*) first arrived in central Florida and one (*Phrynosoma cornutum*) in northern Florida. Four species (*Litoria caerulea, Trachemys scripta elegans, Cosymbotus platyurus*, and *Anolis sagrei*) arrived in both central and southern Florida through multiple releases, and another (*Gekko gecko*) arrived more or less simultaneously in northern, central, and southern Florida.

Although other states have acquired nonindigenous amphibian and reptile species over the years (Fig. 8.2), Florida now has more of these species than any other state (Fig. 8.3). Only Hawaii, an oceanic island system with no native herpetofauna, has even comparable numbers (Wilson and Porras 1983; Eldredge and Miller 1995). Surrounded by water on three sides and by frost on the fourth, southern Florida has some characteristics of an oceanic island and may therefore be more vulnerable to colonization by nonindigenous taxa than are other mainland systems (Myers and Ewel 1990b; Chapter 1 in this volume). The numbers of nonindigenous taxa that have colonized northern and central Florida are similar to those of most other states.

Most of the nonindigenous amphibians and reptiles in Florida are tropical. Thirty-two species originated primarily from tropical regions, two are of temperate origin, and two have wide ranges in both temperate and tropical areas (Table 8.1). This pattern differs from that of Florida's 136 indigenous species (excluding sea turtles) (Ashton and Ashton 1981, 1985, 1988; Conant and Collins 1991; Moler 1992). Twenty-four indigenous species, in 13 genera (*Bufo, Hyla, Gastrophryne, Kinosternon, Sphaerodactylus, Anolis, Cnemidophorus, Rhineura, Rhadinaea, Drymarchon, Tantilla, Micrurus*, and *Crocodylus*), have tropical affinities, whereas 112 species (in 57 genera) are of tem-

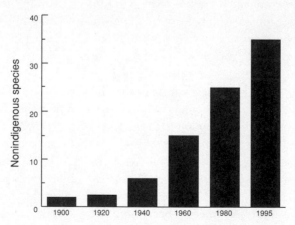

FIGURE 8.2   Accumulation of nonindigenous amphibian and reptile species in Florida over time.

perate origin (Savage 1982). California and Texas, in contrast, two continental states with significant numbers of nonindigenous amphibians and reptiles (Figure 8.3), have been colonized principally by species of north temperate origin (Dixon 1987; M. R. Jennings, pers. comm.).

King and Krakauer (1966) have suggested that the mechanism by which amphibians and reptiles have invaded Florida has changed over the years. The earliest colonizations, prior to about 1930, were by species from nearby islands of the West Indies that immigrated to southern Florida with cargo shipments. During the middle part of the 1900s, additional immigrants from the West Indies were supplemented with intentional and unintentional introductions. Wilson and Porras (1983) have described similar trends.

To determine whether this pattern continues, we examined the numbers of New World and Old World taxa added to Florida's herpetofauna since 1983 and compared these numbers to those of taxa present before 1983. Before 1983, some 22 species were from New World source areas and only 4 were from the Old World (Wilson and Porras 1983). Since 1983, 5 species have arrived from the New World and 5 from the Old World. The proportions before and after 1983 differ significantly (maximum likelihood test; χ^2 = 4.18, d.f. = 1, p = 0.041). This pattern is consistent with continued immigration of Caribbean taxa with cargo traffic supplemented by more global introductions associated with the pet trade.

Because many of the nonindigenous amphibians and reptiles arrived from

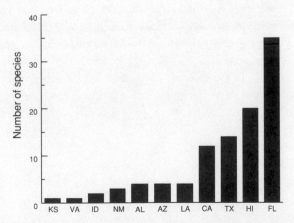

FIGURE 8.3   Comparison of the numbers of nonindigenous amphibian and reptile species in 11 selected states. *Sources:* AL (Alabama), C. Guyer (pers. obs.); AZ (Arizona), R. Reed (pers. comm.); CA (California), M. R. Jennings (pers. comm.); HI (Hawaii), Eldredge and Miller (1995); ID (Idaho), Fichter and Linder (1964), and Linder and Fichter (1970); KS (Kansas), Collins (1993); LA (Louisiana), R. A. Thomas (1994); NM (New Mexico), C. Painter (pers. comm.); TX (Texas), Dixon (1987); VA (Virginia), Conant and Collins (1991), and Mitchell (1994).

the Caribbean and some of these same species appear to thrive in Florida, we explore the relationship between location within the West Indies and probability of colonizing Florida. The Bahamas are close to Florida and the Bahamian herpetofauna has a history of recent overwater dispersal (within the last 80,000 to 100,000 years according to Lind 1969)—two features that should make these taxa more likely immigrants to southern Florida than other Caribbean taxa (Butterfield unpublished data), so we compared the Bahamian and non-Bahamian members of the nonindigenous herpetofauna of Florida. Of the 45 species found in the Bahamas (and Turks and Caicos Islands), 6 have recently colonized Florida. Of the 506 species identified from the remainder of the West Indies exclusive of the Bahamas (Schwartz and Henderson 1991), 12 have colonized Florida. Proportionally more of the Bahamian species reached Florida than did other West Indian species (maximum likelihood test; χ^2 = 12.88, d.f. = 1, $p < 0.0001$). Of the invaders from the Bahamas, 5 of 6 arrived in Florida as immigrants, whereas only 2 of the 12 remaining West Indian species did so (Fisher's exact test, $p = 0.013$). We conclude that proximity to Florida has affected the proportion of nonindigenous West Indian amphibians and reptiles that have colonized the state as well as the methods of transport by which they arrived.

Patterns of Expansion

Twenty-two of Florida's nonindigenous amphibian and reptile species have apparently not dispersed far beyond their sites of arrival. In some cases, insufficient time may have elapsed since colonization. In others, species have colonized habitats surrounded by barriers that limit dispersal. For example, populations of *Ameiva ameiva*, *Anolis cristatellus*, and *Ctenosaura pectinata* introduced on Key Biscayne (Dade County) are still limited to the island. Finally, some species, such as *Gonatodes albogularis*, *Sphaerodactylus argus*, and *Phrynosoma cornutum*, that have had adequate time to disperse have simply failed to do so. These observations suggest that unknown factors such as the genetic makeup of founders or unstudied biotic interactions play a role in colonization success.

Five species—*Trachemys scripta*, *Sphaerodactylus elegans*, *Anolis equestris*, *Basiliscus vittatus*, and *Leiocephalus carinatus*—have undergone limited range expansions since their initial colonizations. Four of these species are restricted to only two or three counties of southeastern Florida. *Trachemys scripta elegans* and *B. vittatus* have spread along many canals in Dade County, whereas *Anolis equestris* has spread through residential areas of Broward and Dade counties and sporadically into the Keys. *Trachemys scripta elegans* has also been introduced into at least two locations in central Florida (Bancroft et al. 1983;

Hutchison 1992). *Sphaerodactylus elegans* is restricted to the lower Keys but has colonized many locations, including some natural areas. Since the introduction of *Leiocephalus carinatus* to Palm Beach County, this species has expanded its range northward and southward along the Atlantic coast.

The remaining nine species have wide, continuous distributions. Eight of these species appear to have expanded their ranges in close association with the movements of humans. The ninth, *Eleutherodactylus planirostris*, may have escaped some of its dependency on humans: it now occurs throughout most of the peninsula, including natural habitats.

We suggest that factors other than time affect range expansion of amphibians and reptiles. Indeed, the mode of arrival in an area may predict a species' success. Species arriving by passive immigration may be better adapted to human disturbances than are species intentionally or accidentally introduced. Immigrants probably comprise a subset of species that are exceptional dispersers, so we predicted that immigrants would be more successful at range expansion than introduced species. Two processes—diffusion and jump dispersal (Pielou 1979; Chapter 1 in this volume)—are thought to contribute to range expansion. Immigrants should be better than introduced species at both of these processes. Immigrants, because they tend to be wide-ranging habitat generalists in their native ranges (Williams 1969), can occupy more habitat patches, and these patches are more likely to be contiguous. Therefore, opportunities for diffusion (the gradual and regular spread of a species' range) are greater. Moreover, immigrants should be more likely to display jump dispersal (long-distance movement outside a species' range), because more suitable patches are available to which they can jump.

In a preliminary examination of this hypothesis, we compared multiple regressions of range and body size on time since arrival for immigrants and introduced species. Area (square root transformed; see van den Bosch et al. 1992) was the dependent variable; number of years since colonization and snout–vent length (SVL) were covariates. Body size was included in the equation because larger animals tend to have larger geographic ranges (Brown 1995). Because exact areas occupied by most species of the nonindigenous herpetofauna are unknown, we estimated this variable by counting the number of counties occupied by each (Table 8.1). Following Means and Simberloff (1987), the Florida Keys and Monroe County were considered separate counties. A significant multiple regression of range on time (adjusted for differences in SLV) was found for immigrants ($F = 9.25$, $p = 0.01$, $r^2 = 0.73$), indicating that these species have consistently expanded their ranges (Figure 8.4). We found no significant relationship for introduced species ($F = 1.80$, $p = 0.19$, $r^2 = 0.14$; Figure 8.4). But because most introductions are recent, these data are insufficient to

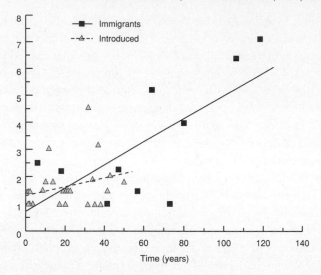

FIGURE 8.4　Regressions of area (number of counties$^{1/2}$) on time (in years) for immigrants and introduced amphibians and reptiles in Florida.

demonstrate that those animals will continue to inhabit restricted ranges or to expand their ranges only slowly over time periods comparable to those available to immigrants.

Invasibility of Natural Habitats

Nonindigenous amphibians and reptiles in Florida are strongly associated with disturbed areas (Wilson and Porras 1983) altered primarily through urbanization or agriculture. In Florida, none of the colonizations originated in natural habitats, and most nonindigenous amphibians and reptiles have remained only in disturbed sites. A few have colonized natural areas, however, as have other vertebrates, so evaluation of potential impacts warrants examination.

Everglades National Park is a good case study. Located at the southern tip of Florida, the park is close to the points of colonization for many nonindigenous species of plants and animals. Of the 36 taxa on our list, 31 occur in two counties of southern Florida (Dade and Monroe) encompassed in part by Everglades National Park (Table 8.2).

To assess the invasibility of the Everglades by nonindigenous amphibians and reptiles, we compiled records from natural-history observation cards archived by park personnel beginning in the late 1940s, from literature records, and from field observations. Fourteen nonindigenous species have been observed at the park. Of these, nine are not known to have reproductive populations within park boundaries. Three probably were introduced intentionally:

TABLE 8.2.

Comparison of Nonindigenous Amphibians and Reptiles Observed
and Established in Everglades National Park
with Nonindigenous Species Established in Surrounding Areas

| Species | County[a] | | Everglades National Park[b] | | |
	Dade	Monroe	Observed	Established Disturbed	Natural
Bufo marinus	X	X	X		
Eleutherodactylus planirostris	X	X		X	X
Osteopilus septentrionalis	X	X		X	X
Trachemys scripta elegans	X				
Gekko gecko	X	X			
Gonatodes albogularis	X	X			
Hemidactylus frenatus		X			
H. garnotii	X	X		X	
H. mabouia	X	X		X	
H. turcicus	X	X			
Sphaerodactylus argus		X			
S. elegans		X			
Anolis cristatellus	X				
A. cybotes	X				
A. distichus	X	X	X		
A. equestris	X	X	X		
A. garmani	X				
A. sagrei	X	X		X	X
Basiliscus plumifrons	X				
B. vittatus	X				
Iguana iguana	X	X	X		
Ctenosaura pectinata	X		X		
Leiocephalus carinatus	X				
L. personatus	X				
L. schreibersi	X				
Ameiva ameiva	X				
Cnemidophorus lemniscatus	X				
C. motaguae	X				
Tupinambis nigropunctatus			X		
Ramphotyphlops braminus	X	X	X		
Boa constrictor	X		X		
Caiman crocodilus	X		X		
Total	28	16	9	5	3

[a] Dade = established in Dade County; Monroe = established in Monroe County.

[b] Observed = observed but not established; Disturbed = established in disturbed areas;
Natural = established in natural areas.

Boa constrictor (1981, Royal Palm Hammock Road), *Tupinambis teguixin* (1994, Royal Palm Hammock), and *Ctenosaura pectinata* (1994, Royal Palm Hammock). Three others appear to have immigrated with human assistance: *Anolis equestris* (1973, 1977, Flamingo), *Ramphotyphlops braminus* (1990, 1994, 1995, Shark Valley, Hole-in-the-Doughnut), and *Anolis distichus* (1978, Royal Palm Hammock). A seventh species, *Caiman crocodilus,* was observed in Taylor Slough during a four-month period in the 1980s, presumably having dispersed through the canal systems surrounding the park. The final two species, *Bufo marinus* (1990, Parachute Key visitor's center, Royal Palm Hammock parking lot) and *Iguana iguana* (1970–1977, 1995, Royal Palm Hammock, Long Pine Key Campgrounds), are exceptional in that they are established along the park boundary. *Bufo marinus* is very common in adjacent agricultural areas but was not observed to breed in the park during intensive fieldwork from 1990 to 1992. In 1995 (W. E. Meshaka, Jr., pers. obs.), *Iguana iguana* was established along canals that border the park but was not known to reproduce in the park.

Five species are established within Everglades National Park boundaries. Two of these, *Hemidactylus garnotii* and *H. mabouia,* are restricted to buildings and immediate surrounding vegetation. The former has been known from the park since 1969 (Kluge and Eckardt 1969) and the latter since 1991 (Butterfield et al. 1993). The presence of these geckos is best explained by human-assisted immigration. Three established nonindigenous species, *Eleutherodactylus planirostris, Osteopilus septentrionalis,* and *Anolis sagrei,* are found in disturbed and natural areas of the park. All three are native to areas close to Florida, namely the Bahamas and Cuba. Given more time, these three species might have reached Florida by natural overwater dispersal. In fact, Lazell (1989) speculates that *E. planirostris* and *O. septentrionalis* (together with *A. carolinensis* and *Sphaerodactylus notatus,* now considered indigenous) dispersed naturally to the lower Keys during pre-Columbian times.

Natural areas of Everglades National Park have thus far resisted colonization by most nonindigenous amphibians and reptiles. Whether other natural areas within Florida will resist colonization remains to be seen. Observations from Everglades National Park suggest that the most likely future colonists of natural areas will be taxa from nearby source areas that have large native ranges and for which seemingly appropriate habitat exists in Florida. On that basis, the list of nonindigenous amphibians and reptiles currently established in Florida and likely to colonize natural areas includes one turtle (*Trachemys scripta elegans*) and seven lizards (*Gonatodes albogularis, Sphaerodactylus argus, S. elegans, Anolis distichus, A. equestris, Leiocephalus carinatus,* and *Phrynosoma cornutum*).

Impacts on Native Herpetofauna

As noted in Chapter 1 of this volume, nonindigenous species can have many different effects on native species, and an understanding of the importance and sometimes even the existence of these effects requires intensive research. Such research has rarely been conducted on the nonindigenous herpetofauna of Florida. Nevertheless, it is worth detailing existing information, as much to indicate which effects seem *not* to occur commonly as to suggest what potential problems should be studied.

Competition

Wilson and Porras (1983) consider competition between indigenous and nonindigenous amphibians and reptiles to be greatly overplayed. In fact, of their two major and eight minor threats to Florida's indigenous herpetofauna, competition with nonindigenous forms was considered to be least important. Although we know of little recent evidence to suggest a stronger case for competition, naturalists seem to believe that *Anolis sagrei* is outcompeting the indigenous *A. carolinensis* and may even extirpate it. This notion has persisted despite the two species' ecological dissimilarity and a long history of co-occurrence in the West Indies (Williams 1969). In fact, in all recent cases where one of these species arrived in an area already occupied by the other, the species arriving later was at least marginally successful (Losos et al. 1993). The observations assumed to indicate competitive exclusion of Floridian *A. carolinensis* by *A. sagrei* are probably better described as a shift in habitat usage by *A. carolinensis* (Collette 1961; Schoener 1975).

Despite our reservations about this *Anolis* example, a comparison of the entire assemblage of anoles in southern Florida with the extensive literature on West Indian anole communities appears to be a fruitful line of research. On the four largest islands of the Greater Antilles (Cuba, Jamaica, Hispaniola, and Puerto Rico), from 7 to more than 35 *Anolis* species co-occur (Roughgarden 1995). These lizards have been divided into six general ecomorphs based on body size, color, modal perch height, and body proportions. Members of different ecomorph categories can co-occur locally (Williams 1983), but on islands where several species occur, members of the same ecomorph are not syntopic (Roughgarden 1995). If a species invades the range of another species of similar size and habitat preference, theory and empirical evidence suggest that the nonindigenous form may establish a static enclave in the range of the indigenous species (Roughgarden 1995), both may undergo shifts in habitat use (Losos 1994), both may undergo shifts in body size as predicted by a character displacement model (Losos 1992a), the larger of the two may decrease in body size and drive the other to extinction as predicted by a taxon-loop model

(Roughgarden 1992), or the resident may evolve away from the invader and the invader evolve into the niche space left vacant by the resident as predicted by a parallel-evolution model (Roughgarden 1995).

The anole assemblage occurring over much of urban Dade County is composed of four ecologically dissimilar species: *A. equestris* (a "crown giant"), *A. carolinensis* (a "trunk-crown" species), *A. distichus* (a "trunk" species), and *A. sagrei* (a "trunk-ground" species). Interestingly, the pattern exhibited by anole assemblages on the larger islands of the West Indies appears to have been recreated in Florida by colonizations by nonindigenous anoles. If the success of anole colonists is governed largely by the processes described by Williams (1983), then two more ecomorphs, a "twig" species and a "grass-bush" species, could fit into the Florida assemblage. It is important to note, however, that the three widespread nonindigenous anoles were also the first three species to be introduced. The other nonindigenous anoles of Florida have been present for short times, occur only in localized populations, and belong to ecomorph categories of one of the more widespread anoles listed earlier. The fates of these later additions offer opportunities for independent tests of hypotheses of community structure generated from the West Indies by Williams (1983), Losos (1992b, 1994), and Roughgarden (1992, 1995). To date, the only examination of these invaders is that by Salzberg (1984), who demonstrated that *A. cristatellus* males reduce the perch height of male *A. sagrei* in syntopy.

Competition might be examined in two other nonindigenous species: *Litoria caerulea* (competing with *Hyla cinerea* and *H. squirella*) and *Cnemidophorus motaguae* (competing with *C. sexlineatus*). Both of these nonindigenous forms presently have small ranges within urban areas of Florida and are allopatric to any known populations of indigenous competitors. But if range expansion occurs, the possibility of competitive exclusion should be examined.

Predation

Nonindigenous species may prey upon indigenous species or vice versa. In Florida, we are aware of only one documented account of a member of the nonindigenous herpetofauna consuming indigenous forms. The Cuban tree frog, *Osteopilus septentrionalis*, preys on *Bufo terrestris*, *Gastrophryne carolinensis*, *Rana sphenocephala*, *Hyla cinerea*, *H. squirella*, and conspecifics (Meshaka 1994). Preliminary data suggest a negative association between the number of *O. septentrionalis* and numbers of *H. cinerea* and *H. squirella*—a pattern explained in part by predation and in part by different habitat preferences (Meshaka, unpublished data; Ashton and Ashton 1988). Other cases undoubtedly exist, but quantification of predatory effects on native populations is lacking. These nonindigenous taxa are generally confined to urban areas where negative effects on indigenous populations are likely to be inconsequential.

Indigenous species preying on nonindigenous species may be more important than the converse. Wilson and Porras (1983) indicate that *Diadophis punctatus* prey on *Eleutherodactylus planirostris, Coluber constrictor* prey on young *Ameiva ameiva,* and *Elaphe guttata* prey on *Anolis sagrei.* Lower Keys populations of *E. guttata* are thought to have increased in urban areas because increasing numbers of *A. sagrei* are available as prey (Wilson and Porras 1983). *Osteopilus septentrionalis* is also preyed upon by *C. constrictor* and *Elaphe obsoleta* (Meshaka and Ferster 1995) as well as barred owls (*Strix varia;* Meshaka 1996) and other indigenous predatory vertebrates (Meshaka, unpublished data). Dalrymple (1994) argues that because the skin of *O. septentrionalis* contains noxious chemicals, few predators on indigenous anurans would be able to eat this nonindigenous frog. He mentions observations of American crows (*Corvus brachyrhynchos*) and black racers (*Coluber constrictor*) attempting to consume *O. septentrionalis* with little success and suggests that these predators were repelled by its skin secretions, but crows readily eat larval, postmetamorphic, and adult *O. septentrionalis* (W. E. Meshaka, Jr., pers. obs.). In general, we conclude that the nonindigenous herpetofauna of Florida is unlikely to affect native forms through direct consumption. Assemblages of indigenous predators of vertebrates might be altered if certain members are better able to exploit the nonindigenous forms as food resources. Contemporary diet studies of indigenous species are needed.

Nonindigenous predators other than amphibians and reptiles are more likely to affect Florida's indigenous herpetofauna. The fire ant (*Solenopsis invicta*) is suspected of affecting indigenous reptiles through predation on eggs and neonates (Mount 1981). Fishes are known to eat amphibian larvae (Wilbur 1984), so a nonindigenous fish, like the walking catfish (*Clarias batrachus*), which is capable of locomotion across land between pools of water and has a very broad diet (Thakur 1978), might dramatically affect larval amphibians. Finally, nine-banded armadillos (*Dasypus novemcinctus*) and feral dogs (*Canis familiaris*) are important predators on indigenous amphibians and reptiles (Causey and Cude 1978; Carr 1982). Other nonindigenous species undoubtedly prey occasionally on Florida's indigenous herpetofauna.

Disease and Parasites

Few studies are available on the effects of disease on natural populations of amphibians and reptiles (Dodd and Seigel 1991), but the transmission of disease from captive or introduced individuals to indigenous animals gives cause for concern. Desert tortoises (*Gopherus agassizi*) from the western Mojave Desert, for example, developed an upper-respiratory-tract disease that proved catastrophic and led, in part, to federal protection. This disease was thought to have been transmitted to wild populations through release of infected captives. A similar disease, suspected to have been transmitted in the same way, has been found in Florida populations of the gopher tortoise (*G. polyphemus*) on Sanibel Island, near Fort Myers (Lee County), and the Tamiami Trail (Dade County) (Dodd and Seigel 1991). Because the same pathogen was found in the two tortoise species, interspecific transfer is worth consideration. We suspect that similar problems could result from transmission of ecto- and endoparasites from nonindigenous amphibians and reptiles to indigenous taxa.

Hybridization

Hybridization with closely related nonindigenous taxa is a potential threat to indigenous species (Chapter 1 in this volume). The only known example in Florida involves *Anolis distichus*. Of 18 recognized subspecies, two and possibly a third are found in Florida. The Florida bark anole (*A. d. floridanus*) was first reported from Miami by Smith and McCauley (1948). Wilson and Porras (1983) consider it indigenous to Florida, but Schwartz (1971) postulates that this subspecies differentiated locally in Florida or represents a western Andros population that became established in Florida. The "Dominican" bark anole (*A. d. dominicensis*) from Hispaniola was first reported by King and Krakauer (1966) from the Tamiami Canal in Miami. A third subspecies, the "Bimini" bark anole (*A. d. biminiensis*) from Bimini in the Bahamas, has been reported from Lake Worth in Palm Beach County (Bartlett 1995b). In Broward and Dade counties, subspecific designation of individuals of these lizards is difficult or impossible because they may share characters of at least two subspecies as a result of hybridization (Miyamoto et al. 1986).

Dalrymple (1994) speculates that *Trachemys scripta elegans,* introduced to Florida by the pet trade, might hybridize with *Pseudemys floridana* or *P. nelsoni,* but no hybrids have been found and, moreover, these same three species co-occur widely in northern Florida without hybridizing. Nevertheless, hybridization associated with escaped pets remains a reasonable concern. Hobbyists keep a wide variety of amphibians and reptiles, including many close relatives of indigenous forms. The biggest threats appear to be to snakes of the genera *Elaphe* and *Lampropeltis,* both of which have several populations with color morphs unique to Florida.

Future Trends

Because the southern tip of Florida is composed of habitats that may be more suitable to tropical Caribbean species than to temperate North American species, the former are likely to continue immigrating. The vegetation of the tropical hardwood hammocks of extreme southern Florida is composed predominantly of species common to the Bahamas and the Greater Antilles (Snyder et al. 1990) and may be of very recent origin (only 5000 years old according to Lodge 1994). Thus we predict that southern Florida's herpetofauna will continue to change from one primarily composed of species of north temperate origin to a tropical fauna dominated by taxa with a long history of dispersal and evolution in the West Indies. In particular, 11 Bahamian species (*Trachemys terrapen, Ameiva auberi, Anolis angusticeps, A. smaragdinus, Sphaerodactylus nigropunctatus, Tarentola americana, Alsophis vudii, Epicrates striatus, Tropidophis canus, Typhlops biminiensis,* and *T. lumbricalis*) are likely to become future immigrants to southern Florida. This list could lengthen if trade barriers between the United States and Cuba are removed.

Wilson and Porras (1983) observe that all nonindigenous amphibians and reptiles that have become established in southern Florida because of the pet industry were at some point imported in large numbers and sold at a relatively low price. The growing popularity of amphibians and reptiles in the pet trade guarantees that such introductions will continue. We predict that such additions will be restricted to static enclaves or will expand slowly in association with human disturbances, but there are no data at present to test this notion. The one possible exception is the release of nonindigenous color morphs of indigenous taxa, which might cause widespread alteration of endemic Floridian morphs.

One group of taxa, not associated with the pet trade, that have immigrated to Florida from areas outside the West Indies include five species of geckos and one fossorial snake, *Ramphotyphlops braminus.* Moreover, the brown tree snake (*Boiga irregularis*) has successfully immigrated to many Pacific islands outside its native range and harmed their ecosystems. Indeed, *Boiga irregularis* has caused the extinction and range reduction of many bird and lizard species on Guam (Savidge 1987; Rodda and Fritts 1992) in addition to lowering the quality of human life (Fritts 1994). Fritts (1994) hypothesizes that *B. irregularis* could become established in Florida and, if so, would threaten native species and negatively affect humans. We see no reason to discount his suggestions, except that snakes are unlikely to become immigrants in the Caribbean region (including southern Florida), perhaps because humans tend to locate and kill snakes when they are associated with cargo transport (Butterfield unpublished data).

With the possible exception of *B. irregularis,* no amphibian or reptile is likely to invade native habitats and alter natural communities as have fire ants (*Solenopis invicta*). We strongly agree with Wilson and Porras (1983), who argue that the biggest threat to the indigenous Floridian herpetofauna is pressure not from nonindigenous amphibians and reptiles but from environmental changes caused by humans. We consider the nonindigenous amphibians and reptiles in Florida a symptom of a much larger problem: uncontrolled human population growth and modification of the environment. Although the release of nonindigenous amphibians and reptiles into the wild should be discouraged, study of this group should be encouraged because much can be learned from them about colonization theory, conservation, and evolution.

ACKNOWLEDGMENTS

We thank Daniel Simberloff for the opportunity to contribute to this project. We also thank Jonathan B. Losos and Paul E. Moler for constructive reviews of this chapter.

9 Nonindigenous Birds

FRANCES C. JAMES

 The status of both the native and the nonindigenous birds of Florida is summarized in two excellent recent books—*Florida Bird Species: An Annotated List* by Robertson and Woolfenden (1992) and *The Birdlife of Florida* by Stevenson and Anderson (1994)—that give many more details about Florida's nonindigenous birds than can be included here. Two other books, by Long (1981) and Lever (1987), review the worldwide history of the introductions of birds beyond their natural geographic ranges. In combination, these references are the best sources of information about who has been releasing birds, their reasons, and the consequences. They do not, however, treat the role of the federal government in regulating the importation of nonindigenous birds into the United States. Because Miami is a major port of entry and many imported birds subsequently escape or are deliberately released in southern Florida, this topic is relevant to the subject of this chapter. An up-to-date account of the ongoing debate about federal regulation of the importation of wild-caught birds for the bird trade is given in the transcript of the recent hearing on the reauthorization of the Wild Bird Conservation Act of 1992. This hearing was conducted on 28 September 1995 by the Subcommittee on Fisheries, Wildlife, and Oceans of the U.S. House of Representatives Committee on Resources. The transcript is available from subcommittee chairman Jim Saxton on request. For the history of the exotic bird trade, see Beissinger and Snyder (1992).

Lists of Nonindigenous Birds

Florida is blessed with a rich native avifauna. The total number of species depends on the criteria chosen for inclusion. Stevenson and Anderson (1994) include 483 species, of which 22 (4.5 percent) are deemed nonindigenous. Robertson and Woolfenden (1992) are more conservative: they list 461 species, of which 11 (2 percent) are nonindigenous. Robertson and Woolfenden exclude 16 species that certainly or presumably reached Florida with human help, subsequently nested in the state, but were not considered to be well established in Florida by the end of 1991. At least one species on this excluded list of 16, the common myna, *Acridotheres tristis,* has been increasing in number in Florida since 1991. Bruce Neville (pers. comm.) believes that if one were to prepare a "hit list," this species should be near the top. Robertson and Woolfenden (1992) have also compiled a list of 119 additional nonindigenous species that are also excluded from their official list of Florida birds. These birds have been observed in the wild in Florida, but they are clearly not established there, and the records are often incomplete. The criteria for distinctions among these lists may be arbitrary, but they are important.

Some Floridians consider exotic birds a pleasing addition to their environment. Most naturalists deplore the presence of these species and consequently have paid little attention to them (Robertson and Woolfenden 1992; Smith and Smith 1993). Exceptions have been Oscar Owre and his students at the University of Miami, Glen Woolfenden and his students at the University of South Florida, and P. William and Susan Smith of Homestead (Dade County). Their focus on this subject clearly shows that we can learn important lessons about avian ecology and population biology from careful study of nonindigenous birds and that without such information we would miss some of the most dynamic events in the ornithology of the state. In this chapter I review the histories of the 11 best-established nonindigenous bird species in Florida (Table 9.1) and comment on a few of the species not included on the Florida species list by Robertson and Woolfenden (Table 9.2). I conclude by outlining the status of the international trade in wild birds and its implications for future introductions of nonindigenous birds in Florida.

Types of Cases

Because bird populations are naturally so dynamic, even more care is required to characterize them than would be required with more sedentary life forms. Species like the cattle egret (*Bubulcus ibis*) and fulvous whistling duck (*Dendrocygna bicolor*), which arrived in Florida on their own, are simply expanding their natural geographic ranges. They are natural invaders in the process of becoming native to Florida and do not qualify as nonindigenous. Species like

Table 9.1.
Eleven Nonindigenous (Exotic) Bird Species That Have Established Breeding Populations in Florida

Common name	Scientific name	Distribution in Florida
Muscovy duck	*Cairina moschata*	statewide
Rock dove	*Columba livia*	statewide
Eurasian collared dove	*Streptopelia decaocto*	statewide
Budgerigar	*Melopsittacus undulatus*	Pinellas and Pasco counties plus other Gulf Coast areas
Monk parakeet	*Myiopsitta monachus*	east coast counties, Tampa/ St. Petersburg area, and elsewhere
Canary-winged parakeet	*Brotogeris versicolurus*	mainly Miami area and Tampa/ St. Petersburg
Red-whiskered bulbul	*Pycnonotus jocosus*	S. Miami
European starling	*Sturnus vulgaris*	statewide
Spot-breasted oriole	*Icterus pectoralis*	Brevard to Dade counties
House finch	*Carpodacus mexicanus*	Leon, Escambia, and Okaloosa counties
House sparrow	*Passer domesticus*	statewide

Source: Robertson and Woolfenden (1992) supplemented by information from Stevens and Anderson (1994).

Table 9.2.
Twelve Exotic Bird Species in Florida That Did Not Have Well-Established Breeding Populations in 1991

Common name	Scientific name	Location of present population in Florida
Common peafowl	*Pavo cristatus*	Brevard County to the Keys
Ringed turtledove	*Streptopelia "risoria"*	St. Petersburg
Rose-ringed parakeet	*Psittacula krameri*	N. Miami, Collier County
Black-hooded parakeet	*Nandayus nenday*	St. Petersburg, Dade County
Blue-crowned parakeet	*Aratinga acuticaudata*	Florida Keys
Red-masked parakeet	*Aratinga erythrogenys*	Dade, Monroe, and Palm Beach counties
Dusky-headed parakeet	*Aratinga weddellii*	Miami Springs
Chestnut-fronted macaw	*Ara severa*	Miami
Orange-winged parrot	*Amazona amazonica*	Dade County
Common myna	*Acridotheres tristis*	Dade, Palm Beach, and Collier counties
Hill myna	*Gracula religiosa*	Biscayne Bay, Coconut Grove
Red-crested cardinal	*Paroaria coronata*	Broward, Dade, and Orange counties

Note: These species are not included by Robertson and Woolfenden (1992) in their list of the 461 species of Florida birds. Nor are the black francolin (*Francolinus francolinus*), red-crowned parrot (*Amazona viridigenalis*), blue-gray tanager (*Thraupis episcopus*), and Java sparrow (*Padda oryzivora*)—species known to have bred in Florida but not to have persisted into the 1990s.

the red-whiskered bulbul (*Pycnonotus jocosus*), which were imported for the pet trade, released, and then became established in the wild, clearly do qualify as nonindigenous. More problematic are cases like the black-bellied whistling duck (*Dendrocygna autumnalis*), whose natural range extends as far north as southern Texas. It wanders occasionally as widely as California, Michigan, Florida, and the Lesser Antilles (Stevenson and Anderson 1994). Because some whistling ducks that escaped from the Crandon Park Zoo in Miami in about 1968 have been breeding nearby for several years (P. W. Smith and M. Wheeler cited by Robertson and Woolfenden 1992), it is not always possible to tell whether the source of the increasing numbers of records scattered through 14 counties of Florida (Stevenson and Anderson 1994) are dispersing wild birds or birds that escaped from captivity and then dispersed. We must simply plead ignorance. Nevertheless, the black-bellied whistling duck seems to be establishing a wild population in Florida. Robertson and Woolfenden (1992) have deemed it a natural invader. The case of the white-winged dove (*Zenaida asiatica*) is similar. This species breeds as close to Florida as Louisiana and the Bahamas and also wanders into many areas of Florida in the nonbreeding season (Stevenson and Anderson 1994). Again, however, well-documented releases of captive birds that have subsequently nested in the wild make it impossible to tell for certain whether the small populations of breeding birds in southern Florida today are derived from wild or captive stock. Robertson and Woolfenden (1992) view the white-winged dove as a natural invader.

Another situation occurs when a nonindigenous bird arrives on its own but has help from human activity in the course of its range expansion. The most dramatic case in Florida is that of the Eurasian collared dove (*Streptopelia decaocto*), which has spread explosively northward through Florida in the last 15 years (Stevenson and Anderson 1994). It is probably a recent immigrant from the Bahamas, where released European birds initiated an expansion (Smith 1987). Humans transported it across the Atlantic Ocean, but it then reached Florida on its own. An interesting complication in this case is that in Pinellas County the Eurasian collared dove has been hybridizing with a wild population of the closely related, presumably domesticated, and possibly conspecific (American Ornithologists' Union 1983), ringed turtledove or Barbary dove (*Streptopelia "risoria"*) (Lever 1987; Stevenson and Anderson 1994; P. W. Smith, pers. comm.).

While the Eurasian collared dove has been spreading northward through Florida and increasing in number, the house finch (*Carpodacus mexicanus*) has been spreading southward. It is a native of western North America and Mexico, but the Florida population derives from birds transported from California to New York and then released (Stevenson and Anderson 1994). These examples show that simple assignment to two classes, immigrant or introduced, can mask interesting variation in the histories of the populations, which sometimes defy full understanding.

When animals are bred in captivity for many generations and then succeed in the wild without human support, they are referred to as feral. Cats, for example, readily establish feral populations in farmland, whereas dogs do not. In Florida, the populations of the Muscovy duck (*Cairina moschata*) and the rock dove (*Columba livia*) are clearly feral. Note that even though they breed in the wild without aid, they are confined to human-altered habitats. The nonindigenous bird species with the worst national reputation as a pest—the introduced European starling (*Sturnus vulgaris*)—is not considered feral because today's population was derived originally from released wild birds.

Exotic Cage Birds

Southern Florida supports a large cottage aviculture industry and many private collections of game birds, parrots, and songbirds. Dealers in exotic pet birds sometimes release large numbers of them to avoid quarantine restrictions on imported birds. Other birds are released by aviculturists because they are unwanted. Some simply escape, or hurricanes destroy outdoor aviaries but not their occupants (Smith 1987). At least 60 species of birds released in Florida have bred in the wild (Robertson and Woolfenden 1992). Sometimes their populations have increased for several generations and then, for unknown reasons, declined. The few species from the bird trade that have succeeded in maintaining self-sustaining populations over the last 20 years are all confined today to suburban areas: the budgerigar (the familiar pet parakeet, *Melopsittacus undulatus*), monk parakeet (*Myiopsitta monachus*), canary-winged parakeet (*Brotogeris versicolurus*), red-whiskered bulbul, and spot-breasted oriole (*Icterus pectoralis*) (Table 9.1). Even in these cases, however, local populations tend to fail for unknown reasons (Lever 1987; Robertson and Woolfenden 1992). The last three species have not spread beyond suburban Miami. Several populations of the budgerigar and the monk parakeet around the state have died out.

Owre and his students (Carleton 1971; Owre 1973; Carleton and Owre 1975) and Robertson and associates (Robertson and Kushlan 1974; Robertson and Frederick 1994) have described the ecological history of urban development along the eastern coast of southern Florida as one that created an environment conducive to the support of released tropical cage birds. In the early 1900s, the Atlantic coastal ridge, which forms the eastern edge of the Everglades basin, was still covered with pineland and tropical hardwood hammock vegetation. The subsequent drainage of the Everglades, dredging of the mangroves, and urban development were accompanied by widespread planting of more than a thousand species of nonindigenous tropical ornamental trees and shrubs.

This artificially produced environment offered nectar, fruit, and seeds at all seasons to alien birds, and these resources were supplemented by food provided at bird feeders in suburban gardens. Thus, for example, Carleton and Owre (1975) found the nonindigenous red-whiskered bulbul roosting in fig and palm trees native to Asia and eating the fruits of the exotic Brazilian pepper (*Schinus terebinthifolius*), loquat (*Eriobotrya japonica*), and various jasmines (Oleaceae).

Introductions of Game Birds

The history of attempts to introduce nonindigenous game birds by private citizens and by state and federal agencies in the United States is mostly one of expensive failures (Phillips 1928 cited by Lever 1987). As with pet birds, the most common result is either immediate failure or modest success and then gradual failure (Phillips 1928 cited by Lever 1987). Even with species that have been successfully introduced, the successes often have followed a long series of failed cases, and what made the difference is often unclear. In the United States, one notably successful case, after many failures, was the introduction in 1882 from China to Oregon of wild-caught ring-necked pheasants (*Phasianus colchicus*). After that example, many states established game farms for the production of animals to be released for sportsmen to shoot. With the possible exception of the black francolin (*Francolinus francolinus*), which was released widely around Florida in the 1960s, none of the numerous introductions of nonindigenous game birds to Florida by the Florida Game and Fresh Water Fish Commission or by sportsmen has resulted in successful establishment. Cox et al. (Chapter 19 in this volume) review the checkered history of the Florida Game and Fresh Water Fish Commission's introductions of nonindigenous game birds to the state.

Potential Threats

I do not view any of the populations of nonindigenous birds in Florida as serious economic or ecological threats. Even so, some Florida birds have caused significant agricultural losses elsewhere, and we should not be blind to potential problems. Dolbeer (1988), for example, summarizes the status of the European starling as a pest in the United States: starlings can cause economic losses to farmers by eating grain and fruit crops; under certain circumstances their droppings can transmit histoplasmosis, a serious human respiratory disease; where their habitats overlap, starlings compete with native birds for nest cavities. These problems have been bigger issues in states where starlings have more massive breeding and wintering populations than they have in Florida, states with large agricultural areas where grain is left in fields or feedlots.

In late summer, the starlings join several species of native North American blackbirds (red-winged blackbirds, *Agelaius phoeniceus;* common grackles, *Quisculus quiscula;* and brown-headed cowbirds, *Molothrus ater*) in attacks on the sunflower seed crop in the Dakotas and the ripening corn in Ohio, Illinois, and elsewhere. Later in the fall and winter, the starlings gorge themselves on rice in Arkansas and Louisiana. The U.S. Fish and Wildlife Service's Office of Animal Damage Control, often in collaboration with state and local governments, used to run a program to control agricultural losses by killing large numbers of birds. One method used in the 1970s was to spray migratory or winter roosts from helicopters with surfactants that caused birds to freeze to death. Between 1974 and 1986, the government killed an estimated 7 million starlings and 26 million blackbirds in 60 winter roosts in the southeastern United States (Dolbeer 1988). These large-scale programs were expensive and, as we shall see, were no more effective than were the quelea control programs in Africa. In both cases, more birds came to the funeral.

Crop Depredation

One instructive agricultural example is the depredation caused in Africa by a small finch called the quelea (*Quelea quelea*). In times of drought, when native grass seed is not available, millions of quelea attack rice, wheat, and millet, but campaigns to eradicate immense roosts of birds by means of traps, poison, dynamite, and even bombs have failed to reduce populations more than temporarily. Although on several occasions pet quelea have escaped to the wild in the United States, they have not become established. We do not want quelea in the North American grain belt, and it is the responsibility of the Fish and Wildlife Service (FWS) to see that they do not get there (Nilsson 1981).

In the 1980s it became clear that the actual losses to farmers caused by birds were not so great as originally supposed (Weatherhead et al. 1982). The birds were often eating grain that had been spilled in the fields—grain that would not have been harvested anyway. In the last decade the FWS policy has changed away from large-scale efforts to kill birds at roosts toward smaller projects aimed at specific properties where damage is most severe. For example, a poison bait approved by the Environmental Protection Agency, called Starlicide, is being used to kill birds in feedlots (Dolbeer 1988). Threats to Florida farmers from depredations by birds, whether native or nonnative, should be handled similarly. In fact, farmers suffer greater losses from rodents, insects, and weeds than they do from birds.

One popular cage bird, the monk or Quaker parakeet, has been treated as an agricultural pest in its native South America. More than 64,000 monk parakeets were imported to the United States for the pet trade between 1968 and 1972

(Banks 1977 cited by Long 1981). By 1974 there were many reports of free-flying birds in California and from Texas to Wisconsin eastward. California and New York quickly established eradication programs. Bucher (1992), however, considers damage to fruit and grain crops by monk parakeets in Argentina and Uruguay to be somewhat exaggerated. Partly because the birds are conspicuous, farmers tend to overstate the damage, which is usually local rather than regional. In South America, the birds are trapped, shot, and poisoned by the thousands (Bucher 1992).

Two species of songbirds that are popular as cage birds, but are classed as agricultural pests in their native India and Java, established small wild populations in southern Florida in the 1960s: the red-whiskered bulbul and the Java sparrow. For unknown reasons, the Java sparrow population in Coral Gables (Dade County) died out in the 1980s. This species is now listed as noxious and can no longer be imported legally (R. Gnam, pers. comm.). The bulbul persists in small numbers in Miami. In the Hawaiian Islands, it eats Natal plums, papayas, strawberries, and mangos (Nilsson 1981). Theoretically the species could pose a threat to berry growers in southern Florida, but the pattern of release, apparent establishment, and then population decline, typical of so many nonindigenous cage-bird species in Florida (Robertson and Woolfenden 1992), now applies to the bulbul as well.

Disease Transmitted to Poultry or Native Birds

The extent of possible harm to populations of native species as a result of diseases introduced by nonindigenous birds is well illustrated by the situation in Hawaii, where avian malaria, transmitted by an introduced mosquito, and birdpox have caused heavy mortality in native birds and contributed to the extinction of some of them (van Riper et al. 1986). Such situations have not been reported in Florida, but because so many birds are accidentally released here, and smuggling of wild birds is a serious problem, the threat is always present.

Wild birds of the parrot family tend to harbor exotic Newcastle disease. The U.S. Department of Agriculture (USDA) screens legally imported birds for this disease and directs a quarantine program. The program was established after a devastating outbreak of exotic Newcastle disease was identified in 1971 in Fontana, California. By the time the disease was contained, two years later, a total of nearly 12 million chickens, turkeys, exotic birds, game birds, and native songbirds had been destroyed. The total cost of this outbreak was estimated to be $56 million. Although recent outbreaks have been contained before such massive losses (Nilsson 1981), the threat is serious. Smuggled birds, and birds removed prematurely from quarantine, can infect both native birds and poultry at any time. Clearly the most direct solution to the problem of nonindigenous avian diseases is to curtail the importation of wild-caught birds.

Changes in the Distribution of Native Birds

Robertson and Woolfenden (1992) report major changes in the last 25 years in the geographic ranges of native birds that breed in Florida. Along with the tremendous pressure on Florida's natural environments from economic development, many species of Florida birds have been increasing their geographic ranges in the state. Robertson and Woolfenden list about 65 species that have extended their breeding ranges, most of them southward and eastward. Examples are the American robin (*Turdus migratorius*), barn swallow (*Hirundo rustica*), blue grosbeak (*Guiraca caerulea*), and indigo bunting (*Passerina cyanea*). Those that are extending their ranges northward and westward are mainly either West Indian species or established nonnatives. Thirty species are less abundant in the state than formerly or have receding geographic ranges. Many of these declines are associated with loss of native habitat like wetlands, pineland, prairie, or scrub. Examples are the large herons and egrets, the endangered red-cockaded woodpecker (*Picoides borealis*), the Florida scrub jay (*Aphelocoma coerulescens*), the eastern bluebird (*Sialia sialis*), the summer tanager (*Piranga rubra*), and the Bachman's sparrow (*Aimophila aestivalis*). None of these changes is clearly attributable to competition with nonindigenous species. Possible exceptions are competition between the rapidly expanding Eurasian collared dove and the native mourning dove (*Zenaidura macroura*) and observations of common mynas attacking purple martins (*Progne subis*) at their nests (B. Neville, pers. comm.).

Although one may deplore the presence of nonindigenous birds in Florida, almost all of them are in human-altered habitats, where they mix only with those native birds that can also survive in such habitats. Are native birds seriously affected by nonindigenous species? The general answer is no.

The Mallard and the Greater Flamingo

Two species that do not qualify for listing in either Table 9.1 or Table 9.2, but could involve nonindigenous lineages of Florida birds, are the mallard (*Anas platyrhynchos*) and the greater flamingo (*Phoenicopterus ruber*). The mallard, of course, is a common native wintering species. In addition feral mallards exist throughout the state all year on farm ponds, canals, and lakes. Stevenson and Anderson (1994) add that there are intermittent breeding records of "wary" mallards (individuals that don't exhibit tame behavior) in Leon, Santa Rosa, Jackson, and Highlands counties. At Lake City (Columbia County), Belle Glade (Palm Beach County), and Zellwood (Orange County), small numbers of mallards associate in summer with resident mottled ducks (*Anas fulvigula*). Apparent hybrids have been noted since 1984 in Highlands and Orange counties. Whether the wary mallards that breed in Florida are wild birds that did not

148 FRANCES C. JAMES

migrate back north in the spring or are originally from captive stock is unknown. Regardless of their origin, their tendency to hybridize with the native mottled duck in peninsular Florida is worrisome. Such introgressive hybridization could eventually destroy the genetic integrity of the mottled duck.

The greater flamingo, flaunted in advertisements for Florida as a tropical paradise and a place to win lottery money, was abundant in extreme southern Florida in the 1800s. Although the records are somewhat equivocal, it probably nested there (Stevenson and Anderson 1994). But even then, most of the flamingos in Florida originated from rookeries in the Bahamas and Cuba, spent the winter in Florida, molted their feathers and grew new ones, and returned (Palmer 1962 cited by Stevenson and Anderson 1994)—a new twist on the "snowbird" label for tourists who winter in Florida. Because there are breeding colonies of flamingos in captivity from which captive birds escape, such as the one at Hialeah Park in Dade County, it is usually impossible to tell whether the flamingos that show up sporadically along the Florida coast now are wild or escapees. Since the 1950s a semidomesticated colony of about 30 greater flamingos has been established at Hialeah (Long 1981; Stevenson and Anderson 1994).

Nonindigenous Birds That Are Well Established

In this section I summarize the status of the 11 species of birds that Robertson and Woolfenden (1992) consider to have well-established nonindigenous populations in Florida in 1991 (Table 9.1). Except as noted, all information is from Robertson and Woolfenden (1992) and Stevenson and Anderson (1994).

MUSCOVY DUCK   The natural range of the Muscovy duck extends from Mexico to Argentina. Domestic stocks are common in Florida as the large black and white patched duck on local ponds. The birds often hybridize with domestic mallards, including the white barnyard "Pekin" duck, producing plumages that do not occur in the wild populations of either species. In fact, most Muscovy ducks in Florida appear to be hybrids. The individuals seen occasionally in natural wetlands or remote coastal areas (Stevenson and Anderson 1994) are not wary and are presumed to be feral. The species has been declining in Dade County over the past few years (P. W. Smith, pers. comm.).

ROCK DOVE   Rock doves in the United States (the "pigeons in the park") and around the world are considered feral because generations ago they were the offspring of domesticated birds (Johnston and Janiga 1995). The original population occupied coastal and inland cliffs from Spain to North Africa and east to India, but rock doves have been held in captivity, possibly since Neolithic times, for food (Lever 1987). They probably came to the United States with early settlers. Johnston and Janiga (1995) describe the few places in Europe where truly wild birds persist and, even there, they found evidence of introgressive hybridization between wild and feral stock. They also discuss the fact that some populations of feral rock doves in Colorado and Oregon have reverted to living on cliffs away from humans. In Florida, however, as in most of the United States, rock doves occur only around farms and bridges and in urban situations, where buildings seem to serve as satisfactory substitutes for the rocky cliff habitat preferred by the original wild populations. As is typical of many feral animals that have a long history of captive breeding, the rock dove breeds in all months of the year.

Rock doves can spread diseases such as psittacosis, cryptococcal meningitis, histoplasmosis, toxoplasmosis, and encephalitis. Their droppings deface buildings, and they can weaken mortar by pecking it for its lime content. In farmland, rock doves eat grain and can compete with chickens for food. The birds can become a nuisance around airports (Lever 1987). In such cases, private companies can be hired to trap and remove them, a policy I view as preferable to poisoning them.

EURASIAN COLLARED DOVE   The Eurasian collared dove is a game bird in the Old World. Since 1930 it has undergone a dramatic range expansion northwestward through western Europe and eastward through northern China and Korea. The eastward invasion may be by natural invaders from western China or Inner Mongolia (Vaurie 1961 cited by Lever 1987) or by birds derived from individuals shipped to China from India (Streseman and Nowak 1958 cited by Lever 1987). The birds can be a local nuisance, especially where grain is being harvested (Lever 1987).

A similarly explosive invasion of collared doves—with similarly complex cases of hybridization with domestic stock of S. "risoria"—is under way in North America. Here the invading collared doves are derived from a release of 50 individuals in the Bahamas in 1974 (Smith 1987). These birds had been brought to the Bahamas by mistake for the pet trade. After they proved to be unsuited for that purpose and some escaped, the owner/breeder released the rest. P. W. Smith discovered the species in Homestead, Florida, in the late 1970s and early 1980s. By 1987, there were at least 1200 pairs in Dade and Monroe counties. By 1993 the species was nesting in Alachua County in the north central area of the Florida peninsula. Collared doves have become common in coastal residential habitats in northern Florida, and there are now records for Alabama, Mississippi, and Louisiana along the coast of the Gulf of Mexico (Stevenson and Anderson 1994) and north to North Carolina (Davis 1995) along the Atlantic coast.

A domesticated strain of a closely related species—the African collared dove (*Streptopelia roseogrisea*), known to aviculturists as the ringed turtledove or Barbary dove (*S. "risoria"*)—has established feral populations in California (Lever 1987) and in Orange and Pinellas counties in Florida (Lever 1987) (Table 9.2). Although the Orange County population died out, the population in St. Petersburg in Pinellas County persists. Smith (cited by Stevenson and Anderson 1994) reports substantial hybridization between the invading collared dove and the smaller ringed turtledove in the Pinellas County population, a phenomenon also occurring in England and elsewhere (Lever 1987).

BUDGERIAR   The budgerigar is the pet parakeet, the most common cage bird in the United States. It regularly escapes or is released in Florida and elsewhere. The species is native to arid regions of Australia, where even wild birds, given rainfall and a food source, can breed at any season. Florida populations are all derived from captive stock, but they retain some of the opportunism of wild stock. In St. Petersburg (Pinellas County), where residents enjoy seeing them in the wild, many released birds were provided with nest boxes. By 1962 thousands of feral birds were nesting in suburban areas. In the late 1970s the species had breeding populations all along the west coast of peninsular Florida, from Dixie County to Collier County, and other populations existed on the east coast near Fort Pierce (St. Lucie County), Port St. Lucie (St. Lucie County), and Fort Lauderdale (Broward County) (Wenner and Hirth 1984). The largest breeding populations occurred from Charlotte County to Citrus County. Flocks of 30 or more occurred as far north as Gainesville (Alachua County) in winter, but the breeding populations did not spread to the state's interior. Then, in the 1980s, numbers declined. Whether or not wild populations will persist in Florida is unknown. None of the many releases of this species outside of Florida have resulted in establishment of wild populations (Lever 1987).

MONK PARAKEET   The monk parakeet is native to southern central South America from Bolivia to Argentina, where it has the reputation—perhaps exaggerated, as we have seen—of a serious agricultural pest. This species occurs widely in parklike habitats in Florida, especially from the central part of the peninsula south. Nesting records for Miami go back to 1969 (Owre 1973). Before Hurricane Andrew, in September 1992, more than 100 pairs nested at Homestead Air Force Base (Dade County), and colonies of a few hundred pairs existed at Miami Beach, Miami Springs, and South Miami (P. W. Smith cited by Stevenson and Anderson 1994). Smaller populations occur in suburban areas along the east coast from Jacksonville in Duval County to Plantation Key in Monroe County. Monk parakeets also nest in the Tampa–St. Petersburg area and elsewhere along the Gulf coast. On the Florida Christmas Bird Counts in

1991, some 933 individuals were reported in thirteen 15-mile-diameter circular areas; 250 of them were in the St. Petersburg area and 204 in Fort Lauderdale. Stephen Pruett-Jones (pers. comm.) estimates that by the mid-1990s there were several thousand monk parakeets in the wild in the United States. Several thousand more occur in a population in San Juan, Puerto Rico (Silver cited by Lever 1987).

Should we be worried about whether the monk parakeet will become a serious pest in Florida? Opinions differ. Its large communal nests sometimes cause short circuits in transformers on utility lines. Stevenson and Anderson (1994) predict that the species will eventually damage our commercial crops. P. W. Smith (pers. comm.), however, finds that local populations in Florida are usually ephemeral, possibly as a result of natural and human-induced nest collapse.

CANARY-WINGED PARAKEET   Canary-winged parakeets, native from Colombia to northern Argentina (Lever 1987), have maintained wild populations in Florida since the 1970s in Sarasota (Sarasota County) and Tampa–St. Petersburg along the Gulf coast and from Fort Pierce to Miami along the east coast. Thousands of birds once roosted in a mixed-species parrot roost in South Miami, but in the 1980s their numbers declined sharply. In 1992 the Dade County Christmas Bird Count included only 234 birds (Stevenson and Anderson 1994). More recent releases of many individuals of the yellow-chevroned form of this species seem to have replaced most of the original population. This form occurs farther south in South America than does the canary-winged form and is not officially recognized as a subspecies (Smith and Smith 1993). This example, which has led to some taxonomic confusion, shows that geographic variation in plumage within the natural range of a species can be compared to the plumage of nonindigenous individuals and can yield clues to the sources of birds sold in the pet trade. We do not know whether the wild population of canary-winged parakeets would be self-sustaining without periodic releases of additional birds.

RED-WHISKERED BULBUL   The red-whiskered bulbul is a common Old World songbird that is resident from India to Vietnam (Lever 1987) and is usually associated with humans (Carleton and Owre 1975). Nonindigenous populations occur in Hawaii, Australia (Stevenson and Anderson 1994), and Mauritius (Lever 1987), as well as in Florida. The Florida population is the only representation of the Old World avian family Pycnonotidae in the Western Hemisphere (Owre 1973). In 1960, five to ten birds escaped from a commercial bird farm in Kendall, in the southern suburbs of Miami (Dade County), and this population has persisted within a 25-square-mile area ever since (Robertson and Woolfenden 1992). Carleton and Owre (1975) have estimated that 250 garrulous birds occupied a communal roost in 1969–1970. It seems unlikely that this species will colonize all of southern Florida as Carleton and Owre (1975) have predicted. In fact, numbers of bulbuls have continued to decline (P. W. Smith, pers. comm.). It now seems more likely that the family Pycnonotidae will be extirpated in the Western Hemisphere.

EUROPEAN STARLING   The European starling is native throughout Europe and east to Lake Baikal. In addition to the immense nonindigenous population that exists from Canada to Mexico, other populations occur in Jamaica, South Africa, Australia, and New Zealand (Lever 1987). The first attempts to introduce the starling into the United States failed. In 1890, however, an eccentric Shakespearean scholar named Eugene Schieffelin released 40 pairs of starlings and, in 1891, 40 more pairs in Central Park in New York City (Phillips 1928 cited by Lever 1987). His plan was to introduce all the birds mentioned by Shakespeare into the United States. By 1898 starlings were common in New York City and on Long Island. Increasing their range in concentric circles, they reached Savannah, Georgia, in 1917, and then the southward movement slowed. Progression through Florida was reflected in records on Amelia Island (Nassau County) in 1918, Leon County in 1924, Sarasota in 1934, Pensacola (Escambia County) in 1935, Miami and Key West (Monroe County) by 1953 (references given by Stevenson and Anderson 1994). The European starling is now a common summer resident in all populated regions of the state. It usually nests in cavities in trees. Migrants supplement the resident population in winter and roost socially at night in large roosts.

SPOT-BREASTED ORIOLE   The spot-breasted oriole is native to the Pacific slope of Mexico and Central America. Ever since escaped cage birds were found to be nesting in the wild in 1949 (Brookfield and Griswold cited by Lever 1987), a small population has persisted in the Miami area, adding a new species to the avifauna of the United States (Long 1981). By the 1960s the population extended from Dade County to Brevard County. Although the birds feed mostly on fruit and nectar from exotic tropical and subtropical plants, the population survived the freezes of January 1977 and the early 1980s. Nevertheless, it now seems to be losing ground. P. W. Smith (pers. comm.) thinks that in 1995 it may no longer occur north of Palm Beach County. Numbers in West Palm Beach, Fort Lauderdale, and Miami are declining.

HOUSE FINCH   The house finch is native to the western United States and Mexico. Because of its status as native to the United States, it cannot be kept legally as a pet cage bird. Some captive birds, probably from the area of San

Diego, California (Aldrich and Weske 1978), were released on Long Island, New York, in 1940 by dealers who had been told that the species was protected by U.S. law (Lever 1987). Since then, the species has been repeating the story of exponentially increasing its numbers and expanding its geographic range in concentric circles, as did the house sparrow and the starling before it. Within a decade the birds had bred as far west as Cleveland, Ohio, and as far north as Brunswick, Maine (Lever 1987). Dispersing individuals arrive in new areas months and even years before nesting occurs. The first Florida record of a wild bird was in Gulf Breeze (Escambia County) in 1983 (R. A. Duncan cited by Stevenson and Anderson 1994). Now the species is an established resident throughout the Panhandle and Big Bend areas of Florida and is rapidly becoming established farther south. This bird is an agricultural pest in Oregon and California, where it is trapped in fruit-growing areas (Lever 1987). Given the high reproductive potential of the house finch, similar problems could be in store for Florida.

HOUSE SPARROW   The house sparrow (*Passer domesticus*), which is native to Eurasia and North Africa, is always found near human habitation. It has been successfully introduced on nearly every continent. The first successful introduction in the United States was in Brooklyn in 1853 after several unsuccessful releases (Silverstein and Silverstein 1974). One reason for the introduction was the hope that it would control larvae of the snow-white linden moth (*Ennomos subsignarius*), which was defoliating trees; another was to make the British feel at home (Lever 1987). Probably supported by subsequent introductions elsewhere in the United States, the species has now spread throughout North America. It had reached Lake City, Florida, by 1882 (Howell 1932) and Homestead by 1930. In Florida, house sparrows do not occur in numbers large enough to cause serious problems. Apparently their numbers peaked at the turn of the century in eastern cities and towns, when spilled grain was everywhere (Phillips 1928 cited by Lever 1987).

Nonindigenous Birds That Are Not Well Established

Of the nonindigenous species that have bred in Florida but have not persisted long enough for Robertson and Woolfenden (1992) to have included them as well established by 1991 (Table 9.2), the one with the most promising future in Florida is probably the common myna, a native of Southeast Asia. Lever (1987) summarizes its status worldwide. In populated areas and farmlands it has been expanding its range naturally southeastward from Thailand to Singapore (Lever 1987). The birds, which are also popular as pets, have been released in many places around the world to serve as scavengers or to control insect pests.

Nonindigenous populations exist in Hong Kong, South Africa, Madagascar, Australia, New Zealand, the Hawaiian Islands, other oceanic islands, and elsewhere. The species is credited with having eliminated pest locusts (*Nomadacris septemfasciata*) in the Mascarene Islands in the Indian Ocean in the 1700s. In the 1800s in the state of Victoria, Australia, many birds were released by the Acclimatization Society. Today it is a locally abundant and aggressive inhabitant of several Australian cities, where it competes mainly with nonindigenous rock doves, house sparrows, and starlings.

In Florida, the common myna was first reported in the Miami area in 1983, and breeding was confirmed in Cocoa Beach and Broward County in the late 1980s (Stevenson and Anderson 1994). It has been increasing around shopping centers and malls in southern Florida (Robertson and Woolfenden 1992), and Stevenson and Anderson (1994) report breeding records for five counties.

The other species in Table 9.2 that is tenuously established in Florida but seems to be increasing in number in the 1990s is the rose-ringed parakeet (*Psittacula krameri*), a native of central Africa and from Pakistan to Sri Lanka. Long (1981) and Lever (1987) discuss its status worldwide. In India the species is regarded as an agricultural pest in fruit orchards and grain fields. Nonindigenous populations exist on Mauritius and in Germany, the Netherlands, Israel, Hong Kong, South Africa, and elsewhere. The species is well established in the British Isles, where in 1983 it occurred in gardens and parks in 50 counties in England, Scotland, and Wales. Although quite common, it has not become an agricultural pest there. The population that became established in southern California was removed, as a precaution, by a control program.

The first published breeding record for Florida was of a pair breeding in North Miami in 1969, but a longer history is suggested by informal information. Additional records are for nesting at Delray Beach (Palm Beach County) in 1972 and in the 1980s from St. Petersburg to Venice (Sarasota County). Stevenson and Anderson (1994) consider the species to be presently established in Dade, Collier, and Dixie counties. This species should be monitored carefully.

The other species in Table 9.2 seem less likely to persist in Florida. The feral common peafowl (*Pavo cristatus*) and ringed turtledove do not thrive in the wild. The black-hooded parakeet (*Nandayus nenday*), blue-crowned parakeet (*Aratinga acuticaudata*), red-masked parakeet (*Aratinga erythrogenys*), dusky-headed parakeet (*Aratinga weddellii*), chestnut-fronted macaw (*Ara severa*), and orange-winged parrot (*Amazona amazonica*) have all nested in the wild. The hill myna, although it is thriving in Vancouver, British Columbia, is showing no signs of similar vigor in Florida. The red-crested cardinal (*Paroaria coronata*) escapes often and may be breeding in limited numbers but is unlikely to persist (Robertson and Woolfenden 1992).

One would think that comparisons of successful, partially successful, and unsuccessful introductions might lead to some insight into the ecology of the establishment of nonindigenous birds. Yet the causes of declines after partial successes are poorly understood.

The International Bird Trade

Until recently, the importation of wild birds for the pet trade was insufficiently regulated in the United States. Mortality of birds in transport and during quarantine has been unacceptably high. Exotic birds have introduced diseases of poultry, and the overharvest of birds in their source countries has contributed to serious population declines, especially in New World parrots (James 1990; Beissinger and Snyder 1992). The United States has been the world's largest importer of exotic birds, accounting for 80 percent of the 1.8 million parrots traded worldwide between 1982 and 1988 (Thomsen and Mulliken 1992).

In 1992, a major step was taken toward solving this conservation crisis: the U.S. Congress passed the Wild Bird Conservation Act (WBCA; P.L. 102-440). It calls for a moratorium on the importation of species thought to be harmed by the trade, the authority to suspend imports, and the development of an approved list of species that can be traded—primarily birds reared in approved foreign captive-breeding facilities. The rules implementing the provisions of the WBCA are still being developed by the FWS Office of Management Authority (Rosemary Gnam, pers. comm.). In effect, the act puts teeth in the United States' participation in the Convention on International Trade in Endangered Species of Wild Fauna and Flora (CITES), to which the United States and more than 100 other countries are signatories. In the future only licensed breeders, zoos, and scientists will be permitted to import wild birds. The aviculture industry will have to rely increasingly on captive-bred stock. Of course, it is difficult to stop the smuggling of wild birds into the United States, but in the near future, wild-caught birds will not be legally imported to be sold as pets unless it has been established that trade is not harming their wild populations. The passage of the WBCA by the U.S. Congress was an important step forward for avian conservation worldwide.

Three years have elapsed since the passage of the WBCA, and it is already up for reauthorization. Aviculturists contend that the regulations established thus far under the act are too restrictive and are urging that the act be weakened to allow substantially more trade. As this chapter goes to press, the future of the WBCA is unclear. If it is reauthorized without serious modification, the numbers of wild-caught birds imported into Miami as part of the commercial pet trade will continue to decline. This will probably mean that fewer nonindigenous birds will escape or be released in southern Florida and that those which are released are likely to have been reared in captivity and even less likely to persist than were their predecessors.

The Overall Picture

The traditional ecological question—whether nonindigenous animals and plants are excluded from native communities by interspecific competition—does not seem to be quite the right question for the birds described here. A better approach might be to ask the extent to which ecological requirements of these birds are fulfilled by the new environment. Another question might address the stages of the life cycles of these birds at which limiting factors operate. Even if the species reproduces in its new environment, does its innate dispersal behavior allow the population to spread?

Southern Florida is like an open zoo for released birds. Guides for ecological tours have commented that nowhere in the world can one see more species of parrots in one day than in the greater Miami area (B. Neville, pers. comm.). Even if they succeed in these habitats for a few generations, most of them fail to persist. We have to admit that we usually do not fully understand why. Diagnosis of why bird populations persist or fail is a difficult procedure. In conservation biology, only a few successful cases are known (Caughley and Gunn 1996). It took a stepwise elimination of several possible causes to deduce that the endangered flightless woodhen (*Tricholimnas sylvestris*) on Lord Howe Island is limited by nest predation by pigs. It took an experimental analysis of both nesting and foraging requirements to reveal why the Seychelles magpie-robin (*Copsychus sechellarum*) is so rare. So perhaps it is not surprising that we cannot answer the "why" questions when it comes to the fates of unplanned introductions. Insightful diagnosis of mechanisms regulating populations of animals and plants, whether they are endangered species or nonindigenous potential pests, is probably the primary challenge to basic research in conservation biology.

ACKNOWLEDGMENTS
I thank P. William Smith for many useful comments on the subject of exotic birds in Florida and for supplying detailed information about several species.

10 Nonindigenous Mammals

JAMES N. LAYNE

 This chapter summarizes the status of nonindigenous mammals that have become established in Florida. I also wish to consider the factors that may account for the different degrees of success of these species, their potential impacts on native biota and communities, and their economic and public health significance. This account is not definitive. Although we can construct a reasonably detailed picture of current status for a few introduced species, data for most species are fragmentary. Furthermore, records for many species are largely or entirely based on sightings, often reported second or third hand. Although this problem may not be serious with such a distinctive and conspicuous mammal as an armadillo, the validity of reported sightings of a species like the jaguarundi is much more open to question. Thus we cannot evaluate the impact of nonindigenous mammals on native biota at present because we lack sufficient knowledge of the life histories and ecology of both the introduced species and the native ones. Finally, even an accurate and complete account of the current status of Florida's alien mammal fauna would give only a snapshot of a dynamic system subject to short-term change as the result of new introductions, extinctions, and expansion and contraction of ranges.

Three major factors contribute to Florida's vulnerability to establishment of nonindigenous mammals. The first is an abundant supply of candidate species, including a large population of exotic wildlife pets, particularly in southern Florida, and numerous tourist attractions, game ranches, and wild-animal importers and dealers. The second factor is a climatic range suitable for both temperate and tropical species. The third is a broad range of natural aquatic, wetland, and upland habitats and human-altered landscapes. To these can be added hurricanes, which may liberate and disperse captive nonindigenous wildlife as well as domestic animals. Hurricane Andrew in 1992 freed large numbers of exotic species, for example, including many primates (Belleville 1994). Despite their potential importance, the only known case in which a hurricane actually facilitated establishment of a nonindigenous mammal in Florida appears to be that of a colony of nutria near Tampa. Hurricanes may also affect established nonindigenous mammals—for example, accumulations of debris and disruption of garbage collection services in the aftermath of a hurricane may benefit Norway and black rats. Hurricanes contributed to severe flooding in southern Florida in 1947 that resulted in a drastic decline in numbers of black-tailed jackrabbits in the Miami area (Layne 1965). Moreover, the population of Mexican gray squirrels on Elliott Key in Biscayne Bay (Monroe County) may have been reduced by Hurricane Andrew in 1992, although census data before and after the hurricane are not available. High water from hurricanes is also undoubtedly detrimental to armadillo populations in low-lying areas.

More than 50 species of free-ranging nonindigenous mammals have been recorded in Florida; still others have undoubtedly escaped notice; and the list does not include another 20 or so species living under semiwild conditions in tourist attractions, ranches, or hunting preserves or, for that matter, the water buffalo used in experiments on aquatic weed control. The known nonindigenous species represent most major taxonomic groups of mammals and regions of the world. A few examples will illustrate the diversity: the binturong, hedgehog, lesser anteater, baboon, chinchilla, mongoose, ferret, ocelot, African lion, Indian elephant, kinkajou, and even a marine species—the California sea lion. Most species are known from a few records of individuals recently escaped or released from captivity. In most cases, if not quickly recaptured or killed, these animals probably do not survive long in the wild. Exceptions include a jaguar shot in 1968 near Felsmere, Indian River County, that had been seen in the area over a two-year period (G. Heinzman, pers. comm.).

Of the nonindigenous mammals that have been recorded as free-ranging in Florida, at least 19 species have succeeded in establishing populations that persisted for at least a few years and another 3 species may have become established. (An additional species appears most likely to have been introduced, although it may have reached Florida naturally.) This account focuses on these 22 species plus four domestic mammals with feral or free-ranging populations in Florida (Table 10.1). Three, and possibly four, of the 22 nondomestic species have apparently been extirpated; all had localized ranges. Eight of the 19 remaining species are widely distributed, and five of these are relatively common.

TABLE 10.1.

Nonindigenous Mammals Known or Believed to Have Become Established in Florida

Species	Origin	Source[a]	Status[b]
Order Chiroptera			
Molossus molossus tropidorhynchus (velvety free-tailed bat)	Caribbean (Cuba)	I	L
Order Primates			
Saimiri sciureus (squirrel monkey)	S. America	A,I	L
Cercopithecus aethiops (vervet monkey)	Africa	I	L
Macaca mulatta (rhesus monkey)	Asia	A,I	L
Order Xenarthra			
Dasypus novemcinctus (nine-banded armadillo)	Western U.S.	A,I,R	Wc
Order Lagomorpha			
Oryctolagus cuniculus (European rabbit)	Europe	U	L
Lepus californicus (black-tailed jackrabbit)	Western U.S.	A	L
Order Rodentia			
Sciurus aureogaster (Mexican gray squirrel or red-bellied squirrel)	Mexico	I	L
Cynomys ?ludovicianus (?black-tailed prairie dog)	Western U.S.	U	Le
Rattus norvegicus (Norway rat)	Asia via Europe	A	Wu
Rattus rattus (black rat)	Asia via Europe	A	Wc
Mus musculus (house mouse)	Asia via Europe	A	Wc
Myocastor coypus (nutria)	S. America	A,I,?R	L
Hydrochaeris hydrochaeris (capybara)	S. America	?A	L?e
Order Carnivora			
Nasua narica (white-nosed coati)	Central–S. America	A	Wu
Canis familiaris (dog)	(domestic)	—	Wc
Canis latrans (coyote)	Western U.S.	A,I,R	Wc
Vulpes vulpes (red fox)	U.S.	A,I,R	Wc
Felis yagouaroundi (jaguarundi)	Central–S. America	I	?Wu
Felis catus (house cat)	(domestic)	—	Wc
Order Artiodactyla			
Sus scrofa (pig)	(domestic)	—	Wc
Axis axis (axis deer)	India	A	Le
Cervus elaphus (elk)	Western U.S.	I	L
Cervus unicolor (sambar deer)	Asia	I	L
Antilocapra americana (pronghorn)	Western U.S.	I	Le
Capra hircus (goat)	(domestic)	—	L

Note: Provisionally included species are indicated by asterisks and questionable species identifications by question marks. Scientific names follow Wilson and Reeder (1993) except for the following species, for which previous, widely used names are retained because of greater familiarity to nontaxonomists (name in Wilson and Reeder 1993 in parentheses): *Cercopithecus aethiops* (= *Chlorocebus aethiops*), *Canis familiaris* (= *Canis lupus*), *Felis yagouaroundi* (= *Herpailurus yagouaroundi*), *Felis catus* (= *Felis silvestris*).

[a] Source of introduction: A = accidental release or escape; I = intentional release; R = natural range expansion a contributory factor; U = nature of introduction unknown.

[b] Status: L = one to a few widely scattered localized populations; Wu = wide distribution but relatively uncommon; Wc = wide distribution and generally common; e = extirpated.

Three of the four domestic species have widespread and abundant feral or free-living populations (Table 10.1).

The 26 nonindigenous mammals here assumed to be currently or previously established in Florida represent seven orders; rodents, carnivores, and artiodactyls predominate. Excluding feral domestic species, the United States ranks first as the source of nonindigenous species, followed by Central and South America. Accidental and intentional introductions have contributed about equally to the nonindigenous mammal fauna of the state, and natural range expansion has apparently played a role in the establishment and spread of three species and possibly a fourth (Table 10.1). Currently extant nonindigenous species comprise 22 percent of all Florida mammal species and 27 percent of the land mammals. In California, introduced mammals make up about 12 percent of all mammals (Lidicker 1991).

Nondomestic Species

In the following list, the 22 species of nondomestic nonindigenous mammals believed to have been established in the wild in Florida are treated first, followed by the four domestic mammals that exist in a free-ranging or feral state.

Molossus molossus tropidorhynchus (Velvety Free-Tailed Bat)

Three populations ranging from about 80 to 300 individuals of this bat were discovered in buildings in the lower Florida Keys (Marathon, Stock Island, Boca Chica) in late 1994 by Phillip Frank (Adams 1996). The subspecies, determined by Karl Koopman of the American Museum of Natural History, occurs in Cuba. Although we cannot rule out the possibility that the Florida population was founded naturally by dispersal of bats from Cuba, circumstantial evidence suggests that it derived from bats released in 1929 in a bat tower on Sugarloaf Key by Richter C. Perky as a mosquito control measure. The tower, which still exists, is located approximately in the center of the presently known range. Information on the species, its geographic origin, and the number of bats released is vague and contradictory. Parks (1973) and informants interviewed by Hoffmeister (1975) claim that no bats were ever released at the tower, whereas Jennings (1958) states that at least 2500 bats, probably Brazilian free-tailed bats (*Tadarida brasiliensis*) from Texas, were released. Stevenson (cited by Lazell and Koopman 1985) also reports that Perky released large numbers of *Tadarida brasiliensis* at the site. The most detailed account of the tower's history (Fix 1966) states that the bats were obtained from Cuba by an associate of Perky's, Steve Singleton. Although the species collected is unknown, *Molossus molossus* is the likely candidate, as it is abundant and widespread in Cuba and elsewhere in the Caribbean and commonly roosts in buildings. Two attempts were made to

establish bats in the tower, but each time they promptly disappeared when released from the tower after several days of confinement to acclimate them.

Given that the present *Molossus* population derives from the Perky introduction, the failure to detect its existence until recently is not surprising in view of
the paucity of knowledge of the bat fauna of the Florida Keys. Jennings (1958)
mentions a report of an unidentified bat on Stock Island in 1950 that may have
been this species; Lazell and Koopman (1985) note that a specimen identified as
Tadarida brasiliensis was collected from a building on Bahia Honda Key in
1975. As the specimen has been lost, the identification cannot be confirmed,
but in light of the general similarity between *Molossus* and *Tadarida*, it could
have been the former.

Saimiri sciureus (Squirrel Monkey)

Squirrel monkeys were introduced at Silver Springs, Marion County, as a
tourist attraction (Maples et al. 1976). There apparently is no record of how
many monkeys were released or when. In the late 1960s, the colony consisted of
12 to 15 individuals living in the hammock at the head of the springs (Layne
1969). By 1975, the monkeys had shifted their range downstream to the Oklawaha River (Marion County) (Maples et al. 1976). Another free-ranging
colony, on the Bartlett Estate in Fort Lauderdale, Broward County, derives from
two pairs released nearby about 30 years ago and numbered 43 individuals in
1988 (Wheeler 1990). A semi-free-ranging colony has also existed since 1960 at
Goulds Monkey Jungle in Miami (DuMond 1967, 1968), and an estimated 500
to 1000 squirrel monkeys from the defunct Masterpiece Gardens near Lake
Wales (Polk County) were living free in the vicinity in the early 1980s. Poor success in efforts to capture them was reported in 1981, and some probably still
persist in the area. A colony in Titusville, Brevard County, founded by 15 individuals that escaped in 1976 from the Tropical Wonderland tourist attraction, is
still in existence (most recent recorded sighting in 1993 or 1994; R. Kirk, pers.
comm.). Still other free-ranging groups of this species probably occur in Florida
or did in the past.

Cercopithecus aethiops (Vervet Monkey)

Hyler (1995) has reported on two troops of vervet monkeys inhabiting Westlake
Park in Dania, Broward County. The population was founded by releases in the
mid-1950s and early to mid-1970s from a failed tourist attraction. As many as
four other primate species are also reported to have lived in the area in the past.
The monkeys inhabit an approximately 3700-ha tract consisting primarily of
mangrove swamp. Censuses of the two groups in 1991 and 1992 give means of
13 and 19 in one group and 23 in the other, including adults, juveniles, and in-
fants. A more recent estimate gives a total of "well over 120" individuals in the

two troops (Belleville 1994). Hyler (1995) found that the principal foods were flowers, fruits, and foliage of red mangrove (*Rhizophora mangle*), black mangrove (*Avicennia germinans*), and white mangrove (*Laguncularia racemosa*), but a variety of other items were eaten, including portions of four other plant species, shelf fungi, and several kinds of invertebrates.

I visited the site in May 1995 and observed no monkeys or even signs in an hour's search along the edge of the swamp and several short transects into the interior. A resident of the mobile-home park bordering the tract told me that he had seen monkeys within the past month. He noted that up to 40 monkeys had been seen at one time in a small grassy plot at the edge of the swamp, where they are fed by people, and that he had seen them go into the residential area across the paved road bordering the wooded tract to feed on fruit in the yards of houses.

Macaca mulatta (Rhesus Monkey)

A well-established population of rhesus monkeys has existed along the Silver River, Marion County, for over 50 years. Maples et al. (1976) state that the source of the colony was two pairs of monkeys released in 1933 during the filming of Tarzan movies. Wolfe and Peters (1987), however, suggest that the introduction predated the filming of a Tarzan movie at Silver Springs and was carried out by Colonel Tooey, manager of the "jungle cruise" boat ride at the time, to increase the wildlife interest of the boat tour. According to this version, a small number of monkeys was released on an island in the river near the main spring prior to 1938, as a 1938 newspaper account of a solitary monkey shot in a nearby town indicated the colony was already established. Six additional monkeys were reportedly released on the river in 1948. In 1961, Jerry Tooey (pers. comm.), then manager of the jungle cruise, stated that the original release, consisting of four pairs from Pakistan, was in 1930 and that three new females were added in 1960. He noted that lone monkeys had been reported from Orange Lake, Lake Eustis (Lake County), and elsewhere in the region.

In the late 1960s when anthropologists from the University of Florida began to study the monkeys, there were two major concentrations—one near the main spring and one below the confluence of the Silver River with the Oklawaha River in the Ocala National Forest (Marion County) (Wolfe and Peters 1987). In 1968 there were 78 monkeys in the upper Silver River area and an unknown number in areas away from the springs (Maples et al. 1976). By 1984, the upper Silver River population consisted of nearly 400 individuals before trapping and removal of 217 monkeys in 1984 and 59 in 1986 (Wolfe and Peters 1987). No estimate was available for the population along the Oklawaha River in the Ocala National Forest, where removal of monkeys had been permitted. Over the years, monkeys have been reported along the Oklawaha River more than 25 miles north (Rodman Dam) and 20 miles south (Starke Ferry Bridge) of the entrance of the Silver River (Maples 1979). The primary habitat of the monkeys is the floodplain swamp forest bordering the river, and a wide range of native plant foods is consumed, even by troops that are artificially provisioned by the tour boats and tourist facility (Maples et al. 1976; Wolfe and Peters 1987).

Another established population of rhesus monkeys has apparently existed in southeastern Florida for more than 40 years. D. F. Austin (pers. comm.) received reports of a colony in Broward County near Dania in 1977. The source was allegedly a tourist attraction that went out of business in the 1950s. This was probably the same colony referred to in a 21 September 1983 *Tampa Tribune* newspaper article stating that 30 to 50 rhesus monkeys inhabiting an area slated for expansion of the Fort Lauderdale–Hollywood International Airport were to be removed. The habitat occupied by the colony is an extensive mangrove swamp with Brazilian pepper (*Schinus terebinthifolius*) and Australian pine (*Casuarina equisetifolia*) on drier sites. In a visit to the area in May 1995, I saw no monkeys in an hour's survey along the periphery of the tract or along several survey lines that had been cut into the interior, but a surveying crew in the area reported seeing monkeys regularly in the early morning along the edge of the swamp and on the paved road bordering it. Indeed, they had sighted four adults and young several days before.

Robert Kirk (pers. comm.) has provided information on a colony in a wooded area in Titusville, Brevard County. It was founded by a group of about 40 or 50 rhesus monkeys, two "bob-tailed macaques" (? pig-tailed macaque, *M. nemestrina*), and squirrel monkeys that escaped from the defunct Tropical Wonderland tourist attraction in Titusville in 1976. They had been placed on an island surrounded by a moat in 1963–1964, but after the attraction was sold and fell into neglect, trees fell across the moat providing an avenue of escape. Monkeys from this group were reported as far as 3 miles from the site in different directions. Some, including the two nonrhesus macaques, were caught and removed soon after escaping. The colony has persisted in the area until the present (the most recent sightings of which Kirk was aware were in 1993 or 1994), and considerable breeding has occurred. The monkeys make regular forays into a nearby mobile-home park for food.

In the lower Florida Keys, rhesus monkeys were stocked on Key Lois (formerly Loggerhead Key) in 1973, and Raccoon Key in 1976, by the Charles River Laboratories to supply animals for medical research (Figure 10.1). Reported populations on the two keys during 1987–1990 were 1322 (Key Lois) and 1473 (Raccoon Key) (Taylor et al. 1994). Individuals have been reported to escape occasionally to other keys (Lazell 1989). A solitary monkey was shot on tiny Little Crane Key in 1994 (T. Wilmers, pers. comm.).

FIGURE 10.1   Rhesus monkeys (*Macaca mulatta*) damaging black mangrove trees (*Avicennia germinans*) on Raccoon Key (Great White Heron National Wildlife Refuge) in the lower Florida Keys. Photo by Wayne Hoffman.

Dasypus novemcinctus (Nine-Banded Armadillo)

The nine-banded armadillo is distributed throughout Florida and is one of the state's most abundant mammals. The history of its spread is the best documented of those of Florida's nonindigenous mammals. The earliest record of the species in the state is of a pair from Texas released near Miami (Hialeah) at the end of World War I; single adults, one a female with young, were killed in the same area in 1922 and 1924 (Bailey 1924a). According to Buchanan and Talmage (1954), a population was established in the Miami area but had disappeared by 1954. Other Texas armadillos were reported to have escaped from a zoo at Cocoa, Brevard County, during a storm in 1924, and single individuals were killed or captured in Flagler and Brevard counties in 1934, 1936, and 1938 (Sherman 1943). On the basis of the known distribution in the early 1940s, at which time armadillos appeared to be well established in a four-county area east of the St. Johns River (Sherman 1943), the primary or sole center of origin of the present introduced population was the coastal areas of Volusia or Brevard County (Fitch et al. 1952; Neill 1952). By 1949, the main distribution had expanded to a 14-county area in the eastern part of the peninsula, and scattered records extended from north central Florida south to Broward County (Newman 1949). Neill (1952) cites many additional records from peninsular Florida west to Wakulla County (the latter considered an introduction rather than natural dispersal by Stevenson and Crawford 1974) but none from the Big Cypress–Everglades region.

On the basis of a mail survey in 1972, Humphrey (1974) showed the introduced population as continuously distributed in peninsular and northern Florida west to the Aucilla River (Jefferson and Taylor counties) except for the Everglades region. He also indicated that the northward expansion of the native range in the western United States, in progress for about 100 years, had reached extreme western Florida, leaving a gap between the introduced and native populations with only a few scattered records. However, records from Everglades National Park between 1959 and 1971 indicate that, although the Everglades region was the last part of the peninsula to be invaded, armadillos did occur there at least sporadically prior to 1972 (Layne 1974). The conclusion that armadillos in western Florida marked the advancing front of natural range expansion from the west is also open to question. Wolfe (1968) attributed the widespread occurrence of armadillos in western Florida by the late 1960s to spread from an introduction near Foley, Alabama, rather than to an extension of the western population, whose eastern range limit at the time was considered to be in central Mississippi. Stevenson and Crawford (1974) agreed. They believed that records in 1972 and 1973 in the Tallahassee area (Leon County) marked the beginning of closure of the remaining gap in the range by natural dispersal from the peninsular population. Thus establishment of continuity of armadillo distribution across the northern Gulf coast apparently involved the merger of three, rather than two, populations, and the zone of contact between the introduced and naturally expanding populations was farther west, in Alabama or Mississippi, rather than in Florida.

Armadillos have apparently not yet reached the Florida Keys, as Lazell (1989) does not mention them, but they have been introduced on a few barrier islands (for example, Sanibel and Captiva in Lee County) (Layne 1974). Armadillos in Florida occur in almost all natural terrestrial vegetation types and in many kinds of human-modified environments (Neill 1952; Layne 1976).

Oryctolagus cuniculus (European Rabbit)

Lazell (1989) states that populations of European rabbits are established on several of the Florida Keys, including West Summerland, Ramrod, Perky, and

Lower Sugarloaf. An early introduction of this species into Florida is suggested by the report of DePourtales (1877) of "a burrowing rabbit" on Rabbit Key located between the mainland and the upper keys.

Lepus californicus (Black-Tailed Jackrabbit)

This species was presumably introduced in the Miami area in the 1930s when animals used in greyhound training escaped (Layne 1965). By 1947, the population was reported to have increased to an estimated 2000 animals and to have expanded westward to the edge of the Everglades. But in that same year, the population was drastically reduced by severe flooding resulting from abnormally high rainfall, two hurricanes, and a tropical depression in quick succession. Jackrabbits are known to have persisted in southern Broward and northern Dade counties until at least 1958 and were still common in the grassy areas between runways and taxiways of the Miami International Airport (Layne 1974). According to C. McArthur of the airport staff (pers. comm.), jackrabbits are still commonly seen and regularly hit by aircraft and vehicles, although the amount of mowed grass habitat has been reduced over the years by runway and other construction. A road-killed specimen observed by H. Kale (pers. comm.) on the Florida Turnpike at the Boca Raton exit in southern Palm Beach County in 1985 appears to constitute the northernmost record in southeastern Florida. It is unknown whether the specimen was from a population in that area or a lone disperser or relocated animal from farther south. Lazell (1989) mentions a *Lepus* specimen of unidentified species collected from Upper Sugarloaf Key, Monroe County, which may have been a black-tailed jackrabbit transported from Dade County. A jackrabbit was observed on the University of South Florida campus in Tampa in the early 1970s by G. Woolfenden (pers. comm.), but there is no evidence that an established population existed in that area. As jackrabbits have been used for greyhound training in a number of different parts of the state, there probably have been other escapes and possibly established populations.

In southern Florida, the species has occurred predominantly in mowed grassland habitats. The fact that the wide, mowed grass shoulders of interstate and other major highways offer corridors for dispersal raises the question of why the species has failed to expand its range northward into the extensive ranchlands of south central Florida, which appear to be ideal jackrabbit habitat.

Sciurus aureogaster (Mexican Gray Squirrel or Red-Bellied Squirrel)

Two pairs of this species from Mexico were introduced onto Elliott Key in Biscayne Bay by J. Arthur Pancoast, an early island property owner, in 1938 (Brown 1969; Tilmant 1980). Schwartz (1952) reported a thriving population in 1951; Brown and McGuire (1969) found the squirrels common in dense subtropical hammock forest habitat throughout the island. In 1972, mean population density in mature subtropical hammock was 2.47 squirrels per hectare (Brown and McGuire 1975). Density estimates ranging from 0.86 to 1.06/ha in different areas of the island in 1977–1978 were reported by Tilmant (1980). The species also reached Adams Key adjacent to Elliott Key by about 1974, probably by swimmming, as the islands are close together. In 1975, an individual was captured while swimming toward Old Rhodes Key across Caesars Creek, and in early 1977 the squirrels were found to have spread to Sand Key, which is narrowly separated from Elliott Key at the north end.

Information on the current status of this species was provided by J. (Bo) Dame, resource manager, Biscayne National Park (Monroe County). Elliott Key was hit by Hurricane Andrew in 1992, and its vegetation was substantially altered, although potential squirrel habitat remained. The species is known to have survived the hurricane on Elliott Key (an individual was sighted in March 1995, and sign believed to be of squirrels has been observed), but no information is available on population size. Although no squirrels have been seen on Adams Key since the hurricane, specific searches for the species have not been conducted.

Cynomys ?ludovicianus (?Black-Tailed Prairie Dog)

Prairie dogs were observed at the Miami International Airport and at other localities in the vicinity in 1973 and 1974 by O. Owre and W. J. Robertson, Jr. (pers. comm.). Despite control measures at the airport during the winter of 1973–1974, prairie dogs were still present as late as May 1974 (R. Thorington, pers. comm.), suggesting that more than a few animals were involved. As prairie dogs and their characteristic burrows are conspicuous, the absence of records in recent years suggests that the population was exterminated or went extinct for other reasons. Although we cannot be sure, the population was probably of the black-tailed prairie dog, the most widespread species in the western United States. According to J. A. Gore (pers. comm.), black-tailed prairie dogs are sometimes sold as pets in Florida and may escape or be released. In 1989 two individuals with burrows were observed at different localities in the Panhandle near Panama City, Bay County.

Rattus norvegicus (Norway Rat)

The Norway rat has long been established in Florida. The earliest published record from the state is apparently that of Allen (1871) from Jacksonville (Duval County) in 1869, but it undoubtedly arrived earlier, although not before the late 1700s (Silver 1927). The Norway rat is assumed to occur throughout the state but has been documented in relatively few localities. It is generally uncommon and less frequent than the black rat in natural habitats away from human-occupied areas.

Rattus rattus (Black Rat)

Except for feral pigs, this species was probably the first nonindigenous mammal to reach Florida. Black rats were almost certainly on the ships of the earliest Spanish explorers to visit the state (Silver 1927) and may have become established during that period. The first documented record was in the late 1800s (Rhoads 1894). The species is widely distributed in Florida, including the Florida Keys and the more isolated Dry Tortugas (Layne 1974) and Marquesas (L. M. Ehrhart, pers. comm.). It is more common overall than the Norway rat. Like the Norway rat and house mouse, black rats are primarily associated with human rather than natural habitats, but they appear to occur more frequently in natural habitat than do the former two species, particularly in southern Florida and the Keys where they are regularly found in a variety of natural and disturbed habitat types (Layne 1974).

Mus musculus (House Mouse)

Although first recorded in Florida by Rhoads (1894), the house mouse undoubtedly arrived much earlier, probably during the early colonial period. It now occurs statewide, including the lower Keys. Self-maintaining populations in the wild appear to be relatively uncommon in Florida, although reported in coastal dunes and in disturbed habitats on barrier islands and the mainland. An unusually large population (about 60 percent capture rate in live traps) was encountered in open weedy-grassy habitat on reclaimed phosphate-mined land in Polk County by B. Toland (pers. comm.). During long-term monitoring of small mammal populations on the Archbold Biological Station in south central Florida (Highlands County), the relatively infrequent appearance of house mice in natural habitats away from buildings has tended to coincide with low populations of native mice (unpublished data).

Myocastor coypus (Nutria)

Nutria have been reported from a number of scattered localities in Florida south to Tampa on the west coast and Vero Beach (Indian River County) on the east coast. The first published record of the species in the wild in Florida was in 1955 (Anonymous 1955), and Griffo (1957) has summarized its status to that time. He recorded nutria farms or records of nutria in the wild in 23 counties. Two known releases for aquatic plant control had occurred in Calhoun and Pinellas counties, and a third was suspected in Putnam County. Individuals assumed to be recent escapees from nearby fur farms were recorded in Hillsborough, Marion, Alachua, and Putnam counties. Five of seven records assumed to represent feral nutria were from Gulf–coastal marsh habitats in Walton, Holmes, Franklin, Dixie, and Levy counties; the two inland locations were in Gilchrist and Hillsborough counties. The best evidence of an established population was in Choctawhatchee Bay (Okaloosa and Walton counties), where small numbers of nutria were reported to have been trapped.

A 1963 survey obtained records from freshwater marsh and swamp and coastal marsh habitats in St. Johns, Alachua, Brevard, Indian River, Santa Rosa, Okaloosa, Jefferson, Escambia, and Walton counties (Wallace 1963). The largest concentrations were apparently in brackish cattail–bulrush habitats along the Indian River in Brevard and Indian River counties and in Walton County (number estimated at 150 animals). Nutria were still present in Indian River County (just north of Vero Beach) in 1972 (W. Haas, pers. comm.). Established populations mapped by Manville (1963) in 1962 occurred along the Gulf coast from Pensacola Bay (Escambia County) to Tampa Bay. Evans (1970) showed four areas with established populations (coastal western Florida, Apalachee Bay area (Franklin and Taylor counties), northeastern Florida, and the mid-Atlantic coast region) as of 1966. The accuracy of these maps is questionable. Although only four years apart, they differ significantly and miss localities known to have established colonies at the time. In neither case were sources of the data cited. Stevenson (1976) noted a colony in Hamilton County near the Georgia line in 1969. In 1971, J. A. Beard (pers. comm.) reported a colony estimated at 20 to 24 animals in a swampy area near Tampa, Hillsborough County. The source of this colony was believed to be 10 to 12 animals that escaped from a nearby nutria farm in 1964 when a hurricane destroyed the pens. In the early 1970s, Brown (1975) found high numbers of nutria inhabiting dairy cattle sewage lagoons in the vicinity of Tampa. Dye (1975) believed that the Hillsborough County population was the only persistent population in Florida at that time. More recent records from Hillsborough County were in 1984 and 1985 (R. J. Callahan, Jr., pers. comm.). A small colony in typical salt-marsh habitat near Aripeka, Pasco County, was located in 1986 (J. L. Wolfe, pers. comm.). A thorough survey of the current status of the nutria in Florida would probably reveal other local populations. Although most of the Florida colonies appear to have originated from escapes or releases, the Gulf coastal populations of the Panhandle might have been founded by natural dispersal from coastal marshes farther west where nutria are abundant.

Hydrochaeris hydrochaeris (Capybara)

J. Wooding of the Florida Game and Fresh Water Fish Commission (pers. comm.) provided information from reliable observers on a small group of capybaras in northern Florida along the Sante Fe River near Brooker, Bradford County, in the early 1990s. The animals may have been escapees from the Bacardi wildlife research facility located in the area. In addition to sightings of the animals in the wild, capybara allegedly from this population was served at a wild-game dinner hosted by a student wildlife group at the University of Florida. The most recent record apparently referable to this introduction is a sighting in late December 1995 by J. Whitehouse (pers. comm.) of a capybara on the bank of a canal in a swamp close to the Santa Fe River near the town of Lulu, Union County. Thus the details of this introduction are sketchy and nothing is known of the breeding status; it appears to have persisted for at least several years.

Nasua narica (White-Nosed Coati)

The number and relatively widespread distribution of reports of coatis in Florida, as well as evidence of breeding, appear to indicate an established wild population. An early appearance of the species in the state is documented by a specimen in the Florida Museum of Natural History from Alachua County in 1928 labeled "escaped captive." In 1971, D. Fertig (pers. comm.) observed an adult with three or four young crossing State Route 70 in Highlands County. Other records from Highlands County include single individuals seen in 1973, 1986, and 1988 and three individuals (of different size) live-trapped and released in 1984. D. Austin (pers. comm.) reported a road-killed specimen on U.S. 27 at Palmdale, Glades County, in 1970, and O. Owre (pers. comm.) observed adults with young in Matheson Hammock County Park, Dade County, in two consecutive years (about 1974 and 1975). J. Lazell informed me in 1973 that a total of more than six coatis had escaped on at least two occasions from the city park in Belle Glade, Palm Beach County, and that some had been trapped around Palm Beach, suggesting a free-ranging population there. In 1988, individuals were observed crossing highways near Fort Lonesome, Hillsborough County, and near Okeechobee, Okeechobee County (K. Lips, pers. comm.). A sighting by a reliable observer also has been reported (year unknown) from near Spring Hill, Hernando County (K. Frohlich, pers. comm.).

Canis latrans (Coyote)

Although present in Florida during the Pleistocene (Webb 1974), the coyote did not persist in the state to pre-Columbian times. The present population is of relatively recent origin, presumably derived from two sources: accidental or intentional introductions within the state and immigration from southern Alabama and Georgia. The latter animals may have been from populations derived from introductions, the naturally expanding population in the Southeast (Paradiso 1966; Hill et al. 1987), or a combination of the two.

The earliest known introductions in Florida, involving 30 animals, were in Palm Beach and DeSoto counties between 1925 and 1931 (Young and Jackson 1951). During the same period, individuals were killed in Collier and Marion counties and on Key Largo, Monroe County (Young and Jackson 1951). A possible coyote was also observed in Everglades National Park in 1959 (Layne 1974). By the late 1970s, coyotes were well established in northern Florida. Information accumulated from various sources during that period by A. Burt (pers. comm.) indicates high populations in some areas of northern Florida— particularly near High Springs in Alachua County, where coyotes were allegedly introduced in the late 1960s. The first confirmation of an established population southward on the peninsula was in 1969 in the Lake Wales area, Polk County, where adults were trapped and pups dug out of a den (Cunningham and Dunford 1970). According to local sources, coyotes had been released in the area for hunting with hounds on at least two occasions: the late 1950s (under the name "black foxes") and in 1968. In addition to known intentional releases, accidental releases of coyotes from fox pens have also occurred (J. Brady, pers. comm.).

By the early 1980s, some 26 specimens, 4 sightings, and 16 areas with frequent, reliable reports had been recorded, revealing that the species was generally distributed throughout the Panhandle and through the central region of the peninsula from Hamilton County to Polk County (Brady and Campell 1983). In 1988, the range extended south to Broward and Collier counties, with frequent reports throughout western Florida and scattered records in the peninsula inward from both coasts south to northern Broward and Collier counties (Wooding and Hardisky 1990). Although Wooding and Hardisky (1990) show no records for Highlands County, a sighting was reported in 1982. Since that time, the number of sightings and localities has increased steadily, indicating that the species is now well established in the county. In a survey of various areas in Highlands County and southeastern Polk County in June 1995, R. McBride (pers. comm.) found coyote tracks widely distributed, indicating a sizable but not "really high" population. Two additional county records in south central Florida since the 1988 survey include a coyote killed in 1991 in Charlotte County (B. Crawford, pers. comm.) and one live-trapped in 1991 in DeSoto County (J. Gingerich, pers. comm.). Coyotes were reported for the first time on Two Rivers Ranch in northern Hillsborough County in 1993 (T. H. Dyer, pers. comm.). Most sightings of coyotes in the south central part of the state have been in areas with extensive improved pastures or native prairie. In his track survey in the region in June 1995, R. McBride (pers. comm.) found that coyotes were more frequent in pasture and citrus grove habitats than in xeric scrub habitats.

Vulpes vulpes (Red Fox)

Although present in the Pleistocene (Webb 1974), the red fox apparently did not occur in Florida in recent times prior to the early 1950s, as it was not included in Sherman's (1952) or any earlier checklist of Florida mammals. Within approximately 30 years after it first appeared, the red fox had become common throughout the state. Like that of the coyote, its spread appears to be the result of immigration from bordering states as well as intentional and accidental local introductions with subsequent expansion. Introductions were probably the primary source of the peninsular population.

Red foxes apparently moved into the Florida Panhandle during the 1950s. In sampling fox populations by trapping in seven Florida counties (Jackson, Jefferson, Leon, Liberty, Marion, Wakulla, Washington) in 1953–1956, Wood (1959) captured red foxes in Washington County in 1953 and Jackson County in 1956. Slightly later, Jennings et al. (1960) noted that red foxes were increasing in parts of western Florida counties but did not give specific locality data. Apparently the first records of the species in the eastern part of the state were from Alachua County (two individuals killed in 1961), Polk County (population established from introductions during 1960–1969), and Orange County (den with young in 1968) (Lee and Bostelman 1969). In 1978, red foxes were reported in Pinellas County and stated to be widespread in central Florida (Edscorn 1976). The first record from south central Florida (Highlands County) was in 1969, and by the late 1970s red foxes had been reported from most other counties in that part of the state (unpublished data). Although red foxes had been observed in Everglades National Park in 1950 and 1970 (Layne 1974), an increasing number of reports came from the Miami area beginning in 1978. Many of the sightings were in developed areas, including the Orange Bowl football stadium (T. Alexander, pers. comm.).

The red fox in Florida occurs in a wide range of natural habitats, including flatwoods, scrub, and hammocks, and in disturbed habitats such as open pastures, citrus groves, and suburbs. In northern Florida, red foxes are less often reported from urban and suburban areas than is the native gray fox, *Urocyon cinereoargenteus* (J. A. Gore, pers. comm.). In south central Florida, they appear to be more prevalent than the gray fox in human-altered habitats such as citrus groves and improved pastures (unpublished data).

Felis yagouaroundi (Jaguarundi)

This species is alleged to have been introduced into Florida by a writer who over a number of years brought back animals from Central and South America and released them in various state parks and other protected areas as well as in the vicinity of his home in Chiefland (Levy County) (Neill 1961, 1977). Just when these releases took place is unknown, but animals believed to be jaguarundis were sighted in Blackwater Forest State Park, Santa Rosa County, between 1934 and 1940 (C. Beck, pers. comm.) and near Arcadia, DeSoto County, about 1907 (H. Stevenson, pers. comm.). The first published report was in 1942 by

Verrill (1942), who was familiar with the species in its native range. He concluded that jaguarundis were present in northwestern Florida on the basis of descriptions by local residents and gave records from the north central region (Fowlers Bluff and Chiefland, Levy County) and south central region (Highlands Hammock State Park, Highlands County). Additional reports between 1942 and 1947 came from around Fort Myers, Lee County (Anonymous 1942), Two Rivers Ranch near Hillsborough River State Park (C. Beck, pers. comm.), and Osceola County (Teagarden 1954). Records compiled by Neill (1961) between 1950 and 1953 indicate that the species was widely distributed throughout peninsular Florida. By the late 1970s, Neill (1977) had accumulated only two additional records and concluded that the population had declined. But additional reports from Everglades National Park (Layne 1974) and elsewhere in the state are available from the 1950s to 1970s, and seemingly reliable reports of sightings in widely separated parts of the state continue to accumulate (unpublished data). Some jaguarundi sightings in Florida may be of animals escaped from captivity. Mount (1984) has noted several apparently reliable sightings of jaguarundis in coastal Alabama, which may signify a westward movement of the species from Florida. Jaguarundis have been reported from a wide range of habitats in Florida, including scrub, flatwoods, hammocks, mangrove–freshwater marsh ecotone, ranchland, and citrus groves.

The status of the jaguarundi in Florida is an enigma. Despite numerous seemingly reliable sightings from most regions of the state over a period of many years, there are no known Florida specimens in scientific collections. Although Neill (1977) presented photographs of the skin and skull of a young road-killed individual he reported collecting in Highlands County about 1954, positive identification is not possible from the photographs, and the whereabouts of the specimen, if it still exists, are unknown. The lack of specimens raises the question of whether the jaguarundi is actually as widespread and numerous as alleged sightings suggest: if it were, at least an occasional specimen shot, trapped, or hit on the road should have come to the attention of wildlife personnel or mammalogists. The absence of such evidence has led some workers to question whether the jaguarundi ever existed in the state (J. Brady, pers. comm.). The fact that jaguarundis are shy and solitary and apparently very wary of traps (Verrill 1942; Daniels 1983) might explain the lack of Florida specimens. Even if a Florida specimen should be collected, the possibility that it was a recently escaped or released captive would have to be ruled out.

Axis axis (Axis Deer)

Allen and Neill (1954) reported that a few individuals of this species escaped in Volusia County around 1930. By the mid-1950s, the deer occurred in four

counties east of the St. Johns River. The subsequent history of this introduction
is unknown, but the population has presumably been extirpated (perhaps by
poaching) for many years. Semiwild herds of this species are presently maintained in the state.

Cervus elaphus (Elk)

The history of an elk introduction in south central Florida was reviewed by
Layne (1993a). Six animals were released on Buck Island Ranch near Lake
Placid, Highlands County, in 1967 or 1968. The herd increased to a maximum
of 28 or 29 individuals in the late 1970s and then declined, presumably as a result of dispersal and poaching. The population has persisted on the ranch and
in the vicinity, however, to the present time. The most recent sighting was of a
herd of ten individuals in December 1993 (L. O. Lollis, pers. comm.). A commercial elk ranch with approximately 60 animals is located in DeSoto County
(Kimball 1993).

Cervus unicolor (Sambar Deer)

Four sambar deer were introduced on St. Vincent Island in the Gulf of Mexico
near Apalachicola, Gulf County, in 1908. By 1940 the population had reached
several hundred (Newman 1948). Numbers declined to less than 50 during
World War II owing to illegal hunting (Newman 1948) and possibly logging
(Lewis et al. 1990), then recovered to about 75 to 100 animals by the late 1960s
(Smith 1969) and around 200 in 1983–1986 (Flynn et al. 1990). The deer
occur in all major habitat types on the island but tend to avoid tidal marsh. They
forage extensively in freshwater marshes and prefer dense woody cover such as
cabbage palm and mixed slash pine–cabbage palm stands and thick oak scrub
when inactive (Flynn et al. 1990). Although the deer are well isolated on the island, at least one reached the mainland (Anonymous 1963). A suggestion to introduce sambar deer from St. Vincent Island into marsh habitats on the mainland apparently was never acted upon (Anonymous 1967).

Antilocapra americana (Pronghorn)

According to T. Hines (pers. comm.), a group of pronghorns was introduced on
a ranch on the east side of the Kissimmee River in Osceola County sometime before 1973. He assumes that they were quickly poached and did not survive, but
pronghorns were periodically reported from this area through the 1970s. Moreover two sightings from the west side of the Kissimmee River on the Avon Park
Bombing Range were reported during this period, one about 1970 and one in
1979 (J. Whitehouse, pers. comm.).

Feral or Free-Ranging Domestic Species

Canis familiaris (Dog)

The status of feral dogs (living completely independent of humans) in Florida is unknown. If any exist, their number is probably low, and they are most likely to occur in less developed parts of the state. I. L. Brisbin, Savannah River Ecology Laboratory (pers. comm.), doubts that feral dog populations exist anywhere in the southeastern United States. Free-ranging dogs, however, those retaining some association with humans, are numerous and occur throughout Florida. In the present context, free-ranging is broadly defined to include both the household pet with limited opportunity to run free and dogs that are essentially free-living but retain a tenuous association with people through scavenging for garbage and food sources associated with human activity. Such dogs have been termed "pariah" dogs by Brisbin (1977). In some cases pariah dogs are morphologically similar to primitive dogs and the Australian dingo. Such a population, termed the "Carolina dog," is known from a remote bottomland forest and adjoining area along the Savannah River on and in the vicinity of the U.S. Department of Energy Savannah River Plant in South Carolina and on the opposite side of the river in central Georgia (Brisbin 1986). A single free-ranging dog resembling the Carolina dog type was captured in northeastern Florida (in Yulee, Nassau County) by Brisbin (pers. comm.), but this dog was killed and the carcass disposed of before material for DNA analysis could be obtained. Campbell (1985) suggests that the heartworm (*Dirofilaria immitis*) may be a limiting factor in feral or free-ranging dog populations in Florida.

Felis catus (House Cat)

Feral cat populations are probably more likely in Florida than feral dogs, and free-ranging cats are common throughout the state. As an example of the potential size of the population, Coleman and Temple (1993) have estimated the number of free-ranging rural cats in Wisconsin to be 1.7 million.

Sus scrofa (Pig)

The presence of feral pigs in Florida dates back to the beginning of the Spanish colonization period in the 1500s (Towne and Wentworth 1950; Hanson and Karstad 1959). Feral pigs have been recorded in all 67 counties of the state; as of 1988, the heaviest concentrations were in the Big Bend region of the Gulf coast and in the south central and southwestern part of the peninsula (Mayer and Brisbin 1991) (Figure 10.2). Degner et al. (1983) have estimated the total state population at more than 500,000. Feral pig populations have multiple origins from domestic stocks, and Eurasian wild boars intended to "improve" the feral

FIGURE 10.2   Recent vegetation damage by wild pigs (*Sus scrofa*) in Cutthroat Seep, Tiger
Creek Preserve, Florida. Photo by Steve Morrison.

stocks for sport hunting have also been introduced in different parts of the state.
Mayer and Brisbin (1991) found evidence of hybridization between domestic
pigs and wild boars in populations on Eglin Air Force Base (Okaloosa and
Walton counties) and portions of Alachua, Polk, Highlands, and Palm Beach
counties; only the Eglin Air Force Base population showed successful establishment of the hybrid type. The hybrid types in the latter population have become
more widely distributed in recent years (I. L. Brisbin, pers. comm.).

Capra hircus (Goat)

Stevenson (1976) says that feral goats were established on St. George Island,
Franklin County. According to J. A. Gore (pers. comm.), the goats were confined to Little St. George Island (Franklin County) by the 1970s and were all removed by 1980. Another feral goat population ranging from about 6 to 20 individuals has existed on Bumblebee Island in Lake Istokpoga, Highlands County,
since at least the 1960s (unpublished data).

Who Succeeds?

Established nonindigenous mammals in Florida differ greatly in their success in expanding their ranges and populations. At one extreme is the black-tailed jackrabbit, which has failed to spread from the general area in which it was introduced some 60 years ago; at the other is the red fox, which has blanketed the state in about half that time. There are no obvious correlations between taxonomic group, geographic origin, circumstances of introduction, body size, life-history traits, or other factors and the success of introduced species in Florida. Thus we are left to conclude that species have done poorly or well for different reasons. Among the less successful species, for example, the tight group cohesion characterizing the social structure of the three primate species may have been a significant factor limiting their spread, whereas poaching pressure may have prevented expansion of the introduced elk (Layne 1993a). There are no obvious reasons preventing expansion of species like the black-tailed jackrabbit and nutria, however, which, on the basis of their life histories and habitat requirements, would be predicted to do well in Florida. The nutria, in particular, despite multiple introductions, abundance of seemingly suitable wetland and aquatic habitats, and a high reproductive potential, has had much less success in Florida than in some areas, such as coastal Louisiana, into which it has been introduced (de Vos et al. 1956).

Among the more successful species, large home-range size and long-distance dispersal tendencies were probably key factors in promoting rapid spread of the coyote and red fox, whereas small home-range size and low dispersal may have contributed to the equally rapid expansion of the armadillo by facilitating settlement of artificially relocated individuals (Layne and Glover 1977). Armadillos are conspicuous and easily captured by hand, and there are numerous anecdotal accounts of people catching them in one locality and releasing them elsewhere, including one killed in upstate New York in the early 1960s (unpublished data). Such relocations probably were a major factor in the rapid spread of the species in peninsular Florida and might also have contributed significantly to the range expansion of the native population in the west.

A relatively high reproductive rate probably aided the spread of the coyote and red fox, as well, whereas a different reproductive trait, delayed implantation, may have played a role in the success of the armadillo. Delayed implantation in the armadillo results in a normal gestation period of 8 or 9 months, which under conditions of stress can be extended to as much as 24 months (Storrs et al. 1988). Presumably prolonged gestation would aid establishment of new populations by increasing the probability that naturally dispersing or relocated females were pregnant and would give birth in the new locality. The single sex of the usual identical quadruplets produced by armadillos, however, would limit population foundation by single pregnant females to those with male litters.

Land-use changes have presumably contributed to the spread of some non-indigenous species in Florida by providing more favorable habitat conditions or corridors for dispersal. Movement of armadillos into the Everglades and Big Cypress regions, for example, was probably facilitated by drainage and the construction of levees and dry road shoulders through wet areas. Conversion of dense forest areas and wetlands into pastureland, citrus groves, and other more open habitat types may have favored coyotes and red foxes.

Parasites may also influence establishment and survival of nonindigenous mammals. For example, the absence in south central Florida of the helminths *Parelaphostrongylus tenuis* and *Elaeophora schneideri,* which do not significantly affect deer but are detrimental to elk, may have contributed to the persistence of the introduced elk in that region (Layne 1993a).

Ecological Impacts

Nonindigenous mammals are known or suspected to affect native biota in various ways, mostly negative, but for the most part, the exact nature of these effects and their overall importance have not been documented through actual research. Effects of nonindigenous mammals on native species or communities may occur through habitat destruction, predation, disease and parasites, genetic effects, or competition.

Habitat Destruction

Feral pigs, through their rooting, have a more destructive impact on natural habitats than any other exotic Florida mammal. Intensive rooting in wet areas can result in considerable destruction of vegetation and slow recovery. Rooting in mesic hammocks may cause drying of the litter layer to the detriment of soil microorganisms as well as larger invertebrates and such vertebrates as salamanders and shrews. Adverse effects of pig activity on soil microarthropods in Hawaiian rain forest was demonstrated by Vtorov (1993). Pigs may also have an indirect deleterious impact on natural vegetation through dissemination of exotic plants. Jensen and Vosick (1994) suggest that pig rooting may facilitate the spread of the tropical soda apple (*Solanum viarum*), and one of the avenues of dispersal of hydrilla (*Hydrilla verticillata*) is by shedding of the resistant turions in the feces of pigs that have fed on the plants. The artificially stocked rhesus monkeys on Raccoon Key have severely degraded the mangroves and other vegetation (N. Goodyear, pers. comm.), presumably to the detriment of nesting birds and probably also the endangered silver rice rat (*Oryzomys argentatus*) known to occur on the island before introduction of the monkeys. There is less evidence of direct damage to natural habitats by other nonindigenous mammals. Limited destruction of marsh vegetation ("eatouts") by nutria has been reported (Wallace 1963), and rooting by armadillos has been claimed to disrupt the organization and productivity of the leaf-mold stratum of forest soils (Carr 1982). Foraging by goats on Bumblebee Island in Lake Istokpoga at times of high population has resulted in marked destruction of natural vegetation, including the climbing dayflower (*Commelina gigas*), listed as threatened by the Florida Department of Agriculture and Consumer Services (unpublished data). The Mexican gray squirrel was found to be primarily responsible for damage to native thatch palms (*Thrinax morrisii* and *T. radiata*) on Elliott Key in Biscayne National Park (Tilmant 1980). In 1977–1980 approximately 67 percent of the palms were damaged and over 40 percent were dead or dying. Black rats were also a potential cause of palm damage.

Predation

Nonindigenous wild species such as the armadillo, coyote, red fox, coati, and jaguarundi prey on a wide variety of native invertebrate and vertebrate animals. The potential impact of predation by nonindigenous mammals on endangered and other at-risk species in Florida is of particular concern. The armadillo is the only species whose food habits in Florida have been studied in detail. Under most conditions it probably has no serious effect on native invertebrate or vertebrate populations, although it may be a significant predator of the scrub lizard (*Sceloporus woodi*), sand skink (*Neoseps reynoldsi*), and other small reptiles endemic to the Florida scrub habitat (Layne 1976). Holler (1992) has cited the red fox as a potential threat to the Perdido Key beach mouse (*Peromyscus polionotus trissyllepsis*), and according to J. A. Gore (pers. comm.) red foxes and coyotes are also important predators of sea-turtle eggs and of hatchlings and chicks of ground-nesting birds, particularly colonially nesting birds such as least terns (*Sterna albifrons*) and black skimmers (*Rynchops nigra*). L. N. Brown (pers. comm.) found partially eaten shells of the Florida tree snail (*Liguus fasciatus*), a state-listed Species of Special Concern, in two nests of black rats he examined on Key Largo in the 1970s, but whether or not predation by rats is a significant cause of mortality in tree snail populations is unknown.

Among feral or free-ranging domestic mammals, pigs consume small vertebrates and dogs prey on a wide variety of native species ranging from gopher tortoises (*Gopherus polyphemus*) (Douglass and Winegarner 1977; Causey and Cude 1978) to deer. House cats, by virtue of their presumably larger population, probably have an even greater overall impact on native vertebrates than dogs. In addition to such prey as birds and small rodents, they regularly kill shrews, moles, frogs, snakes, and lizards, which they often do not consume. Predation by cats may have played a significant role in the reduction or actual extinction of Sherman's short-tailed shrew (*Blarina carolinensis shermani*) (Layne 1992) and in the decline of populations of beach mice (*Peromyscus polionotus*) in dune habitats along the Gulf and Atlantic coasts (Bowen 1968; Humphrey and Barbour 1981; Frank 1992). In 1976, A. Litman (pers. comm.) reported that wild house cats had become very abundant on north Key Largo, Monroe County, in the range of the endangered Key Largo wood rat (*Neotoma floridana smalli*) and Key Largo cotton mouse (*Peromyscus gossypinus telmaphilus*). According to P. Frank, Florida Game and Fresh Water Fish Commission (pers. comm.), free-ranging cats are still numerous in that area, and the common practice by local residents of putting out food for raccoons, which also attracts cats, may contribute to the high population. Although their food habits have not been studied, the cats are potential predators on the wood rats and cotton mice. Because of their persistence in pursuing prey at low densities (Pearson 1964), cats may be particularly efficient in eliminating native small mammals and other vertebrates from small remnant natural habitats in developed areas.

On the positive side, some nonindigenous mammals may serve as a food source for native carnivores in Florida. The native species most dependent on an introduced species is the Florida panther (*Felis concolor coryi*): feral pigs are a major component of its diet (Maehr et al. 1990). Pigs also are included in the diet of the black bear (*Ursus americanus*), alligator (*Alligator mississipiensis*), and bobcat (*Lynx rufus*). Prey remains from barn owl (*Tyto alba*) nests on Sanibel Island indicate that the black rat is a major component in the diet of the owls on the island (unpublished data). Bobcats (Wassmer et al. 1988), long-tailed weasels (*Mustela frenata*) (Layne 1993b), and indigo snakes (*Drymarchon corais*) (unpublished data) are also among the known predators of black rats in Florida. House mice are included among the prey of barn owls (Trost and Hutchinson 1963) and American kestrels (*Falco sparverius*) (unpublished data). Armadillos are taken by panthers, bears, and alligators, and road-killed specimens are a food source for native scavengers such as the crested caracara (*Caracara plancus*), black vulture (*Coragyps atratus*), and turkey vulture (*Cathartes aura*). Armadillos were the most abundant (24 percent) of 185 road-killed vertebrates observed being fed on by caracaras in south central Florida (unpublished data.). Considering the probable decline from historic levels of carrion supply as a result of more efficient ranching practices, screwworm control, changing land-use patterns, and other factors, the great number of armadillos killed on Florida highways probably represents an increasingly important food resource of caracaras and vultures.

Disease and Parasites

Nonindigenous species may vector diseases and parasites that affect native mammals. Red foxes, coyotes, and cats, for example, carry rabies. The reappearance in 1994 in hunting dogs of the canine rabies strain not seen in the state since the 1950s has been attributed to introduced coyotes, according to an article in the *Tampa Tribune* of 5 March 1995. Cats are particularly likely candidates for transmission of rabies to native mammals, as they have the highest incidence of rabies of any domestic species, at least in part because of the difficulty of enforcing laws requiring vaccination, and are often numerous in suburban or rural areas where they come in contact with wildlife (Krebs et al. 1995). Feral or free-ranging cats may also introduce feline panleucopenia virus into populations of bobcats and possibly Florida panthers (Wassmer et al. 1988), and pseudorabies virus, widespread in feral pigs in Florida, has been implicated in the death of at least one Florida panther (Glass et al. 1994). The source of the hookworm *Ancylostoma pluridentatum,* which infects Florida panthers and bobcats in Collier County, the only known North American locality for this parasite (Forrester 1991), may have been captive-bred panthers of apparently mixed ancestry involving South American animals that were released in Everglades National Park between 1957 and 1967 (O'Brien et al. 1990). Florida panthers have apparently acquired infections of trichinosis from feeding on feral pigs (Forrester 1991). A western North American flea, *Hoplopsyllus glacialis affinis,* typically associated with rabbit hosts, is well established on cottontail and marsh rabbits in Florida and may have been introduced on black-tailed jackrabbits in southern Florida (Layne 1971).

Genetic Effects

Hybridization with native species is another potential impact of nonindigenous mammals. The only possible case in Florida involves the European rabbit (*Oryctolagus cuniculus*) and the endangered Keys marsh rabbit (*Sylvilagus palustris hefneri*) in the Florida Keys, where the two occasionally coexist. Lazell (1989) says there was no evidence of hybridization in these situations, but his remark implies that hybridization might be possible. As hybridization with the coyote was the primary threat to the red wolf (*Canis rufus*) during its last days in the wild, the rapid spread and increasing abundance of coyotes in mainland Florida appears to have foreclosed the possibility that the red wolf could be reintroduced there (Layne 1978). Hybridization between two nonindigenous species in Florida—coyote and dog—may also have implications for native wildlife. These species hybridize freely and produce fertile hybrids, which at least in some regions can potentially backcross to either parent in the wild (Gipson et al. 1975). Individuals of mixed ancestry may differ morphologically or behaviorally from pure coyotes. The relatively large size of the coyotes in New England, for example, which have been termed "wild canids" or "New England canids," is apparently attributable to introduction of dog and possibly wolf genes into the population early in its history (Silver and Silver 1969). Theoretically, morphological or behavioral differences of hybrids might be reflected in differences in food habits or other aspects of ecology and life history, resulting in impacts on native species different from those of either parental species.

Competition

Competition with native species for food or other resources is presumed to be one of the most important impacts of nonindigenous mammals, although actual evidence of such effects and their magnitude is largely lacking. Feral pigs in Florida are considered major competitors with deer (*Odocoileus virginianus*), turkeys (*Meleagris gallopavo*), bears, raccoons (*Procyon lotor*), and many other mammals and birds for acorns and other mast. On the basis of habitat overlap and food habits elsewhere in their ranges, red foxes may be significant competitors of the native gray fox. Competition for prey may also occur between coyotes and bobcats. Competition from nonindigenous species may be an especially serious threat to some of Florida's threatened and endangered native rodents. Goodyear (1992) concludes that the endangered silver rice rat of the lower Florida Keys has been competitively excluded or greatly reduced on some keys by black rats. Competition with the black rat was also considered a possible cause of relatively low density of the Key Largo wood rat (Hersh 1981). Competition with house mice may contribute to the decline of beach-mouse populations (Humphrey and Barbour 1981), but Breise and Smith (1973) conclude that in disturbed inland or beach habitats where both species occur, their relative abundance is due to an interaction between availability of homesites for the house mice and predation by cats, to which beach mice are more vulnerable. Thus the effect on beach mice by house mice may be indirect, although with the same result.

Economic Impacts

Although the focus of this volume is on the relationships of nonindigenous species to Florida's natural ecosystems, the economic and public-health impacts of nonindigenous mammals, which are arguably greater than those of any other group of exotic organisms except plants and insects, warrant brief mention.

Negative Values

Pig rooting in improved pastures is a serious agricultural problem, and in some situations pigs also invade residential areas and root up lawns. Armadillos, too, damage lawns and landscape plantings with their digging. Black rats are a significant pest in sugarcane fields (Samol 1972) and may damage citrus and other crops. Armadillo burrows are hazardous to horses and riders on cattle ranches and may weaken earthen dams and levees. The introduced rats and the house mouse are serious pests in urban, suburban, and rural areas through their consumption and contamination of foodstuffs and stored products and damage caused by gnawing and burrowing. Red foxes, coyotes, and free-ranging dogs take a toll of domestic livestock. An estimated 200 calves were killed in Florida by coyotes in 1991 (Emerson 1994). Coyotes also prey on sheep in Florida. An extreme example of the magnitude of such predation is a report from a Highlands County sheep rancher of the loss of 200 lambs and 50 ewes to coyotes in a single year (J. Hendrie, pers. comm.). Group hunting behavior of coyotes and dogs allows them to attack larger domestic animals, such as adult cows and horses. In some areas of the state, coyotes also do significant damage to watermelon crops by eating the melons (Emerson 1994).

Moreover, nonindigenous mammals are known (or potential) vectors of a number of diseases or parasites affecting humans or domestic animals. Perhaps the most serious of these is brucellosis. Feral pigs are a major reservoir for the disease, which is a threat to domestic swine, cattle, and humans (van der Leek et al. 1993). Wild pigs also are involved in transmission of pseudorabies and trichinosis, have been recorded as seropositive for eastern equine encephalitis, and are a potential reservoir of African swine fever should it be introduced into Florida (Forrester 1991). Swine lungworms carried by feral pigs transmit swine influenza and hog cholera virus, which constitute a threat to domestic pigs (Forrester et al. 1981). Red foxes, coyotes, and feral or free-ranging dogs and cats are important potential sources of rabies transmission to humans and domestic animals. Armadillos are believed to be involved in amplification and transmission cycles of St. Louis encephalitis virus (Day et al. 1995) and have been recorded as seropositive for leptospires (Motie et al. 1986) and arboviruses (Bigler et al. 1975). Wild armadillos with leprosy infections have been recorded in Texas, Louisiana, Mississippi, and Mexico. Although monitoring to date has revealed no Florida cases, suggesting that the founders of the introduced population were leprosy-free, infections could enter the state through the expanding population from the west (Storrs and Burchfield 1985). The armadillo is also a potential reservoir of Chagas disease (Forrester 1991). Murine typhus has been found to be endemic in Norway rats and black rats in some Florida localities (Rickard and Worth 1951). The simian herpes B virus present in the Silver Springs rhesus monkey population poses a significant threat to humans (Chapman 1994), and a number of cases of injuries to people from attacks by the monkeys are on record. Risks of attacks from wild primates are increased in situations where they are habituated to humans by being fed. In the case of the vervet monkeys in Dania, this danger led to passage of a county ordinance prohibiting feeding of wild animals. Forrester (1991) lists additional parasites and diseases carried by nonindigenous mammals that are of known or potential significance to humans or domestic animals.

Positive Values

Feral pigs are an important game species in Florida and elsewhere in the United States and are also used for commercial purposes. Degner et al. (1983) have estimated the recreational and commercial value of feral pigs at about $8.3 million per year. Although considered a nuisance by authorities, feral pigs in some state parks are an attraction to visitors and are thus favored by concessionaires who conduct wildlife tours. The rhesus monkeys at Silver Springs presumably also enhance its appeal to tourists. Red foxes provide economic benefits from sport hunting with hounds. Armadillos consume pest species such as love bug larvae (*Plecia nearctica*) and fire ants (*Solenopsis invicta*), and their burrows provide a refuge for various small native wildlife species. They are used to a limited degree for their meat as well, mainly in outdoor cookery, and for the manufacture of novelties such as baskets made of the dried "shell." The armadillo is important as a laboratory animal for the study of leprosy and development of a leprosy vaccine, and a considerable number of animals have been captured from the leprosy-free population in south central Florida for shipment to medical research laboratories in other countries. The finding by Jackson et al. (1972) that the dermal ossicles are strong accumulators of strontium-90, presumably via the food chain, suggests that the armadillo may also be a valuable bioindicator for monitoring the level of radioactive fallout. Mayer and Brisbin (1991) note that some feral species' populations may warrant conservation as sources of genes that have been lost from modern domestic stocks and cite such a pig population in southwestern Florida in the vicinity of the Peace (Polk to Charlotte countries) and Caloosahatchee rivers. Feral pigs in this region exhibit the mule-footed condition and a gray adult-striped juvenile color pattern (Mangalitza type) that may reflect morphological traits of the earliest introductions of pigs in Florida.

The Question of Control

As it is not feasible for practical as well as financial reasons to eliminate or even significantly reduce nonindigenous mammals such as the red fox, coyote, and armadillo once they have become widespread and abundant, emphasis in Florida is on reducing or regulating potential sources of exotic mammals—such as the pet trade, accidental importation, deliberate introductions, game ranches, and tourist attractions—and responding rapidly to eliminate any free-ranging individuals or newly established populations before they gain a foothold (Quinn 1994b). Public sentiment for certain mammals, particularly primates, may impede efforts to remove them from the wild. In the case of the widespread and abundant nonindigenous mammals, control efforts have been limited to local populations that pose significant environmental or economic threats.

The feral pig is the most frequent target for control on both public land (Girardin 1994; Hardin 1994; Nelson and Richards 1994; Chapters 16 and 18 in this volume) and private lands. Pigs trapped in state parks and other lands are frequently released in public hunting areas. Between 1960 and 1976, for example, 2848 pigs were stocked in the J. W. Corbett Wildlife Management Area in Palm Beach County (Mayer and Brisbin 1991). As a matter of official policy, armadillos, pigs, and other nonindigenous species are removed from state parks when possible. Coyotes, armadillos, feral pigs, and feral or free-ranging dogs and cats on ranches and other private lands are often trapped or shot as opportunity affords, and more intensive control efforts may be instituted when actual loss of livestock is involved. Despite its introduced status in the state, the red fox is legally protected along with the native gray fox in deference to fox hunters, who prefer it because it gives their hounds a longer chase. In recent years, there has been considerable controversy over the rhesus monkeys at Silver Springs: some state agencies advocate removal; various public-interest groups vigorously oppose such action.

Prospects

Given the many factors determining the chance of arrival and success or failure of an exotic species in a new environment, any attempt to predict future additions to the list must be highly speculative. Nonetheless, their popularity as pets, the known escapes or releases, and the evidence of persistence of some individuals in the wild suggest that the ferret (*Mustela putorius*) and ocelot (*Felis pardalis*) may eventually establish wild breeding populations in Florida. I am aware of accounts of ocelot sightings in the wild in the Miami area in 1958, Alachua County in 1962, and Highlands County during the 1970s—and there are undoubtedly more. Ferrets were captured in Highlands Hammock State Park (Highlands County) in 1978 and Jonathan Dickinson State Park (Martin County) in 1979 (R. Roberts, pers. comm.). The latter animal may have been one that was lost by a park visitor eight months before. Other sightings of ferrets in the wild in Highlands and DeSoto counties occurred in 1973 and 1987.

Of all the nonindigenous mammals that have appeared in Florida, the species most likely to become a serious pest if ever established is the mongoose

(*Herpestes auropunctatus*), which has had a devastating effect on the indigenous faunas of the Caribbean islands on which it was introduced (de Vos et al.
1956). Nellis et al. (1978) reported an introduction of the species in the Miami
area in 1977. A young female mongoose was captured on Dodge Island, Port of
Miami, in February, and two others were alleged to have been killed a month before. Another claimed to have been seen was not reported again after May, when
intensive trapping was terminated. The source of these animals may have been
a pregnant female transported in a fruit shipment from the West Indies. Van
Gelder (1979) has reviewed records of introductions of the species and suggests that it is less successful at becoming established in mainland habitats than
on islands.

ACKNOWLEDGMENTS

I thank the many people cited in this chapter who have provided me with observations on nonindigenous mammals in Florida over a 40-year period. The
chapter profited greatly from the comments and suggestions of J. Brady,
C. Belden, J. Wooding, I. L. Brisbin, Jr., and J. A. Gore. I am also grateful to the
editors of this volume for the opportunity to present this synthesis of information on Florida's nonindigenous species and for their careful editing of the
manuscript.

11 Nonindigenous Marine Invertebrates and Algae

JAMES T. CARLTON AND
MARY H. RUCKELSHAUS

Although we know little about nonindigenous marine and estuarine invertebrates and algae of the southeastern United States, including Florida, the patterns of documented biological invasions in other regions of North America and the world suggest that numerous nonindigenous species from the warmer waters of South America, the eastern Atlantic Ocean, and the Pacific Ocean historically became established in Florida through human transport. Briefly reviewed here are the transport mechanisms that may have brought—or continue to bring—nonindigenous marine organisms to Florida's coastal waters. Because this is the first formal attempt to document the diversity, history, and biogeography of the introduced marine organisms of Florida, we can provide only a preliminary assessment here.

Mechanisms of Introduction

Attempts by the Spanish to colonize Florida, beginning in the early sixteenth century (Natkiel and Preston 1986), no doubt led to the transport and probable introduction 500 years ago of the first nonindigenous species of fouling and boring organisms on and in ship bottoms. Accompanying these organisms were mobile animals within the fouling masses and in the deep holes made by boring shipworms (teredinid bivalve mollusks) and boring gribbles (limnoriid isopods) in ship hulls (Allen 1953; Carlton 1985, 1987; Carlton and Hodder 1995). The same vessels that transported fouling organisms externally and were riddled with boring organisms internally carried rock, sand, and other "solid" or "dry" ballast that held foreign species of semiterrestrial talitrid amphipods (shore hoppers), maritime insects, and beach, dune, and marsh plants (Carlton 1992c).

The modern manifestation of shipborne dispersal of nonindigenous marine species is the transport of organisms in ships' ballast water (Carlton 1985; Carlton and Geller 1993)—a viable transport mechanism for virtually any planktonic or smaller nektonic species in the water at the time of ballasting that has been linked to numerous invasions since the 1970s (Carlton and Geller 1993), including the transport of red-tide-causing dinoflagellates (Hallegraeff and Bolch 1991, 1992). Ballast water is carried by ships on both cargo legs (in small volumes) and ballast legs (in much larger volumes) for trim, stability, and other purposes. Ballast water may be released as a vessel travels along a coast, as it enters a port and moves in harbor channels toward a dock, or in the dock itself. The Port of Miami, for example, receives large volumes of ballast water from general cargo and passenger traffic (Carlton et al. 1995).

Other human-mediated transport mechanisms may move species to new localities. These include the transplantation of marine animals and plants for mariculture (aquaculture), along with their diverse associated biota, and release of organisms initially imported for use as aquarium stock, fish bait, or human food (Carlton 1992a, 1993; Carlton and Scanlon 1985). All these mechanisms may have facilitated transport of nonindigenous species to Florida.

Transport to Florida waters of warm-water marine invertebrates, algae, protists, and fish from hundreds of ports, harbors, and anchorages around the world has been going on for five centuries. Thus we cannot conclude, like Winston (1995), that because the oldest record of the bryozoan *Zoobotryon verticillatum* in the Caribbean "goes back almost 200 years," it "could well be native" to that region. It is therefore challenging to distinguish species introduced by shipping from species that some biogeographers and systematists would recognize as "naturally" cosmopolitan. Scores of species of marine organisms in Florida coastal waters—particularly among the sponges, hydroids, flatworms, nemerteans, polychaetes, oligochaetes, bryozoans, gammarid amphipods, copepods, shipworms, and ascidians—are often described as occurring in Caribbean Florida, eastern Atlantic, Pacific, Australasian, and Indian Ocean waters.

Such taxa are traditionally considered "native" to Florida (and to the Caribbean biota in general), but often there is little or no evidence to support either this assignment or their recognition as introductions. Thus it is logical to view many of the cosmopolitan species in the Florida marine biota as cryptogenic (Carlton 1996)—that is, species whose status as introductions is not resolved. Paleontological, archaeological, historical, biogeographic, systematic, genetic, and other forms of evidence (Chapman and Carlton 1991, 1995) must be mustered to classify cryptogenic species as either native or nonindigenous.

Introduced Marine and Estuarine Invertebrates

There are no systematic studies of the nonindigenous marine and brackish-water invertebrates of Florida waters. Direct comparisons of early species lists with modern ones (with due consideration for taxonomic pitfalls and collecting biases) from Florida bays, estuaries, and lagoons would undoubtedly reveal a host of species not recorded by previous workers. In turn, after those species are set tentatively aside that (because of their size, habitat, or rudimentary taxonomic status) may easily have been overlooked by earlier workers, the remaining species would undoubtedly yield a fruitful harvest of taxa needing reexamination.

We offer here examples of some shipborne introductions (Table 11.1). It is important to emphasize that the 20 species listed as established introductions in Table 11.1 (of a total of 39 species, the rest of which are cryptogenic or of unknown status) are only examples. Thus Table 11.1 cannot be summarized as saying that "20 introduced species of marine invertebrates are known from Florida." This number is a *minimum;* the true number of introduced species is probably at least twice that. In Coos Bay, Oregon, for example, a literature review suggested the presence of 11 introduced marine invertebrates (Carlton 1979a), but on-site studies for three years raised the number to 60 (J. Carlton and C. Hewitt, unpublished data). Winston (1995) has noted that 58 percent (21 of 36 species) of bryozoans from the Indian River Lagoon (Brevard County) in eastern Florida are "circumtropical or cosmopolitan," commenting that shipping activity has "resulted in the augmentation of the original fauna to its present level." Only 6 of these 21 species, the clearest cases, have been listed in Table 11.1; it seems improbable that the remaining 15 taxa are all native.

Thirteen species of introduced fouling invertebrates are listed in Table 11.1. In some Florida fouling communities, introduced barnacles, tube worms, bryozoans, and ascidians are common (Mook 1983; McPherson et al. 1984). Although the ecological interactions between these and native species have not been examined in Florida waters, one can predict competitive, predatory, and disturbance interactions (Carlton 1989, 1992d). Introductions have also occurred in habitats as diverse as supralittoral communities (the European shore isopod *Ligia exotica*); marshes (the European snail *Ovatella myosotis*); intertidal, nonfouling assemblages (the Asian sea anemone *Haliplanella lineata* and perhaps the West African pulmonate limpet *Siphonaria pectinata*); soft-bottom infaunal communities (the polychaete worm *Boccardiella ligerica*); and sublittoral coral reefs (the Indo-Pacific black-lipped pearl oyster, *Pinctada margaritifera,* which, although not necessarily established, suggests the vulnerability of even this habitat to invasions).

Table 11.1.

Examples of Introduced and Cryptogenic Marine
and Estuarine Invertebrates of Florida

Species and status	Origin and mechanism of introduction	Reference[a]
PLATYHELMINTHES		
Taenioplana teredini (C)	Pacific?/S	Riser (1970, 1974)
A predatory flatworm found inside the burrows of the shipworms *Teredo furcifera* and *Bankia fimbriatula*.		
CNIDARIA		
Hydrozoa (hydroids)		
Cordylophora caspia (+)	Eurasia/S	Wurtz and Roback (1955)
Anthozoa (sea anemones)		
Haliplanella lineata	Asia/S	Schick and Lamb (1977)
The well-known orange-striped sea anemone (in most literature as *H. luciae*) often common on intertidal hard substrates.		
CRUSTACEA		
Cirripedia (barnacles)		
Balanus reticulatus	Indo-West Pacific/S	Spivey (1979)
Balanus amphitrite	Indo-West Pacific/S	Henry and McLaughlin (1975); Bingham (1992)
Balanus trigonus	Indo-West Pacific/S	Zullo (1992)
All three barnacle species are common in Florida fouling communities.		
Isopoda (isopods)		
Sphaeroma walkeri	Indian Ocean/S	Miller (1968); Conover (1979); Carlton and Iverson (1981); Kensley et al. (1995)
A fouling and occasionally boring isopod.		
Sphaeroma terebrans (+)	Indian Ocean/S	Richardson (1897); Rehm and Humm (1973); Simberloff et al. (1978); Ribi (1981, 1982); Estevez (1994)
A wood-boring isopod; large populations occur in mangroves.		
Limnoria pfefferi (C)	Indo-Pacific/S	Menzies (1957); Kensley and Schotte (1989); Cookson (1991)
A wood-boring isopod widely distributed in the Indo-Pacific Ocean that appears to have been introduced into the region between the Bahamas and Belize, including Puerto Rico and the U.S. Virgin Islands. Closely related species are found primarily in the Indo-Pacific or Pacific Ocean.		
Limnoria saseboensis (C)	Pacific?/S	Menzies (1957); Kensley and Schotte (1989); Cookson (1991)
A wood-boring isopod known from Japan, Fiji, and Florida but whose closely related sibling species occur largely in the Pacific (Schotte 1989; Cookson 1991).		

[a]Reference cited establishes the presence of the species in Florida—not necessarily its recognition as introduced. Pending a thorough review of the history of Florida introductions, no attempt is made here to determine the earliest Florida record of the species in question.

TABLE 11.1.   *(Continued)*

Ligia exotica	Europe/S	Van Name (1936); Kensley et al. (1995)

Amphipoda (amphipods)

Chelura terebrans (+)   Europe?/S   Bousfield (1973)

A wood-burrowing amphipod occupying old *Limnoria* (gribble) burrows.

Tanaidacea (tanaids)

Zeuxo maledivensis (+)   Pacific/S?   Sieg and Winn (1981)

Decapoda (crabs and shrimps)*b*

Charybdis helleri   Indo-Pacific/BW   Lemaitre (1995)

An Indo-Pacific portunid (swimming) crab apparently transported from the Eastern Mediterranean to the Caribbean and South America in the 1980s and from there to the Indian River, Florida, in 1995.

ANNELIDA

Polychaeta (polychaete worms)

Hydroides elegans   Australasia?/Indian   Zibrowius (1971)
Ocean?/S?

A common serpulid tube worm that builds calcareous tubes in fouling communities.

Boccardiella ligerica (+)   Europe?/S?   Kravitz (1987)

Mud-dwelling, tube-building spionid worm.

MOLLUSCA

Gastropoda (snails)

Siphonaria pectinata (C)   Mediterranean/S   Voss (1959); Carlton (1992b)

Intertidal limpet-like snail; see Carlton (1992b) for status.

Ovatella myosotis   Europe/SB   Carlton (1992b)

A small salt-marsh and drift-line snail.

Ercolania fuscovittata (−)   California/S   Mikkelsen et al. (1995)

A California nudibranch that Mikkelsen et al. (1995) consider not established.

Bivalvia (clams, mussels)

Mytella charruana (−)   Eastern S. America/   Lee (1987); Carlton (1992b)
BW or S

Charru mussel; occurred in large numbers in a power-plant intake at Jacksonville in 1986; the population may have been ephemerally established through release of larvae in the ballast water of an oil tanker from Venezuela.

Pinctada margaritifera (−)   Tropical Pacific/BW?   Chesler (1994)

This is the common black-lipped pearl oyster of the tropical Pacific Ocean. A total of three confirmed living specimens have been found by SCUBA divers in relatively deep sublittoral waters on coral reefs off the east coast of Florida, 80 km north of Miami at Boca Raton and Boynton Beach. These specimens—collected in 1990 (not 1992 as reported by Chesler 1994), 1993, and 1994—all measured approximately 19–20 cm in diameter (Chesler 1994; J. R. Chesler, pers. comm.; M. R. Bukstel, pers. comm.). Their similar size suggests the release of a similar cohort, perhaps as larvae in ballast water.

---

b The saber crab, *Platychirograpsus spectabilis* (= *P. typicus*) (Grapsidae), one of Florida's better-known invasions from Mexico by way of log importations (Marchand 1946), is generally regarded as a freshwater species and not treated here, but its local distribution in Florida suggests that it may occupy oligohaline or mesohaline waters of Tampa Bay on migratory excursions (D. Camp and W. Price, pers. comm.).   *continues*

TABLE 11.1. *(Continued)*

Martesia striata (C)   Indo-Pacific?/S   Turner and Johnson (1971)

> The origin of this wood-boring bivalve remains to be worked out, but its now global distribution parallels that of the large and probably Indo-Pacific guild of wood-boring sphaeromatid isopods, limnoriid isopods, and shipworms. Turner (1955) notes fossil material from Japan, and it appears to have been redescribed at least twice from Florida (as *Martesia americana* from Fort Dade and as *M. funisicola* from Lake Worth) (Turner 1955).

Bankia carinata (–, C)   Indo-Pacific?/S   Turner (1971)

> Florida records based in part on a population at Fort Pickens, Pensacola, Florida, described as *Bankia caribbea* Clench and Turner, 1946.

Bankia fimbriatula (C)   Pacific?/S   Riser (1970); Turner (1971); Hoagland (1986)

Lyrodus bipartitus (C)   Pacific?/S   Turner (1971); Hoagland (1986)

Lyrodus medilobatus (–)   Indo-Pacific/S   Mikkelsen et al. (1995)

> An Indo-Pacific shipworm that Mikkelsen et al. (1995) indicate was collected (as *L. mediolobatus*) in the Indian River region but has not become established.

Lyrodus massa (C)   Pacific?/S   Turner (1971); Hoagland (1986)

Teredo bartschi (C)   Pacific?/S   Turner (1971); Hoagland and Turner (1980); Hoagland (1986)

Teredo clappi (C)   Pacific?/S   Turner (1971); Turner and Johnson (1971)

> The type locality of this shipworm is a ship's keel in Key West, Florida, but it has been recorded predominately from the Indo-Pacific and Pacific oceans, suggesting the probable origin of the ship in question.

Teredo furcifera (C)   Pacific?/S   Riser (1970); Turner and Johnson (1971)

ECTOPROCTA (BRYOZOANS)

Ctenostomata

     Victorella pavida (+)   Europe/S   Mook (1983)

     Zoobotryon verticillatum (C)   Europe?/S   Mook (1983); Bingham (1992)

     Sundanella sibogae (+)   Pacific?/S   Winston (1995)

Cheilostomata

     Conopeum "seurati" (C)   Europe/S   Winston (1982, 1995)

     Cryptosula pallasiana (+)   Europe?/S   Winston (1982, 1995)

     Watersipora subovoidea (+)   Indo-West Pacific?/S   Mook (1983); Winston (1995)

CHORDATA

Ascidiacea (sea squirts)

     Styela plicata (+)   Asia/S   Mook (1983); Bingham (1992)

     Botryllus schlosseri   Europe/S   Mook (1983)

     Botryllus niger (C)   E. Atlantic?/S   Mook (1983)
     (= *Botrylloides nigrum*)

> All three sea squirts are seasonally common in Florida fouling communities.

Note: Mechanisms of introduction: BW = ship ballast water and sediments; S = ship fouling and boring communities; SB = ship solid ballast (rocks, sand). Status codes: + = identified here as introduced; – = not known to be established; C = cryptogenic.

The Mangrove-Boring Isopod *Sphaeroma terebrans*

The case history of the wood-boring sphaeromatid isopod *Sphaeroma terebrans* in Florida illustrates several fundamental challenges in invasion biogeography and ecology—including the greater difficulty of enumerating nonindigenous species in marine than in terrestrial ecosystems and the fact that even relatively well known species may not be recognized as nonindigenous. *Sphaeroma terebrans* was described first from Brazil (as *S. terebrans* by Bate in 1866) and later from Florida (as *S. destructor* by Richardson, 1897), and it now occurs intermittently from South Carolina to Brazil (Estevez and Simon 1975; Harrison and Holdich 1984; Carlton unpublished observations). Its "historical" presence and its apparently early "widespread" distribution make it easy to assume that the species is native to South American and southern North American waters, but it is not uncommon for nonindigenous species to be first described or redescribed from a region into which they have been introduced. Well-known examples are the first descriptions of the gammarid amphipod *Gammarus tigrinus* from the North Sea, although it is native to North America (Bousfield 1973), and of the tube-building serpulid worm *Ficopomatus enigmaticus* (= *Mercierella enigmatica*) from France, although it is native to Australia (Carlton 1979b). The shipworm *Teredo clappi*, although it has a Florida type locality (actually a ship's hull), is another probable example (Table 11.1). Ironically, *S. terebrans* was described simultaneously by Bate (1866) from its home waters under a different name, *Sphaeroma vastator*, reflecting the breadth of morphological variation in this species.

Sphaeroma terebrans is widespread through the Indo-Pacific and the Indian Ocean (Harrison and Holdich 1984; Carlton unpublished observations). It is native to those waters and was introduced in the bored hulls of wooden ships in the nineteenth century to tropical and subtropical Atlantic waters (Carlton unpublished observations). This conclusion may surprise some Florida workers, but all discussions of the ecological role of this isopod in Florida omit reference to its evolutionary or biogeographic history, even though earlier isopod workers had already noted that ships probably played a role in its global distribution (Miller 1968).

Carlton (unpublished observations) details the evidence for the introduced status of *S. terebrans* in the Atlantic Ocean and predicts that red mangrove (*Rhizophora mangle*) prop roots from Atlantic paleoestuarine habitats, if available as fossil, subfossil (archaeological), or pre-1850 museum material, will be found to bear no evidence of sphaeromatid isopod bore holes. Florida collections may be particularly fruitful in this regard, as opposed to the more isolated and less well-studied sites in South American or African waters. Carlton (unpublished observations) further predicts that molecular genetic (allozyme and mtDNA) studies will confirm that Atlantic *S. terebrans* differ little if at all from their parental Indo-Pacific stocks.

The evidence that *S. terebrans* is introduced can be briefly summarized as follows. First: Evolutionarily, *S. terebrans* originated in the Indian Ocean region; it is a member of a group of closely related *Sphaeroma* species that are or were found only there, and it has no morphological relatives in the Atlantic Ocean (Harrison and Holdich 1984; Carlton unpublished observations). Second: Biogeographically, outside its broad distribution throughout the Indo-Pacific and Indian Ocean regions, *S. terebrans* occurs in the western Atlantic from South Carolina to Brazil and in the eastern Atlantic from Liberia to the Congo River. It is absent from the eastern Pacific. Like all peracarid crustaceans, it has nonplanktonic larvae, and neither it nor any other coastal marine invertebrate has ever been recorded from drifting wood or any other substrates that have been clearly adrift in the open ocean even for several weeks (as judged, for example, by the age of attached pelagic lepadomorph barnacles). Natural transport from Pacific or Indian Ocean waters to the Atlantic Ocean would involve a transit of many months if not years through oligotrophic, high-salinity, UV-exposed waters. Moreover, *S. terebrans* is not a plate-tectonic (for example, Tethys Sea) relict—no such relict shallow-water marine animal or plant in the Florida-Caribbean region has remained so genetically similar to allopatric populations. The most parsimonious explanation for the distribution pattern of this isopod is that it was introduced in historical time. Third: *Sphaeroma terebrans* is well known to bore into ship bottoms, so a clear mechanism was available for human-mediated transoceanic and interoceanic dispersal. Fourth: Other wood-boring *Sphaeroma* species of the Indian Ocean have demonstrated a propensity for global human-mediated dispersal, so the phenomenon is not unusual in the group. For example, *Sphaeroma annandalei* was introduced into Brazil, *Sphaeroma quoyanum* into Pacific North America, and *Sphaeroma walkeri* into Australia, Hawaii, Hong Kong, and the Atlantic Ocean (Carlton and Iverson 1981; Harrison and Holdich 1984; Morton 1987). *Sphaeroma terebrans* was transported out of the Indian Ocean region to the western Atlantic about a century before *S. walkeri* arrived in South or North America (Carlton and Iverson 1981).

Sphaeroma terebrans occurs in decaying wood, hard-packed sand, and other solid substrates (Harrison and Holdich 1984; Carlton unpublished observations). It also commonly bores into the living prop roots of mangrove trees. Extensive but patchy damage to red mangrove trees in Florida has been noted by Rehm and Humm (1973) and Rehm (1976). An earlier literature debating the ecological and implied evolutionary relationships between mangroves and this isopod (Rehm and Humm 1973; Simberloff et al. 1978; Ribi 1981, 1982) has been reviewed by Perry and Brusca (1989). Rehm and Humm (1973) believed that isopod-bored roots broke off at the high-water line and failed to reach the ground, thus reducing the stability (and productivity) of the mangrove ecosystem. Simberloff et al. (1978), in contrast, suggested that the boring activities of *Sphaeroma* might increase root-tip branching and thus prop-root proliferation. Ribi (1982) concluded that *S. terebrans* may both prevent mangroves from expanding toward the water (by preferentially boring into peripheral root endings) and increase the density of established roots, thereby increasing mangrove stability and potentially helping to reduce soil erosion. In Costa Rica, Perry and Brusca (1989) found that *Sphaeroma peruvianum*, a mangrove-boring isopod, causes a 50 percent decrease in aerial-root growth rate (due to breakage at the root tip caused by boring activity) and that stimulation of new root-tip growth does not compensate for the growth loss. Net root production in bored roots was 62 percent below that in unbored ones. Boring does result in regeneration of multiple sets of root tips but not in a greater number of roots entering the ground per unit time and thus does not increase the stability of trees during storms.

Although considerable work remains to be done on the distributional ecology and spatial extent of mangroves before its arrival, it is clear that *S. terebrans*, where abundant for a long time, may have a substantial impact on the distributional patterns and biomass of mangroves. Rehm (1976) concluded that red mangroves on the west coast of Florida are confined to the upper portion of the intertidal zone by the boring activities of *S. terebrans*, so red mangroves may have extended farther down the shore before its arrival or sometime before the 1850s in coastal Florida.

Shipworms and Gribbles

A number of tropical and subtropical species of wood-boring shipworms and gribbles, noted earlier as common inhabitants of now largely extinct wooden vessels, are said to occur both in Florida (and greater Caribbean and South American waters) and in the Pacific and Indian oceans—a situation reminiscent of the distribution of *S. terebrans*. Shipworm studies (Turner 1966, 1971) and gribble literature (Menzies 1957), although often noting the association of these taxa with wooden ships and piers, have generally implied that these taxa have been primarily distributed by means of natural floating wood.

The contrast, however, between the undocumented arrival of such wood arriving with living shipworms in the western Atlantic from the Pacific or Indian oceans and the documented transits of thousands of wooden ships between these waters moves us to identify certain taxa that are more widespread throughout the Pacific and Indian oceans and less so in the Atlantic as cryptogenic species (Table 11.1). We predict, however, that once morphological, detailed biogeographic, historical, and evolutionary-genetic studies are undertaken on these taxa, they will be found to be introduced to the Atlantic Ocean. If this prediction proves correct, then wood boring in Florida waters has changed dramatically: Hoagland (1986) notes that the shipworms *Lyrodus massa, L. floridanus, L. bipartitus, Teredo furcifera,* and *T. bartschi* and the boring clam *Martesia striata* "were found together in mangrove wood from the Ft. Pierce Inlet (St. Lucie County), Florida"; all but the second species are listed in Table 11.1.

Introduced Marine and Estuarine Plants

Even less information is available on nonindigenous marine plants in Florida's waters than on the marine invertebrates. No cases of invasion have been documented, almost surely because data are inadequate. Modes of possible introduction are numerous and lead us to believe that multiple introductions have probably taken place over the past several centuries. But unless introduced species have a pronounced ecological impact—like toxic dinoflagellate blooms (Hallegraeff and Bolch 1991, 1992) or drastic changes in community composition (Meinesz et al. 1993; de Villele and Verlaque 1995)—they can easily go unnoticed. Even if a tiny fraction of the introductions of marine plant species have resulted in successful establishment of a nonindigenous taxon, a thorough comparison of present-day floras with species lists from the past (Taylor 1960; El-Sayed et al. 1972; Dawes 1974), modified by the caveats noted earlier, should identify a number of nonindigenous marine plant taxa in Florida.

Although no nonindigenous marine plants are known to be established in Florida, a number of notoriously invasive marine plant species occur in Atlantic waters. In Table 11.2 we list macroalgal species that have already invaded habitats outside their native ranges and whose present distributions put them close to Florida. The phytoplankton undoubtedly has nonindigenous members as well, but because almost nothing is known about the microalgae, we focus on macroalgal species. Almost all the taxa listed function normally under a broad range of environmental conditions—for example *Sargassum muticum* can grow and reproduce sexually at temperatures ranging from 10 to 25°C (Norton and Deysher 1989; Ogawa et al. 1990)—and the temperature requirements of all algae on the list would allow them to survive in Florida (Walford and Wicklund 1968; Dawes 1974).

Species with several dispersive life-history stages have shown especially rapid rates of invasion throughout their nonindigenous ranges. For example, vegetative fragments and sexually and asexually produced spores all contribute to dispersal and establishment in *Codium fragile tomentosoides,* which has spread along the Atlantic coast 450 km south to North Carolina over about 30 years (Searles et al. 1984; Carlton and Scanlon 1985). Moreover, taxa such as *S. muticum* that are monoecious and self-compatible should be especially good colonizers of new habitat (Neushul et al. 1992).

A number of the macroalgae in Table 11.2 could have dramatic ecological effects if introduced into Florida. *Caulerpa taxifolia, Codium fragile tomentosoides, Sargassum muticum, Undaria pinnatifida,* and *Polysiphonia breviarticulata* have brought about changes in physical environmental characteristics such as light and nutrient levels (Givernaud et al. 1989; Kapraun and Searles 1990), a decrease in the abundance of other local species of algae and sea grasses (de Villele and Verlaque 1995), and shifts in the community composition of herbivorous invertebrates and fishes (Boudouresque et al. 1994) in areas where they have invaded.

The escape and subsequent spread of *Caulerpa taxifolia* in the Mediterranean Sea led to one of the most thoroughly studied invasions by a marine plant—and the documented ecological impacts are unnerving. What makes the

TABLE 11.2.

Examples of Marine Macroalgae with the Potential to Invade Florida Waters

Species	Origin	Present (invaded) distribution	Mode of establishment[a]
CHLOROPHYTA			
Caulerpa taxifolia	Pacific, tropical Atlantic, Red Sea	Mediterranean, Gulf of Mexico	S, F
Caulerpa scapelliformis	Pacific	West Indies, Mediterranean	S, F
Codium fragile tomentosoides	Japan	W. Atlantic, Mediterranean, E. Pacific, New Zealand	S, F, A
PHAEOPHYTA			
Cystoseira fimbriata	S. Europe	Bermuda	?
Sargassum muticum	Japan	E. Pacific, W. Atlantic	S, F
Undaria pinnatifida	Japan	Atlantic, Mediterranean, Australia, New Zealand	S
RHODOPHYTA			
Antithamnion nipponicum	Japan	Mediterranean, W. Atlantic	S
Bonnemaisonia hamifera	Pacific	W. Atlantic	S
Polysiphonia breviarticulata	E. Atlantic	W. Atlantic, West Indies	S, F
Prionitis sp.	Pacific	Texas	?

Sources: Caulerpa taxifolia (Taylor 1960; Meinesz and Hesse 1991; Boudouresque et al. 1992, 1994; Meinesz et al. 1993, 1994; Directorate General for Environment 1994; Verlaque 1994); *Caulerpa scapelliformis* (Taylor 1967); *Codium fragile tomentosoides* (Carlton and Scanlon 1985; Fletcher et al. 1989; Yotsui and Migita 1989; Bird et al. 1993); *Cystoseira fimbriata* (Taylor 1960); *Sargassum muticum* (Rueness 1989; Espinoza 1990; Fernandez et al. 1990; Givernaud et al. 1990); *Undaria pinnatifida* (Hay 1990; Floch et al. 1991); *Antithamnion nipponicum* (Verlaque and Riouall 1989); *Bonnemaisonia hamifera* (Dixon and Irvine 1977; South 1984; Breeman 1988; Breeman and Guiry 1989); *Polysiphonia breviarticulata* (Kapraun and Searles 1990); *Prionitis* sp. (Wynne 1993).

a Modes of establishment: S = sexually produced spores; F = vegetative fragments; A = asexually produced spores.

C. taxifolia story so useful is that the source and time course of the introduction and spread are known. In 1984, this siphonous green alga was first discovered in Mediterranean waters in front of l'Aquarium du Musée Océanographique de Monaco (Meinesz et al. 1993). The initially observed group of plants consisted of a patch 1 m in diameter at 12 m depth. Five years later, *C. taxifolia* covered 5 ha of the nearshore benthos. After almost a decade of an exponential rate of spread—and despite a massive eradication effort—it was estimated to cover in excess of 1300 ha by late 1993 (Meinesz et al. 1993; Vaugelas et al. 1994).

Like other invasive taxa, *C. taxifolia* has shown much higher densities and rates of spread in its nonindigenous range than throughout its native distribution (Boudouresque et al. 1994). The main ecological impact of the introduction is degradation of extensive sea-grass (*Posidonia oceanica*) meadows and decline of their fish and invertebrate inhabitants (Boudouresque et al. 1992, 1994; de Villele and Verlaque 1995). The alga produces toxic secondary compounds (terpenes) that have deterrent effects on fish, sea urchins, microalgae, and bacteria and that also appear to contribute to its ability to displace sea-grass patches. High dispersal capability (*C. taxifolia* can spread rapidly from vegetative fragments in addition to sexually produced spores; Meinesz et al. 1993), a robust physiological tolerance to a range of environmental conditions (Gayol et al. 1995), and production of secondary compounds noxious to fouling organisms, herbivores, and other macroflora combine to make *C. taxifolia* a formidable invader.

Stemming the Tide of Invasions

What lessons can we extract from carefully studied invasions of marine plants such as *Caulerpa taxifolia* or from the impacts of previously overlooked invasions by marine invertebrates like *Sphaeroma terebrans?*

First, the ecological consequences of an invasion can be dramatic. And in most cases studied thus far, they have not been adequately understood until after the nonindigenous species is well established. We do not have enough information on most invading marine organisms to know whether their ecological impacts can be predicted (and thus avoided before the invasion has gotten out of control). What we need is far more basic biological and ecological research on the impacts of nonindigenous species in marine and estuarine communities. Genetic analyses (Francis et al. 1989; Goff et al. 1992; Geller et al. 1994; Van Oppen et al. 1995) and experimental studies addressing survival of nearshore algae and invertebrates under oceanic (high seas) conditions will be critical to our ability to stem future introductions.

A second important message we may take from the histories of *Caulerpa taxifolia* and *Sphaeroma terebrans* is that the potential and realized ecological consequences of establishment of nonindigenous marine organisms in Florida waters should be ample motivation for a careful evaluation of the safeguards in place designed to prevent such introductions. There is also abundant evidence for harmful social and economic consequences of invasions by nonindigenous marine organisms—ranging from impacts on recreation in nearshore areas (for example, *Polysiphonia breviarticulata* blooms; Kapraun and Searles 1990) to economic ramifications (for example, the extensive depredations of fouling and boring organisms; Woods Hole Oceanographic Institution 1952; Turner 1966) to effects on human health (for example, lethally toxic introduced dinoflagellates; Hallegraeff and Bolch 1991, 1992).

Of the numerous possible mechanisms that would facilitate the transport and release of nonindigenous marine organisms into Florida, all have been implicated in invasions by nonindigenous species in other areas. Carlton (1989, 1992a) has reviewed many examples among marine invertebrates. Increasing mariculture activities in Florida in particular could lead to accidental releases. Indo-Pacific penaeid shrimp, for example, including the tiger shrimp *Penaeus monodon* and the white shrimp *Penaeus vannamei* (Wenner and Knott 1992), released from shrimp aquaculture facilities have been trawled by commercial fishermen in southeastern U.S. waters.

There are many likely sources of marine plants nonindigenous to Florida waters:

- Fouling on ships' hulls (for example, *Undaria pinnatifida* sporophytes, which have survived a four-week ocean voyage; Hay 1990)

- Transport in ballast water (for example, *U. pinnatifida*, Sanderson 1990; numerous dinoflagellate species, Hallegraeff and Bolch 1991, 1992)

- Arrival with aquaculture packing material or on bivalve spat for commercial culture (for example, *U. pinnatifida, Antithamnion nipponicum, Sargassum muticum,* Rueness 1989; Verlaque and Riouall 1989; Floch et al. 1991)

- Escape from field aquaculture (for example, *U. pinnatifida* and S. *muticum,* Floch et al. 1991; Neushul et al. 1992)

- Escape from aquaria (for example, *Caulerpa taxifolia,* Meinesz et al. 1993)

Regulation of the potential sources of marine plant and animal introductions into Florida is not coordinated by a single agency: it comes, rather, through a number of indirect avenues (Brown 1994). For plants, for example, the Florida Department of Environmental Protection has regulatory control over permit requests for transplanting or introducing marine species for restoration, mitigation, or commercial purposes under Florida Statutes Chapter 370.081 (Futch and Willis 1992; D. Schmitz, M. Durako, and S. Adams, pers. comm.). At present, commercial sales of marine plants to public aquaria or hobbyists are not regulated. Indeed, the Florida Department of Agriculture disseminates a booklet entitled "Florida Aquatic Plant Locator" (Florida Department of Agriculture and Consumer Services 1993) that lists commercial sources of freshwater and marine plants for sale throughout the state. The most recent edition contains a number of supposedly native marine macroalgae for sale—including two *Caulerpa* species—and four nonindigenous phytoplankton taxa. There are no regulatory mechanisms to prohibit future sales of nonindigenous taxa (N. Coile, pers. comm.).

Prospects

As is typical of most subtropical and tropical shores around the world, the history, diversity, and ecological consequences of biological invasions remain poorly known in Florida. Centuries of shipborne transport have obscured the natural patterns of distribution of many neritic species, leading to the common view that many species are "naturally" widely distributed, if not cosmopolitan, and to overlooking of many, and perhaps in Florida most, introductions. That such overlooked invasions are not restricted to obscure taxa is well illustrated by the case history of the introduced mangrove-boring isopod *Sphaeroma terebrans*.

That future invasions by nonindigenous marine organisms will occur in Florida seems inevitable, as does the need to stem this tide. These invasions will occur by the "natural" progression of invasive species now approaching Florida—such as the nonindigenous South American brown mussel *Perna perna*, which is making its away along the Gulf of Mexico coast (Hicks and Tunnell 1993)—and by the release of organisms through the mechanisms noted earlier. To understand the magnitude of past marine invasions in Florida (and thus their ecological, economic, and social consequences) and to lay the foundation for preventing future invasions, we need basic biological and ecological research in three areas:

- Systematic studies at both the community level (to detect changes in species composition over time and thus to identify candidate nonindigenous species) and the species level (to determine probable areas of endemicity, dispersal tracks, and the timing of introductions)

- Genetic and ecological research and studies on the temporal and spatial patterns of human-mediated dispersal mechanisms—such as ballast water and mariculture—all designed to pinpoint the most likely sources of current and future invasions and identification of some of the species potentially involved

- Experimental studies on the ecological impacts of nonindigenous marine and estuarine organisms

Such studies will provide better information about the past history of invasions, how they have altered Florida's coastal ecosystems, and, as with *Sphaeroma,* how our perceptions and understanding of these ecosystems are in turn influenced. This work, combined with research on probable modes of introduction and the anticipated effects of nonindigenous species both on "natural" communities and on those that have already sustained invasions, is a necessary precursor to formulation of robust strategies for the prevention of future introductions and for their control if they have already become established.

ACKNOWLEDGMENTS

We thank John R. Chesler and Mary R. Bukstel for sharing their collection data and experiences for the Pacific pearl oyster *Pinctada* in Florida waters, David Camp and Wayne Price for helpful discussions on the biology of the crab *Platychirograpsus,* Rafael Lemaitre for access to his data on the crab *Charybdis,* and Henry Lee for information on the appearance of the Venezuelan mussel *Mytella.* Comments by two anonymous reviewers, and by Daniel Simberloff, Michael Beck, Dennis Hanisak, Mike Durako, and Clinton Dawes, permitted us to clarify and expand upon certain points.

PART III

Managing Nonindigenous Species: Strategies and Tactics

12 Ecological Restoration

John M. Randall, Roy R. Lewis III,
and Deborah B. Jensen

Efforts to control nonindigenous species in natural areas
are rarely, if ever, undertaken solely to eliminate or reduce
their populations. The ultimate goals, rather, whether
stated or not, are to allow native species and communities
to replace the pest and improve habitat for wildlife. In
most natural areas, pest control is therefore best viewed as
an integral part of a comprehensive restoration program.
Ecological restoration is here defined as "returning an
ecosystem to a close approximation of a pre-existing condition (usually that prior to an identifiable disturbance or
stress)" as opposed to "returning to a pristine condition"
(Lewis 1990; National Research Council 1992). This
perspective—emphasizing what is being managed *for*—
helps to ensure that ends and means do not become con-
fused and that pest-control activities do not degrade populations and communities of native species more than the
pest does (Schwartz and Randall 1995). It also reflects a
lesson many restorationists have learned through failed
projects: impacts of nonindigenous pests must be evaluated during planning and then addressed if the success of
native-species plantings and other actions is to be ensured
(Baird 1989; Berger 1993).

Ecological restoration becomes necessary when a disturbance or impact damages a site in a way likely to be
irreversible without intervention (Cairns 1986). Many aggressive nonindigenous species invasions meet these criteria, but the type of intervention necessary depends on
the factors that allowed the invasion and the changes it has
caused. For example, certain nonindigenous species invade sites that do not seem to have been otherwise

disturbed. These invasions can sometimes be addressed directly by removal or sharp reduction of the pest population and through measures to prevent its reestablishment. Effective prevention entails systematic checks for new incursions, barriers or fences, and cooperation with neighboring landowners to eliminate nearby sources of invasion. At the other extreme are invasions made possible only by disturbances or large-scale alterations of ecosystem processes. Changes in hydrological regimes and suppression of wildfire, for example, have promoted nonindigenous species invasions in many biological communities in Florida and elsewhere. (See, for example, Loope and Dunevitz 1981a; Ewel 1986; Moyle 1986.) Alterations of this sort are often the root cause of a suite of problems, including nonindigenous species invasions, that at first appear unrelated. Restoring natural processes, when possible, is the logical first step in these situations and may deprive invasive nonindigenous species of their competitive advantages, leading to reduction or elimination of their populations without further direct action. Of course, real restoration projects are usually more complicated than these hypothetical situations, involving multiple disturbances, fragmented landscapes, and invaders that cause secondary impacts, kill plantings of native species, and prevent reestablishment of natural processes.

Restoration projects planned without attention to these complexities and especially to the impacts and reproductive biology of nonindigenous species often fail, wasting effort and money (Clark 1994; Thayer and Ferriter 1994). In some cases they actually result in an increase in nonindigenous species populations (Girardin 1994). For example, hydrological manipulation and soil disturbance undertaken to prepare sites for native species plantings often encourage colonization by aggressive plants that may outcompete native species (Hobbs 1991; Hobbs and Huenneke 1992). Small "piecemeal" projects often fail when their areas are reinvaded from large sources of propagules on nearby sites that were not adequately considered and addressed.

Successful control of nonindigenous species is far more likely for projects that adopt a regional approach and address possible sources of reinvasion. One method is to use a "target" and "bull's-eye" model, in which the "bull's eye" is the restoration site and the "target" is an area surrounding it, a buffer zone free of nonindigenous species. The target may include both public and private lands. In some cases it may be possible to expand the target treatment area with time so that nonindigenous species are methodically removed from an increasing area around the restoration site (see Lewis 1994: fig. 4.2f). Experience indicates that once the problem is explained, private landowners may cooperate by removing nonindigenous species from their property or allowing public agency workers (or volunteers) to do so. Maintaining a buffer can reduce long-term costs of keeping the bull's-eye free of nonindigenous species. If significant impediments to controlling nonindigenous plants in the target area are identified during planning, however, there are two options: the entire effort may be abandoned, or additional funds can be budgeted for policing the restoration site in the future so that incipient reinvasions can be located and eliminated. Undertaking this kind of analysis will help to eliminate "feel good" control programs destined to fail in the long term.

Success of restoration projects that include nonindigenous species control should be judged by more than reduction in nonindigenous species numbers or areal cover. Although it is often difficult to determine what measures are most appropriate, they should be based on the project's ultimate goals and capable of indicating whether those goals have been attained (Schwartz and Randall 1995). Measuring the success of restoration projects, and of natural-area management in general, has been the subject of lively interchange in the ecological and conservation biology literature. For example, a recent issue of the journal *Restoration Ecology* dedicated to the Kissimmee River restoration contains several articles that propose a variety of measures ranging from population parameters of plants, insects, and fish to ecosystem process parameters (Dahm et al. 1995; Harris et al. 1995; Toth et al. 1995; Trexler 1995).

Florida's ecosystems have suffered greatly from nonindigenous species invasions, but they have also inspired some of the most ambitious and innovative restoration projects yet attempted. Here we describe several of these projects to illustrate the variety of problems nonindigenous species invasions cause during restoration and the necessity of including programs to identify and manage such species as integral parts of restoration projects.

The Hole-in-the-Doughnut

One of the most ambitious and expensive restoration projects on Florida's public lands to date was undertaken on a small portion of the Hole-in-the-Doughnut, a roughly 4000-ha area of previously farmed land in Everglades National Park (Doren et al. 1990). The aptly named Hole-in-the-Doughnut was surrounded by, but not included in, Everglades National Park when the park was established in 1934. Portions of the area were purchased by the park as they were abandoned by farmers, and the final piece was added in 1975 when all farming in the Hole-in-the-Doughnut ceased (Doren et al. 1990). Much of the area was quickly invaded by aggressive nonindigenous plants after cultivation ceased, especially where farmers had "rock plowed" the substrate. Rock plowing, a technique developed in the 1950s, crushes the upper 10 to 20 cm of limestone bedrock to create an artificial soil better suited for crops (Ewel et al. 1982). It created soil where there had been little or none before and changed what had been low-nutrient anaerobic conditions to higher-nutrient, aerobic conditions (Orth and Conover 1975). These changes are believed to have increased the sites' susceptibility to invasion and dominance by nonindigenous plants, particularly Brazilian pepper (*Schinus terebinthifolius*) (Ewel et al. 1982; Doren et al. 1990). Portions of the Hole-in-the-Doughnut that were cultivated, but not rock plowed, reverted primarily to native vegetation in which small, scattered areas were dominated by Brazilian pepper (Ewel et al. 1982).

Upon acquiring the Hole-in-the-Doughnut, the U.S. National Park Service recognized it faced the related problems of reestablishing native vegetation and preventing invasion by exotics (Doren et al. 1990). Information was not available on successional trends on abandoned rock-plowed land in southern Florida, but it was hoped that native species would soon push out invasive nonindigenous species. A monitoring program determined that Brazilian pepper instead colonized and then eliminated herbaceous vegetation. This species reached low densities 5 to 10 years after abandonment and then increased in density by up to twentyfold per year until it achieved nearly continuous cover. After roughly 20 years, the stands began to thin, leaving fewer, larger plants (Loope and Dunevitz 1981b; Doren and Whiteaker 1990a). The invasion continued to spread into adjacent areas that had never been rock plowed as Brazilian pepper seed rained down on them (Ewel et al. 1982).

Studies in other areas also found that nonindigenous species which invade and become established in disturbed areas can, with time, spread into adjacent undisturbed habitats. This spread may result from "infection pressure"; at least a few of the large numbers of propagules shed into adjacent areas year after year will land in open sites under favorable conditions, even in habitats where appropriate sites and conditions are rare (Salisbury 1961; Ewel 1986). Moreover, natural selection on populations of invaders in disturbed areas may, in time, yield more competitive and invasive individuals (Baker 1965). Where these phenomena are suspected, nonindigenous species should be controlled on sites that might otherwise be of little conservation value in order to prevent them from becoming "staging areas" for the invasion of more valuable habitats nearby (Ewel 1986).

Attempts to restore portions of the Hole-in-the-Doughnut began in 1972. During the following decade a variety of methods were employed, but most were ineffective in the long term. Prescribed fire was effective for the control of smaller Brazilian pepper trees surrounded by heavy accumulations of fuel but ineffective for larger trees or for preventing invasions of large areas (Doren and Whiteaker 1990b). One partially successful project was the creation of artificial wetlands: soil was removed from one area and mounded in another to create uplands (LaRosa et al. 1992). In this case, increased hydroperiods and removal of the soil in the wetlands appeared to prevent establishment of Brazilian pepper (Doren et al. 1990). Another project involving removal of artificially created soils from sites outside the park that had been illegally rock plowed also demonstrated that this technique could prevent reestablishment of Brazilian pepper (Dalrymple et al. 1993).

To evaluate the effects of substrate removal systematically—particularly the relative importance of soil removal and increased hydroperiods to vegetation dynamics—park researchers initiated a project on a rock-plowed site in the Hole-in-the-Doughnut in 1989 (Doren et al. 1990). The existing vegetation, dominated by Brazilian pepper, was bulldozed from a 24.3-ha area and burned. Soil, originally created by rock plowing, was scraped down to the bedrock limestone on an 18.2-ha section of the site and removed from the park. On the remaining 6.1 ha, the soil was partially removed, leaving roughly 2 cm above the limestone bedrock (Doren, pers. comm.). Succession was allowed to proceed, and vegetation, soil mycorrhizae, and water levels were monitored. Within the first year, wet-prairie vegetation was regenerating on the entire site and 54 percent of the species present were classed as wetland species. Brazilian pepper rarely got established where all the soil had been removed but did so far more often where it had been partially removed (Doren et al. 1990). In fact, Brazilian pepper densities on the latter section were similar to those found on newly abandoned rock-plowed sites (Doren et al. 1990). The partial-removal site will probably be dominated by Brazilian pepper in a few years unless further control measures are taken.

Although the project demonstrated that removal of rock-plowed substrate encourages the establishment of wet-prairie vegetation on large acreages, it was also labor intensive and costly. Approximately 68,700 m^3 of rock-plowed substrate was removed from the 24.3-ha site in 3437 dump-truck loads at a total cost of $640,000 (Doren et al. 1990), borne entirely by a private developer as mitigation for the development of wetlands elsewhere in Dade County. In the autumn of 1995 the National Park Service began preparing a request for proposals to restore a 200- to 400-ha section of the Hole-in-the-Doughnut with this technique. Such restoration of the entire 2000 ha of rock-plowed land is anticipated to take 20 years and cost an estimated $37,000 per hectare plus expenses for research and monitoring that may total more than $20 million (Doren et al. 1990; R. Doren, pers. comm.).

Endangered Lands in Hillsborough County

Another restoration project, on a 264-ha site near Tampa Bay, was stalled by Brazilian pepper invasions. Unfortunately, the invasion was ignored in its early stages, when it could have been controlled with relative ease, while plans for the site were developed. The site was purchased by the Hillsborough County Endangered Lands Acquisition and Protection (ELAP) Program in 1991 for $2 million (Beatley 1991; Lewis 1994: figs. 4.2a–c) and lies 5 km north of a

1050-ha private habitat restoration effort undertaken by Tampa Electric Company. When the site was purchased, row crops and recently abandoned limestone and shell quarries occupied 144 ha; mangrove, tidal marsh, and saltern habitats occupied the remaining 120 ha. The ELAP Program does not normally buy such disturbed lands, but local interests, including the National Audubon Society and the Surface Water Improvement and Management (SWIM) Program of the Southwest Florida Water Management District, persuaded the county to purchase it with the goal of restoring it to native or seminative plant communities. The SWIM Program committed a total of $1.65 million to the project, including matching grants (SWIM 1990, 1992).

Restoration efforts bogged down early in the planning phase when disagreements about "what to restore" arose. Some argued that upland areas should be restored to a mosaic of the upland and freshwater wetland communities that had existed there historically. Others wanted to scrape soils from the uplands in order to reduce their elevation and create mangrove forests. The Florida Game and Fresh Water Fish Commission mediated a settlement stipulating that most upland areas on the site should be restored to upland and shallow freshwater plant communities (Lewis 1994: fig. 4.2d).

The county intended to keep the agricultural fields cultivated to keep them free of invasive nonindigenous plants, but during the debate the farmers elected not to renew their leases. As no money was available to manage the fields, they were quickly revegetated by a mixture of nonnative and native "old field" plant species including Brazilian pepper, dog fennel (*Eupatorium* spp.), and indigo (*Indigofera* spp.). Brazilian pepper dispersed into the fields from unmowed drainage strips along their edges just as it had in smaller fields on the property abandoned earlier.

Immediate implementation of a control program could have prevented further spread of the Brazilian pepper and other nonindigenous species until action on the restoration project was initiated. But SWIM decided that although *wetland* restoration was an appropriate use of the $1.65 million, control of nonindigenous species on uplands was not—particularly if the species was an invader of marginal wetlands like Brazilian pepper. Two years passed while the project cooperators applied to various sources for funds to control the pest plants. Meanwhile, the county used the limited funds and labor at its disposal to conduct small herbicide applications and prescribed burns on the site, but populations of Brazilian pepper and other invasive nonindigenous plants continued to expand. At the same time, the restoration project was on hold and none of the $1.65 million was spent. In July 1994, Hillsborough Community College was granted funds to implement a Brazilian pepper and Australian pine (*Casuarina* sp.) control program on much of the site *outside* the limits of the farm fields. By autumn 1996, the invasions on the old fields still had not been addressed in an organized fashion.

Blowing Rocks Preserve

A series of restoration projects designed to remove a variety of nonindigenous plants and restore native vegetation has been carried out at The Nature Conservancy's Blowing Rocks Preserve on Jupiter Island (Martin County). The preserve encompasses 30 ha and has 1.5 km of shoreline on the Atlantic Ocean and the Indian River Lagoon. It is an important nesting site for the federally listed endangered leatherback sea turtle (*Dermochlys coriacea*) and the federally listed threatened loggerhead (*Caretta caretta*) and green (*Chelonia mydas*) sea turtles. The property was dominated by nonindigenous plants when The Nature Conservancy acquired it in 1968. An invasion by Australian pines (*Casuarina equisetifolia*) had severely altered the dynamic dune ecosystem along the beach. This species can establish itself rapidly, increasing in height by up to 5 m and producing over 10,000 wind-borne seeds each year. There was concern that, by shading the beach, the trees lowered the temperature of the sand where incubating sea turtle eggs were buried, promoting the production of a higher ratio of males to females, which is detrimental to population increase. Moreover, exposed roots of the Australian pines trapped and entangled adult sea turtles when they came ashore to nest.

A goal of restoring native vegetation to the entire site was established, but the work was divided into smaller projects. Each focused on a specific nonindigenous species population or portion of the preserve and depended heavily on volunteer labor. This approach has been extremely successful at this small site located near urban areas that are a source of numerous volunteers but would almost surely be inadequate on large, remote sites.

The first project at Blowing Rocks, initiated in 1985, involved cutting down Australian pines in a 2.5-ha area along the dunes and coastal strand (Steve Morrison, pers. comm.). The Australian pines were introduced on the island as windbreaks and landscape plants in 1916 and by 1985 formed a nearly monospecific stand with little or no understory. Two years later, The Nature Conservancy contracted Post, Buckley, Schuh & Jernigan, Inc. (PBS&J) to assist with restoration of a 6-ha area on the east side of the preserve. PBS&J used chainsaws, a backhoe, and other heavy equipment to remove Australian pines and other exotics. An on-site incinerator was used to burn cut brush and trees in order to minimize dispersal of their seeds and eliminate biomass.

Because of the invasiveness of Australian pine, Brazilian pepper, and the other nonindigenous species on the site, cleared areas were revegetated to deter reinvasion. Between 1985 and 1987, volunteers helped plant more than 60,000 individuals of 85 native species on the 6-ha restoration site (D. Gordon, pers. comm.). Sea oats (*Uniola paniculata*) were planted on the seaward side of the first dune, and mature saw palmettos (*Sabal palmetto*) taken from the site of a planned development nearby were planted in the dunes with other native species. Mangrove seedlings were planted in the wetlands, and native grasses and herbaceous species were established in adjacent areas. The site was policed to ensure quick removal of any emerging nonnatives.

In 1991 a project to restore a 15-ha site on the western side of the preserve was begun. Like the earlier restoration effort, it involved removal of large numbers of aggressive exotics followed by plantings of native species. This time, however, much of the labor was provided by volunteers who met monthly from October through May. Volunteers also collected native plants from nearby areas slated for development and helped propagate native plants from local stock in the preserve's nursery.

Direct costs associated with the restoration from 1985 through 1988 were $252,707, not including staff time, operating expenses, or the value of the volunteer labor. The preserve added a volunteer coordinator, a restoration coordinator, and a native-plant nursery in 1991 for an additional cumulative expense of $111,360. In 1990, the preserve began to document the volunteer hours spent on the restoration effort. Since 1990, some 1810 volunteers have worked a total of 8021 hours on the project. This volunteer time is valued at $114,700 ($14.30/hour based on information from Thousand Points of Light Foundation).

Volunteers continue to help maintain restored areas by removing nonindigenous species that appear there (N. J. Byrd, pers. comm.). Because populations still persist outside the preserve, vigilant monitoring for reestablishment will remain important. Species they encounter frequently include lather leaf (*Colubrina asiatica*, a woody vine in the Rhamnaceae), and *Wedelia trilobata*, a composite introduced as a ground cover.

Restored Phosphate Mines

Large restoration projects have been undertaken on some sites mined for phosphorus in Florida. More than 104,900 ha was mined, primarily in the central Florida mining district, which covers 526,000 ha (Albin 1994). Although mine owners are not legally required to reclaim lands mined before 1 July 1975, about 60 percent of the roughly 60,300 ha mined before this date qualifies for reclamation funding through the state's nonmandatory reclamation program. Mine owners are legally required to reclaim lands mined after this date (roughly 44,600 ha; Albin 1994). Reclamation is here defined as "rehabilitation of land disturbed by resource extraction to a beneficial use while providing for the protection of the environment" (Florida Statute 378.402(2); A. B. Whitehouse, pers. comm.). The result of reclamation may be ecological restoration, creation of new habitats that did not exist on the site (often referred to as "replacement"), or some combination of the two. Phosphorus-mine reclamation projects are required to attempt hectare-for-hectare, type-for-type, restoration of wetlands, rehabilitation of at least 10 percent of damaged upland forests, and replacement of lost wildlife values (Albin 1994). Legal standards for these projects require that indigenous species be used for revegetation except where noninvasive agricultural crops, grasses, and temporary ground covers are planted.

Lands mined before 1975, where reclamation is not required, often harbor invasive nonindigenous species, and areas that have been colonized for long periods often harbor huge nonindigenous seed banks (Albin 1994). Such sites are often close to more recently mined areas and serve as seed sources for newly disturbed or reclaimed lands. State inspectors commonly find cogon grass (*Imperata cylindrica*), lantana (*Lantana camara*), castor bean (*Ricinus communis*), Brazilian pepper, hydrilla (*Hydrilla verticillata*), water hyacinth (*Eichhornia crassipes*), water lettuce (*Pistia stratiotes*), and tropical soda apple (*Solanum viarum*) invasions on large expanses of mined and reclaimed land (Albin 1994). Of these, cogon grass is generally the most serious biological threat to successful reclamation. It thrives in the heavily disturbed and open habitats characteristic of newly reclaimed areas such as sand tailings, overburden (material overlying a mineral deposit), and clay fill. It is extremely difficult to kill and responds positively to prescribed burns (Snyder et al. 1990).

Water hyacinth and water lettuce are commonly found in mine water-recirculation systems, but these sites are usually altered during reclamation (Albin 1994). Hydrilla is common in lakes constructed on reclaimed phosphate mines. Control of these invasions is coordinated with the Florida Department of Environmental Protection's Bureau of Aquatic Plant Management when hydrilla levels become excessive.

Several other nonindigenous plants are relatively minor problems on reclaimed sites (Albin 1994). Lantana occurs throughout central Florida but does not appear to affect reclamation sites substantially. Castor bean is found in concentrated locations but is easily treated and replaced by native species. Brazilian pepper appears to be spreading from the region east of the Peace River (Polk to Charlotte counties) westward into the phosphate district. It has been successfully treated and eliminated in a few areas but will probably become a serious problem in larger, ongoing wetland restoration projects. Tropical soda apple has thus far been a relatively minor problem but is spreading rapidly in the mining district. It may have been introduced at several sites in either seed mixtures or hay bales used to control erosion (Albin 1994).

The Kissimmee River and Lake Okeechobee

The Kissimmee River (Osceola and Okeechobee counties) restoration, the largest river restoration ever initiated anywhere, is intended to restore roughly 70 km of river channel and 11,000 ha of wetlands (Koebel 1995). When the Kissimmee was channelized and isolated from its floodplains by work carried out between 1962 and 1971, several invasive nonindigenous plant species colonized the canal and remnant river course (Koebel 1995). The most troublesome are a familiar threesome: water hyacinth, water lettuce, and hydrilla. Of these, hydrilla is the only rooted, submerged species and the only one likely to persist in much of the restored river (Toth et al. 1995). Reestablishment of continuous flow is expected to flush the other two species from most of the river, because both are generally free-floating and unattached, but they will probably remain abundant in slack-water areas such as oxbows and other backwater habitats (Toth et al. 1995). Nonindigenous plants that invaded the Kissimmee River floodplain while it was isolated from the river and drained include Brazilian pepper, guava (*Psidium guajava*), cogon grass, and Bahia grass (*Paspalum notatum*). Toth et al. (1995) note that all of these species are intolerant of prolonged inundation and will probably persist only in the wet prairie zone at the margins of the restored floodplain, where hydroperiods will be relatively short. Active control programs may be necessary there.

The fish fauna of the Kissimmee was relatively depauperate before the river was channelized, and, although the process severely disturbed fish populations, no fish species are known to have been lost, and relatively few nonindigenous fishes have invaded (Trexler 1995). Four nonindigenous fishes are present—the walking catfish (*Clarias batrachus*), blue tilapia (*Oreochromis aureus*), grass carp (*Ctenopharyngodon idella*), and common carp (*Cyprinus carpio*)—but none of them is abundant. In contrast, aquatic habitats further south in Florida have been invaded by many nonindigenous fishes (Trexler 1995). Because nonindigenous fishes are generally more successful in altered habitats (Ross 1991), Trexler (1995) suggests the Kissimmee restoration may help prevent invasions that the unrestored river would not have been able to resist.

The Kissimmee River empties into Lake Okeechobee, which has become eutrophic in recent decades as a result of nutrient loading from agricultural runoff. Modeling studies indicated that a reduction in nutrient loading, particularly phosphorus loading, will be needed to return the lake to a stable and desirable trophic level (Davis and Reel 1983). In the early 1980s the South Florida Water Management District examined a variety of strategies and chose to attack the problem in two ways: by reducing nutrient inputs and by removing nutrients from the lake itself (Mericas et al. 1990). In 1987, the Lake Okeechobee Aquatic Weed Harvesting Demonstration Project began physical removal of large quantities of aquatic vegetation in an effort to reduce concentrations of nutrients significantly. Aquatic macrophytes were mechanically cut and removed from over 150 ha near the northern shore of the lake in 1987–1988 (Mericas et al. 1990). Hydrilla, a nonindigenous species, made up the bulk of the material removed (more than 95 percent). Careful records were kept on the acreage cleared and the wet weight of the vegetation removed, and samples of it were analyzed for phosphorus, nitrogen, potassium, magnesium, and calcium. As these data indicated that on average each harvesting machine in use removed 11.7 kg of phosphorus from the lake per day (Mericas et al. 1990), in its earlier stages this project was viewed as a success. It later became clear, however, that mechanical harvesting simply could not remove enough nutrients to produce the desired effect (A. Ferriter, pers. comm.), so the project was discontinued.

Impacts of Vines

Vining species are among the most troublesome invaders in Florida's natural areas and throughout much of the United States east of the Plains. Despite their common morphology, invasive vines display a variety of life histories and ecological characteristics and differ in both their impacts on restoration projects and their responses to management actions. As the following examples indicate, some vines are especially troublesome following natural disturbances like hurricanes, some are easily controlled with fire, whereas others apparently increase after fires or burn dangerously and promote spot fires.

Nonindigenous vines quickly overtopped many tropical hardwood hammocks damaged by Hurricane Andrew and were assumed to be the greatest threat to recovery of the hardwood stands. Five species particularly aggressive in hammocks managed by the Natural Areas Management Division of the Dade County Parks Department were sewer vine (*Paederia cruddasiana*), air potato (*Dioscorea bulbifera*), two jasmines (*Jasminum dichotomum* and *J. fluminense*), and wood rose (*Merremia tuberosa*) (McHargue 1994). Park managers began systematically removing these and other nonindigenous vines from Matheson Hammock, Snapper Creek Hammock, Castellow Hammock, and the hammock at the Charles Deering estate roughly six months after the storm. The vines are cut and piled where they cannot reroot, and the exposed stumps are treated with herbicide. A study conducted in three of these hammocks indicates that, as anticipated, recovery and establishment of native species was much better on plots from which nonindigenous vines were removed (C. Horvitz, pers. comm.). Additional work in Castellow Hammock on plots established before the storm, at sites that supported high and low densities of nonindigenous vines, indicates that poststorm vine densities are correlated with prestorm densities (Molnar and Randall, unpublished data). Areas invaded before the storm are now blanketed with wood rose and jasmine, whereas areas with low densities before the storm are regaining native cover rapidly. The control plots for these studies are virtually the only areas from which nonindigenous vines have not been systematically removed at least once on the four hammocks listed here and on several other hammocks managed by the county.

Skunk vine (*Paedaria foetida*) invaded sandhill forest on the Janet Butterfield Brooks preserve in Hernando County but has apparently been brought under control with prescribed burns (Gann and Gordon 1994). Fire had been excluded from the site historically, allowing hardwoods to dominate. In 1992 prescribed burns were initiated to promote the return of pine dominance and reduce skunk vine densities. Prior to burning, monitoring indicated a mean of 13.2 skunk-vine stems around each tree sampled. The first burn was conducted in early March, and a month later no surviving skunk vine stems were found around sample trees. One year later, a mean of 0.61 stem per tree was recorded, a statistically significant decrease ($p < 0.05$) of 95 percent (Gann and Gordon 1994). During the same period densities of vines around trees in unburned control areas did not change significantly. Treated areas were again burned in late winter of 1993, resulting in a further 52 percent decrease in skunk-vine numbers around sample trees (Gann and Gordon 1994).

Burns conducted in late winter are effective against skunk vine apparently because its perennial stems are vulnerable to fire at that time (Gann and Gordon 1994; Gordon, pers. comm.). Prescribed burns at Janet Butterfield Brooks preserve also appear to have reduced densities of some native vines with perennial stems but may be promoting the invasive nonindigenous air potato, which has stems that die back each winter at this latitude. Air potato's stems emerge later in the spring and may actually benefit when prescribed burns conducted earlier in the season reduce competition from other vines.

An invasion by climbing fern, *Lygodium microphyllum,* in Jonathan Dickinson State Park (Martin County) interferes with the prescribed burn program implemented to restore communities there. This vining fern produces dense clusters of finely divided fronds that are highly flammable when dry. It climbs to the canopy in cypress swamps, providing "ladders" for fire and thus promoting crown fires in communities where they had been rare (Roberts and Richardson 1994). Flaming clusters of the lightweight fronds sometimes break away from the vines and blow to ignite other canopy trees (J. B. Smith, pers. comm.). Research is currently under way to identify the most appropriate methods to control this species.

Impacts of Nonindigenous Animals

We were unable to find well-documented cases in which control of nonindigenous animals was carried out in the context of a restoration program or where

12 . ECOLOGICAL RESTORATION

they hindered restoration. Although negative impacts by several nonindigenous animals are widely recognized, impacts on native fauna at restoration sites may have been overlooked because most projects emphasize rehabilitation of native plants and vegetation and monitor these more closely than the fauna.

Effects of feral pigs (*Sus scrofa*) are obvious in many of Florida's natural areas, where their rooting disturbs vegetation and soils on sites covering several hectares. The impacts of such rooting on restoration efforts in Florida are not clear, but there are anecdotal reports that it changes plant species competition on some sites. For example, feral pig rooting after prescribed burns in seeps and depression wetlands that supported populations of rare cutthroat grass (*Panicum abscisum*) at Tiger Creek preserve (Polk County) near Lake Wales led to the replacement of this species with the more common native red root (*Lachnanthes caroliniana*) (D. Gordon, pers. comm.). Clearer evidence is available from outside Florida. An exclosure study in South Carolina found that feral-pig herbivory was responsible for crop failure in a naturally regenerating stand of logged longleaf pine, *Pinus palustris* (Lipscomb 1989). After two growing seasons, fenced areas had the equivalent of 200 longleaf pine seedlings per hectare large enough to resist ground fires; the unfenced areas had three. Impacts of feral pigs are especially severe in Hawaiian forests, which had no mammalian grazers before the arrival of humans. Exclosure studies there demonstrated that pig digging and herbivory affect nutrient cycling, reduce regeneration by many native woody plant species, facilitate invasions by nonindigenous plants, devastate soil microarthropod communities, and reduce the diversity of native bird and terrestrial insect species (Spatz and Mueller-Dumbois 1975; Loope and Scowcroft 1985; Stone 1985; Vitousek et al. 1987; Stone et al. 1992; Vtorov 1993). In fact, removing pigs and preventing their return are the most important components of most forest restoration programs in Hawaii (A. Holt, pers. comm.).

Scattered populations of the nonindigenous rodent nutria (*Myocastor coypus*) are present in northern and central Florida (Lowery 1974). We found no evidence that they significantly depress tree regeneration in the state, but in Louisiana, where they are widespread, nutria herbivory severely depresses or prevents bald-cypress (*Taxodium distichum*) regeneration (Blair and Langlinais 1960; Conner and Toliver 1987, 1988; Conner and Day 1989; Conner and Flynn 1989). Nutria kill bald-cypress seedlings by clipping them near the soil surface or pulling them up to feed on the bark of the taproot and root collar. They caused heavy mortality in three bald-cypress plantings, even when "Vexar" plastic guards were installed around the seedlings (Conner and Tolliver 1987). The patterns of damage suggest that the intensity of herbivory is related to the proximity of nutria resting mounds. The combined impacts of hydrological changes and nutria herbivory are, in fact, so severe in Louisiana's wetlands that bald-cypress swamps may slowly fade from the state as older trees die and fail to replace themselves (Conner and Tolliver 1990). Bald-cypress and pond-cypress (*T. distichum* var. *nutans*) stands in Florida, however, appear to be replacing themselves vigorously, even at recently clearcut sites (Ewel et al. 1989).

The imported red fire ant (*Solenopsis invicta*) and a few other nonindigenous insects have had devastating impacts on natural areas in Florida (Porter et al. 1988; Tschinkel 1993), but again there is little information on their role in restoration work. There is concern, however, that the cactus moth (*Cactoblastis cactorum*), which recently became established in Florida, may interfere with plans to reestablish populations of a rare prickly pear cactus, *Opuntia spinosissima*, in the Keys. The cactus moth was used with spectacular success as a biological control agent against nonindigenous species of prickly pear (*Opuntia* spp.) in Australia and on several Caribbean islands. It apparently arrived in Florida either by spreading unassisted to the Keys, where it was first detected in 1989 (Habeck and Bennett 1990), or by accidental introduction on cacti shipped from the Dominican Republic to a nursery near Homestead (Dade County) (Pemberton 1995). Once in the state, it quickly became a pest of native *Opuntia* species (Habeck and Bennett 1990; Johnson and Stiling 1996) including *O. spinosissima*, which has just one known U.S. population comprising only 11 plants. In spring 1995 a project to restore *O. spinosissima* to three additional sites on Big Pine Key and Key Largo was announced. Plants propagated at the Center for Plant Conservation at Fairchild Botanical Garden will be planted at the three sites with funding from the U.S. Fish and Wildlife Service (D. Gordon, pers. comm.). These may have to be protected from the cactus moth by cages (as are all the adults in the extant wild population). If so, the plants may need to be pollinated artificially because the cages will probably exclude pollinators.

Prospects

Florida is home to some of the most severe nonindigenous species invasion problems and some of the largest and most ambitious restoration projects in the Western Hemisphere. Not surprisingly, nonindigenous species invasions have figured prominently in many restoration projects in the state. In some cases, restoration was initiated specifically to reverse or control an invasion; in others, restoration initiated to address some other disturbance was derailed by invasions. There is abundant information about impacts of nonindigenous plants on restoration of upland and wetland habitats in Florida but surprisingly little about their impacts on restoration of aquatic habitats. Even less information is available on how nonindigenous animals affect restoration of any habitat in the state.

The examples presented here illustrate the need for evaluation of potential impacts of nonindigenous species invasion before a restoration project is begun so that appropriate actions can be planned and implemented as quickly and efficiently as possible. The examples also make clear that intensive effort, long-term commitment, and large sums of money are often necessary to control invasions successfully. This chapter and others in this volume point to the necessity of taking an integrated approach to control of nonindigenous species at local, regional, and continental scales. An integrated approach will require the establishment of programs to prevent or slow introduction of new invaders and to detect and treat new invasions quickly, before explosive spread occurs (Hobbs and Humphries 1995). Reducing disturbance where possible will be important, as disturbance is widely recognized as one of the key factors influencing the invasibility of natural areas (Hobbs 1991; Hobbs and Huenneke 1992). It will also require an assessment of the value of particular sites and their degree of degradation and potential invasibility so that management priorities can be established to ensure more efficient use of limited labor and funds (Hobbs and Humphries 1995). Without this approach, we have little chance of maintaining healthy native communities or effectively restoring damaged sites for the long term and therefore little chance of ensuring that many native species can continue to evolve with our ecosystems.

13 Eradication

DANIEL SIMBERLOFF

 Campaigns to eliminate every individual of an invasive nonindigenous species are seductive but controversial (Perkins 1989). On the one hand, continuing costs of maintenance control (when such control is even possible) would be obviated. On the other hand, the prospect may seem technologically impossible and, moreover, the attempt may be enormously expensive and harm nontarget species (Newsom 1978).

Certainly the long campaign to eradicate the introduced fire ant (*Solenopsis invicta*) from Florida and the rest of the South was an expensive disaster (Davidson and Stone 1989). In 1957, Congress authorized $2.4 million to eradicate the ant, but problems immediately surfaced (Lofgren 1986). Not only did the initial heptachlor applications kill wildlife and cattle, but the entire project had been developed without provision for monitoring human health and wildlife (Davidson and Stone 1989). Florida withheld matching funds for the campaign, and heptachlor was abandoned. Researchers next developed mirex bait. Not only did fire ants rapidly reinvade areas where they had been eliminated, but trace residues of mirex were discovered in many nontarget organisms, including competitors and predators of fire ants. These findings led to cancellation of mirex registration by the U.S. Environmental Protection Agency in 1977. The costs of the applications up to that date were ca. $200 million—and the range of fire ants expanded severalfold during the eradication campaign (Davidson and Stone 1989).

Often a management program is explicitly aimed at

eliminating a species completely—an "eradication" campaign—but the methods are exactly those that would have been used simply to reduce a population to an economically or ecologically acceptable level. If such a reduction is achieved and maintained, the campaign can be considered a success even if the pest is not eradicated. Tamarisk (*Tamarix ramosissima*), for example, has been reduced to an acceptable level at The Nature Conservancy's Coachella Valley Preserve in California by a combination of mechanical and chemical means and is maintained at this level by the same means (Randall 1993). This is a successful project even though complete eradication has not been achieved. There was no conflict between an eradication campaign and a program of maintenance control.

The real problem arises when, as is often the case, different means would be used for eradication than for maintenance control (Dahlsten 1986). The chemical blitz to eliminate fire ants would certainly conflict with cultural controls and biological controls that have been suggested as possible palliatives for this invasion, even if none of these controls could completely eradicate the ant.

Successful Eradications in Florida

Many critics of eradication (such as Dahlsten 1986) argue that eradication of insect pests can succeed only if the nonindigenous species is detected early. Some eradications in Florida do appear to validate this contention. An infestation of the Asian citrus blackfly (*Aleurocanthus woglumi*) was discovered in Key West in 1934 and never spread beyond this island during a $200,000, three-year spray campaign using a mixture of paraffin oil, whale oil, soap, and water. The last individual was seen in 1937 (Hoelmer and Grace 1989). The insularity and small size of Key West aided this effort, as did the destruction of the railroad bridge by a hurricane in 1935. The Overseas Highway was not started until 1937, so traffic of humans and commodities between Key West and the mainland was greatly reduced. It is instructive that the same species was discovered in 1976 in a much larger area centered on Fort Lauderdale (Broward County). An eradication campaign was again attempted, but the area was too large, and low-level infestations remained. In 1979, the eradication effort was abandoned in favor of a program of maintenance control, or "containment" (Hoelmer and Grace 1989).

Insularity played a key role in the famous eradication campaign against the screwworm fly (*Cochliomyia hominivorax*), a cattle pest first seen in the United States in 1935 (Baumhover et al. 1955; Dahlsten 1986). The technique of releasing sterile males to cause wild females to mate fruitlessly was first field-tested on Sanibel Island (Lee County), where apparent eradication led to a similar trial on the 440-km^2 island of Curaçao. Eradication there led to the eradication campaign in the Southeast, and the technique has been used in other such programs, like that against the Mediterranean fruitfly (medfly, *Ceratitis capitata*) in California.

The giant African snail, *Achatina fulica* (Figure 13.1), has been introduced into many countries in Asia as well as numerous islands in the Pacific and Indian oceans and recently the West Indies. It is a serious agricultural pest in many locations, and predatory snails introduced as potential biological-control agents have extinguished numerous endemic snail species. (See the references in Civeyrel and Simberloff 1996.) It was eradicated in Florida, and the campaign is inspirational (Mead 1979). In 1966, a boy returning to Miami from Hawaii smuggled in three live specimens, and his grandmother liberated them in her garden. In September 1969, the resulting infestation was brought to the attention of the Florida Division of Plant Industry, leading to an immediate survey and emergency plans. Within ten days, the Commissioner of Agriculture announced that the snail was established and appealed for public cooperation in reporting and eliminating it. A brochure was posted, and a copy (150,000 in all) was mailed to every residence in certain zip codes. A total of 133 private properties were initially quarantined, but three subsequently discovered infestations

FIGURE 13.1   Giant African snail (*Achatina fulica*) feeding on a plant in southern Florida in the early 1970s. Photo by Florida Department of Agriculture and Consumer Services.

led to the quarantine of 438 properties (about 42 blocks). Worse, within days a second infestation was discovered 40 km north in Hollywood (Broward County).

An eradication campaign was mounted that used hand picking plus a granulated chemical bait. There were frequent, meticulous square-foot-by-square-foot surveys, and by 1971 only 46 snails were uncovered during six months (compared to 17,000 in the previous 16 months). Then a single live adult snail was found in the Hollywood infestation 17 months after the last live specimen had been found. Less than a month later, a third major infestation, probably two to three years old and containing over 1000 live snails in one block, was found 5.6 km southwest of the original site. The entire block was quarantined, and a substantial buffer zone was surveyed and treated. Nine months later, a fourth infestation (perhaps three years old) was found 3.6 km north of the original one, then a fifth one of similar age along the Biscayne River Canal near 135th St., about 1 km north of the original infestation.

Despite the disappointment engendered by these newly discovered infestations, the Division of Plant Industry persisted in its campaign. By 1973, over 18,000 snails had been collected (plus eggs), but in the first half of that year, only three snails were found, at two sites. By April 1975, no live specimens had been found for almost two years and the campaign, which cost over $1 million, was judged successful, although frequent surveys continued, as did baiting and carbaryl drenches, for many months. The snail has not been found again in Florida, although several pet stores in the state have recently been found selling a related large African snail (Thomas 1995). The Florida eradication was a model for the only other successful eradication of *Achatina fulica,* a campaign in a town in Queensland, Australia (Colman 1978).

Not all successful eradication campaigns are of such locally distributed pests. The citrus canker (*Xanthomonas campestris* pv. *citri*), a bacterium, was found in southern Dade County in 1912, having arrived in the Gulf States around 1910 (Merrill 1989). It had spread as far as Texas by 1914. A strict quarantine, chemical controls, and a major educational effort to keep citrus workers from spreading it led to successful eradication. By 1916, it was mostly reduced to local outbreaks, and the last infestation in Florida was in 1927; no new infestations were seen through 1984. The cost of the eradication campaign was about $2.5 million.

For both the African snail and the citrus canker, key aspects of the biology of the invader suggested that eradication might be possible. For the canker, five factors appeared crucial: the bacterium relies heavily on humans for transport; it has a restricted host plant range (citrus) and the host is introduced; it does not survive long away from its host; authorities reacted quickly once the bacterium was discovered; and growers strongly supported the effort, including the quarantine and destruction of infected fruit and plants (Merrill 1989). For the snail, the fact that it does not self-fertilize was crucial (Mead 1979). Had the biology of the pest been inappropriate, these campaigns could have been expensive disasters like the effort to eradicate the fire ant.

The repeated campaigns to eradicate the medfly in California are enormously expensive and highly controversial, as they entail aerial spraying of residential neighborhoods with malathion. Just the routine monitoring for this insect by the California Department of Food and Agriculture costs over $7 million annually, and several more millions are expended for monitoring if an outbreak is detected (J. R. Carey, pers. comm.). Although some researchers claim that these eradication campaigns are successful (Saul 1992; Voss 1992), the fact that the "reinvasions" crop up in the same neighborhoods, plus aspects of the spread of recorded flies, seem convincing evidence that the medfly is actually established in southern California and the reinvasions are simply outbreaks of a low-density, persistent population (Carey 1992a, 1992b).

The same insect has apparently been totally eradicated from wide areas of Florida at least once, however, and perhaps twice (Rohwer 1958). In 1929, the medfly was first found in Orlando and rural areas of central Florida, eventually spreading to 20 counties. Nevertheless, in a "scorched earth" 18-month, $7 million campaign (Ayers 1957) with 6000 employees, the insect was eradicated. Measures included a strict quarantine for which the National Guard manned roadblocks, destruction of produce and plants, trapping, and insecticide spraying. That the eradication succeeded is evidenced by the fact that the insect was not seen again in Florida until 1956. Then it surfaced in 28 counties, including Miami, as well as other urban areas like St. Petersburg and Tampa, and rural citrus areas. A roadblock-enforced quarantine was again established, while some 800 workers destroyed fruit, sprayed with an aerial malathion bait (Clark and Weems 1989), and trapped much more extensively than in 1929. The cost was over $10 million. Although this campaign was declared successful, the situation may be the same as in California. The medfly reoccurred occasionally in the 1960s, each time leading to a mini-eradication campaign, was not seen at all in the 1970s, and surfaced rarely in the 1980s (Clark and Weems 1989).

Keys to Eradication

It seems quite likely that nonindigenous vertebrate species can be eradicated if their habitat is sufficiently restricted and the efforts are adequately supported. The South American nutria (*Myocastor coypus*), for example, present in Florida and implicated in local damage there (Chapter 10 in this volume), has been an environmental scourge in Louisiana and England. After half a century in England, where numbers peaked at 200,000, successful eradication was achieved in

1989, although strict monitoring continued for several more years (Gosling 1989). Keys to the success were intensive scientific study of the natural history and population dynamics of this rodent, the fact that it is restricted to wetlands and the margins of water bodies, and an innovative and assiduous trapping program. Both the British Ministry of Agriculture and local drainage boards funded this eradication campaign; just the years 1980–1989 cost about $5 million. Surely a similar effort could succeed in Florida, where the populations are small and local (Chapter 10 in this volume), although reinvasion from fur farms would remain a constant threat.

Similarly, several isolated populations of introduced fish species in Florida have been eradicated (Courtenay et al. 1986; Chapter 7 in this volume), including *Serrasalmus humeralis,* the pirambeba, and *Cichlasoma trimaculatum,* the three-spot cichlid. New Zealand scientists have eradicated various combinations of twelve mammal species—ranging from the house mouse (*Mus musculus*), black rat (*Rattus rattus*), Norway rat (*R. norvegicus*), and rabbit (*Oryctolagus cuniculus*) through feral domestic animals such as pigs and goats—from numerous small islands (Veitch and Bell 1990).

Cox et al. (Chapter 19 in this volume) argue that eradication of feral pigs in Florida is probably technically impossible, but they present no evidence and, it seems, no one has seriously considered such a project. No doubt such an endeavor would be very difficult, but there is no evidence that the current method of attempted maintenance control through public hunts effectively limits ecological damage (Chapter 20 in this volume).

Certain nonindigenous plant species have far more devastating ecosystem-wide impacts in Florida than do any of the nonindigenous animals (Chapters 3 and 23 in this volume). There is no list of successful pest plant eradication projects, for Florida or anywhere else, analogous to those listed here for animals. A key hindrance to terrestrial plant eradication is the fact that seeds can remain viable in the soil for over a century.

A long-term project in the Carolinas to eradicate witchweed (*Striga asiatica*), an African root parasite of several grass plant crops, has reduced its range there from about 162,000 ha to 23,000 ha (Westbrooks 1993). This project has entailed herbicides, soil fumigation to kill seeds, and regulating the movement of soil-contaminated equipment and crops likely to carry soil (Westbrooks 1993). Very precise habitat requirements for witchweed seeds suggest that at least this life-history stage is more vulnerable than in most other pest plant species, but complete elimination is far from assured. Of seven other plant eradication programs that the U.S. Department of Agriculture Animal and Plant Health Inspection Service has mounted in cooperation with other federal and state agencies, only that for Asian common wild rice (*Oryza rufipogon*) in a 0.1-ha part of the Everglades National Park appears to have succeeded (monthly surveys continue), although ranges or densities of several of the other species have been reduced. Worldwide, few nonindigenous plant species have been completely eradicated from refuges (Macdonald et al. 1989).

The aquatic plant *Hydrilla verticillata* (Chapter 3 in this volume) has been the target of a massive eradication campaign (R. O'Connell, pers. comm.) in California, where it first invaded in 1976, probably arriving with commercial aquarium plants from Florida. Infestations in ponds, lakes, and canals have been eradicated in 10 of 17 counties. The 64-ha Lake Murray, for example, was cleared over a 16-year period by chemical and mechanical means, including a massive drawdown and the use of divers with suction dredges. This lake and other infested ones were quarantined, and public access was temporarily prohibited to prevent the infestation from spreading. Along with other means, biological control by grass carp has been used in canals (but not lakes), and the 1000 km originally infested has been reduced to about a kilometer. Elsewhere, however, Clear Lake (about 17,000 ha), one of the largest natural bodies of water in California, has been invaded and presents an even greater problem, particularly as its popularity as a bass-fishing lake precludes its being quarantined. Thus, if uncontrolled, the Clear Lake infestation can serve as a source of propagules for many other lakes and rivers. The outcome of the Clear Lake campaign cannot yet be predicted. In fiscal year 1994–95, the entire California anti-hydrilla project cost about $1.9 million, of which 64 percent was for the Clear Lake campaign.

In Thurston County, Washington, Eurasian water milfoil (*Myriophyllum spicatum*), discovered in 1987 in 130-ha Long Lake, has been the subject of an intensive five-year eradication campaign (Thurston County Department of Water and Waste Management 1995). After a combination of intensive chemical treatment with a systemic herbicide and mechanical treatment of small residual infestations, the project appears to be close to complete success.

Although eradication of substantially distributed invasive plant pests has usually failed, it is not necessarily a hopeless dream. New technology may transform control of both plant and animal nonindigenous species—for example, the control of weeds by genetic engineering of herbicide resistance into crops is already approved by the U.S. Department of Agriculture for cotton and soybeans (Anonymous 1995b; Kareiva et al. 1996). Numerous other ways can be imagined in which genetic engineering of either a pest or its enemy can greatly reduce if not completely eradicate the pest (Simberloff 1990; Anonymous 1992). No doubt at least some of these approaches will prove problematic (Tardieu 1995), but one or more may well allow eradications that are today barely contemplated.

Eradication campaigns, even when technically feasible, can run afoul of political and social pressure if a nonindigenous species, even an invasive pest, develops its own constituency (D. Simberloff and M. Tebo, unpublished data).

Attempts in Florida and the Northeast to eradicate the monk parakeet, for example, a potential agricultural pest (Chapter 9 in this volume), produced objections from bird lovers. And the residents of Gulf Stream (Palm Beach County), Florida, objected to the removal of Australian pine, arguing that "extensive research conducted by the Town clearly demonstrates that . . . Australian pines [do] not diminish the natural coastal environment." In fact, Australian pine devastates the natural coastal environment (Johnson 1994; Chapter 3 in this volume), so public education is as essential for eradication as are technology and funding.

ACKNOWLEDGMENTS

I thank Ross O'Connell for extensive information on the campaign to eradicate *Hydrilla* in California.

14 Maintenance Control

JEFFREY D. SCHARDT

Florida is home to some 3500 species of plants, as many as 25 percent of which are nonindigenous (Ward 1989). Although most nonindigenous plants have little impact on Florida's ecosystems, 60 species are listed by the Florida Exotic Pest Plant Council (1995) as highly invasive, capable of rapidly dispersing and disrupting natural ecosystems.

Wetland and upland invaders such as melaleuca (*Melaleuca quinquenervia*) and Brazilian pepper (*Schinus terebinthifolius*) have long been recognized as threats to native plant and animal diversity (Workman 1979; Schortemeyer et al. 1981). Although parcels of land have been cleared of these invasive species, no statewide management policy has been developed and management funding is both insufficient and inconsistent. Cleared sites, therefore, often are recontaminated by seeds or rootstock or from adjacent infestations (Clark 1994; Lewis 1994).

The nonindigenous floating water hyacinth (*Eichhornia crassipes*), in contrast, once Florida's worst aquatic weed problem, has been brought under control statewide. As recently as 1960, more than 50,000 ha of water hyacinth clogged Florida's navigable waterways (U.S. Congress 1965). But since 1990, fewer than 800 ha of the weed have existed in Florida's 500,000 ha of public fresh waters (those accessible by public boat ramp) (Schardt and Ludlow 1995). This reduction rests on four essential principles: the adoption of a management concept known in Florida as "maintenance control"; the designation of a lead government agency responsible for implementing the management policy statewide; a sufficient, dedicated management-funding source; and an ample, well-trained labor force.

This chapter summarizes the steps taken to bring water hyacinth under control in Florida and contrasts that effort with the difficulties encountered trying to suppress a more recently introduced aquatic invader, hydrilla (*Hydrilla verticillata*). The need for a coordinated effort and a financial commitment is further demonstrated by the early success in the state's large-scale effort to control melaleuca on Lake Okeechobee (Figure 14.1).

Water Hyacinth

Water hyacinth was introduced into Florida as a horticultural curiosity in the early 1880s (Tabita and Woods 1962) and quickly spread across the state, aided by farmers who mistakenly believed it would make good cattle fodder (U.S. Army Corps of Engineers 1973). By the end of the nineteenth century, water hyacinth so disrupted logging and steamboat traffic on the St. Johns River and its tributaries (Figure 14.1) that the 55th U.S. Congress, through the Rivers and Harbors Act of 1899 (Chap. 425, Sec. 1), authorized the U.S. Army Corps of Engineers to crush, divert, or remove water hyacinth from these waterways. At the same time, Florida acted to curtail the spread of water hyacinth by enacting Chapter 4753 (now Florida Statute 861.04) prohibiting the placing of water hyacinth in streams and waters of the state and prescribing penalties for violations.

At the turn of the century, only the Corps of Engineers attempted water

FIGURE 14.1   Location of sites discussed in this chapter.

hyacinth control in Florida. Passive devices such as log booms and fence barriers were placed across infested creeks and streams to prevent discharge of floating water hyacinth plants into the St. Johns River. Physical labor was employed to break up small mats that collected among snags and pilings, allowing these mats to float downstream into the Atlantic Ocean (Zeiger 1962). Inorganic chemicals were tested but were rejected because they were ineffective or toxic to nontarget organisms (Buker 1982). As a result, water hyacinth control was conducted almost exclusively by mechanical means until the late 1940s (Figure 14.2).

Mechanical apparatus eased but did not solve the water hyacinth problem. Even the most efficient machines could not operate effectively in shallow water along shorelines and therefore left a continuous source of plant material for re-infestation (Zeiger 1962). The sheer mass and volume of the plant impeded success. One hectare of water hyacinth weighs between 191 and 415 metric tons (wet weight) and has an average volume of 168 m^3 (U.S. Army Corps of Engineers 1973). Finally the growth rate of water hyacinth, which has a doubling

FIGURE 14.2   Conveyer 1 removing water hyacinth from the Caloosahatchee River at LaBelle, Florida, 1939. Photo from U.S. Army Corps of Engineers, Jacksonville, Florida.

time of approximately two weeks under optimum growing conditions (Penfound and Earle 1948; Mitchell 1976), was so great that machines could not keep pace despite round-the-clock operations in some cases (Wunderlich 1967).

The phenoxy herbicide 2,4-D was discovered in 1942. After several years of field evaluation, the Corps of Engineers found water hyacinth to be highly susceptible to 2,4-D and began using this compound in Florida waters by 1948 (Buker 1982). In 1952, using federal funding, the Florida Game and Fresh Water Fish Commission (FGFC) began a limited state program to control water hyacinth (Woods 1963). Until this point, water hyacinth had been managed for commercial interests (navigation) and human welfare (flood and mosquito control). Managing it to conserve ecosystems became a concern late in the program. In 1955, the Florida legislature appropriated approximately $250,000 for water hyacinth control. The FGFC and Corps of Engineers worked in tandem using 2,4-D and the newly available herbicide diquat along with an array of mechanical harvesters. Despite having the same tools used in Florida's current management program, the FGFC and Corps of Engineers were only able to slow, not stop, the spread of water hyacinth.

In 1965, the Corps of Engineers completed a study on control of additional plant species not included in the Rivers and Harbors Act of 1899. This five-year pilot program—the Expanded Project for Aquatic Plant Control—tested the feasibility of control and eradication of water hyacinth and other objectionable species from the waters of eight states. An important finding was that eradication of this invader was not feasible, but management was possible if done consistently (U.S. Congress 1965).

During the early 1960s, several additional state and federal agencies became involved in water hyacinth control in Florida, including the U.S. Soil Conservation Service, the U.S. Fish and Wildlife Service, the Central and Southern Flood Control District, and various local governments including Lee, Dade, Broward, Hillsborough, and Polk counties. At this point, however, the various agencies were pursuing different goals. The Corps of Engineers, for example, wished to maintain navigation, the FGFC to enhance fish and wildlife habitat, local governments to promote flood control and mosquito abatement. Water hyacinth in public waters was sometimes controlled by more than one government entity or not controlled at all. Because many of Florida's public waters are interconnected, unmanaged waters served as contamination sources for downstream waters with control programs. Despite extensive management efforts, the program remained fragmented. As a result, the Corps of Engineers estimated a water hyacinth population in 1960 exceeding 51,000 ha in Florida public waters and projected an additional 20,000 ha unless improved management strategies were instituted (U.S. Congress 1965).

Clearly a coordinated state effort and a consistent funding source were necessary to alleviate the state's aquatic plant problems. The Florida legislature resolved one problem by supplying consistent funds. In 1969, it enacted Florida Statute 371.171 (now 327.28) to provide funding from boat registration revenues. Additional funding was provided in 1971 through the creation of Florida Statute 213.11 (now 206.606), which transferred gasoline taxes generated by motorboats to the Aquatic Plant Trust Fund. The legislature then designated the Department of Natural Resources (now the Department of Environmental Protection, FDEP) the state's lead agency for aquatic plant control by enacting Florida Statute 372.925 (now 369.20), the Florida Aquatic Weed Control Act.

The FDEP was authorized to coordinate the activities of all government and private groups engaged in aquatic plant control as well as to direct the research and planning related to these activities. The FDEP was chosen to lead the program because of its multiple-use orientation toward the state's natural resources. Most important, in 1974 the legislature, recognizing that water hyacinth could not be eradicated but could be managed, mandated that nonindigenous aquatic plants, particularly water hyacinth, be controlled under maintenance programs. A maintenance program is defined in Florida Statute 369.22(2)(d) as "a method for the control of non-indigenous aquatic plants in which control techniques are utilized in a coordinated manner on a continuous basis in order to maintain the plant population at the lowest feasible level as determined by the department."

The concept of maintenance control of nonindigenous vegetation in Florida's natural waterways was developed by the Corps of Engineers upon completion of a massive water-hyacinth control project—Operation Clean Sweep—on the St. Johns River in 1973 (Buker 1982). Previously the species had been allowed to reach problem levels before control measures were implemented. These efforts resulted in the death of large amounts of floating plant biomass, which led in turn to environmental damage by detrital loading (Figures 14.3 and 14.4). The only way to avoid environmental disturbances associated with such operations was to prevent water hyacinth from reaching large population sizes. Joyce (1985) has reported that, compared to allowing water hyacinth to cover a body of water before it was controlled, maintaining cover at 5 percent or less reduced herbicide use more than twofold, reduced sedimentation up to fourfold, and reduced dissolved oxygen depressions.

Water hyacinth management along the Suwannee River, which divides peninsular Florida from the western Panhandle (Figure 14.1), exemplifies the benefits of maintenance control (Figure 14.5). FDEP records indicate that the species covered more than 1000 ha of the Suwannee River during 1974. Marshes and creeks were clogged, and fringes sometimes more than 50 m wide lined the river channel. To preserve navigation, more than 2900 kg (active ingredient) of the herbicide 2,4-D was applied in 1974. Several thousand metric

FIGURE 14.3   Massive water hyacinth buildup on Lake Rousseau, 1965. Photo from U.S. Army Corps of Engineers, Jacksonville, Florida.

FIGURE 14.4   Lake Rousseau shortly after water hyacinth control, 1965. Photo from U.S. Army Corps of Engineers, Jacksonville, Florida.

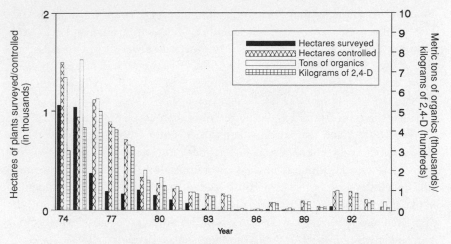

FIGURE 14.5   Maintenance control of water hyacinth on the Suwannee River, 1974–1994.

tons (dry weight) of organic detritus was deposited as a result, in addition to that from leaf and root material naturally sloughed from live plants.

Maintenance control operations were begun during 1976 to reduce the standing crop of water hyacinth on the Suwannee River. Amount of plant material killed, herbicide use, and management costs increased initially but decreased along with the tonnage of organic sedimentation produced as water hyacinth coverage was reduced. By the early 1980s, maintenance control had been achieved along the Suwannee River. Occasionally water hyacinth growing in marshes and sloughs that are inaccessible to management crews is flushed into the river by flooding, as it was in 1991, when coverage rose from less than 1 ha to approximately 40 ha. Coverage was returned to about 2 ha by the end of 1992. Once water hyacinth competition was removed and control operations were scaled back, diverse communities of native plants returned to the banks and marshes along the river.

The FDEP has designated water hyacinth the top management priority of the 18 nonindigenous plants detected in Florida public waterways (Schardt 1992). Funding is allocated for its management before any other aquatic plant control activity, and the approximately 50 federal, state, and local government field crews that manage plants in Florida public waters are instructed to give its management the highest priority.

Although hydrilla was nearly 60 times more abundant than water hyacinth in Florida waters in 1994 (unpublished FDEP plant survey records), water hyacinth has greater potential to cause problems. Because it is not attached to the substrate, it is easily moved by wind and water currents. A seemingly innocuous stand in an isolated cove, therefore, can be transformed into a navigation or flood hazard during a brief storm or if left unmanaged for only a short period. The 800-ha population on Lake Okeechobee, for example, quadrupled to 3200 ha in five months when herbicide control operations were suspended and replaced with mechanical harvesters in 1986. Two years and $2 million were spent restoring maintenance control conditions in Lake Okeechobee by means of herbicides (unpublished FDEP control records).

Since 1982, the FDEP has monitored public waters in Florida (Schardt and Ludlow 1995). Each year aquatic plants are inventoried as part of the FDEP's program to find and contain (or eradicate) nonindigenous aquatic plants at low cost before they become established and require long-term maintenance programs. Annual surveys also provide up-to-date information for development of statewide priorities for distribution of limited funding. Finally, surveys allow the FDEP to monitor trends in aquatic plant communities, especially those that may be related to nonindigenous plants or plant management programs.

Maintenance control of water hyacinth was achieved in Florida public waters during the early 1990s (Figure 14.6). This once ubiquitous plant, although present in about 65 percent of Florida's 450 public bodies of water, has been reduced to a minor component of their flora through integration of physical, mechanical, herbicidal, and biological controls at an average annual cost of $2.7 million between 1990 and 1994 (for management of 13,400 ha of water hyacinth mixed with another nonindigenous floating plant, water lettuce, *Pistia stratiotes;* FDEP 1995). Although host-specific insects and pathogens have been released in Florida to increase water hyacinth leaf mortality and to reduce plant size and reproductive vigor (Goyer and Stark 1981, 1984; Center and Van

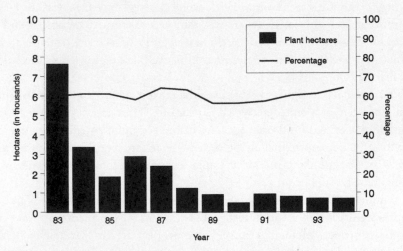

FIGURE 14.6 Hectares of water hyacinth and percentage of public waters in which water hyacinth was reported in Florida, 1983–1994.

1989; Grodowitz et al. 1991), their role in population reduction is difficult to assess. These agents may take months or years to reach densities that will reduce population size (Harley 1990). Because water hyacinth grows so quickly in Florida, management must include herbicides and mechanical devices.

Hydrilla

Hydrilla was deliberately introduced by the aquarium industry during the early 1950s (Schmitz et al. 1991). Absence of any screening at ports of entry, ignorance of hydrilla's invasive potential, and inadequate monitoring in Florida waters allowed it to expand across the state for more than a decade. Expansion was aided into the early 1990s by sportsmen and managers, who exploited hydrilla as fishery habitat (Montalbano et al. 1979; Moxley and Langford 1982; Leslie 1994).

As with water hyacinth, the first hydrilla-related problems were reported not as environmental disturbances but as flood-control threats (in South Florida drainage canals) and navigational restrictions (in the Crystal River, Figure 14.1). Hydrilla was initially misidentified as a species of elodea (Blackburn et al. 1967), and managers became concerned when herbicides effective in elodea management did not produce satisfactory control (L. Bitting, pers. comm.).

When the plant was correctly identified as hydrilla in 1968, the experience gained in the water hyacinth management program was employed against this new invader. Research funding was authorized for identification of vulnerable stages in hydrilla's life cycle. Field trials were conducted to ascertain effectiveness of various biological, chemical, mechanical, and physical controls. Possession of hydrilla was regulated by federal law (Federal Noxious Weed Act) and state administrative code (Chap. 62C-52, F.A.C.). Government agencies in Florida organized to manage hydrilla under the coordination of the FDEP.

Research confirmed field observations that hydrilla grows extremely fast (Haller 1978), competes intensely with native vegetation (Haller and Sutton 1975), and has multiple reproductive capabilities for rapid and broad distribution (Haller 1978). Research also suggested that once hydrilla became established, eradication would be difficult because it forms underground tubers that resist control (Brunner and Batterson 1984). Thus long-term maintenance programs would probably be necessary to prevent it from completely overgrowing Florida's shallow waters.

In contrast to the water hyacinth case, the public and even some fish and wildlife managers were not convinced that hydrilla posed a serious statewide threat. Relatively few public waters were affected by the middle 1970s, and they were managed sufficiently to sustain navigation and most recreational activities. As hydrilla entered new systems, especially those with little submersed vegetation, water clarity often improved, sport fisheries were enhanced (Porak et al. 1990), and waterfowl were concentrated on this easily accessible food source (Johnson and Montalbano 1984). Large-scale control efforts often met with opposition from fishing enthusiasts and the managers charged with administering Florida's $1.5 billion per year freshwater fishing industry (*Tallahassee Democrat* 1975; Colle 1982).

Opposition usually lasted until hydrilla overgrew the waters. Then local support for control soared—particularly for restoration of recreational uses and property values. Statewide public support remained low, however, because when only a few water bodies were affected, many nearby alternatives were available. This style of crisis management was contrary to Florida law but was unavoidable because conflicting arguments, for and against control, sent confusing messages to lawmakers resulting in inadequate and inconsistent funding. Despite the existence of technology adequate to manage it for as little as $250 per hectare per year in 1995, hydrilla continued to expand across Florida and fill public waters.

The statewide infestation was held below 21,500 ha until the early 1990s (Schardt and Ludlow 1995). But when funding failed to keep pace with hydrilla's growth, the aquatic invader expanded within infested waters and was transported into adjacent waters. When funding is increased, hydrilla infestation can be held to a low level. When funding is reduced, control is lost, hydrilla expands, and a higher level of funding is necessary to regain the previous level of control. This upward trend has accelerated in Florida in the 1990s (Figure 14.7). FDEP survey records indicate that hydrilla coverage in public waters nearly doubled from about 20,250 ha in 1992 to 40,000 ha in 1994. Whereas $2.5 million was required to manage hydrilla in 1985, more than $14 million was needed in 1995—but less than $5 million was available (FDEP 1995). Hydrilla coverage and management costs are expected to continue increasing until a sufficient, consistent, funding source is provided.

Management of hydrilla infestations in Florida benefits from an economy of scale. More than 150 small populations, ranging from trace amounts to coverage of 40 ha, were designated by the FDEP for control in 1994 at a cost of approximately $450,000, or an average of $3000 per water body. Projected costs for the remaining 43 infested water bodies totaled more than $13 million, or an average of approximately $300,000 each (FDEP 1995). Between 1990 and 1994, more than 40 new hydrilla infestations were detected by FDEP inspectors, mostly at public boat ramps (FDEP 1995). Many appear to have been eradicated before tubers could be established. Others are under intensive maintenance to prevent spreading. Several grew out of control. Most notable among these is the hydrilla population in 3000-ha Lake Weohyakapka (Polk County) (Figure 14.1). Hydrilla was first observed there, scattered along the shoreline, by FDEP inspec-

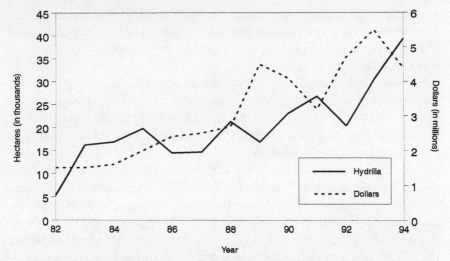

FIGURE 14.7   Hectares of hydrilla reported in Florida public waters vs. dollars spent to manage hydrilla from 1982 through 1994.

tors in 1992. By August 1994, hydrilla covered nearly 2100 ha of the lake. Appropriate management would have cost approximately $750,000 in 1995, but the money was not available. Today Lake Weohyakapka has become nearly unusable as a recreational resource.

Hydrilla management is expensive, but failure to manage it is more so. In one of the few economic-impact studies relating to hydrilla infestations, Milon et al. (1986) found that Orange Lake and Lake Lochloosa (Alachua County) of north central Florida (Figure 14.1) generated economic activity each year of approximately $10 million when surfaces were free of hydrilla and water hyacinth. Colle et al. (1985) have estimated a 90 percent loss in revenues from declining use of Orange Lake by anglers when it was covered by hydrilla.

The explosive expansion of hydrilla in Florida public waters between 1992 and 1994 offers compelling evidence that eradication of new infestations and maintenance control of established hydrilla populations are the most environmentally and economically sound methods for dealing with this species. It is now incumbent upon federal, state, and local agencies working in unison with the public to secure ample, consistent funding for the research and management needed to bring hydrilla under control.

Melaleuca

Melaleuca was deliberately introduced into Florida from Australia during the early twentieth century. Seeds were intentionally broadcast by airplane in the everglades region of southern Florida during the mid-1930s (Austin 1978a). Melaleuca has since spread to become a dominant component of the southern Florida landscape, covering nearly 182,000 ha in 1993 (Thayer and Ferriter 1994) (Figure 14.8).

Despite this expansion and warnings from ecologists regarding problems such as loss of plant and animal diversity (Austin 1978a; Schortemeyer et al. 1981; Ostrenko and Mazzotti 1981) and increased fire hazards (Schortemeyer et al. 1981; Flowers 1991), melaleuca was still available in the early 1990s at nurseries for ornamental plantings.

After more than a decade of efforts, a federal law (the Federal Noxious Weed Act) and a state measure (Florida Statute 369.251) were passed in 1993 prohibiting the sale of melaleuca. A broad program was begun by public agencies and private groups to educate the public about this invasive plant (Figure 14.9). Government agencies and private groups have been controlling melaleuca throughout southern Florida for many years, but there has been little coordination (Clark 1994). Consequently, although some areas have been cleared of melaleuca—such as Sanibel Island (Lee County), Everglades National Park (Doren and Jones 1994), and areas of the Big Cypress National Preserve (T. Pernas, pers. comm.) (Figure 14.1)—seed banks and adjacent dense melaleuca forests present a continuous source for reinfestation.

Although various groups have conducted active melaleuca control programs for as long as three decades, much of the management information is anecdotal (D. Thayer, pers. comm.). The Exotic Pest Plant Council has compiled control

FIGURE 14.8   Dead melaleuca trees in Lake Okeechobee, 1995. Photo by Don Schmitz.

FIGURE 14.9   Sign describing melaleuca (*Melaleuca quinquenervia*) management efforts in Lake Okeechobee, Florida. Photo by Don Schmitz.

data (Laroche 1994), and the South Florida Water Management District has co-ordinated field trials on chemical, mechanical, and physical melaleuca management techniques. The U.S. Department of Agriculture has initiated overseas evaluation of biological controls.

No government entity has yet emerged to coordinate statewide melaleuca management efforts, but in 1993 the Florida legislature directed the FDEP to spend $1 million per year on melaleuca control. Although this money was diverted from the hydrilla maintenance program, it has provided an opportunity to apply the maintenance control philosophy to melaleuca management on a large scale. Previous funding for melaleuca control was inconsistent and often grossly inadequate (U.S. Congress 1993). Large-scale control programs have been difficult to initiate because follow-up (maintenance control) funding was not guaranteed. Without follow-up maintenance, melaleuca would regrow quickly from seed and adjacent infestations.

In 1994, the South Florida Water Management District appropriated funds to match with state revenues for maintenance control of melaleuca on two sites,

Lake Okeechobee and the Everglades Conservation Areas to the northeast of Everglades National Park (Figure 14.1). These sites were selected because of their definable boundaries, ecological importance, and infestation, over thousands of hectares, with dense melaleuca forests as well as sparse outlier populations. Two sites were chosen so that if environmental conditions precluded operations in one area, funding could be transferred to the other to retain trained field crews and take full advantage of funding while it was available (D. Thayer, pers. comm.).

As the control strategy for Lake Okeechobee and the conservation areas is the same, only Lake Okeechobee is discussed here. To prevent further encroachment into the Okeechobee marsh, outlier trees and small stands of melaleuca were attacked first. Control was achieved by hand pulling or girdling and treatment of the exposed cambium (Laroche 1994). This labor-intensive work progressed toward the dense monospecific forests of melaleuca, which were more economically treated with foliar herbicides applied by helicopter (Laroche 1994).

Latitude–longitude coordinates are recorded for each outlier or group of melaleuca trees. These sites are reinspected every three to six months, and any missed plants or regrowth are controlled before plants can mature and produce seeds (A. Ferriter, pers. comm.). Large trees are killed but, to reduce management costs, not removed. They usually fall and begin to decompose in the humid climate of southern Florida within a few years after being killed (D. Thayer, pers. comm.). The methods of herbicide application have little impact on nontarget vegetation. Resurgence of native flora from the seed bank has been extensive enough on Lake Okeechobee that expensive artificial revegetation has not been necessary.

Approximately 2400 ha of melaleuca existed on Lake Okeechobee in 1993, spread over some 12,200 ha of marsh. After two years of maintenance control techniques, more than 1400 ha of melaleuca has been eliminated over an area of about 8100 ha (A. Ferriter, pers. comm.). Most of the labor-intensive outlier control was achieved during the first two years. The Okeechobee program was originally projected to take about 5.5 years and cost approximately $10 million. Managers now anticipate that if current environmental conditions persist, maintenance control may be achieved in five years at a cost of $5 million. About $50,000 will be needed annually to control regrowth.

Lessons

The most important lessons provided by water hyacinth, hydrilla, and melaleuca management are that preventing establishment of invasive, nonindigenous species and early intervention after these species are introduced are much more attainable and cost-effective than management after their dispersal and establishment. Periodic inventories and timely control of newly discovered invasive, nonindigenous species allow managers to be proactive in protecting public lands. Maintenance control of even widely dispersed and seemingly unmanageable invaders is possible with a well-researched, well-staffed, well-coordinated, and well-funded management commitment.

Water hyacinth infestations were successfully reduced in Florida's public waters once the state designated a lead agency and committed consistent, ample funding and a well-trained, well-staffed labor force to achieving maintenance control.

All the steps that led to solution of the water hyacinth problem have been followed in the hydrilla management program except commitment of sufficient funding: funding for hydrilla control has been both inconsistent and inadequate since the middle 1980s. As a result, hydrilla expanded from fewer than 15,000 ha in 1986 to nearly 40,000 ha in 1994. The austerity efforts of governments at all levels involved with Florida's hydrilla management program have had the perverse effect of increasing the amount of funding that will ultimately be needed to attain maintenance control of hydrilla.

Melaleuca covered 4.5 times as much Florida public land in 1994 as did hydrilla, but most is in remote locations and poses little economic hardship to the public. Like hydrilla, melaleuca is manageable on a large scale by means of current technology, as evidenced by the control results on Lake Okeechobee. But because of poor public support and intense competition for funding with social programs, statewide maintenance control of melaleuca is not foreseeable in the near term.

15 Biological Control

TED D. CENTER, J. HOWARD FRANK,
AND F. ALLEN DRAY, JR.

 Biological control is a scientific discipline based primarily on ecological theory but relying also on ethology, taxonomy, physiology, genetics, biochemistry, and other disciplines. It derives from the premise that populations of many organisms are held at low, noninjurious levels by their natural enemies and postulates that natural enemies can therefore be used to control specific plant and animal pests (Huffaker et al. 1976). Biological control often hinges on some aspect of dependency, or host specificity, between the pest and the biological control agent (bioagent). In most cases this association has developed over evolutionary time. Wilson and Huffaker (1976) define biological control as "the role that natural enemies play in the regulation of the numbers of their hosts, especially as it applies to animal and plant pests." The following list defines some of the key terms important to understanding biological control:

> *adventive:* not native to the environment under discussion; alien, exotic, nonnative, or nonindigenous
> *euryphagous:* using a wide variety of food hosts; also called polyphagous
> *generalist:* a herbivore, predator, parasite, or parasitoid that uses many hosts, often unrelated ones
> *host:* an organism from which a second organism gains shelter, food, or both
> *monophagous:* using a single food host
> *parasite:* an organism living in or on another organism from which it derives nourishment

parasitoid: a parasite living within its host only during immature stages, eventually killing the host

specialist: a herbivore, predator, parasite, or parasitoid that uses only one host or a few closely related hosts

stenophagous: using a limited number of food hosts

In this chapter we restrict our discussion to density-dependent, self-sustaining interactions among populations of organisms achieved by the importation, release, and establishment of nonindigenous bioagents. This is classical biological control, also known as inoculative biological control. Density dependence describes the case in which the pest's mortality rate increases with its density. The pest population increases in a favorable environment until corresponding increases in repressive forces (natural enemies, competition for food or shelter) are induced. These repressive forces cause a population decline, then ease, allowing the pest population to increase again, beginning the cycle anew. The result is fluctuation of the pest population around a relatively stable average level. This level is lower when more repressive forces are present.

Despite popular misconception, the goal of a biological control program is never eradication. If the pest species is eradicated, species that exploit it exclusively will be eradicated as well. In other words, the bioagents would disappear along with the pest. Bioagents often start by reducing pest outbreaks. Theoretically, as the host declines to low population levels the bioagent has more difficulty finding it, the bioagent population is reduced more than the pest's, and a low-level steady state results. From that point on, the predator or parasitoid prevents outbreaks of pests rather than reducing outbreak populations (Murdoch et al. 1985). Effective biological control is often difficult to demonstrate because to see it in action one must observe pest outbreaks not happening. For example, one species introduced into Florida for water hyacinth (*Eichhornia crassipes*) control in the late 1970s was referred to in the popular media as a "super bug." When water hyacinth was not quickly eliminated, the headlines read "super bugs not so super," even though this insect killed 65 percent of newly produced juvenile plants, thus preventing infestations from growing larger. This lack of appreciation for the effects of bioagents stems from unrealistic expectations of their capabilities.

Background

Biological control has been practiced for hundreds of years. Early attempts sought relief from pests without concern for nontarget effects and lacked rigorous verification of efficacy. Examples include the domestication of cats by

Egyptians to protect grain stores from rodents and use of predacious ants by Chinese citrus growers (ca. A.D. 900) and Yemeni date growers (ca. A.D. 1200) to protect their trees (Doutt 1964; Simmonds et al. 1976; Coppel and Mertins 1977). Linnaeus used more exacting scientific methods when he tested the predacious ground beetle *Calosoma sycophanta* as a control for orchard pests during the mid-eighteenth century (Hagen and Franz 1973; Coppel and Mertins 1977).

The nineteenth century saw an increase in the number of biological control projects, including the first attempt against a weed. Local control of the prickly pear cactus *Opuntia vulgaris* in India (1795) by an immigrant eriococcid bug, *Dactylopius ceylonicus*, led to the purposeful introduction of this insect to other areas of India during 1836 and Sri Lanka during 1865 (Julien 1992). This attempt was highly successful. In 1840, Boisgiraud continued Linnaeus' earlier work with *C. sycophanta* by using this beetle to suppress gypsy moths (*Lymantria dispar*) on poplar trees in France (Hagen and Franz 1973). In Italy, Villa reduced levels of several garden pests (*Cetonia, Pieris, Forficula*) in 1844 with predacious beetles (Hagen and Franz 1973).

Pathogens of pests were also investigated during the nineteenth century. Metchnikoff, for example, conducted encouraging field trials in Russia during 1884 on the green muscardine fungus (*Metarhizium anisopliae*), released against the wheat cockchafer (*Anisoplia austriaca*) (Coppel and Mertins 1977). Vertebrates used as control agents during this period included *Bufo marinus* (giant toad) against white grubs of sugarcane in Martinique and mongoose (probably *Herpestes javanicus*) against rats in the West Indies (Simmonds et al. 1976; see also Hoagland et al. 1989).

A watershed event for biological control came in 1889 with a United States project that successfully controlled cottony-cushion scale (*Icerya purchasi*). This pest was devastating California's citrus industry until the vedalia beetle (*Rodolia cardinalis*), imported from Australia, brought it under complete control (Caltagirone and Doutt 1989). The pace then began to accelerate and involve more discriminating use of agents (Coppel and Mertins 1977). In Hawaii, several bioagents controlled the sugarcane leafhopper *Perkinsiella saccharicida* (Coppel and Mertins 1977). In Australia, the moth *Cactoblastis cactorum* and *Dactylopius* spp. bugs controlled prickly pear cacti (*Opuntia* spp.; Julien 1992) (Figure 15.1). Many nations began experimenting with herbivorous fish to control aquatic plants during the early twentieth century (Clausen et al. 1978; Julien 1992).

The modern period of biological control (roughly 1950 to the present) is characterized by increasingly rigorous preintroduction protocols to protect against attack of nontarget species. For example, biological control agents of

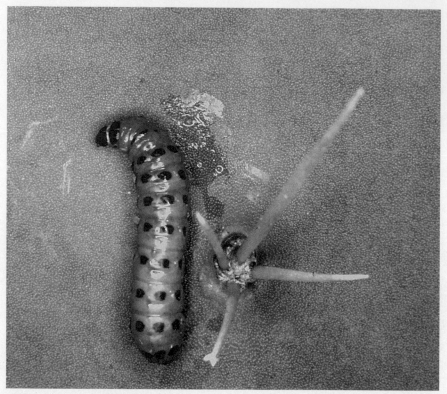

FIGURE 15.1   *Cactoblastis cactorum* larva on *Opuntia stricta*. Photo by Peter D. Stiling.

alligatorweed (*Alternanthera phyloxeroides*) were screened during the early 1960s against 14 to 30 plant species in six to eight families (Buckingham 1990). In contrast, biological controls of hydrilla (*Hydrilla verticillata*) were screened during the mid-1980s against 41 to 51 plant species in 18 to 27 families.

Recent successes abound. Alligatorweed, once a severe problem in the southeastern United States, is now suppressed by a flea beetle (*Agasicles hygrophila;* Figure 15.2), a moth (*Vogtia malloi*), and a thrips (*Amynothrips andersoni*), from South America (Cofrancesco 1991). *Hypericum perforatum* (common St. John's-wort, Klamath weed) is controlled in the northwestern United States by three European beetles (*Chrysolina quadrigemina, C. hyperici,* and *Agrilus hyperici*) (Julien 1992). Infestations of *Salvinia molesta* (karibaweed, giant water fern) are reduced in Australia and Papua New Guinea by a South American weevil (*Cyrtobagous salviniae*) (Room et al. 1985). The Chinese grass carp or white amur, *Ctenopharyngodon idella,* successfully controls aquatic weeds in many countries (Julien 1992), but consumption of nontarget

FIGURE 15.2   Damage by alligatorweed flea beetle (*Agasicles hygrophila*) on alligatorweed (*Alternanthera phyloxeroides*). Photo by Larry Nall.

plants can be problematic. Pathogenic fungi used as biological controls include *Puccinia xanthii* from the United States, which controls *Xanthium strumarium* (noogoora burr, common cocklebur) in Australia (Julien 1992). Numerous other examples have recently been published (Hoffman 1991; Nechols et al. 1995).

Pest insects, too, are successfully managed by means of biological controls. The yellow clover aphid (*Therioaphis trifolii*) is controlled in California by the parasitoid wasps *Praon palitans*, *Trioxys utilis*, and *Aphelinus semiflavus* (van den Bosch et al. 1964). In Florida and Texas, citrus blackfly (*Aleurocanthus woglumi*) infestations are reduced by the parasitoids *Amitus hesperidum* and *Encarsia opulenta* from India and Pakistan (Dowell et al. 1979; Sailer 1981). In Canada, the winter moth (*Operophtera brumata*), whose larvae feed on hardwood foliage, was controlled by parasitoids (*Cyzenis albicans* and *Agrypon flaveolatum*) from Europe (Embree 1966). Gambusia for mosquito control is the most widely used fish bioagent (Wilson 1965). Clausen et al. (1978) cite additional examples of successful biological control of pest insects.

How Does Biological Control Work?

The following list outlines the stages in a classical biological-control project:

- Select the target: Investigate its biological suitability as a target. Identify conflicts of interest. Interview proponents and opponents of controlling the proposed target. Conduct an economic and ecological cost-benefit analysis.

- Locate potential biological controls: Identify the native range of target species. Survey the native range of natural enemies. Determine each natural enemy's potential to inflict substantial stress on the target.

- Determine specificity: Examine the host range of natural enemies in native areas. Import the best candidates into U.S. quarantine for screening against native species related to the target, native ecological analogs to the target, and economically important species potentially at risk. Drop nonspecialists from further consideration.

- Obtain permission to release biological controls: Report findings from specificity trials to federal and state agencies responsible for approving biological control agent releases. Reply to requests for clarification or additional host trials.

- Establish biological controls in the target's adventive range: Develop laboratory colonies of approved agents using progeny from quarantine colonies or fresh genetic stock from the agent's native range. Test various release strategies to determine which is the most effective. Use this strategy to establish persistent populations of agents at field sites. Enhance the natural dispersal of agents where appropriate.

- Determine efficacy: Measure the ability of agents to reduce host numbers. Effects may include direct death of individuals, reduced reproductive success, slowed growth, and greater susceptibility to other repressive factors.

- Incorporate biological controls into management plans: Use various technology-transfer methods to ensure that resource managers are knowledgeable about biological control agents and their use. Investigate interactions between biological controls and other control methods to determine the best strategies for integrating alternative approaches.

It is important at the outset to determine whether attempting to control a particular weed or pest insect is in fact in the public interest. It may be difficult to justify a project if the potential economic, ecological, and aesthetic benefits are perceived as minimal (Harris 1991). Such a perception may arise from the difficulty of assessing direct economic damage or when the damage is borne by a small or politically impotent segment of the population. Infrequently visited natural areas or sparsely populated private lands may suffer tremendous damage but not generate enough public impetus for a control program. Habeck et al. (1993) have proposed that a biological control project becomes economically worthwhile when expected savings exceeds about $60,000.

Moreover, the proposed target may not be regarded as a pest by everyone. Australian pine (*Casuarina* spp.), for example, is valued by Florida's beachgoers for shade but is disliked by conservationists because of its invasive attributes. Because eradication may be impossible should public sentiment change (Harris 1988), conservative actions and exhaustive inquiry (DeBach 1964) are necessary in resolving conflicts of interest: introduction of a bioagent is generally irreversible after it is established. Biological control must be in the best public interest, and all concerns must be considered. The process is therefore highly regulated in the United States, at both the state and federal levels, although coordination and consolidation of these regulations would improve their effectiveness (U.S. Congress 1993).

Determining whether biological control is the best approach can be difficult, even after control is deemed appropriate. Eradication programs using chemical or physical methods may be prudent for newly arrived pests, but if the pest is likely to reinvade or eradication cannot be guaranteed, then biological control may be warranted—especially if the pest has been controlled elsewhere by a readily available bioagent. A typical weed program requires 11 to 13 years just to get the bioagents to the field (Andres 1977). Biological control should always be investigated as a potential strategy when the pest infests large areas. Not only are chemical or physical controls likely to be temporary and more expensive in this circumstance, but they also require a perpetual investment of human labor, whereas biological control programs are typically self-sustaining (Andres 1977). Best of all is to integrate biological, chemical, and physical controls in order to capitalize on the strengths of each.

Selecting the Target

The best targets for biological control are nonindigenous species with few native relatives. Nonindigenous species with many native congeners are more difficult to suppress. The contrast between tropical soda apple (*Solanum viarum*), from South America, and melaleuca (*Melaleuca quinquenervia*), from Australia, illustrates the difference. No native members of the genus *Melaleuca* or its subfamily, the Leptospermoideae, occur in the United States. A few members of its family, the Myrtaceae, are native to Florida, but none is closely related to *Melaleuca*. An organism introduced from Australia that was host-specific to at least the subfamily level would be unlikely to attack any native plant species in Florida. In contrast, the genus of tropical soda apple, *Solanum,* contains 33 species native to the United States, 27 of which are native to the Southeast (Soil Conservation Service 1982). The genus and family (Solanaceae) also include numerous economically important species: tomato, eggplant, potato, red pepper, tobacco, petunia, and others (Bailey 1971). Obviously, any South American insect introduced to control tropical soda apple must be highly specific. The elevated level of screening thus required raises the cost.

Insects and plants have historically constituted the greatest proportion of biological control targets (but see van Driesche 1994). These can be broadly classified as either pests of agriculture and horticulture or pests of native species and natural areas, although there is frequent overlap. Most biological control insects established in Florida (34 of 42) were released against insect pests of agriculture and horticulture. The remaining eight were released against invasive nonindigenous aquatic plants, which affect natural areas and agriculture.

Nonindigenous insects are arriving at alarming rates. Two destructive pests, the Mediterranean fruit fly (*Ceratitis capitata*) and the oriental fruit fly (*Dacus dorsalis*), were intercepted 304 and 2469 times, respectively, between 1985 and 1990 (Carruthers and Coulson pers. comm.). Estimates of the rate at which pests invade the United States vary, but Sailer (1983) has reported that one new major pest becomes established every three years. (See also Carruthers and Coulson pers. comm.) Frank et al. (Chapter 5 in this volume) report that at least one major pest and ten other species become established in Florida annually.

Nonindigenous species can have broad regional effects—such as destruction of American elms by the Dutch elm disease (*Ophiosoma ulmi*). This fungus is almost certainly of Asian origin but probably arrived from Europe prior to 1930 on elm veneer logs (Daughtrey 1993). The disease was vectored by another nonindigenous species, the European elm bark beetle (*Scolytus multistriatus*). These pests illustrate the tremendous potential for pathogens as control agents when the affected species are nonindigenous instead of natives (see Charudattan and Walker 1982). Frank et al. (Chapter 5 in this volume) summarize current knowledge on nonindigenous insects established in Florida.

The number of potential plant invaders entering the United States each year is staggering. Nearly 363 million nonindigenous plants, representing hundreds of species, were imported in 1993 through the port of Miami alone (D. R. Thompson, pers. comm.), so it is not surprising that many species escape cultivation to become established in native communities. Although some displace dominant native plant communities in "pristine" areas, most invade disturbed ecosystems. Natural areas can be protected from disturbance by creation of preserves, but not necessarily from invasion (Lonsdale 1992b).

Examples of troublesome nonindigenous plants are abundant. Melaleuca devastates valuable wetlands in Florida (Hofstetter 1991; Bodle et al. 1994; Chapter 3 in this volume). The neotropical floating water hyacinth blankets open water surfaces of lakes and rivers with dire consequences. Hydrilla forms dense beds in aquatic sites, usurping the water column and displacing other submersed aquatic vegetation. As a result of such invasions, aquatic communities lose their diversity and become monospecific, often lacking phytophagous consumers and producing detritivore-based faunal assemblages. (See, for example, Hansen et al. 1971 and Dray et al. 1993.)

Virtually every nonindigenous pest is vulnerable to biological control: the more fully a pest is studied, the more diverse its natural enemies are found to be (Wilson and Huffaker 1976). A biological control effort should therefore be preceded by a thorough study of the factors regulating the pest, and only host-specific bioagents should be introduced. Examples of scientific introductions of bioagents to control organisms other than insects, weeds, or plant pathogens are rare. Cats and mongooses were introduced in many places to control rodents, but by individual people who lacked training in biological control. The *Myxoma* poxvirus and the rabbit flea introduced into Australia to control the European rabbit illustrate the possibilities for vertebrate control (DeBach and Rosen 1991).

Selecting the Bioagent

Animals and pathogenic fungi that eat or parasitize a broad range of hosts or prey are called generalists (or polyphagous or euryphagous). The gypsy moth is a good example: it feeds on more than 100 tree species (Tietz 1972). The grass carp too is polyphagous—it feeds on a wide range of aquatic plants (Sutton and Vandiver 1986)—but most of the successful biological-control introductions have been of specialists, that is, highly host-specific species.

Species that exploit a single host are called monophagous and are the most desirable bioagents. Stenophagous organisms (specialists) exploit a few taxonomically related hosts. Because related hosts are often absent in the adventive range of the pest, the specialist attacks only a single species in that range and functions as a monophage. Host-specific species are commonly the most reliable control factors holding a given species at low densities in stable habitats. Huffaker et al. (1976) note that generalists contribute to overall community balance and serve as ancillary control factors when specialists fail. They also suggest that biological control of a widely distributed pest throughout its range will probably require an array of highly specific species.

Some workers suggest that generalists provide the best control, but the idea of using them has been criticized because of the danger that nontarget hosts might be attacked (Howarth 1983; Miller and Aplet 1993). The generalist predatory snail *Euglandia rosea,* for example, released against the giant African snail *Achatina fulica*, caused the extinction of several precinctive Hawaiian tree snails, including *Achatinella mustelina* (Hadfield and Mountain 1980). Most researchers believe that the safety of nontarget species is paramount, so they generally advocate using only specialists as control agents.

Some workers (Pimentel 1963; Carl 1982; Hokkanen and Pimentel 1984; Harris 1993) propose that natural enemies of species closely related to a pest are more promising, particularly for native pests, than natural enemies of the pest itself. The bioagent could be "any organism that has not had a major role in the evolutionary history of the host" (Lockwood 1993). Examples of successful "new associations" are limited to bioagents of the Cactaceae (Goeden and Kok 1986; Greathead 1986), so researchers (Huffaker et al. 1971; Carl 1982; Goeden and Kok 1986; Greathead 1986) have disputed the validity of this "new association" hypothesis. Andres (1981b) discusses the subtleties and nuances involved in this approach. Some argue that it should not be ruled out; others insist that native biota should never be subjected to biological control (Pemberton 1985b; Harris 1993; Lockwood 1993).

The use of vertebrates as bioagents is fraught with difficulties. Mongooses (*Herpestes auropunctatus*) released in Hawaii to control rats ravaged native, ground-nesting birds (Funasaki et al. 1988). Ferrets, stoats, and weasels caused similar declines in New Zealand when introduced to control rabbits (Simberloff 1992). Vertebrates make poor bioagents because most have broad host ranges. The grass carp, for example, introduced in Florida to control hydrilla, has occasionally been effective (Sutton and Vandiver 1986) but has often eliminated desirable native vegetation as well.

Managing the Bioagent

Some of the procedures cited earlier—searching for natural enemies, ranking candidates, determining their host fidelity, and establishing these agents in the pests' adventive range—have been discussed by others (Harris 1973, 1989, 1991; Wapshere 1975, 1981; Hokkanen and Pimentel 1984; Goeden and Kok 1986; Waage 1990) and need not be reiterated here. Strategies for postrelease management of these agents, however, have been discussed less often. The most common strategy is to rely on the bioagent to find the pest and to achieve autonomously the necessary abundances. This strategy provides the advantage over chemical and mechanical pest control that little effort is expended placing the agent in contact with the pest. DeBach (1964) notes that this strategy accounts for most of the successes. Its main advantage, however, is also its principal weakness. Because the bioagent becomes freely available after establishment, no profit incentive drives private exploitation. Thus biological control depends on public support and is underused.

What Are the Risks?

On the one hand, critics of biological control point to careless, unscientific introductions of vertebrates—like the mongoose in Jamaica and the cane toad in Australia—to express concerns about safety. On the other, advocates sometimes exaggerate its virtues, leading to unduly optimistic expectations. Public concerns over use of environmental toxicants, such as herbicides and pesticides, have contributed further to the view that biological control is an environmental panacea. Conversely, because there are few examples demonstrating ecological risks, critics often raise the specter of ecological disaster by citing unsupported anecdotes and cases in which vertebrates have been released by nonprofessionals (Howarth 1983; Miller and Aplet 1993). Biological control, like many sciences, has appropriate uses and inappropriate uses (Pemberton 1985b).

The risks fall into three main categories. Two of them—called "economic spillover" and "environmental spillover" by Tisdell et al. (1984)—have been sources of public controversy. The third is uncertainty. "Economic spillover" results when the bioagent autonomously searches for the pest over a wide geographic range after limited introductions. Although it is generally considered advantageous, it is disadvantageous if the person receiving the "benefit" does not want it. *Echium plantagineum* in Australia is a good example. The Australian Commonwealth Scientific and Industrial Research Organisation (CSIRO) planned to release several bioagents to control this rangeland weed because it is toxic to livestock and displaces valuable forage, but a few graziers use it as sheep forage during droughts, and some beekeepers value it (Delfosse and Cullen 1981). The project proceeded only after prolonged legal battles and passage of the Australian Biological Control Act of 1984 when a special commission determined that the program would have a benefit-to-cost ratio of 9:1. This litigation delayed release of the bioagents for nearly a decade (Delfosse 1990), and even then not all of the issues were resolved. Beneficiaries stood to gain more than opponents would lose, but the question of income distribution was ignored. The gainers did not compensate the losers (Tisdell et al. 1984). Issues like these can be resolved only through the political process.

"Environmental spillover" is attack by a bioagent on a nontarget species. The invasion of Florida by the cactus moth *C. cactorum* provides a good example. This South American stenophage was released as a "new association" in Australia to control *Opuntia inermis* and *O. stricta,* which originated from the Gulf coast of North America. It was a spectacular success. The Commonwealth Institute of Biological Control introduced this moth into the Leeward Islands in 1957 (Simmonds and Bennett 1966), but possible collateral effects on native Caribbean *Opuntia* were not considered. The moth either dispersed or was introduced throughout the Caribbean, eventually reaching Florida, where it attacked native cacti (Tuduri et al. 1971; Habeck and Bennett 1990; Pemberton 1995). *Opuntia stricta* is native to Florida, where populations may now be jeopardized. The rare semaphore cactus (*O. corallicola*), which occurs only in the Florida Keys, is also endangered (Simberloff 1992). The consequences could be even more serious if *C. cactorum* enters states with many native *Opuntia* species. (See Chapter 5 in this volume for further discussion.) This example illustrates the need to consider broad regional implications before introducing bioagents. Releasing *C. cactorum* in Australia was appropriate, even though the insect was not entirely host-specific, because no native *Opuntia* occur there. It was an inappropriate organism to release in the Caribbean region, however, where native *Opuntia* do occur.

Howarth (1983) has claimed that biological control of exotic insects in Hawaii extirpated many native insects, resulting in turn in the demise of some native birds, but Funasaki et al. (1988) point out that most parasitoids implicated in the decline of native Hawaiian Lepidoptera were not introduced deliberately. Lai (1988) notes that there are no data showing that bioagent introductions contributed to these extinctions. Neither do any data indicate that they did not. In point of fact, there are no data.

Less equivocal examples of environmental spillover appear to be lacking. Most of the frequently cited examples, including many discussed by Howarth, stem from early projects lacking rigorous safety protocols. Nonetheless, some agents introduced to control pest insects do attack agents introduced to control weeds (Howarth 1983; R. Friesen, pers. comm.), thus preventing their establishment or limiting their effectiveness.

Another concern related to environmental spillover is that an agent might "switch" hosts. Agriculture provides many examples in which a stenophagous native insect broadened its host range to include a nonindigenous host. The apple maggot fly (*Rhagoletis pomonella*), for example, fed exclusively on hawthorne prior to the introduction of the apple into the United States but now accepts the latter as an alternate host (Feder 1995). The Baltimore checkerspot butterfly (*Euphydryas phaeton*) specialized on turtlehead (*Chelone glabra,* a U.S. native) prior to the introduction of plantain (*Plantago lanceolata*). Some populations of the Baltimore checkerspot in the northeastern United States now use plantain exclusively (Bowers et al. 1992). We know of no investigations demonstrating this phenomenon in the reverse, however. That is, despite the fact that "biological control introductions are experiments on a grand scale" (Greathead 1986) that provide ample opportunity for imported insects to switch from their exotic hosts to native plants, no such host range extensions have been reported (Harris 1993).

Uncertainty of success is the third major risk. All releases are experiments in which neither agent establishment nor pest suppression is guaranteed. Risk-averse public agencies and landholders (such as farmers) are often willing to pay a premium to reduce the risk of uncertainty. If biological control is effective, but not as reliable as chemical control, these public agencies and landholders are likely to continue to use chemicals.

What about Resistance?

Just as insects routinely develop resistance to insecticides, pests might develop resistance to bioagents. As noted in the cactus moth example, however, biological control normally remains effective for many years. Resistance might, in fact, be offset by parallel evolution of increased virulence of the bioagent (Tisdell et al. 1984). Such evolution by bioagents is illustrated by the flea beetles (*Chrysolina* spp.), which were introduced into northwestern North America to control Klamath weed (*Hypericum perforatum*). *Chrysolina quadrigemina* and *C. hyperici* quickly reduced this weed by 98 percent in California, but *C. quadrigemina* initially did not thrive in humid areas of British Columbia. It adapted to the local climate within a few years, however, and now provides better control in that region (Peschken 1972; Andres et al. 1976; Harris and Maw 1984). In contrast, the European rabbit developed resistance to the *Myxoma* poxvirus that was introduced into Australia to control it (Davis et al. 1976). Evolution of resistance to bioagents can thus occur but is probably of little overall significance.

What Are the Benefits?

In general, benefits are derived only from successful biological control programs, though one might reasonably argue that the knowledge gained from "failed" projects is beneficial. Success, in turn, can be defined on two distinct levels: economic and ecological. A project might be considered successful when project costs are less than the benefits gained, though the ratio may be difficult to determine (DeBach 1964; Hokkanen 1985). The aesthetic value of conservation areas, for example, cannot easily be translated into economic terms (Dennill and Donnelly 1991). Weed-induced losses are difficult to estimate, too, as are savings realized from biological control. Simultaneous implementation of other control measures simply complicates the problem.

Ecological benefits derived from the project may be difficult to measure, as well. A small reduction in pest equilibrium density may, for example, be considered a complete success in one situation, whereas a large reduction may be insufficient in another (Hokkanen 1985). Ehler and Andres (1983) say that "a project may be successful, ecologically speaking, when the natural enemy establishes itself, affects the host, and enters into a balance with its host and the environment. Unfortunately, this balance may not always result in the desired level of economic control."

Past Successes

Hokkanen (1985) has reported that nearly 3000 bioagent introductions have occurred worldwide during the past 100 years. These introductions represent nearly 1000 agents released against 200 pest insects and weeds. There have been over 525 biological control attempts against pest plants—an attempt denotes the introduction of one control agent into one country against one target weed (Julien 1982)—but only 117 of these attempts were rated as successful, so 60 to 75 percent were failures. Of the 51 arthropods introduced into North America for weed control, 38 became established and 15 of these were effective (that is, reduced weed abundance somewhere within their range; Ehler and Andres 1983). Complete biological control of insect pests resulted from about 16 percent of the projects attempted, but some ecological success resulted from 58 percent of the attempts (Hall et al. 1980). Overall, complete success as a proportion of total success has steadily risen since the 1930s as knowledge of ecological processes has increased (Hokkanen 1985). Failures most often result from inadequate information about the system or from lack of persistence in attempts (van den Bosch 1968).

Frank and McCoy (1993) report the release of 151 insect species in Florida since 1890 for use against insects and weeds. About 28 percent of these agents became established. Thirty-four were released against insect pests of agriculture and horticulture, the other eight against aquatic weeds (Frank and McCoy 1993).

Cost-Benefit Analyses

The benefits derived from biological control vary with the potential productivity of the pest-free resource and the degree of control realized (Andres 1981a). About 3.5 scientist-years (SY) are required to study and develop a weed-feeding insect (Andres 1981c). (A scientist-year equals the time of one scientist plus technical, administrative, and logistical support, currently about $300,000 per SY based on USDA estimates.) When a complex of weed-feeding insects is needed against a particular target (as with melaleuca), 20 SY may be required (Harris 1993). Fortunately, clearance of the first agent often reduces the time required to produce additional agents—important because host screening constitutes a major project expense. Evaluation of the agent's performance is increasingly regarded as a vital component in biological control programs (U.S. Congress 1993), even though it can easily require five SY or more when the effects are subtle and manifested over a wide geographical range.

A complete program typically costs $4 to $6 million for a weed target (Harris 1979, 1993). If the project is a complete success, however, enormously larger savings are realized. Annual benefits from classical biocontrol programs exceed $180 million in the United States, and this estimate includes only those programs (four on weeds and five on arthropods) for which reliable cost-benefit data are available (Coulson and DeLoach in press). Hays (1992, cited by Coulsen and Delouch, in press) attributes to classical biological control over $2 billion in savings to U.S. agriculture during the past decade alone. In contrast, the total cost of all federal and state research on classical biological control from 1888 to 1976 has been estimated at a mere $20 million, and annual U.S. expenditures for all (not just classical) biological control programs totaled only $21.5 million in 1987 (Coulson and DeLoach in press).

Milon et al. (1986) have reported that hydrilla infestations on just two Florida lakes caused $10 million annually in recreational-industry losses. The state of Florida spent $1.2 million in 1989 to apply herbicides to 5000 acres of a 15,000-acre hydrilla infestation at Lake Istokpoga (Highlands County). Florida's public waterways harbor about 80,000 acres of hydrilla; attaining maintenance-control levels would require an annual expenditure of about $14.5 million (Chapter 14 in this volume).

A chemical eradication effort directed against citrus blackfly in southeastern Florida cost $15 million in 1976. A biological control program was successfully implemented (both economically and ecologically) after this effort failed. The biocontrol program cost $2.2 million but saves $9.3 million annually (in 1980 dollars). This program was founded on research in Mexico and Texas, without which it would have cost more (Frank and Bennett 1990). The return on the investment is better than 400 percent annually.

When the use of chlordane was banned in the 1970s, cattle ranchers were left without a pesticide to control mole crickets. State-funded research led to release of three bioagents (a wasp, a nematode, and a fly) during 1983–1987. They provided partial (ecological) success by 1991. Florida's budgetary difficulties in 1991 led to dismantling of this program before additional bioagents could be evaluated. Meanwhile, mole crickets caused losses of more than $30 million to the turf industry in 1986 (Sailer et al. 1994).

Biological control of water hyacinth can be very cost-effective (Center et al. 1990). Records from the Louisiana Wildlife and Fisheries Commission document a statewide reduction of water hyacinth from about 1.2 million acres to some 300,000 acres after introduction of biological control agents. The immediate value of a reduction of this magnitude would be about $25 million based on the minimum cost for one herbicidal treatment. The accumulated value to date is nearly $350 million. The actual saving is much less, however, because the vastness of the infestations limited herbicidal control to less than 1 percent of the total acreage. Herbicide (2,4-D) usage on water hyacinth simultaneously declined from 15,000 gallons per year to 2000 gallons per year—yet a higher percentage of the total infestation was treated, thus producing more effective overall control. This example clearly illustrates the principle that "a region which does not use man's money and energy unnecessarily to perform tasks which nature can perform for free will have an economic advantage " (Bayley and Odum 1973 cited by Andres 1981a).

Florida's Experiences

In Florida, biological control projects against weeds of natural areas, particularly weeds of aquatic and wetlands habitats, currently outnumber those against insect pests. Projects focusing on pest insects are discussed by Frank et al. (Chapter 5 in this volume). Plant species that have been subjected to biological controls include alligatorweed, water hyacinth, hydrilla, water lettuce (*Pistia stratiotes*), Eurasian water milfoil (*Myriophyllum spicatum*), melaleuca, Brazilian pepper (*Schinus terebinthifolius*), and milkweed vine (*Morrenia odorata*).

Alligatorweed was the first aquatic weed to become the target of biological control in the United States. Research conducted in Argentina identified more than 40 insects attacking alligatorweed, a South American native (Coulson 1977). Three of these were released in Florida and elsewhere: the alligatorweed flea beetle (*Agasicles hygrophila*), the alligatorweed thrips (*Amynothrips andersoni*), and the alligatorweed stem borer (*Vogtia malloi*). (See Vogt 1960, 1961; Maddox 1968, 1970; Maddox et al. 1971; Brown and Spencer 1973.) This project has been so successful that in 1981 fewer than 1000 problem acres of alligatorweed were reported in the United States; over 97,000 acres were reported for 1963, prior to release of the insects (Cofrancesco 1991).

The water-hyacinth weevils *Neochetina eichhorniae* and *N. bruchi* were first released in the United States in 1972 and 1974, respectively (DeLoach 1976; DeLoach and Cordo 1976; Center 1982). The water-hyacinth moth *Sameodes albiguttalis* was first released in 1977 (Cordo and DeLoach 1978; DeLoach and Cordo 1978; Center and Durden 1981). Although these agents have occasionally elicited dramatic reductions in water-hyacinth coverage (Center et al. 1990), the effects frequently are more subtle and do not result directly in plant mortality. Instead, high bioagent abundances slow growth (Center and Van 1989). Populations therefore contain fewer, smaller plants; they expand at a lower rate; and they are more susceptible to other causes of mortality (Center 1987; Center and Van 1989). Such causes include herbicides, frost, and *Cercospora rodmanii,* a fungal pathogen of water hyacinth that induces leaf necrosis and secondary root rot (Mitchell et al. 1994). Water-hyacinth bioagents have become established throughout the Southeast and California.

The South American weevil *Neohydronomus affinis* was released against water lettuce in Florida in 1987 (Thompson and Habeck 1989; Dray et al.

1990). At some sites, rapidly increasing weevil populations caused dramatic declines (Dray and Center 1992). This bioagent has established itself throughout peninsular Florida, Louisiana, and Texas (Grodowitz 1991; Grodowitz et al. 1992). A second water-lettuce insect used in Florida is the Asian water-lettuce moth (*Spodoptera pectinicornis*). This insect was first employed in Thailand to eliminate water-lettuce infestations in rice paddies (Suasa-Ard and Napompeth 1982). Persistent populations have not yet become established in Florida.

Four insects have been released to control hydrilla in Florida. Releases of the Indian tuber weevil (*Bagous affinis*) and the Asian hydrilla fly (*Hydrellia pakistanae*) began in 1987 (Center et al. 1990; Buckingham 1994). The tuber weevil is best suited to waterways that experience periodic droughts. Nascent populations gained temporary footholds at one waterway in Florida and another in California, but evidence for long-term persistence is lacking (Center 1992a; Grodowitz et al. 1993). The fly is established in Florida, Texas, Alabama, and Georgia (Center 1992b; Grodowitz et al. 1993). High densities of *H. pakistanae* have been associated with declining hydrilla populations in Florida and Alabama (Grodowitz et al. 1993). Two Australian insects, the hydrilla fly (*Hydrellia balciunasi*) and the hydrilla stem weevil (*Bagous hydrillae*), were released in 1989 and 1991, respectively (Center 1992b). Both became established in Texas (Center 1992a; Grodowitz et al. 1993), and the weevil has persisted at one Florida site for several months. A weevil (possibly another *Bagous* species) and a root-feeding beetle are being studied in China for use against hydrilla (Buckingham 1992). An African tip-feeding midge from Lake Tanganyika (Pemberton 1980; Markham 1986) also has potential as a bioagent for control of hydrilla. Although much is already known about insect herbivores of hydrilla from various regions of the world, much remains to be investigated.

Early investigations of Eurasian-water-milfoil insects in eastern Europe (Lekic and Mihajlovic 1971) and Pakistan (Baloch et al. 1972) produced no useful bioagents. More recently, Creed and Sheldon (1991, 1992, 1994) studied the weevil *Euhrychiopsis lecontei*. The fungus *Mycoleptodiscus terrestris,* isolated in Massachusetts during 1979, has also been studied for use against both Eurasian water milfoil and hydrilla (Gunner 1983; Shearer 1992, 1993; Theriot et al. 1993). Chinese insects of possible use against Eurasian water milfoil include stem-feeding *Eubrychius* and *Bagous* weevils, a flower-feeding *Phytobius* weevil, a *Hydrellia* leaf-mining fly, and a tip-feeding *Cricotopus* midge (Buckingham 1992, 1993). *Bagous, Phytobius, Eubrychius,* and *Hydrellia* have been imported into U.S. quarantine for further studies (C. A. Bennett 1993; Buckingham 1992, 1993).

Two Australian insects that attack melaleuca, the tip-feeding weevil *Oxyops vitiosa* and the sawfly *Lophyrotoma zonalis,* are under study in U.S. quarantine (C. A. Bennett 1993; Balciunas et al. 1994). Other candidates under investigation in Australia include a gall fly (*Fergusonina* sp.), three sap-feeding bugs (*Boreioglycaspis melaleucae, Eucerocoris suspectus,* and *Pomponatius typicus*), and a stem-boring cerambycid beetle (Habeck et al. 1994). South American insects being investigated for use in Florida against Brazilian pepper include the thrips *Liothrips ichini,* the sawfly *Heteroperryia hubrichi,* and several leaf tiers from the genus *Episimus* (Habeck et al. 1994).

A fungal pathogen, *Phytopthora palmivora,* was developed for use against milkweed vine during the 1970s and early 1980s (Ridings et al. 1977). Other pathogens have been or are being investigated. (See Charudattan and Walker 1982; Mitchell et al. 1994.)

What Does the Future Hold?

Because biological control has been used less frequently in Florida than in California and Hawaii, public perceptions of its safety, economy, and efficacy are still being formed. The public must be educated about its goals and practices so they can make informed decisions. Garnering adequate support, both political and financial, has always proved difficult:

> Although classical biological control celebrated its 100th anniversary last year [1989], it has never received appropriate attention in the United States. During the first fifty years, it was plagued by bureaucratic rivalries and shifting priorities within U.S.D.A. and between federal authority and the states. By 1950, it was overtaken by the rush to chemical controls, and although work has proceeded, it has been only through the persistent efforts of a small group of biological control researchers. Their work may suffer neglect again if the more glamorous pursuit of genetic engineering is allowed to overshadow work on population dynamics and the basic interactions among organisms. Biological control should never again take a back seat to other forms of pest control research. Consistent funding and coordinated regulatory support would provide the vital start to make biological control central to United States pest control programs. [Hinkle 1992:127]

This statement applies as well to Florida as to the United States as a whole.

Even though Florida is one of the three states most affected by nonindigenous species, unlike Hawaii and California it has no policy, no consistent source of funding, and no lead agency for dealing with these pests. Schmitz et al. (1993) note, in fact, that $98 million was spent in Florida during 1980–1991 for chemical and mechanical control of aquatic weeds alone. Funding for classical biological control was meager in comparison: programs directed at aquatic weeds received less than 4 percent of this amount over the same 12-year period.

Delfosse (1990) estimates that there is a 50-year backlog of projects in Australia. Harris (1991) estimates a similar backlog in Canada. These are probably conservative estimates and pale in comparison to Florida, where nonindigenous species comprise "about 15 % of Florida's plant species, 16 % of its fish species, 22 % of its amphibian species, 42 % of its reptile species, 5 % of its bird species, and 23 % of its mammalian species" (Frank and McCoy 1992). The Florida Exotic Pest Plant Council has listed nearly 150 invasive nonindigenous plant species (Habeck et al. 1994). Many of these will eventually require biological controls. Several are already increasing rapidly: melaleuca, Brazilian pepper, the Australian pines (*Casuarina equisetifolia, C. glauca,* and *C. cunninghamiana*), Japanese climbing fern (*Lygodium japonicum*), carrotwood (*Cupaniopsis anacardioides*), Chinese tallow tree (*Sapium sebiferum*), and skunk vine (*Paederia foetida*). Similar numbers of insect pests probably exist. Thus, even if it were possible to halt further invasion by nonindigenous species, it is unlikely that the need for biological control in Florida would soon end.

In 1987, the National Academy of Sciences reported that "biological control can and should become the primary method used in the United States. . . . The need for alternatives to complement or replace chemical control dictates placing an increasing emphasis on biological control research and development." This recommendation will never be fully implemented until the science of classical biological control gains the support it deserves.

PART IV

Managing Nonindigenous Species: Policy and Implementation

16 Management in National Wildlife Refuges

Mark D. Maffei

The U.S. Fish and Wildlife Service (FWS) manages close to 400,000 ha of land in Florida (Table 16.1). From the hardwood hammocks of St. Marks National Wildlife Refuge in North Florida to the tropical islands of Key West National Wildlife Refuge, it manages a wide range of biological communities. Among the management goals for virtually all of these lands is the maintenance or restoration of the function, structure, and species composition of the native ecosystem.

The FWS has a national mandate to protect fish and wildlife and their habitats and has, within the last few years, begun to take an ecosystem approach to its mandate (U.S. Fish and Wildlife Service 1994). Because the invasion of native communities by nonindigenous species significantly threatens the systems managed by the FWS, its land managers devote substantial resources to reducing their impact. Nonindigenous trees are controlled by fire, herbicides, or mechanical removal. Aquatic weeds are controlled with herbicides or water management regimes designed to limit their spread. Controlling nonindigenous animals may require sacrificing native species—as when rotenone is used to remove carp from lakes and ponds—but is necessary because invasive nonindigenous species are one of the greatest threats facing the management of FWS lands.

All the national wildlife refuges (NWRs) in Florida have been invaded by nonindigenous life forms. Some invasions are of little consequence; others pose a serious threat. Some nonindigenous plants and animals found on NWRs in Florida are listed in Tables 16.2 and 16.3. The effort to control any of these species varies from refuge to refuge and depends on the threat it poses.

TABLE 16.1.

National Wildlife Refuges in Florida, September 1993

Refuge	Size (ha)
Archie Carr NWR (Indian River and Brevard counties)	21
Arthur R. Marshall Loxahatchee NWR (Palm Beach County)	59,500
Caloosahatchee NWR (Lee County)	16
Cedar Keys NWR (Levy County)	337
Chassahowitzka NWR (Citrus County)	12,322
Crocodile Lake NWR (Monroe County)	2656
Crystal River NWR (Citrus County)	27
Egmont Key NWR (Monroe County)	133
Florida Panther NWR (Collier County)	9465
Great White Heron NWR (Monroe County)	77,932
Hobe Sound NWR (Martin County)	397
Island Bay NWR (Charlotte County)	8
J. N. "Ding" Darling NWR (Lee County)	2165
Key West NWR (Monroe County)	84,335
Lake Woodruff NWR (Volusia County)	7913
Lower Suwannee (Levy and Dixie counties)	20,299
Matlacha Pass NWR (Lee County)	207
Merritt Island NWR (Brevard County)	55,977
National Key Deer NWR (Monroe County)	3318
Okeefenokee NWR (Baker County)	1489
Passage Key NWR (Pinellas County)	26
Pelican Island NWR (Indian River County)	1908
Pine Island NWR (Lee County)	222
Pinellas NWR (Pinellas County)	13
St. Johns NWR (Brevard County)	2532
St. Marks NWR (Wakulla and Jefferson counties)	26,547
St. Vincent NWR (Franklin County)	506

Source: U.S. Fish and Wildlife Service (1993).

Nonindigenous Plants

Although most nonindigenous plants pose little threat to NWR lands, some species aggressively invade undisturbed areas or outcompete native species—including threatened and endangered ones (Duever et al. 1979; Grow 1984)—for disturbed areas. They crowd out native plants (Myers 1983) on which wildlife depend, alter ecosystems by increasing evaporation, and poison and irritate wildlife and nearby plants (as do Brazilian pepper and melaleuca; Austin 1978a; Williams 1980). In Florida alone, control and eradication of nonindigenous plants on national wildlife refuges in fiscal year 1993 exceeded $200,000.

Melaleuca quinquenervia (melaleuca) is one of the most problematic alien

TABLE 16.2.
Partial List of Nonindigenous Plant Species
Found on National Wildlife Refuges in Florida

Scientific name	Common name
Acacia auriculiformis	earleaf acacia
Alternanthera philoxeroides	alligator weed
Brachiaria mutica	para grass
Casuarina equisetifolia	Australian pine
Casuarina glauca	scaly-bark beefwood
Clerodendrum speciosissimum	Java glorybower
Colocasia esculenta	wild taro
Colubrina asiatica	lather leaf
Cupaniopsis anacardioides	carrotwood
Dioscorea bulbifera	air potato
Egeria densa	Brazilian elodea
Eichhornia crassipes	water hyacinth
Hydrilla verticillata	hydrilla
Ipomoea aquatica	water spinach
Leucaena leucocephala	lead tree
Lonicera japonica	Japanese honeysuckle
Lygodium japonicum	Japanese climbing fern
Melaleuca quinquenervia	melaleuca
Myriophyllum spicatum	Eurasian water milfoil
Panicum repens	torpedo grass
Paspalum notatum	bahia grass
Pistia stratiotes	water lettuce
Psidium guajava	guava
Salsola kali	Russian thistle
Salvinia rotundifolia	water spangles
Sapium sebiferum	Chinese tallow tree
Scaevola taccada var. *sericea*	scaevola
Schinus terebinthifolius	Brazilian pepper
Sesbania punicea	purple sesban
Sesbania vesicaria	bladderpod
Sonchus asper	prickly sow thistle
Sonchus oleraceus	common sow thistle
Thespesia populnea	seaside mahoe
Vitex trifolia	vitex
Wisteria sinensis	Chinese wisteria
Yucca aloifolia	Spanish bayonet

TABLE 16.3.

Partial List of Nonindigenous Animal Species Found
on National Wildlife Refuges in Florida

Scientific name	Common name
Anolis sagrei	brown anole
Astronotus ocellatus	oscar
Belonesox belizanus	pike killifish
Bufo marinus	marine toad
Cervus unicolor	sambar deer
Cichlasoma bimaculatum	black acara
Clarias batrachus	walking catfish
Columba livia	rock dove
Dasypus novemcinctus	armadillo
Eleutherodactylus planirostris planirostris	greenhouse frog
Felis catus	feral cat
Felis yagouaroundi	jaguarundi
Hemidactylus turcicus turcicus	Mediterranean gecko
Leiocephalus carinatus armouri	northern curly-tailed lizard
Marisa cornuaurietus	golden-horn marisa
Molothrus bonariensis	shiny cowbird
Oreochromis aureus	blue tilapia
Osteopilus septentrionalis	Cuban tree frog
Passer domesticus	house sparrow
Pomacea bridgesi	spike-topped apple snail
Rattus norvegicus	Norway rat
Rattus rattus	black rat
Sphaerodactylus elegans	Indo-Pacific gecko
Streptopelia risoria	ringed turtledove
Sturnus vulgaris	European starling
Sus scrofa	pig
Tilapia mariae	spotted tilapia

plants on NWRs in Florida. It imperils NWR resources in southern Florida and is capable of becoming a problem in central Florida as well. The species thrives in freshwater marshes and upland areas (Woodall 1980; Myers 1984; Hofstetter 1991; Chapter 3 in this volume)—habitats well represented on NWR lands— and is a severe problem on A.R.M. Loxahatchee, National Key Deer, Crocodile Lake, Hobe Sound, and Merritt Island NWRs, where it is aggressively invading the marsh communities. It has been removed from J. N. "Ding" Darling NWR. Because of its impact on the freshwater marshes of the Everglades, this tree threatens the endangered Everglade kite (*Rostrhamus sociabilis*) and wood stork (*Mycteria americana*); indeed, it is a threat to all organisms that depend on marshes (Austin 1978a; Mazzotti et al., 1981). Since implementation of a control program (Maffei 1991) in April 1992, more than a million melaleuca trees have been removed from A.R.M. Loxahatchee NWR alone. Control programs for this plant are also under way at Crocodile Lake, Florida Panther, Merritt Island, National Key Deer, and Hobe Sound refuges.

Coastal refuges are seriously affected by Australian pine (*Casuarina equisetifolia*). Beaches and dunes are rapidly invaded by this tree, the canopy of which shades out native vegetation. This tree is a particular problem on refuges, such as Hobe Sound, Merritt Island, and Archie Carr NWRs, with beaches used as nesting sites for sea turtles, all of which are federally listed as endangered or threatened species. When a female sea turtle encounters the roots of the Australian pine while excavating her nest, she abandons her nesting attempt. The large tangles of roots exposed when the trees topple, as they often do, have trapped and killed both adult and hatchling sea turtles. Control of Australian pine is under way at Merritt Island, Hobe Sound, J. N. "Ding" Darling, Egmont Key, Crocodile Lake, and Florida Panther NWRs.

Brazilian pepper (*Schinus terebinthifolius*) has invaded Merritt Island, National Key Deer, Crocodile Lake, Hobe Sound, A.R.M. Loxahatchee, Egmont Key, Key West, Great White Heron, J. N. "Ding" Darling, and Florida Panther NWRs. Control efforts are under way at all these sites. On Florida Panther NWR, shading caused by dense stands of Brazilian pepper and Australian pine kills plants used by white-tailed deer (*Odocoileus virginianus*), which are an important food of the endangered Florida panther (*Felis concolor coryi*).

Other nonindigenous plants for which control efforts are under way on refuges in Florida include the Chinese tallow tree (*Sapium sebiferum*) at St. Vincent NWR; lather leaf (*Colubrina asiatica*) at National Key Deer Refuge; water hyacinth (*Eichhornia crassipes*) at Lake Woodruff and A.R.M. Loxahatchee refuges; lead tree (*Leucaena leucocephala*) at National Key Deer, Crocodile Lake, and Key West refuges; and Japanese climbing fern (*Lygodium japonicum*) at Hobe Sound.

Nonindigenous Animals

The dangers of nonindigenous animals that invade national wildlife refuges in Florida are difficult to assess. Although it is relatively easy to determine the extent to which nonindigenous plants can invade native areas, the impact of nonindigenous animals on native biological communities, and on those native species with which they compete directly, is often less obvious. The Norway rat (*Rattus norvegicus*), for example, found throughout the state, may directly compete with the endangered Key Largo wood rat (*Neotoma floridana smalli*) and cotton mouse (*Peromyscus gossypinus allapatocola*) in the Florida Keys NWRs, but the direct impact is difficult to measure.

The impacts of certain nonindigenous animals are less obscure. Feral pigs

(*Sus scrofa*) threaten both plant and animal communities. They destroy the nests of sea turtles, gopher tortoises (*Gopherus polyphemus*), indigo snakes (*Drymarchon corais couperi*), shore and wading birds, and other species, and they prey on native reptiles and amphibians. Their rooting disrupts the soil, causing dune destabilization or damage to wetlands and other plant communities. They compete with native animals for acorns, an important food resource. On some refuges, such as St. Vincent and Florida Panther NWRs, the pig is a reservoir for pseudorabies virus, which threatens large carnivores such as the endangered Florida panther and is deadly to the endangered red wolf (*Canis rufus*). Feral pigs are present on A.R.M. Loxahatchee, Chassahowitzka, Lower Suwannee, Merritt Island, National Key Deer, St. Marks, and St. Vincent NWRs. Hunting to control populations occurs on Lower Suwannee, St. Marks, and St. Vincent refuges. Pigs are taken on Merritt Island NWR by a contract hunter and by authorized refuge personnel.

The armadillo (*Dasypus novemcinctus*), considered nonindigenous by some (Chapter 10 in this volume), affects native life forms. Armadillos excavate sea-turtle nests and feed on the eggs. Like feral pigs, they disrupt soils and can kill seedling plants. Armadillos are found on NWRs throughout Florida except in the Florida Keys. No refuge removes armadillos systematically, but on Hobe Sound and St. Vincent NWRs, they are removed when encountered.

Many other nonindigenous species that may affect native species are found on Florida NWRs. The marine toad (*Bufo marinus*) can poison larger animals that attempt to eat it. The brown anole (*Anolis sagrei*) appears to be displacing the native green anole (*A. carolinensis*) in the Florida Panther NWR (J. Krakowski, pers. comm.). Nonindigenous fish species, such as oscars (*Astronotus ocellatus*), may be competing with native sport and nongame fishes for nest sites and food (Chapter 7 in this volume). Oscars and other nonindigenous fishes also prey on juveniles of native fishes and on small nongame species. Fire ants (*Solenopsis invicta*) compete with and displace native ants throughout Florida and may be affecting ground-nesting birds and burrowing animals (Chapter 5 in this volume). On St. Vincent NWR, nonindigenous sambar deer (*Cervus unicolor*) damage or kill trees on which they rub their antlers.

Future Needs

Management and control of nonindigenous species of plants and animals on NWRs in Florida is generally paid for out of a refuge's base operations and maintenance funding. But refuge budgets are simply inadequate for the control of invasive nonindigenous species. Thus control programs for these species on national wildlife refuges can be characterized as "when time allows" programs. Refuge managers facing the choice between spending scarce dollars on controlling invasive plants and animals (which may not yet have a substantial impact on the refuge's resources) and completing other refuge projects and maintenance generally opt to delay work on nonindigenous species. When these species become a problem, however, the cost of removal is much greater than it would have been had there been early control (Chapters 14 and 23 in this volume).

On some NWRs, the impact of nonindigenous species is so great that special funding is provided. A.R.M. Loxahatchee NWR has special funding for control of melaleuca, for example, and Crocodile Lakes NWR for control of Australian pine, Brazilian pepper, and melaleuca. Funding to monitor and control potential problem plants and animals should be available before a crisis develops. Research on the ecosystems of national wildlife refuges is needed so that potential impacts of nonindigenous species can be determined before they occur. Control efforts can then be initiated more efficiently and economically.

Finally, interpretation and education are needed. The support of the public is necessary if efforts to prevent the introduction of nonindigenous species, or to control those already present, are to succeed. Only when the value of protecting and maintaining native ecosystems is explained will the public understand the damage caused by nonindigenous species and support the efforts of the FWS and other land management agencies to control them.

17 Plant Management in Everglades National Park

R OBERT F. D OREN AND D AVID T. J ONES

Everglades National Park, a World Heritage Site and Biosphere Reserve, encompasses more than 600,000 ha and is the only subtropical wilderness in the continental United States. The park is located in southern Florida at the southern terminus of the vast wetland complex known as the Everglades (Figure 17.1). Established in 1947 to preserve the unique biological resources of the area, the park contains a variety of habitats within its boundaries: shallow-water marine habitat (240,000 ha), saltwater wetland forests and marshes (192,000 ha), freshwater marshes and prairies (162,000 ha), and upland pine and tropical hardwood forests (6000 ha). The distribution of vegetation is largely controlled by the hydrological regime, surface geology, and overlying soil type. Natural disturbances (fires, freezes, hurricanes, sea level changes) and human activity (drainage, development, introduction of nonindigenous plants) also have powerful effects on vegetation patterns. Remarkably, this complex ecological wilderness lies within a hundred miles of 3.5 million people.

Of the 850 plant species reported from the park, 221 (26 percent) are nonindigenous in origin (Whiteaker and Doren 1989). This number includes more than half of all the nonindigenous species known to be established in southern Florida. Indeed, Loope (1992) considers the park among the four U.S. national parks worst affected by nonindigenous species. Many of the nonindigenous species were planted by settlers before the park's establishment. Natural disturbances that are part of the southern Florida environment have allowed weedy

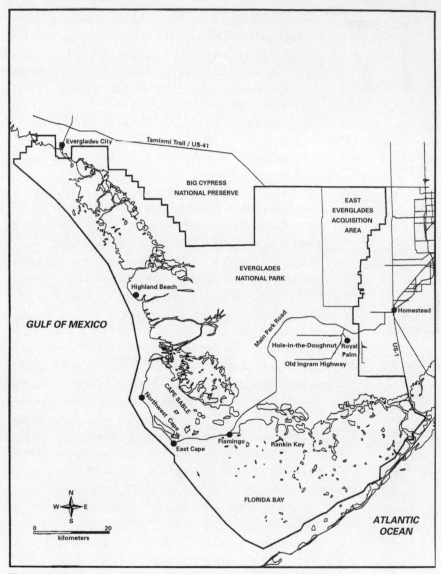

Figure 17.1 Map of Everglades National Park and environs.

species to become established, an effect amplified by human activities. The most successful nonindigenous species are so well adapted to an altered habitat that they outcompete native species (Ewel et al. 1982).

Management of nonindigenous plants in the park has developed in response to laws, general directives, and policies. Under the National Park Service (NPS) Organic Act of 1916, the NPS is charged with management of the parks to "conserve the scenery and the natural and historic objects and the wildlife therein and to provide for the enjoyment of the same in such manner and by such means as will leave them unimpaired for the enjoyment of future generations." NPS policy states that nonindigenous species will be managed "up to and including eradication . . . whenever such species threaten park resources . . . [and] high priority will be given to [nonindigenous species] that have a substantial impact on park resources . . . " (National Park Service 1988, 1991).

Management of nonindigenous plant species is given high priority in the park's Resource Management Plan (Everglades National Park 1982) and is articulated in the Everglades *Exotic Plant Control Handbook* (Doren and Rochefort 1983), which establishes guidelines, priorities, and methods for controlling nonindigenous plants in each of the park's four districts: Pine Island, Flamingo, Florida Bay, and Northwest. Work is shifted among these districts according to other work assignments and funding.

Historical Perspective

The first efforts at plant eradication in the park began before the park's establishment and were directed at a native species—wild cotton (*Gossypium hirsutum*)—in an effort to control the spread of the cotton boll weevil *(Anthonomus grandis;* LaRosa et al. 1992). In the years after the park was established, little attention was given to nonindigenous species or their effects on the environment. Many plants were associated with abandoned homesites or dwellings of both native Indians and Europeans. The earliest control work focused on aquatic weeds in the Royal Palm Pond. After Hurricane Donna in 1960, park staff noticed an increase in populations of Australian pine (*Casuarina* spp.), and the first control was attempted in 1963. Australian pine trees continued to spread, however, after Hurricane Betsy in 1965 (LaRosa et al. 1992). In 1969, the first nonindigenous plant management plan for the park was directed at Australian pine (Klukas 1969).

The first formal, comprehensive, nonindigenous plant management plan for the park, written in 1973 and updated in 1977, included over 100 additional pest plants not previously reported. The highest-priority actions were to control Australian pine and to eradicate melaleuca (*Melaleuca quinquenervia*). The latest update, in 1988 (Whiteaker and Doren 1989), listed 221 species of nonindigenous plants in the park. The update provides information on current distribution, potential to spread and invade native plant communities, and management approaches.

Australian pine, melaleuca, and Brazilian pepper (*Schinus terebinthifolius*), three of the park's most disruptive and widespread nonindigenous plant species, have been the focus of most of the management effort. Two additional species, however, lather leaf (*Colubrina asiatica*) and shoebutton ardisia

(*Ardisia elliptica*), are becoming increasingly widespread in the park and are now the target of some control measures.

Australian Pine

Casuarina equisetifolia and *C. glauca,* two of eight species of Australian pine introduced into Florida, are found in the park (Figure 17.2). The former species invades many vegetation types but is restricted in its distribution by long hydroperiods. It survives best on well-drained sites and tolerates brackish soils and sea spray, colonizing open sand and shell beaches and coastal prairies. Scattered individuals can be found in higher-elevation mangrove stands and in the interior of keys in Florida Bay. The latter species has invaded saw-grass (*Cladium jamaicense*) marshes and southern Florida slash-pine (*Pinus elliottii* var. *densa*) forests but is most common along roadsides and berms and in burned tree islands. As a result of its prolific root suckering and widespread root system, it often forms compact, dense stands.

Stands of *Casuarina* are detrimental to wildlife and affect certain threatened and endangered species in the park. Loggerhead turtles (*Caretta caretta*) and green sea turtles (*Chelonia mydas*) require gently sloping beaches with soft sand for successful nesting. Because Australian pines impede nesting, beaches dominated by *Casuarina* are rarely used by these turtles. Australian pines have displaced native beach-stabilizing vegetation, allowing wave action to erode sand adjacent to roots, resulting in a steeper embankment.

As early as 1956, individual trees were reported at several locations in the park: in the southeastern corner near Card Sound, along the Ingraham Highway to the mangrove zone, on several keys in Florida Bay, and along the Gulf coast beaches (LaRosa et al. 1992). No action was taken until 1963, when post-hurricane surveys revealed rapidly increasing populations on Cape Sable and Highland beaches. From 1963 to 1970, mechanical measures (tree cutting and uprooting) and chemical control were carried out. Prescribed burns killed scattered trees within prairies but were ineffective in dense stands, because of a lack of fuel in the herbaceous understory.

By 1970, Australian pine covered several thousand hectares, prompting a control program that extended from 1971 through 1978. During this period, 86,300 trees were treated in the park interior (including Highland Beach and the northern shore of Florida Bay). Because of budget constraints, only 12,000 trees were treated from 1979 through 1985. A survey in 1983 estimated that dense stands covered 7200 ha in the southeastern corner of the park.

Since 1989, Australian pine has been treated in the East Everglades Acquisition Area (East Everglades) of the park. This 42,400-ha area, located in the northeastern corner of the park (Figure 17.1), is characterized by seasonally in-

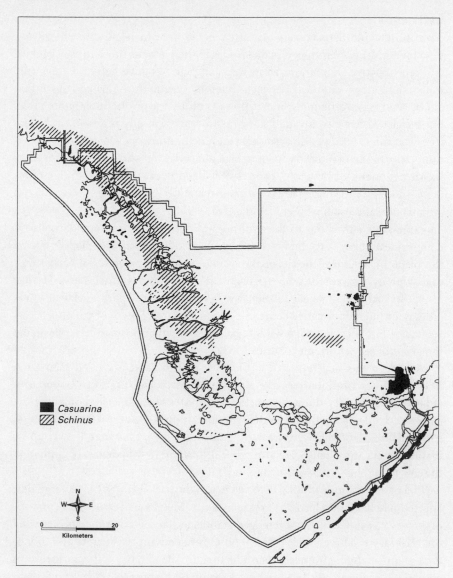

Figure 17.2 Map of distribution of Australian pine (*Casuarina*) and Brazilian pepper (*Schinus*) in Everglades National Park (based on 1987 data).

undated saw-grass and muhly (*Muhlenbergia* sp.) prairies with scattered tree is-lands. Australian pine is widespread throughout the exposed pinnacle rock ("rocky glades") portions of the area and remains a threat to the slightly elevated sites that once supported bayhead or tropical hammock vegetation that was re-moved by severe wildfires (Schomer and Drew 1982).

Melaleuca

The potential for displacement of native vegetation by melaleuca may be greater than for any other introduced plant species in the park. Its numerous vegetative and reproductive adaptations allow it to compete in native habitats, especially transition zones (ecotones) between different vegetation types and disturbed areas. Cypress–prairie and pine–cypress ecotones seem the most vulnerable, but melaleuca has also invaded saw-grass marshes, muhly prairies, slash-pine forests, tropical hardwood hammocks, and the buttonwood–mangrove association. Desiccation of its seed capsules, brought on by freezing, drought, fire, herbicide treatment, or breakage, can result in seed release. It is the timing of release—under wet or under dry conditions—that determines the success of seed germination and establishment (Bodle et al. 1994).

Large-scale alterations to the hydrological regimes of southern Florida and concomitant changes to the fire regime have also increased the distribution of this plant. In fact, melaleuca is, on the average, able to increase numbers of stems each year by a factor of ten, especially after wildfires (Laroche and Ferriter 1992). In addition to its effects on native vegetation of the park, melaleuca provides poor habitat for native fauna: the expansion and growth of this species essentially eliminates nonavian wildlife habitats, and it appears to be undesirable forage for deer (Schortemeyer et al. 1981).

Melaleuca was first reported in the park in 1967 and occurred as isolated trees near park headquarters and along the eastern and northern boundaries (LaRosa et al. 1992). Treatment before the mid-1970s consisted of felling the trees and applying herbicide to the remnant stump. Between 1979 and 1984, some 8300 individuals were treated, mostly seedlings that were pulled up by hand. Larger individuals were girdled or frilled and herbicides were applied in the cuts (the "hack-and-squirt" method).

The primary concentration of melaleuca in the park is in the East Everglades: approximately 28,000 ha of the area is infested. Melaleuca occurs in the prairies of this area as single plants, even-aged stands, and monotypic forests of several age classes. Aerial reconnaissance in 1993 revealed approximately 200 monotypic stands, ranging from less than 1 ha to more than 20 ha, in the northeastern corner of the area alone. Because its expansion in this area threatens native plant communities, intensive efforts to control its spread have been carried out there since the mid-1980s.

Brazilian Pepper

Brazilian pepper can rapidly colonize disturbed areas and persist through later successional stages (Ewel et al. 1982). As a pioneer species in areas undergoing secondary succession, it grows rapidly, produces many seeds that are consumed and dispersed by animals (especially birds, both native, particularly American robins, *Turdus migratorius,* and nonindigenous), and sprouts readily. Seedlings can establish themselves and survive in open areas, as well as under dense canopies. The fact that few other plant species fruit during the winter, when Brazilian pepper seeds are dispersed, may explain the success of this plant in southern Florida (Ewel 1986).

Brazilian pepper has invaded many habitats in the park, including saw-grass marshes, muhly prairies, tropical hardwood hammocks, coastal hardwood hammocks, slash-pine forests, and the salt-marsh–mangrove ecotone (Figure 17.2). Outliers and small to medium-sized populations on the edges of the sandy soils between East Cape and Northwest Cape threaten nesting habitat for the gopher tortoise (*Gopherus polyphemus*), a threatened species in Florida. It rarely grows on sites flooded longer than three to six months and hence is rarely found in marshes and wet prairies. It thrives on disturbed soils created by natural disruptions—such as hurricanes—and is especially invasive in areas affected by human activity, particularly abandoned farmlands, roadsides, and canal banks. In southern Florida, farming practices, especially rock plowing, alter the substrate and allow Brazilian pepper to outcompete native species on these sites (Meador 1977) and alter successional vegetation patterns (Loope and Dunevitz 1981b). Brazilian pepper does not compete with melaleuca for sites: they differ in their relationship to fire and water regimes and in their degree of dependence on human modification of natural conditions (Ewel 1986).

Although Brazilian pepper was noted to be invading mangrove areas around Everglades City in 1961, control was not attempted there (LaRosa et al. 1992). Most efforts at controlling it have been concentrated in the Hole-in-the-Doughnut—an area of abandoned farmland (4000 ha) in the middle of the park (Figure 17.3)—and adjacent pinelands. Rock plowing in half of the area altered the substrate, resulting in conditions favoring Brazilian pepper establishment, that is, higher nutrient levels and aerobic conditions (Doren et al. 1990). Attempts at restoring the native wetland communities in the Hole-in-the-Doughnut were initiated in 1972 and included mowing, disking, rolling, chopping, and bulldozing (Koepp 1979) (Figure 17.4).

Experiments with use of herbicide and fire on Brazilian pepper have also been conducted in the Hole-in-the-Doughnut. Herbicide applied directly to basal bark proved the most effective means of killing Brazilian pepper and had minimal impact on the surrounding vegetation (Ewel et al. 1982; Doren et al. 1991b). Repeated fires may slow down the invasion rate, but they do not exclude Brazilian pepper establishment. Although invasion progresses with or without fire (Doren et al. 1991a), prescribed burning at three-year to seven-year intervals has severely restricted its establishment within the park (Loope and Dunevitz 1981b).

Figure 17.3 A monospecific stand of Brazilian pepper (*Schinus terebinthifolius*) in the Hole-in-the-Doughnut area of Everglades National Park. Photo by Don Schmitz.

Current Management Practices

The strategy employed by the park in managing its most serious nonindigenous plant pests, primarily Australian pine and melaleuca, is to focus on individual plants and outliers rather than dense stands (Moody and Mack 1988). Concentrated efforts to control the spread of these two species in the East Everglades, an area of significant invasion, began in the mid-1980s with the establishment of the East Everglades Exotic Plant Control Project by the multi-agency East Everglades Resource Planning and Management Committee (DeVries 1995). With the assistance of the Florida Exotic Pest Plant Council (EPPC), the park developed the control project and acquired funding from several cooperating state agencies.

The treatment season for melaleuca and Australian pine in the East Everglades extends from December to May. For melaleuca, the most isolated, distant seed trees are treated first, and treatment progresses toward the center of the infestation. Control is achieved by annual treatment of regrowth and recruits at

Figure 17.4 Aerial view showing the creation of artificial wetlands and uplands in the Hole-in-the-Doughnut, Everglades National Park, through the removal of Brazilian pepper (*Schinus terebinthifolius*) and soil. Photo by Don Schmitz.

previously treated sites. Surveillance is conducted by helicopter for outliers, and low to medium densities of melaleuca are treated as found. Locations of high-density, monotypic stands are recorded via the Global Positioning System for subsequent foliar spraying. Field data (location, number of resprouts, number of stems treated, environmental conditions, seed tree heights) are entered into a geographic information system to create project maps. The methods used to control melaleuca include mechanical means (pulling seedlings by hand), chemical or herbicidal treatment (cut stump, basal bark, and foliar), and physical measures (prescribed burn or prolonged inundation after cutting). Larger trees are felled with chain saws and the exposed cambium treated with a herbicide (Arsenal 50%). Monotypic stands of seedlings are either cleared with brush cutters or treated with a herbicide (Arsenal 1.5%) by foliar application. Australian pine is controlled by pulling of seedlings and treatment of larger stems with a herbicide (Garlon 4) by the basal bark method.

Total treatment cost for the control of melaleuca and Australian pine in the

East Everglades for the ten years from 1986 through 1995 was about $1.2 million. Table 17.1 provides a detailed breakdown of annual costs. During this period, approximately 4.5 million melaleuca and 100,000 Australian pine stems were treated by hand pulling, by cut stump and basal-bark method, and by foliar application (DeVries 1995).

The removal of Australian pine from the keys in both Florida Bay and Dry Tortugas National Park is carried out when time and manpower permit. These areas include critical habitat for endangered animal species, such as nesting beaches for sea turtles (at Dry Tortugas keys) and American crocodiles (*Crocodylus acutus*) (at Florida Bay keys). These areas have high priority for control of Australian pine, which can eliminate nesting habitat.

Control of Brazilian pepper in the Hole-in-the-Doughnut (Figure 17.4) is part of one of the park's largest wetlands restoration efforts. A study report by Doren et al. (1990) indicates that the complete removal of disturbed substrate from abandoned, rock-plowed lands, and the subsequent increase in hydroperiod, has altered secondary successional patterns in favor of native wetland vegetation and the concomitant exclusion of Brazilian pepper. This technique will be used in the Hole-in-the-Doughnut on all former wetlands that were invaded by Brazilian pepper after abandonment of the farmlands, an area of about 2000 ha.

Many areas of the west coast mangrove forests in the park have been damaged by hurricanes (Donna in 1960, Betsy in 1965) and freezes (in 1977 and 1981) and have subsequently been invaded by Brazilian pepper (Armentano et al.

TABLE 17.1.

Treatment Costs and Labor Hours for *Melaleuca* and *Casuarina*
Treatment in the East Everglades: 1986–1995

Year	Flight cost ($)	Herbicide cost ($)	Cost per year ($)	Labor hours, NPS[a]	Labor hours, non-NPS[a]
1986	2261	3	2624	32	32
1987	12,965	187	13,152	224	336
1988	25,495	556	26,051	708	48
1989	55,198	442	55,640	926	442
1990	70,451	788	71,239	1419	112
1991	111,053	7744	118,797	2755	72
1992	146,295	7400	153,695	4942	64
1993	93,553	11,540	105,093	2252	1160
1994	110,000	11,358	121,358	4509	0
1995	145,000	9000	154,000	2789	0
Total	772,271	49,018	821,649	20,556	2266

[a] "NPS" Indicates National Park Service personnel.

1995). An unpublished study by the park in 1987 reveals that some 42,000 ha of west-coast mangrove and interior coastal prairie habitat within the park contained this plant (Figure 17.2). Removal of Brazilian pepper from mangrove ecosystems is difficult: mechanical removal disturbs mangrove community substrate, which favors Brazilian pepper establishment; the foliar application of herbicides in mangrove forests could kill nontarget species. Removal by hand and direct injection of herbicides have eliminated this species, but only on a small scale. Further research is urgently needed to determine the best means of removing large infestations of Brazilian pepper within mangrove communities.

Other nonindigenous plant species, such as lather leaf and shoebutton ardisia, are problems in the park and becoming increasingly widespread. The control of these plants is carried out sporadically, if at all, because of labor, time, and financial constraints. Lather leaf (*Colubrina asiatica*), native to tropical Asia, has rapidly spread through coastal vegetation around parts of Florida Bay and on certain keys (Rankin Key, for example) and is becoming a highly visible feature of the Everglades landscape. By virtue of its climbing habit, lather leaf forms dense carpets of growth covering buttonwood, mangroves, and associated coastal vegetation. The fan palm (*Thrinax*) and mahogany hammocks along the Florida Bay coast in the park are particularly vulnerable to the destructive effects of lather leaf. Its potential to spread onto the keys throughout Florida Bay and other coastal areas is high because its buoyant seeds can be dispersed by water. In the few areas of the park where lather leaf is managed, such as Flamingo area and Rankin Key, herbicides (Garlon 4) are used to treat it.

Shoebutton ardisia (*Ardisia elliptica*) originated in Southeast Asia and has invaded disturbed hardwood hammocks and abandoned agricultural lands in the Royal Palm area of the park. It also occurs in the Flamingo area, where it was first reported in 1995 (Seavey and Seavey, unpublished data). It comprises a major part of the understory in the Brazilian pepper-dominated sections of the Hole-in-the-Doughnut. Seavey and Seavey (1994) predict that habitats with high humidity and an intact overstory, such as tropical hardwood hammocks, adjacent to a seed source are especially vulnerable to invasion by this species; pine forests, inland and coastal prairies, and mangrove forests appear to be the least susceptible. Treatment, which has been concentrated in Paradise Key at Royal Palm, is by mechanical means (hand pulling) and chemical methods (Garlon 4 herbicide).

Prospects

The control of melaleuca in the East Everglades is currently the only funded nonindigenous-species management priority in the park. It has been funded since 1987 through compensatory off-site wetland mitigation and contributions from several state agencies. Specific funding for other projects is still lacking. The eradication of Australian pine, for example, will require significant resources to eliminate the large, dense stands along the eastern boundary and southeastern corner of the park. Removal and control of Brazilian pepper within the Hole-in-the-Doughnut will be funded through a Dade County mitigation bank. Brazilian pepper control in the salt-marsh–mangrove ecotone along the west coast needs to be addressed.

Biological control methods are an important component of an integrated pest-plant management approach. Such methods are expensive and will take a long time, but they could result in eventual elimination or significant reduction of continued manual and chemical control in the region and the park.

All management programs should be supplemented with ongoing monitoring for evaluation. An extensive and permanent series of plots is needed throughout the park in order to monitor the status of nonindigenous species by plant community. These plots would serve as the basis for planning control actions and providing critical information on long-term trends and site vulnerability to invasion.

Regional coordination between affected agency land managers has been excellent, and the park continues to cooperate in all such interagency efforts to manage nonindigenous plants. The EPPC has provided one of the most effective forums for the exchange of ideas and conflict resolution concerning nonindigenous species in the state. Regional control plans continue to be developed by EPPC for high-priority species such as melaleuca (Laroche 1994) and Brazilian pepper. Moreover, some funding and management work have been channeled to the park by other agencies through the promotion of coordinated efforts by the EPPC.

18 Management on State Lands

Florida has one of the nation's most intensive land-acquisition programs. Between 1974 and 1994, through various legislative initiatives, the state acquired over 2.1 million hectares of land at a cost of over $1 billion (Office of Environmental Service 1995). Coupled with nonsovereign state landholdings, these purchases have brought significant portions of Florida's undeveloped lands into state ownership and management.

Many lands purchased for conservation or recreation purposes are acquired with established populations of nonindigenous plants and animals, and all are subject to invasion from nearby sources. As a result, state land managers are on the front line of defense against nonindigenous species in Florida and have become acutely aware of the risks to the state's native plants and animals. Moreover, the thousands of miles of rights-of-way in Florida's highway system must be monitored and maintained to reduce the threat of nonindigenous species introductions.

State Agencies

The majority of state lands under active management in Florida are administered by the Florida Park Service, the Game and Fresh Water Fish Commission, and the Division of Forestry. Maintenance and protection of highway rights-of-way are administered by the Florida Department of Transportation.

Florida Park Service

The Florida Department of Environmental Protection's Division of Recreation and Parks manages 143 properties totaling approximately 868,260 upland hectares and

212,506 submerged hectares (Office of Park Planning 1995). Fewer than 90 of the parks, recreation areas, preserves, reserves, special feature sites, museums, ornamental gardens, and trails that comprise the Florida Park Service (FPS) are staffed full-time. All are available for public access of some type.

The mission of the FPS is "to provide resource-based recreation while preserving, interpreting and restoring natural and cultural resources" (Florida Park Service 1994a). FPS policies call for identification and removal of nonindigenous species. Although removals are tracked annually by park and by district, there is currently no mechanism for tracking the costs associated with combatting invasive nonindigenous species. Removal of nonindigenous species is frequently associated with the FPS commitment to restoring altered natural systems and hydrological regimes, but removal efforts have occasionally been slowed or stopped where there has been significant conflict with recreational needs or public sentiment. At John U. Lloyd Beach State Recreation Area (Broward County), for example, the removal of Australian pines (*Casuarina equisetifolia*) has been slowed significantly by complaints from beachgoing visitors who object to the resulting loss of shade.

Game and Fresh Water Fish Commission

The Florida Game and Fresh Water Fish Commission (FGFC) is to some degree responsible for management and public use on approximately 14 million hectares in the state (Boyter 1994). The Types I and II wildlife management areas make up a system that protects resources and provides recreational and educational opportunities for both consumptive and nonconsumptive users. Type I wildlife management areas are generally those on which the FGFC provides both resource management and public use and include most lands acquired by the FGFC through state land purchase programs. Type II areas are owned and managed by forestry companies, water management districts, and the U.S. Air Force. The FGFC maintains a minimal management presence on these lands but does provide some law enforcement (Boyter 1994). The FGFC's mission is "to manage freshwater aquatic life and wild animal life and their habitats to perpetuate a diversity of species with densities and distributions that provide sustained ecological, recreational, scientific, educational, aesthetic and economic benefits" (Boyter 1994).

The FGFC has an especially complex and challenging management relationship with nonindigenous species. Because it regulates nonindigenous animals, including the occasional legal introduction, it cannot simply designate all nonindigenous species for removal but must instead adhere to certain guidelines:

- Approval of the executive director must be obtained before initiating a "hands-on" study outside laboratory conditions of a species not native to Florida that is not already in public or private lands or waters.

- Species not native to Florida will be evaluated to determine negative and positive impacts on native fish, wildlife, their habitats, and people prior to introduction into the wild.

- Only those species not native to Florida that are clearly shown to have positive benefits to humans and no significant negative impacts to native fish, wildlife, and their habitats will be considered for introduction.

- Where practical and necessary, established populations and isolated occurrences of species not native to Florida will be monitored, controlled, used, or eliminated.

- Importation of species not native to Florida will be as provided by sections 372.26 and 372.265 of the Florida Statutes (Boyter 1994).

Division of Forestry

The Florida Department of Agriculture and Consumer Services' Division of Forestry (FDOF) has some management responsibility on more than 2.3 million hectares of public land. It is the lead managing agency for 32 state forests totaling 1.3 million hectares and manages more than 454,664 additional hectares through special agreements with other state agencies, several counties, and water management districts (Hardin 1994).

The FDOF's mission is "to protect and manage Florida's forest resources through a stewardship ethic to assure these resources will be available for future generations" (Hardin 1994).

FDOF policies call for identifying and removing nonindigenous species, but management and control are emerging issues as species such as cogon grass (*Imperata cylyndrica*) become an exponentially increasing problem. Although FDOF districts are aware that invasive nonindigenous species exist, no districts or state forests have conducted systematic assessments. Most occurrences have been observed during normal management activities (Hardin 1994).

Department of Transportation

The Florida Department of Transportation (FDOT) is responsible for safety, maintenance, and vegetation protection on Florida's rights-of-way. The FDOT's goal is a high-quality transportation system. The FDOT's Environmental Policy for State Transportation Facilities states: "the Department will cooperate with the State's efforts to avoid fragmentation of habitat and wildlife corridors by including these considerations in all phases of its operations" (Caster 1994). The FDOT's nonindigenous removal program is driven by three considerations: interference with the safe operation of motor vehicles on state roads; concern for the increased costs associated with nonindigenous pest plant control and removal; and protection and replacement of native trees and other vegetation (Caster 1994).

Impacts on Native Species

All three land management agencies and the FDOT report extensive nonindigenous plant species infestations on the lands they manage. Each reports higher concentrations in southern Florida and different primary invasives in different geographic regions of the state. Subtropical coastal areas of Florida have been singled out as particularly threatened.

Few summary data are available on nonindigenous plant removal on state lands. FPS records for 1994 indicate that 58 state parks treated 1.5 million individual nonindigenous plants of various species, as well as 4 metric tons of mostly elodea (*Egeria densa*) and hydrilla (*Hydrilla verticillata*), and cleared 152.46 hectares of such species as kudzu (*Pueraria montana*), cogon grass, and air potato (*Dioscorea bulbifera*). Statewide, 94 nonindigenous plant species were individually reported as being removed from parks, together with an additional 25 species grouped as "other" (Florida Park Service 1994).

In 1994, the FDOF estimated that over 5436 hectares in six state forests are infested with nonindigenous plant species and that infestations are acknowledged but unestimated in three additional forests (Hardin 1994). Moreover, a myriad of grasses can invade rights-of-way and challenge the FDOT's commitment to stopping the spread of nonindigenous plants on transportation corridors. Torpedo grass (*Panicum repens*), Johnson grass (*Sorghum halepense*), smut grass (*Sporobolus jacquemontii*), crowfoot grass (*Dactyloctenium aegyptium*), and vassey grass (*Paspalum urvillei*) are examples (Caster 1994).

Australian pine (*Casuarina equisetifolia*) is of particular concern in southern Florida and along the southern coasts of the state: in 1990, Australian pines were found in 30 state parks (Stevenson 1990); in 1994, some 84,628 Australian pines were removed from state parks (Florida Park Service 1994b).

Recent surveys of upland coastal vegetation in southern Florida (Johnson and Muller 1993) indicate that Australian pine has colonized large sections of the remaining undeveloped portions of barrier islands along both the Atlantic and Gulf coasts. On the Gulf coast from Pasco to Collier counties, Johnson and Muller (1993) identified 60.2 miles of coastline as undeveloped (out of a total coastline length of 178 miles)—of which 20.1 miles, or 33 percent, were heavily invaded by Australian pine. Similarly, on the Atlantic coast from Indian River County to northern Dade County, 46 percent of the undeveloped barrier island coast is heavily invaded (Johnson 1994). Australian pines have also been found to interfere with sea-turtle nesting activities in the Florida Keys, where beaches are typically narrow and pine growth can occur close to the water line (Duquesnel 1994).

Another serious threat to southern Florida and the southern coasts is Brazilian pepper (*Schinus terebinthifolius*). It differs from Australian pine in that it does not require bare soil to invade (Duquesnel 1994). Brazilian pepper can become established in native vegetation, or it can form monospecific stands on disturbed sites (Johnson 1994). It is consistently present in the coastal forest understory on outer barrier islands on both the Atlantic and Gulf coasts, from Brevard and Pasco counties southward (Johnson 1994a). The plant was present in 42 state parks in 1990 (Stevenson 1990). In 1994, some 161,431 individual Brazilian pepper plants were removed from state parks (Florida Park Service 1994b). It is a threat on wildlife management areas (Boyter 1994) and is widespread, if not dense, on DuPuis Reserve (Martin and Palm Beach counties) and Picayune Strand (Collier County) state forests (Hardin 1994). It causes maintenance and safety problems for the FDOT, as it grows through right-of-way fences and eventually lifts them clear of the ground, allowing wildlife and humans onto high-speed, limited-access highways (Caster 1994).

Cogon grass is widespread and occurs in extremely thick stands along rights-of-way in central and northern Florida. The FDOT reports 900 large monospecific stands accounting for more than 741 hectares of right-of-way (Caster 1994). In 1994 FPS staff treated approximately 59 hectares of monospecific stands and many unreported smaller stands (Florida Park Service 1994). The FDOF initiated a five-year work plan in 1993 to eliminate cogon grass in state forests (Hardin 1994). It is presently considered a threat to nine state forests. The effects of cogon grass on native species have been observed and reported by the FDOF and FGFC (Hardin 1994; Weimer 1994) and are of concern to the FGFC (Boyter 1994). Anecdotal information suggests that gopher tortoises (*Gopherus polyphemus*) abandon areas infested with cogon grass. Loss of the gopher tortoise results in the disappearance of the numerous species that depend on gopher-tortoise burrows, including gopher frogs (*Rana aesopus*), eastern indigo snakes (*Drymarchon corais*), and scarab beetles (Scarabaeidae) (Myers 1990). Where cogon grass inhibits longleaf-pine regeneration, endangered red-cockaded woodpecker (*Picoides borealis*) nesting and foraging may be affected. Moreover, the woodpecker's nest trees are vulnerable to the intense fires that can occur in cogon grass ground cover (Hardin 1994). The FDOF reports that most of the Withlacoochee State Forest (Citrus, Hernando, Sumter, and Pasco counties) may be infested with cogon grass within a generation (Hardin 1994).

Melaleuca (*Melaleuca quinquenervia*) poses a serious threat to Florida's highway rights-of-way. Along the Homestead Extension of Florida's Turnpike (Dade County), for example, melaleuca is the most serious nonindigenous invader, continually interfering with the operation and maintenance of drainage conveyances. Because of limited staff, equipment, and funding, the FDOT removes melaleuca trees only from drainage conveyances, public areas such as rest stops, structures, and mitigation areas (Caster 1994). Melaleuca is one of the most invasive plants in the state parks of southern Florida (Florida Park Service 1994b) and has infested approximately 247 hectares of Golden Gate State Forest, where the highly fragmented nature of state holdings and lack of an ownership map compound the problem (Hardin 1994).

Skunk vine (*Paederia foetida*) and air potato threaten hammocks on public lands (Hardin 1994), and ardisia (*Ardisia crenulata*) is becoming the dominant understory plant in hammocks at both the Lake Overstreet addition to Maclay State Gardens (Leon County) (E. Johnson, pers. comm.) and Withlacoochee State Forest, where, along with skunk vine, it threatens the federally listed Cooley's water willow (*Justica cooleyi*) (Hardin 1994).

Other species reported as being particularly invasive include Chinese tallow tree (*Sapium sebiferum*), described as one of the most invasive of northern Florida's nonindigenous plants (Weimer 1994); tropical soda apple (*Solanum viarum*), spreading through pastures and open areas of central Florida (K. Alvarez, pers. comm.); and beach naupaka (*Scaevola taccada*) on southern Florida coasts (Duquesnel 1994). Table 18.1 lists endangered and potentially endangered plants from southern Florida state parks that are threatened by invasive nonindigenous species.

State land managers have few summary removal data and little understanding of the impacts of nonindigenous animals on the lands they manage. Removal records either are not maintained at all or exclude important aquatic invaders such as blue tilapia (*Oreochromis aureus*). FPS records indicate that some 1778 nonindigenous animals were removed from 25 state parks in 1994, but the costs are unrecorded (Florida Park Service 1994).

Feral pigs are the major nonindigenous animal threat to state lands. The FPS reports 1205 individuals removed from state parks in 1994, including 886 from Myakka River State Park (Sarasota County) alone (Florida Park Service 1994b). They are major pests in Seminole, Withlacoochee, and Goethe state forests (Levy County) and are present in others (Hardin 1994). On wildlife management areas, populations are controlled to some degree through liberal hunting regulations (Boyter 1994).

Feral pigs impose an obvious and extensive toll on native ground cover through their rooting habits (Hardin 1994). In the Panhandle, they may threaten Appalachian relict species such as columbine (*Aquilegia candensis* var. *australis*), bellflower (*Campanula americana*), and mayapple (*Podophyllum peltatum*) (Ludlow 1994). They pose a threat to humans through the potential transmission of brucellosis, trichinosis, and pseudorabies, which the FGFC has

TABLE 18.1.

Endangered and Potentially Endangered Plants Threatened by
Nonindigenous Plant Species in Southern Florida State Parks

Scientific name	Common name
Acacia choriophylla	tamarindillo
Cordia sesbestena	geiger tree
Jacquemontia reclinata	beach clustervine
Jacquinia keyensis	joeweed
Argusia gnaphalodes	sea lavender
Okenia hypogaea	burrowing four-o'clock
Pseudophoenix sargentii	buccaneer palm
Remirea maritima	beach star
Scaevola plumieri	inkberry
Suriana maritima	bay cedar
Encyclia tampensis	butterfly orchid

Source: Duquesnel (1994).

also determined to have been transmitted to endangered Florida panthers (*Felis concolor coryi*), (Boyter 1994).

Armadillos (*Dasypus novemcinctus*) were the second most removed nonindigenous animals from state parks in 1994; 409 were removed statewide (Florida Park Service 1994b), and they are well established on many lands in public ownership (Boyter 1994; Hardin 1994). The rooting habits of armadillos affect ground-cover species in much the same ways as do those of pigs. Domestic and feral dogs (*Canis familiaris*) and cats (*Felis catus*) are present on most public lands, particularly near campgrounds and other high-use areas (Florida Park Service 1994b; Hardin 1994). Cats are voracious predators; a single cat may kill hundreds of birds in one year (D. Bryan, pers. comm.). They may also take a toll of threatened rodents such as the Anastasia beach mouse (*Peromyscus polionotus phasma*) (Bard 1994) and the Key Largo wood rat (*Neotoma floridana smalli*) (Duquesnel 1994). Dogs disrupt resting and feeding patterns of various species and are known predators of gopher tortoises and wild turkeys (*Meleagris gallopavo*).

Conflicts and Challenges

For public land managers, the removal and control of nonindigenous species is only one element of a diverse set of resource management policies. None of the agencies is able to focus primarily on this objective, and each attempts to control invasive species as part of its daily management activities. Some major removal projects have been carried out through mitigation and grants and where natural forces have destroyed nonindigenous species and offered restoration opportunities. At Cape Florida State Recreation Area (Dade County), for example, an extensive stand of Australian pine and other upland nonindigenous species was completely destroyed by Hurricane Andrew in August 1992. Earlier attempts by the FPS to remove the Australian pines had been largely thwarted by local opposition to removal of any shade tree. But given the opportunity afforded by Andrew, a complete restoration of natives is now under way.

In another example of conflict between nonindigenous species control and public policy or sentiment, the FGFC deals with feral pigs by imposing liberal hunting regulations on them in wildlife management areas. The agency believes it has effected a fair measure of control of pig populations through regulated hunting (Boyter 1994); other public land managers, however, advocate pig removal rather than management (Florida Park Service 1994b; Hardin 1994). At Silver River State Park, local residents and the Florida legislature combined to prevent the FPS and the FGFC from removing nonindigenous rhesus monkeys, even though they are infected with a disease potentially lethal to humans. On all managed lands, the pervasive cats and dogs that threaten native wildlife can frequently belong to neighbors. In areas of high visitation, such as campgrounds and picnic areas, removal of cats and dogs can be an extremely touchy proposition.

Most public land managers currently have no formal mechanism for tracking nonindigenous plant removal costs. According to a detailed survey conducted by the FPS in 1989, state park costs are between $250,000 and $350,000 annually. In fiscal 1993, the FGFC spent $200,000 on vegetation control, a large percentage of which was spent controlling nonindigenous species (Boyter 1994). The FDOF reported specific removal projects totaling $11,905 in 1993 but estimated that $400,000 would be needed over the next several years to control cogon grass in Withlacoochee State Forest alone (Hardin 1994).

Including mowing, manual elimination, and chemical removal, the FDOT estimates that over $7 million was spent on nonindigenous plant control on rights-of-way in fiscal 1993–1994. In addition, $4.5 million was allocated for fiscal year 1996 for the removal of Australian pine and melaleuca and their replacement with more desirable species (Caster 1994).

The actual costs and management impacts of removing nonindigenous plants on public lands are extremely difficult to measure. Much of the removal occurs as part of daily work duties and is not reported. In areas where nonindigenous plants have been removed, surveys and hand pulling of invading seedlings are necessary to prevent reinvasion, and these removals are rarely reported. Nonindigenous plant removal is often part of natural-systems restoration projects but is rarely listed as a separate budget item.

As Florida's lawmakers begin to shift their emphasis from land acquisition to land management over the next decade, the need for a systematic approach to the issue of nonindigenous invaders will be critical. Land managers, including the FDOT, agree that separate accounting and budgeting processes are important to future planning and management of nonindigenous plant control (Boyter 1994; Caster 1994; Hardin 1994).

What future steps can be taken by the legislature? The key steps are those that would specifically prohibit the introduction or planting of invasive nonindigenous plants. Clear policies governing the introduction of new species and prohibiting the use of state funds to plant nonindigenous species are needed, as are tough laws governing sale and use. The short-term economic impact of such legislation would be more than offset by the savings associated with reducing control costs.

Actual removal costs associated with nonindigenous animals on state lands are relatively low. State parks frequently employ contractors to remove pigs, which has helped to reduce staff time devoted to their control, but even so there is a significant drain on the workforce.

Additional research is needed on the environmental impacts associated with nonindigenous animals. Moreover, the incidence of nonindigenous microorganisms and their impact is poorly understood, although the Florida Department of Agriculture and Consumer Services' Division of Plant Industry has expressed interest in using state lands as research sites for compiling lists of insect and microorganism species.

Public education and understanding of the threats posed by nonindigenous animals, particularly pets, would go a long way toward solving some of the most sensitive problems associated with nonindigenous animals on public lands. On remote public lands, where no local animal control service exists, removal of dogs or cats can be difficult. Land managers must work to gain the cooperation of neighbors in order to minimize removal needs and control domestic animals.

19 Management by Florida's Game and Fresh Water Fish Commission

JAMES A. COX, LT. THOMAS G. QUINN,
AND H. HUGH BOYTER, JR.

 Fish and wildlife agencies in the United States have played a paradoxical role in managing nonindigenous species. On the one hand, they are instrumental in controlling the import, release, and spread of many nonindigenous species—for example, an early piece of U.S. conservation legislation, the Lacey Act, was initiated and supported by wildlife agencies in part because it reduced the chances that nonindigenous species would become established (Lund 1980). On the other hand, fish and wildlife agencies are also responsible for the establishment of nonindigenous species. The frequency of such introductions has declined dramatically in recent years, but the practice continues and is rarely free of controversy.

Florida's fish and wildlife agency, the Florida Game and Fresh Water Fish Commission (FGFC), was created in 1944 and given the regulatory and executive powers of the state with respect to wild animal life and freshwater aquatic life. The agency is directed by a five-member commission appointed by the governor, which has traditionally consisted of conservationists, large landowners, and hunting and fishing enthusiasts. The agency's stated goals are to "manage freshwater aquatic life and wild animal life . . . to perpetuate a diversity of species with densities and distributions that provide sustained ecological, recreational, scientific, educational, aesthetic, and economic benefits" (FGFC 1994a).

Law Enforcement

The United States is the largest trader of wildlife in the world, importing over $1 billion in live wildlife and

297

wildlife products each year (Traffic 1990). Florida's proximity to tropical regions and the use of Miami as a major port of entry have made the state one of the world's primary distribution centers for nonindigenous species. This trade provides an obvious means by which many nonindigenous species have become established (Temple 1992).

The FGFC has developed numerous rules to regulate the import, sale, and storage of nonindigenous species in hopes of minimizing the chances for nonindigenous species to become established. In 1972, the FGFC was perhaps the first state agency in the nation to adopt caging standards for the display and transport of wildlife. A general prohibition against nonindigenous species is embodied in Section 39-4.005 of the Florida Administrative Code:

> It shall be unlawful for any person to possess, transport, or otherwise bring into the state or to release or introduce in the state any wildlife or freshwater fish that is not native to the state, unless such person shall first secure a permit from the Commission. Such permits shall be granted only after duly authorized agents have made such investigation and inspection of the wildlife or freshwater fish as may be deemed necessary.

Thus the FGFC regulates the possession and sale of nonindigenous and native wildlife through a permitting process that requires inspection of storage facilities and shipments. Permits are required for zoos, tropical fish farms, wildlife importation, commercial American alligator (*Alligator mississipiensis*) farming, venomous reptile trade, personal pet ownership, private hunting preserves, falconry, wildlife rehabilitation, and other such activities. Appendix A at the end of this chapter lists additional rules concerning the possession of wildlife. At present, more than 10,000 individuals or corporations are permitted to exhibit wildlife, to trade and sell wildlife, or to possess native or nonindigenous wildlife as personal pets. Permits are renewed and reviewed annually, and many applications are denied for legal and administrative reasons (if, for example, the applicant has not demonstrated an ability to operate a wildlife rehabilitation facility). Persons conducting regulated activities without a permit may be charged with a misdemeanor (up to six months in jail and $500 fine).

The number of permits granted is increasing as a result of Florida's role in the nonindigenous species trade and its expanding human population. The increases shown in Table 19.1 are conservative estimates, however, because permits are held primarily by commercial entities that deal in hundreds or even thousands of specimens, not by individual owners or even low-volume dealers. Although there are problems associated with legal and illegal owners of small numbers of nonindigenous species, the greater threat probably lies with large-volume dealers engaging in illegal activities. Large-volume dealers can hold and

TABLE 19.1.
Number of Active FGFC Permits for Possession
or Sale of Wildlife in 1980 and 1993

License type	1980	1993	% Increase
Retail fish dealer[a]	549	3660	567
Hunting preserve	67	94	40
Exhibition and sale of animals	544	3723	584
Personal pet	76	194	155
Venomous reptile	182	403	121
Alligator farm	n/a	61	0

[a] In 1991 this license was combined to include resident retail dealers as well as exotic and resident wholesale dealers.

exchange hundreds of potential propagules—and tighter restrictions on dealers in recent years have caused them to seek novel, and often illegal, solutions to satisfy a growing trade.

Oversight of FGFC permit compliance is administered by the Inspections Section within the Division of Law Enforcement, established in 1971 and consisting of nine wildlife inspectors—clearly too few to handle the number of permit holders. Each inspector would have to visit five permitees each day (in addition to other duties) in order to inspect all permitees in a year. Moreover, only one inspector is assigned to cover the Miami airport (Dade County) and the broader region that includes 12 other counties. Because these inspectors are a first line of defense against the illegal importation and release of nonindigenous fish and wildlife in Florida, additional staffing seems imperative.

The nonindigenous fish trade is particularly difficult to oversee and control: the number of dealers operating in Florida has doubled since 1980; prohibited species are hard to identify (Table 19.2); and a huge volume of material must be inspected. From 1992 through 1995, some 7858 shipments of tropical fish arrived at the Miami airport. Most of these shipments contained 25 to 500 containers of fish. The single inspector working the Miami region would have to inspect at least 60 containers per hour per day to check these shipments.

Some importers attempt to exploit the shortage of inspectors by including prohibited or restricted fishes in large shipments, by declaring a much smaller number of restricted fishes than are actually being shipped, or by intentionally mislabeling boxes. Recent examples include mislabeling of electric eels (family Electrophoridae) as "common catfish," piranhas (family Serrasalminae) as pacus (*Colossoma* spp.), and parasitic candiru catfish (family Trichomycteridae) as kuhli loaches (*Acanthophthalmus kuhli*). The last example included over 3000 individual fish that were destined for distribution all over Florida. Candiru catfish can parasitize humans and could easily thrive in Florida waters (Courtenay 1994).

TABLE 19.2.
Prohibited and Restricted Freshwater Fishes

Common name	Family, genus	Species
PROHIBITED FISHES		
African electric catfishes	Malapteruridae	all species
African tigerfishes	subfamily Hydrocyninae	all species
Air-breathing catfishes	Clariidae	all species
Air-sac catfishes	Heteropneustidae	all species
Candiru catfishes	Trichomycteridae	all species
Freshwater electric eels	Electrophoridae	all species
Lampreys	Petromyzonidae	all species
Piranhas and pirambebas	Serrasalminae	all species
Snakeheads	Channidae	all species
Tilapias	*Oreochromis*	all species except *O. aureus,* *O. mossambicus,* *O. niloticus*
Trahiras or tigerfishes	Erythrinidae	all species
Green sunfish	*Lepomis*	*L. cyanellus*
RESTRICTED FISHES		
Bighead carp	*Aristichthys*	*A. nobilis*
Bony tongue fishes	Osteoglossidae	all species except *Osteoglossum bicirrhosum*
Dorados	*Salminus*	all species
Freshwater stingrays	Potamotrygonidae	all species
Grass carp	*Ctenopharyngodon*	*C. idella*
Nile perches	*Lates*	all species
Silver carp	*Hypophthalmichthys*	*H. molitrix*
Snail or black carp	*Mylopharyngodon*	*M. piceus*
Tilapias	*Oreochromis*	*O. aureus,*[a] *O. mossambicus,* *O. niloticus*
Walking catfish	*Clarias*	*C. batrachus*

Note: Prohibited fish are considered dangerous to public health and welfare or harmful to Florida's aquatic waterways. Permits to import, possess, or tranship these species are generally granted only for research or public viewing where adequate quarantine conditions are satisfied. In the case of restricted fishes, permits may be issued for import providing quarantine facilities are adequate to prevent escape or accidental release into state waters.

[a] *Oreochromis aureus* has become established in central and southern Florida and may be possessed, cultured, and transported there without a permit.

Fish identification skills and an understanding of the intricacies of the tropical fish trade are clearly necessary for control of these illegal shipments. A new inspector assigned to the Miami region in October 1987 seized 296 shipments in 1987 but 2665 in 1988 as a result of improved identification skills and a better understanding of illegal practices. This inspector was transferred, in October 1989, and seizures declined to 110. If 1988 was typical, as many as 2400 illegal shipments may have passed through the Miami airport in 1987 and 1989.

Confiscations of illegal fishes have also led some dealers to attempt to establish captive breeding populations in aquaculture ponds rather than relying on illegal imports, a practice uncommon until recently. Unconfirmed reports suggest that some fish wholesalers are attempting to breed electric catfish and tilapia (*Tilapia* spp.), and at least one Florida importer has successfully bred red-bellied piranha (*Serrasalmus nattereri*) and offered the fish for sale. Piranha appear to be popular in today's commercial market, and the FGFC has seized 12 specimens to date.

The Humane Society of the United States reports that more than 500,000 wild birds are illegally imported into the country each year, and the U.S. Department of Justice estimates that another 150,000 are smuggled into the United States from Mexico (Mulliken and Thomsen 1990). With the implementation of the Wild Bird Conservation Act by the U.S. Fish and Wildlife Service (FWS), an increase is anticipated in the number of birds being smuggled into Florida for breeding purposes or to be sold directly to pet owners. The number of bird breeders, both legal and illegal, will also surely increase in response to the shortage of nonindigenous birds likely to result from tighter regulations.

The FGFC has record-keeping requirements for bird dealers in Florida, but inspection staff cannot follow up on all transactions. Current regulations require that "persons transferring any live non-native bird to another shall maintain documentation for a period of 24 months following such sale or transfer. The documentation shall include name, address, date of sale, number and species sold" (Florida Administrative Code Section 29-6.006). The FGFC inspectors have noticed increases in the number of complaints involving stolen birds and the smuggling of species listed by the Convention on International Trade in Endangered Species of Wild Fauna and Flora.

Within the last three years, the importation of mammals listed by the convention has increased by 18 percent at the Miami airport (Traffic 1991, 1993). The FGFC has seized more than 300 illegally imported mammals over this same period. Judging from the number of seizures, smaller species such as lemurs (family Lemuridae), hedgehogs (*Erinaceus algirus*), civets (*Viverra* spp.), and agoutis (*Dasyprocta* spp.) appear to be the most popular illegal imports. Between 1987 and 1993, however, wildlife inspectors also seized 54 cougars (*Felis concolor*). In the case of cougars and other large mammals, simply locating a suitable site to house the seized evidence is a troublesome task. The list of large mammals seized in recent years includes bears (family Ursidae), bison (*Bison bison*), rhinoceroses (family Rhinocerotidae), lions (*Panthera leo*), tigers (*Panthera tigris*), several species of deer (family Cervidae), and chimpanzees (*Pan troglodytes*).

Arrests associated with illegal possession of primates (Table 19.3) typically occur when a complaint is received about an escaped primate. Such complaints averaged about one per month a few years ago but today average nearly one per week. In 1989, some 17 primate escapes were documented; by 1993, this figure had increased to 41. As this increasing trend continues, the chances that a new population of nonindigenous primate will become established seem to be growing. Capturing primates is difficult, too, especially in an urban setting when lethal force is necessary.

Human empathy for primates (Kellert 1979) and public opinion can also make control or eradication efforts difficult (Temple 1990). A population of rhesus monkeys established on the Silver Springs River in the early 1930s by a local tourist attraction seeking to enhance the appeal of its "jungle cruises" had expanded by the early 1980s to nearly 400 individuals (Chapter 10 in this volume). The Florida Department of Environmental Protection purchased much of the land surrounding the tourist attraction in the early 1990s, established a state park, and announced that it planned to remove the monkeys from state-owned lands because the animals posed a threat to human safety and were not a native species. A strain of herpes B virus that can be lethal to humans was detected in the monkey population, and at least 25 attacks had been documented by 1984, several involving serious bites (FGFC 1992).

Opposition to the planned removal quickly surfaced in the form of the "Friends of the Silver Springs Monkeys." The group included the president of the local Audubon Society, persons concerned about animal rights, and research anthropologists from a local university. The group circulated petitions calling for a halt to the "death sentence" imposed on the monkeys. It also helped to draft legislation to protect the monkeys, which was introduced in the Florida House of Representatives by the local legislator in 1993. The bill did not pass, but it did receive considerable media coverage and has probably had an indirect effect on other projects to remove charismatic mammals.

Trade in nonindigenous species of reptiles and amphibians has increased sharply, and Traffic (U.S.A.) (1993) has recorded a 33 percent increase in the number of reptile shipments into Miami from 1987 to 1992. The popularity of captive frogs, salamanders, lizards, snakes, and turtles has also led to the formation of large "herp expositions" that attract many buyers and sellers. In 1990, a weekend herp exposition in Orlando attracted more than 200 dealers from

TABLE 19.3.

Seizures and Arrests by the FGFC for Prohibited or
Restricted Nonindigenous Wildlife Violations: 1985–1993

Year	Seizures		Arrests	
	Fish	Mammals	Primates	Other
1985–1986	38	11	2	19
1986–1987	296	22	4	24
1987–1988	2665	21	1	51
1988–1989	5370	48	6	82
1989–1990	110	41	24	194
1990–1991	6	32	26	247
1991–1992	704	nd[a]	nd[a]	318
1992–1993	48	53	34	193

[a] No data.

many countries; another in August 1994 attracted more than 300. Attendance at both expositions exceeded 10,000, and thousands of dollars and specimens changed hands during these weekend events.

Large-volume dealers operating at these expositions also commonly sell species indigenous to other parts of the United States, some of which are protected by other state wildlife agencies. A recent arrest made at a herp exposition included the seizure of 26 Carolina pygmy rattlesnakes (*Sistrurus miliaris*), a protected species in some states. Other popular trade species found on the threatened and endangered species lists of other states included the Gila monster (*Heloderma suspectum*), bog turtle (*Clemmys muhlenbergi*), and Eastern indigo snake (*Drymarchon corais couperi*), indigenous to Florida and also listed as a threatened species (Wood 1995). The FGFC has placed uniformed officers at these shows to curb illegal activities, but staff reports and documented seizures suggest that most illegal transactions take place behind closed doors. A covert effort may help to curtail these illegal sales.

The use of Florida as a distribution center for species listed by other states requires the FGFC to cooperate with other federal and state fish and wildlife agencies. Some cases involve multistate transactions originating in Florida—as when the FGFC assisted California in the investigation of an illegal shipment of young alligators. Other cases involve transactions that end in Florida—as when the FGFC assisted Oklahoma in tracking down smuggled Gila monsters purchased by Florida dealers. Often Florida serves as a temporary holding area for rare reptiles shipped from other countries. A long-term operation conducted in cooperation with the FWS found that caiman (*Caiman crocodilus*) were smuggled into Florida in boxes marked as tropical fish and then shipped to dealers elsewhere in the United States. This operation resulted in the seizure of 350 caimans and a federal indictment leading to a \$25,000 fine. Another investigation uncovered an attempted shipment of Galápagos tortoises (*Geochelone* spp.) through Florida.

Large snakes and venomous snakes and reptiles present special hazards to human safety, to Florida's environment, and to wildlife inspectors. The effects of the release of the brown tree snake (*Boiga irregularis*) on the avifanua of Guam (Savidge 1987) is a stunning example of the potential problems caused by nonindigenous snakes. The number of venomous reptile permits granted in Florida increased by 121 percent from 1980 to 1993 (Table 19.1). There has also been an increase in the number of complaints involving escaped pythons (*Python* spp.) and boas (family Boidae). During summer months, wildlife officers in southern Florida respond to complaints of loose boids about once a week. Most of these snakes exceed 3 m, and one well-published incident involved an 8-m python living under a house.

Increases in the number of commercial crocodile farms have created another potential avenue for nonindigenous species to enter Florida. FGFC inspectors have noticed a trend for farms to raise nonindigenous crocodilians for sale and distribution nationwide. The accidental or intentional introduction of nonindigenous crocodiles into Florida's environment could threaten native alligator populations (by predation) as well as human safety (Quinn 1994a).

Other activities requiring permits (falconry and wildlife rehabilitation) seem less likely to lead to the establishment of nonindigenous species. A possible exception is private game preserves and farms where many nonindigenous species, primarily big game mammals and upland game birds, may be housed and bred for private or commercial purposes (Dickey 1976). Current regulations (Florida Statutes, 372.16) stipulate that such operations must cover less than 640 acres and be fenced in such a manner that domestic game cannot escape and wild game on surrounding lands cannot enter.

Planned Introductions

The Florida Administrative Code includes broad prohibitions against the import of nonindigenous species, but it also states that "nothing in this rule shall prohibit the Commission or its duly authorized agents from bringing into the state or releasing or introducing any wildlife or freshwater fish."

The FGFC maintains the authority to release nonindigenous and native wildlife or freshwater fish. Such planned introductions were not uncommon in the early decades of the agency, when much of the focus was on game-species management. The scope of these planned introductions may seem enormous if judged by today's standards, but many were strongly encouraged by recreational hunters (Bump 1951, 1968; Bump and Bohl 1964) and academicians (such as Craighead and Dasmann 1974), as well as by federal funds designated specifically for research on the introduction of nonindigenous species (Bump and Bohl 1964).

Variations on a "vacant niche" hypothesis were used as a basis for these introductions, as was the potential expansion of recreational hunting and fishing opportunities. Craighead and Dasmann (1974) argued that post-Pleistocene extinctions of numerous North American ungulates left ecological niches that could be filled by nonindigenous game species. In contrast, Martin (1976) proposed that nonindigenous fishes would "utilize new ecological niches created by man (e.g., reservoirs, canals, and thermal discharge areas)." This argument also supported bird introductions (Bump 1951). Cornwell (1970) suggested that some bird introductions might have another theoretical benefit: the introductions might lead to a discovery of why "southern marshes now are lacking in significant numbers of breeding waterfowl."

The historical emphasis on nonindigenous species introductions is not surprising in view of the successful introduction of the ring-necked pheasant (*Phasianus colchicus*) into North America. This Asian species was first released in the United States in the late 1800s, but it was not until the early to mid-1900s that it became widely established in western states (Bump and Bohl 1964). By 1964, pheasants constituted about 35 percent of all the nonmigratory, upland game birds harvested annually in the United States (Bump and Bohl 1964). Not surprisingly, there has been considerable interest in releasing other nonindigenous game species that might match the pheasant's success, particularly in the Southeast where pheasant introductions had largely failed (Bump 1968). Moreover, pheasants appeared to use ruderal habitats not occupied by native species, and this habit supported the vacant niche theory used to support other attempted releases (Bump 1951). Closer examination, however, has shown that, at least in Illinois, pheasant populations have expanded into natural areas and may be contributing to the decline of the greater prairie chicken (*Tympanuchus cupido*) (Vance and Westemeier 1979).

Following the successful introduction of the pheasant, wildlife agencies throughout the nation attempted to establish other nonindigenous species, particularly during the 1950s and 1960s. Those in many western states focused heavily on introducing nonindigenous big-game species such as oryx (*Oryx gazella*), ibex (*Capra ibex*), and Barbary sheep (*Ammotragus levuia*) (Lund 1980), but few such attempts were made in Florida. One, an effort to establish the pronghorn antelope (*Antilocapra americana*), is perhaps the most noteworthy because the program demonstrates the influence that a few people can exert in decision-making processes. Antelopes were released in the Everglades region at the request of some influential state officials and contrary to the recommendations of FGFC staff. A commercial airline transported the antelope from Colorado and heavily publicized its involvement in the venture. Antelopes were sighted for a few years after the release, but the experiment ultimately failed. Elk (*Cervus elaphus*) and sambar deer (*Cervus unicolor*) were released on several private hunting preserves (Chapter 10 in this volume), but most releases of big-game species appear to have been initiated by individual hunting-preserve owners, not the FGFC. The most extensive mammal importation program supported by the FGFC dealt with white-tailed deer (*Odocoileus virginianus*), an indigenous species. Deer were translocated from other regions of the United States, to stimulate "hybrid vigor" in Florida's deer population (G. Spratt, pers. comm.).

Two programs, both cooperative ventures between the FGFC and the U. S. Fish and Wildlife Service (FWS), were established specifically to introduce nonindigenous bird species. An Exotic Waterfowl Program created in the mid to late 1960s sought to assess the suitability of certain South American ducks for introduction in Florida. The FGFC assigned a waterfowl biologist to the FWS for 18 months, who traveled to Argentina to look for candidates for introduction (Cornwell 1970). One result was the release of 60 rosy-billed pochards (*Netta peposaca*) on Payne's Prairie (Alachua County) in the late 1960s (Hutt 1967; Floyd 1971), which did not become established (Robertson and Woolfenden 1992; Stevenson and Anderson 1994). The FGFC also released Muscovy (or royal) ducks (*Cairina moschata*) in several areas in the late 1960s (Elliot 1967), but these introductions were also unsuccessful (Robertson and Woolfenden 1992; Stevenson and Anderson 1994).

A second bird program, the Foreign Game Bird Introduction Program, focused on upland species, but little preliminary research appears to have been conducted before releases occurred (Kellog et al. 1978). The program persisted until the mid-1970s (Nelson 1963; Kellog et al. 1978) and included releases of ring-necked pheasant (Nelson 1963), red jungle fowl (*Gallus gallus*) (Keller 1963), black francolin (*Francolinus francolinus*) (Murry 1963), and chukar (*Alectoris chukar*). Only the releases of black francolin in southern and central Florida appear to have resulted in a self-sustaining population (Robertson and Woolfenden 1992; Stevenson and Anderson 1994). Sight records of this species persisted into the early 1980s, but the population has now apparently vanished (Robertson and Woolfenden 1992).

Robertson and Woolfenden (1992) mention two additional unsuccessful introductions, perhaps involving the FGFC, which we can neither support nor contradict: the release of the helmeted guinea fowl (*Numida meleagris*) around Lake Okeechobee in the 1960s and that of the Egyptian goose (*Alopochen aegyptiaca*) around Tampa at about the same time.

Opposition to the introduction of nonindigenous waterfowl and upland game species became more common in the late 1960s and 1970s (Weller 1969; Scanlon 1976), and most of the commission's work on introducing nonindigenous bird species began to wane. Two exceptions were an attempt to establish a nonmigratory, breeding population of Canada geese (*Branta canadensis*) and efforts to increase the range of the nonindigenous white-winged dove (*Zenaida asiatica*). Both programs continued until the late 1970s. Releases of Canada geese made in several northern Florida counties led to the establishment of persistent populations that appear to be expanding (Stevenson and Anderson 1994). In most instances, these birds occupy ruderal pasture lands (J. Cox, pers. obs.). The white-winged dove became established in southern Florida after a release in the early 1960s by a bird breeder. The population expanded quickly northward along the Atlantic coast, and in 1976 the FGFC approved a program to enhance the spread of this potentially new game species. Adults were captured in Dade County and transported and released in agricultural areas in northern and central Florida.

Although introductions of game birds and mammals have largely subsided, introductions of nonindigenous game fish continue. The FGFC has primarily focused on nonindigenous piscivorous fish that could help to control other nonindigenous fish species and also enhance sport fisheries in urbanized areas (Shafland 1986).

Research on nonindigenous piscivorous species has led to the recent introductions of two species of cichlid fishes (*Cichla ocellaris* and *C. temensis*) collectively called the peacock cichlid or "peacock bass" (Shafland 1995; see also Chapter 7 in this volume). These are the first exotic species that have been legally released into Florida waters (Shafland 1993). Cichlids were first evaluated for introduction into Florida in the mid-1960s (Ogilvie 1966). Four species were imported from Venezuela and released in a lake in central Florida, but none survived. A more extensive release program was initiated in the late 1980s, when canals in Dade County were stocked with 20,000 peacock cichlid. The formal review period leading up to this release lasted approximately four years (Shafland 1994). The introductions were performed to enhance sports fisheries in an urban setting and to control populations of other nonindigenous fishes that dominate many canal systems (Shafland 1995). Research conducted before the release indicated that salinity and water temperatures would restrict the range of both species to existing canal systems (Moody 1986). Both cichlid species have become established and appear to feed heavily on spotted tilapia (*Tilapia mariae*), a nonindigenous species that constitutes 70 percent of total fish biomass in some canals (Shafland 1995).

Research on nonindigenous piscivorous species has also led to the release of a hybrid *Morone* bass (the "sunshine bass," *Morone chrysops* × *M. saxatilis*).

This release may not be considered a nonindigenous introduction (Shafland and Lewis 1984) because the range of the striped bass (*Morone saxatilis*) includes portions of northeastern Florida (Ware 1995). The releases are noted here because these hybrid fish grow more quickly, have higher survival rates, and are more tolerant of warmer waters (Ware 1995) than native *Morone* and because the ecological impacts of releasing large numbers of hybrid fish (approximately 2 million annually) are not fully understood. Release of *Morone* hybrids has enhanced sports-fishing opportunities in many areas of Florida (Ware 1974), and the hybrids appear to feed primarily on native shads (*Dorosoma* spp.) that are generally abundant in Florida waters (Ware 1974).

Some nonindigenous fish have been investigated by FGFC for their use in controlling nonindigenous weeds (Martin 1976). In the 1960s, several *Tilapia* species seemed to offer promise in this arena (Smith 1970; Ware et al. 1975). The blue tilapia (*Oreochromis aureus*) was introduced for study at several artificial ponds in Hillsborough County in 1961. At the end of the study, the FGFC decided the species would be undesirable (Smith 1970), but by this time the species had been caught by local residents and transported to other lakes and was quickly spreading (Smith 1970). Smith (1970) and Shafland (in press) indicate that this escape was the result of illegal hijackings; Courtenay (Chapter 7 in this volume) lists this as a purposeful introduction by the FGFC. In any event, the species has become abundant in some lakes, and a commercial fishery is now based on blue tilapia (Langford et al. 1978). The FGFC has also constructed more secure fisheries research facilities to reduce the chances that future experiments will go awry (Shafland 1979).

The FGFC also released the Mozambique tilapia (*Oreochromis mossambicus*) into an isolated lake in Polk County in 1963 to test its ability to control weeds. The species devoured most of the aquatic vegetation, but it preferred native to nonindigenous plants (Smith 1970). No further releases were made, and the species does not appear to have spread (Chapter 7 in this volume).

Research on the grass carp (*Ctenopharyngodon idella*) has been performed in the hope that it would control nonindigenous weeds (Ware et al. 1975). Political and public pressures to release this species were high in the mid-1970s (Anonymous 1975), but the FGFC refused to release the carp until a mechanism for controlling the naturalization of the species was developed. In fact, the agency was accused of dragging its feet by people who wanted permits to use grass carp before a control was developed (Anonymous 1975). By the late 1970s, sterile fish could be produced through induced triploidy, either through hybrid crosses or through direct manipulation (Buck 1979). Sterile carp have since been released with some success (Courtenay 1994), but problems have also arisen. Carp are voracious herbivores and can eliminate both target and non-target plants for many years where they are released (Leslie et al. 1987). Among the nontarget plants often eliminated are species important to waterfowl, wading birds, and other wildlife (Brakhage 1987). It has also proved difficult to establish a stocking density that produces predictable results. Current research on triploid carp focuses on the development of economical stocking and removal techniques (FGFC 1995).

Finally, two recent programs overseen by the FGFC and the FWS might also be construed as introductions of nonindigenous species or subspecies into Florida but may represent the first of many introductions designed to save rare species close to extinction. In 1993, six subadult whooping cranes (*Grus americana*) were released on the Three Lakes Wildlife Management Area (Osceola County) in an attempt to establish a nonmigratory population. Nesbitt (1978) indicates that whooping cranes probably bred in Florida at one time, but Robertson and Woolfenden (1992) and Stevenson and Anderson (1994) find the evidence inconclusive. Debate over this introduction and its potential effects on native species (such as the Florida sandhill crane, *Grus canadensis floridanus*) has been relatively minor, perhaps because of the imperiled status of whooping cranes and the likelihood that they once bred in Florida.

In 1995, eight female cougars from Texas were released in southwestern Florida to offset the effects of inbreeding in the Florida panther (*F. concolor coryi*) (Maehr and Caddick 1995), and at least one Texas female has produced offspring in the wild. The degree of inbreeding in the Florida panther has been inferred from low genetic variability, small population size, and certain physical ailments (Roelke et al. 1993). The threat of inbreeding has been assessed by means of genetic and demographic models (Seal and Lacy 1989; Roelke et al. 1993), which suggest the population is in imminent danger of extinction unless new genetic stock is introduced. This introduction has faced greater criticism than the introduction of whooping cranes. Maehr and Caddick (1995) have reviewed the demographic data for Florida panthers and question the evidence for inbreeding depression. On the basis of a review of other translocations, Maehr and Caddick (1995) also suggest that this introduction should have been preceded by more closely controlled studies.

Management and Research

The FGFC has a long history of managing lands harboring nonindigenous species. In 1952, for example, it was one of the first agencies in the nation to begin a program to control water hyacinth by means of chemical herbicides and mechanical harvesters (Woods 1963). Today the agency is involved in the management of approximately 2.3 million ha of land owned by state and federal entities.

Southern and central Florida present the greatest management challenges.

FGFC staff in southern and central Florida report problems with cogon grass (*Imperata cylindrica*), melaleuca (*Melaleuca quinquenervia*), Bahia grass (*Paspalum notatum*), Brazilian pepper (*Schinus terebinthifolius*), tropical soda apple (*Solanum viarum*), downy myrtle (*Rhodomyrtus tomentosus*), Chinese tallow tree (*Sapium sebiferum*), Australian pine (*Casuarina* spp.), and climbing fern (*Lygodium* spp.). In 1993, the FGFC spent approximately $200,000 on nonindigenous vegetation control. Other management costs are hidden under standard activities such as prescribed burning and water management.

The feral pig (*Sus scrofa*) is perhaps the most controversial nonindigenous species at the center of many land-management decisions. Damage caused by feral pigs is well documented (Bratton 1975; Van der Werff 1982; Hoeck 1984; Scott et al. 1986), and pigs are removed from many state parks and state preserves (Chapter 18 in this volume). FGFC management of feral pigs is complicated by the use of the species in both agriculture and recreational hunting. The FGFC defines "wild hogs" as feral pigs in a "wild hog area" that cannot be legally claimed to be domestic. A "wild hog area" is defined in turn as a designated area on which hunting of feral pigs is permitted by commission rules (FAC, 39-1.004.84 and 39-1.004.85). Wild hog areas include state wildlife management areas, state forests, water management district lands, state reserves, and other areas. Pig populations there are controlled primarily through public hunts (Chapter 20 in this volume).

Feral pigs are sought by some 60,000 hunters each year (unpublished FGFC hunter survey data) and rank second in popularity only to white-tailed deer. In 1994, according to the FGFC, the estimated harvest was 63,400 pigs. The current harvest limit set by the FGFC is one pig per day per person, and hunting is allowed only during seasons and in manners established by the commission (FAC, 39-21.004). Current regulations also specify a minimum shoulder height of approximately 35 cm, allowing pigs to reach breeding age before they are taken. The FGFC does not stock feral pigs, but pigs are stocked on some privately managed hunting areas.

Although the current regulations applied to wild hog areas have not reduced populations to the degree that some land managers wish (Girardin 1994; Hardin 1994), few data are available to show the extent of pig damage on public lands in Florida and, moreover, pigs are rarely the sole source of ecological disturbance. Hardin (1994) says that pigs are "clearly a serious pest" in many state forests, but he offers no data on the extent of pig damage or comparison of pig damage to that created by timber-management practices or off-road vehicle use. Similarly, Glisson (Chapter 18 in this volume) lists feral pigs as a serious problem at Myakka River State Park (Sarasota County), a facility that also receives approximately 1.4 million human visitors annually. A comparison of pig damage with that created by human visitors might help in decisions regarding the allocation of limited management funds. Finally, Girardin (1994) cites pig damage to artificial flood-control levees (not natural areas) and an unspecified decrease in the production of longleaf pine (*Pinus palustris*) seedlings and other herbaceous vegetation as a basis for concern.

There are, of course, published studies that clearly document the damage caused by feral pigs (Bratton 1975; Katahira et al. 1993), but many stem from work conducted on national parks, where public hunting is forbidden. These studies also indicate that comprehensive pig eradication is an expensive undertaking when left exclusively to land managers: Katahira et al. (1993) estimated the time needed to eradicate 175 feral pigs in Hawaii at 20 hours per animal, not including equipment costs. These estimates, derived for a forested mountain region, are probably high, but because Florida has one of the largest feral pig populations in the Southeast (Wood and Lynn 1977), total eradication on large management areas would probably cost hundreds of thousands of dollars. Belden and Frankenberger (1977), however, found that feral pig populations on some management areas in Florida were severely depleted by hunting pressures, and special hunts have been developed to help control pigs there. Feral pigs in Hawaii were controlled most effectively by hunting with dogs combined with perimeter fencing (Katahira et al. 1993).

In the late 1980s, the FGFC released six sterilized feral pigs within the ranges of two Florida panthers to assess whether pigs could be used to enhance the prey base for panthers (Maehr et al. 1989). Officials were concerned about panthers south of Alligator Alley (Interstate 75) in Collier County that appeared to suffer from malnutrition and a small prey base. Although most of the pigs were eventually killed by predators, panthers clearly accounted for only one of the kills. The experiment was deemed unsuccessful unless large numbers of pigs were to be released (Maehr et al. 1989).

The FGFC operates an Exotic Fish Laboratory in Boca Raton (Palm Beach County) that collects basic information on the distribution, habits, and general biology of nonindigenous fishes in Florida (Shafland 1979). It is also responsible for documenting the effects of nonindigenous fishes on the environment, developing new techniques for managing them, and reviewing species for possible introduction. The laboratory covers some 10 ha, consists of 17 experimental ponds, and has a closed water system. The laboratory is surrounded by a 5-m levee and a 2.5-m chain-link fence and has withstood at least one Category 3 hurricane without incident.

In addition to introducing nonindigenous species to enhance sports fishing, as described earlier, staff at the Exotic Fish Laboratory have been instrumental in eliminating nonindigenous fish populations considered dangerous to Floridians or Florida's environment (Shafland, in press; Chapter 17 in this volume). A reproducing population of piranha (*Serrasalmus humeralis*), for example, was discovered during a routine inspection of a tourist attraction and eliminated with rotenone in 1977 (Shafland and Foote 1979). The fishes had been introduced by the owners of the attraction in 1963 or 1964. Shafland (in press) lists nine other established populations of nonindigenous fishes that have been eliminated by the FGFC, by habitat alterations, or by natural causes, but some established fish populations have grown to the point where elimination is deemed virtually impossible (Shafland, in press).

The FGFC also operates a wildlife research lab and provides grant funds to support numerous research projects. Although most of the research focuses on indigenous species rather than nonindigenous ones, occasionally there is some overlap, as in recent studies on the effects of feral house cats (*Felis catus*) on beach-mouse populations (Frank and Humphrey 1992; Gore and Schaefer 1993). Several contracted research projects have focused specifically on problems associated with nonindigenous species, including studies on the effects of fire ants (*Solenopsis invicta*) on the herpetofauna of the Florida Keys and techniques for managing melaleuca.

Lessons and Prospects

Florida's diverse human population includes many people interested in possessing nonindigenous species as pets, cultivating them for food, shade, and color, and hunting them for recreation. Management of nonindigenous species will therefore be a factor on Florida's public lands for many decades to come, and the difficulties will only increase as the state's human population expands (Figure 19.1). The most difficult choices will probably be in allocating limited resources among control of nonindigenous species and the many other important management programs (such as fire and people management). In many cases, the most cost-effective approach will be one that prevents new alien species from becoming established and maintains established nonindigenous species at densities where minimal ecological damage occurs.

Wildlife inspectors, and the regulations they enforce, represent an important line of defense against the establishment of new nonindigenous species in Florida. More inspectors are needed to regulate the trade in nonindigenous fish and wildlife in Florida and its projected expansion over the next few years (Traffic 1990). The FGFC has repeatedly sought funding for additional staff, but the Inspections Section staff has added only three positions since the program's inception in 1971. More training might prove helpful, as well, particularly in identification of nonindigenous fish.

Minor changes to FGFC regulations might provide a low-cost means of controlling additional nonindigenous species. The agency maintains a list of fishes prohibited on the basis of their threat to humans and Florida's environment

PROTECT
Our Natural Resources

DON'T RELEASE EXOTICS

Foreign animals will compete with our native fish and wildlife for food and space. Some may present a threat to human safety, while others may become agricultural pests or public nuisances. State law prohibits the release of any non-native wildlife!

FLORIDA
GAME AND FRESH WATER FISH COMMISSION

The exotic species illustrated above are: marine toad (*Bufo marinus*); monk parakeet (*Myiopsitta monachus*); spotted tilapia (*Tilapia Oreochromis mariae*).

FIGURE 19.1   Florida Game and Fresh Water Fish Commission poster warning the public not to release exotic animal species into the environment.

(Table 19.2), for example, and similar lists should be developed for the many nonindigenous mammals, reptiles, amphibians, and birds that have proved to be destructive. Several species of primates and large mammals are generally not permitted as personal pets (Appendix A), but general prohibitions might be developed for these and other species. The brown tree snake, for example, has devastated bird communities on Guam (Savidge 1987) and should be listed as a prohibited species in Florida. A review of the scientific literature should be undertaken, too, and used to expand the list of prohibited species.

The FGFC has moved quickly to eradicate certain nonindigenous species that pose a threat to human health (Shafland and Foote 1979), and the feasibility of eradicating other species that pose a threat to the environment should be explored. Populations of nutria (*Myocastor coypus*), for example, appear to be localized in Florida (Chapter 10 in this volume) and have been eliminated from some areas of the state (Brown 1974). Given the damage that expanding nutria populations have caused in coastal habitats in Louisiana (Lowery 1974), eradication could be cost-effective (Chapter 13 in this volume). Biologists should periodically review the feasibility of eradicating small, isolated populations of nutria and make recommendations to FGFC and other agencies.

In many instances, nonindigenous species appear to have spread beyond the point where they can be eradicated. Feral pigs, for example, have occupied Florida for nearly 500 years (Belden and Frankenberger 1977), and it seems unlikely that they can be permanently removed from many public lands. The use of this species in agricultural operations poses additional problems. Because concern over the damage they cause is increasing, new research is needed that focuses on the impacts of feral pigs in areas where their populations are controlled through public hunts (Chapter 20 in this volume). Finally, new technologies may hold promise for controlling feral pigs and other nonindigenous species that have become abundant (Chapter 13 in this volume). Periodic reviews should assess the application of new technologies to problem species in Florida.

The FGFC participated in or sanctioned the release of dozens of nonindigenous species during the 1950s, 1960s, and 1970s, but since 1980 the peacock cichlids have been the only clearly nonindigenous species released. The agency estimates that recreational activities associated with these releases have an annual economic value exceeding $1.4 million (FGFC 1995) and believes the popularity of the peacock bass is growing among anglers in southern Florida (Huttenmeyer 1994; Lampton 1994), where they account for as much as 33 percent of the fishing activity on some canals (FGFC 1994b). This release has had economic and recreational benefits in line with the stated goals of the agency, but the potential ecological costs cannot yet be fully measured (Chapter 7 in this volume).

Huenneke (1988) argues that our ability to predict the success or failure of an introduction has greatly improved, and the FGFC's most recent introductions appear to confirm this assessment (Shafland 1995). The FGFC and other fish and wildlife agencies, as well as other agencies concerned with biological control (Chapter 15 in this volume), may be willing to take certain risks associated with the introduction of nonindigenous species if the chances of harm appear to be slight and the potential benefits seem significant. The guidelines used in such decisions have become much more strict than in similar work in previous decades (Shafland 1995)—and, moreover, the decisions appear to be less influenced by public opinion. In some of the introductions discussed earlier (pronghorn antelope and rhesus monkey, for example), just a few influential people or groups played major roles in key decisions.

Optimism about planned introductions could be quickly dashed, of course, by an unforeseen ecological catastrophe. Even the most intensive research leading up to a planned introduction cannot be complete because a thorough understanding of both the introduced species and the host ecosystem is rarely available. It would be simple to call for a moratorium on the release of any nonindigenous species, but government agencies reflect the interests of a diverse populace, and many Floridians are still keenly interested in maintaining nonindigenous species for recreational reasons. Much of the progress to be made in the control of nonindigenous species therefore probably lies in public education, particularly regarding the risk of ecological catastrophes (Temple 1990). As Temple (1990) notes, most people do not recognize the difference between a nonindigenous and an indigenous species or realize the damage that invaders can inflict. Florida's most destructive nonindigenous population, therefore, will probably continue to be the 14 million people derived from foreign ancestries.

ACKNOWLEDGMENTS

We thank T. Breault, B. Hartman, R. Kantz, F. Montalbano, and T. O'Meara for helpful comments on earlier versions of this chapter.

Appendix A. Regulations from the Florida Administrative Code Pertaining to Nonindigenous and Native Wildlife in Florida

All individuals not previously permitted to possess Class 1 or Class 2 wildlife (listed below, Rule 39-6.002, Florida Administrative Code) must qualify for a permit by meeting the following criteria as specified by Rule 39-6.0011, F.A.C.:

(a) Submit a notarized statement that the construction of the facility, its cages, and enclosures, is not prohibited by county ordinance, and, if within a municipality, municipal ordinance.

(b) Applicants shall demonstrate no less than one (1) year of substantial practical experience (to consist of no less than 1,000 hours) in the care, feeding, handling, and husbandry of the species for which the permit is sought, or other species, within the same biological family (except ratites, which shall be in the same biological suborder), which are substantially similar in size, characteristics, care and nutritional requirements to the species for which the permit is sought. This requirement shall not apply to applicants for permits to possess ostriches when possessed for propagation purposes, and not for public exhibition.

(c) Shall not have been convicted of a violation of captive wildlife regulations for three (3) years prior to application for such permit.

SECTION 39-6.002. Categories of Captive Wildlife.

1. F.G.F.C. hereby established the following categories of wildlife:

CLASS I WILDLIFE		CLASS II WILDLIFE	
Chimpanzees	Black caimans	Cheetahs	Wolverines
Gorillas	Komodo dragons	Caracals	Honey badgers
Gibbons		African golden cats	American badgers
Drills and mandrills	CLASS II WILDLIFE	Temminck's golden	Old World badgers
Orangutans	Howler monkeys	cats	Binturongs
Baboons	Uakaris	Fishing cats	Hyenas
Siamangs	Mangabeys	Ocelots	Dwarf crocodiles
Gelada baboons	Guenons	Clouded leopards	Alligators, caimans (ex-
Snow leopards	Bearded sakis	Coyotes	cept American alliga-
Leopards	Guereza monkeys	Gray wolves (including	tors)
Jaguars	Celebes black apes	wolf × domestic hy-	Ostriches
Tigers	Idris	brids which are 25%	Cassowaries
Lions	Macaques	or less domestic dog)	
Bears	Langurs	Red wolves (including	CLASS III
Rhinoceroses	Douc langurs	wolf × domestic hy-	WILDLIFE
Elephants	Snub-nosed langurs	brids which are 25%	All other wildlife not
Hippopotamuses	Proboscis monkeys	or less domestic dog)	listed herein, except
Cape buffalos	Servals	Asiatic jackals	those for which a
Crocodiles (except	European and Cana-	Black-backed jackals	permit is not required
dwarf and Congo)	dian lynx	Side-striped jackals	pursuant to rule 39-
Gavials	Cougars and panthers	Indian dholes	6.0022 F.A.C.
	Bobcats	African hunting dogs	

2. Except as provided elsewhere, Class I wildlife shall not be possessed for personal use.

3. Persons possessing any captive wildlife for purposes of public display or sale shall obtain a permit as specified elsewhere.

4. Persons possessing Class II wildlife as personal use wildlife shall purchase a permit as provided elsewhere.

5. Persons possessing Class III wildlife as personal use wildlife shall obtain a no-cost permit.

SECTION 39-6.0022. Possession of wildlife in captivity permits.

2. No permit shall be required to possess the following wildlife for personal use, unless possession of a species is otherwise regulated by other rules of the F.G.F.C.

Reptiles	Moles; shrews	Hamsters	Button quail
Gerbils	Rabbits	Parrots	Prairie dogs
Amphibians	Squirrels, chipmunks	Finches	Chinchillas
Shell parakeets	Ferrets	Myna birds	
Rats and mice	Guinea pigs	Toucans	
Canaries	Cockatiels	Doves	

20 Management in Water Management Districts

Amy Ferriter, Dan Thayer, Brian Nelson, Tony Richards, and David Girardin

The Florida Water Resources Act of 1972 launched a significant change in the state's approach to natural resource management, creating the state's five water management districts. District boundaries are based on natural, hydrological basins rather than political boundaries, and the districts are responsible for most water issues in the state— from providing flood protection and water-supply protection to restoring and managing natural ecosystems. The five districts are the South Florida, Southwest Florida, St. Johns River, Suwannee River, and Northwest Florida water management districts (Figure 20.1).

The South Florida (SFWMD), Southwest Florida (SWFWMD), and St. Johns River (SJRWMD) districts have nonindigenous species control programs in place. Nonindigenous species management within the districts keeps navigation channels open, provides drainage for floodwater abatement, keeps water control structures and pumping stations operational, enhances fish and wildlife habitats, reduces destruction of beneficial native plant communities, and improves aquatic recreational activities. Within each district, these programs can be divided into control of nonindigenous aquatic plants, control of nonindigenous wetland and upland plants, and control of nonindigenous vertebrates.

Control of Nonindigenous Aquatic Plants

Although many foreign plant species are established within district waterways, none has caused more severe environmental problems than water hyacinth (*Eichhornia*

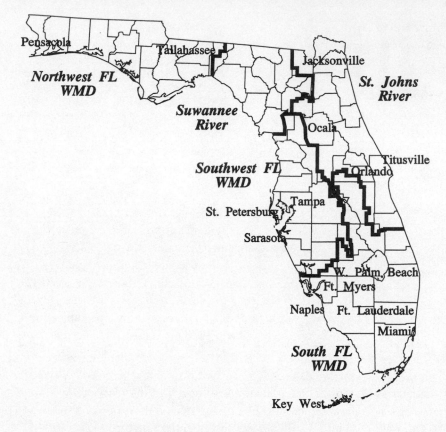

Figure 20.1 Florida's water management districts.

crassipes), water lettuce (*Pistia stratiotes*), and hydrilla (*Hydrilla verticillata*). Because they grow rapidly and can produce a large amount of biomass near the water surface, these species have often altered and replaced entire native vegetational communities, degraded water quality (especially dissolved-oxygen levels), increased sedimentation rates, reduced fish and wildlife habitat, hindered recreational use, and increased flood potential (Schmitz et al. 1993).

Large amounts of plant biomass within a flood control system will reduce the system's functioning during periods of increased flow. If enough vegetation blocks water control structures or jams against bridge pilings, these structures may function as dams, causing damage to the structures and leading to upstream flooding. Water hyacinth, water lettuce, and hydrilla populations are easily swept downstream in large mats.

Most plant control operations involve nonindigenous aquatic plants. Much of the management program is funded through cooperative agreements with the U.S. Army Corps of Engineers and the Florida Department of Environmental

Protection (FDEP). A recent dramatic increase in hydrilla statewide has severely limited management. The hydrilla management program in public lakes is funded through the FDEP-administered Cooperative Aquatic Plant Management Program. Even though approximately $39 million was spent managing hydrilla in Florida waters between 1980 and 1993 (Chapter 14 in this volume), hydrilla coverage in Florida public lakes and rivers has increased drastically during this period. In 1982, hydrilla infested approximately 5300 ha; by 1993, this number had soared to 30,400 ha. As funding has not kept pace with this expansion, effective management has not been possible. Instead, hydrilla control in Florida's lakes has become a sort of salvage effort designed to allow limited access and recreation. Some $670,000 was spent on aquatic plant control during 1993 by the SWFWMD; in 1994, more than $1.3 million was spent in the SJRWMD and about $4 million in the SFWMD.

The primary control method is the application of aquatic herbicides. Physical control techniques (drawdowns) are rarely feasible because most district waters are free-flowing and cannot be contained or released as required. Mechanical harvesting is often used in areas adjacent to potable water intakes, in fast-flowing waters, and in other situations where herbicides are ineffective or cannot be used. Elsewhere harvesting is usually not feasible (Thayer and Ramey 1986). Harvesters are slow, making the process both costly and time-consuming. Moreover, populations of water hyacinth and water lettuce persist in backwater areas too shallow to float a harvester, and the plants grow around trees, logs, stumps, and rocks that block harvesters. If these populations are neglected, water hyacinth can quickly reinfest an entire waterway.

The water management districts also use biological control. Triploid grass carp (*Ctenopharyngodon idella*) have been used successfully in canals to control hydrilla and hygrophila (*Hygrophila polysperma*). The only limitation to using more grass carp in canals has been the requirement for fish barriers. Research continues on barriers that do not interfere with flow requirements. Insects have been imported and released on—or have naturally dispersed to—many water bodies within the state (Chapter 15 in this volume). Except for alligatorweed (*Alternanthera philoxeroides*), which is held to an acceptable level by biological control agents, it is difficult to assess the degree, if any, to which other nonindigenous plant species are affected by biological control. Water hyacinth, water lettuce, and hydrilla expand rapidly whenever nonbiological control methods are curtailed.

Control of Nonindigenous Wetland and Upland Plants

The rapid spread of invasive nonindigenous plant species in Florida's wetlands and uplands has displaced native vegetation, disrupted natural ecosystem functions, and reduced available habitat for native plant and animal species (Chapter 3 in this volume). Although public waters have been managed through a well-coordinated statewide maintenance-control program, there is no comparable eradication or management program for invasive nonindigenous plant species in wetland and upland public lands.

The three most troublesome such species are melaleuca (*Melaleuca quinquenervia*), Brazilian pepper (*Schinus terebinthifolius*), and Australian pine (*Casuarina* spp.). The results of a recent (1993) aerial survey of southern Florida (from the north rim of Lake Okeechobee south to Florida Bay) are shown in Figure 20.2. This survey was flown again, and extended north to Orlando, in the spring of 1995, but the data have not been analyzed as of this writing.

The SFWMD initiated an intensive melaleuca management program in November 1990 in the Everglades Conservation Areas (ECAs) supported by Florida Power & Light mitigation funds of $500,000 for three years. Melaleuca is currently managed in the ECAs and the marsh area of Lake Okeechobee. Moreover, programs have been initiated on many Save Our Rivers lands (critical water-resource lands) managed by the SFWMD. The FDEP contributed $400,000 in fiscal year 1994, and $1 million in fiscal year 1995, to the melaleuca control effort in the ECAs and Lake Okeechobee, bringing the SFWMD funding level for melaleuca management to $2 million in fiscal year 1995. All control operations in this program are carried out by private contractors.

By 1995, all seed-bearing melaleuca trees had been eliminated from the ECAs south of Alligator Alley (Interstate 75 between Naples and Fort Lauderdale; approximately 162,000 ha). Seedling control continues in these areas. In Lake Okeechobee, control of mature trees and seedlings in the outer marsh area has been emphasized. To date, nearly all melaleuca control has entailed treating individual trees, which is labor intensive. Each tree must be completely girdled and the exposed cambium treated with herbicide. Moreover, particularly in the ECAs, trees are spread over such a large area that access must be by helicopter. Research is continuing into the use of aerial application of herbicides. Since 1990, the SFWMD has supported U.S. Department of Agriculture research on biological control of melaleuca with insects. The annual contribution in recent years has been $150,000.

A variety of government agencies and private landowners are currently pursuing melaleuca control for purposes including management of natural areas, compliance with water use permit requirements, and range improvement. Amounts and sources of funding vary considerably, as do treatment methods. Melaleuca control strategies in southern Florida must be integrated to achieve any sort of success; a ranking system for treatment areas is needed to prevent a piecemeal approach and to allow coordination between government agencies

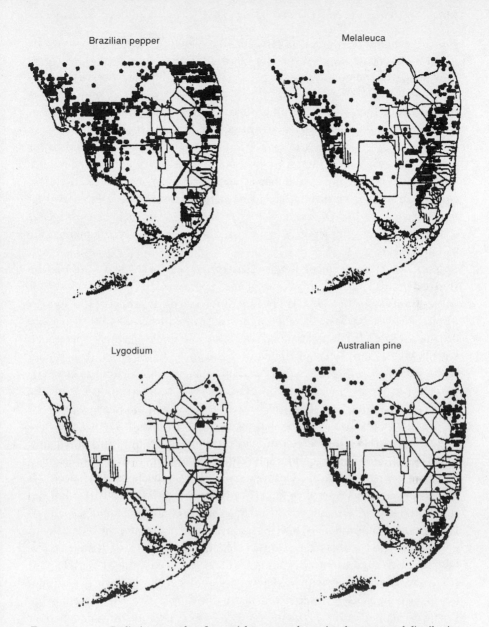

Brazilian pepper

Melaleuca

Lygodium

Australian pine

FIGURE 20.2   Preliminary results of an aerial survey to determine the extent and distribution of four nonindigenous species in southern Florida. Observation lines were established across the region; observation points were documented from fixed-wing aircraft and a global positioning system.

and private landowners (Chapter 22 in this volume). This system could be based on such criteria as potential for realizing control, the effects of hydrological disturbance on nonindigenous plant invasion, and wildlife populations.

The SWFWMD governing board adopted Procedure 61-9, Control of Terrestrial Exotic Flora and Fauna on District-Owned Lands, in 1994 to establish control programs for nonindigenous species on district-owned lands. The mere presence of such a species does not automatically make it a candidate for control. Because funding and workforce are limited, the program concentrates on species that grow and spread rapidly and can disrupt or replace native plant communities. Highest priority is accorded to the control of infestations that threaten areas of high resource value: habitat for threatened or endangered species, remnant examples of natural Florida communities, areas of high public use, and the like.

The Water Management Lands Trust Fund, established in 1981, is the source of funds for control of nonindigenous species on SWFWMD's Save Our Rivers and Preservation 2000 lands. Operations on other properties are funded through district and basin ad valorem funds. Approximately $24,000 was spent on the control of invasive upland and wetland nonindigenous species on SWFWMD lands in 1994—particularly cogon grass (*Imperata* spp.), air potato (*Dioscorea bulbifera*), Brazilian pepper, and melaleuca. Melaleuca has been discovered on newly purchased property in the southern portion of the SWFWMD. A control strategy has been determined and is budgeted for 1997.

Brazilian pepper is rapidly expanding its range throughout Florida (Kushlan 1990). This invader interferes with natural colonization by native plant communities and forms monospecific stands that reduce habitat diversity. Brazilian pepper is probably the most widespread upland nonindigenous plant species. It is reported as a major problem for SFWMD, SWFWMD, and SJRWMD. Because the seeds are distributed by birds, herbicide control is temporary at best. In addition to conventional herbicide controls, fire may reduce the biomass and invasive potential of this plant (Doren et al. 1991a), but its long-term effectiveness is poorly understood. Beginning in fiscal year 1994, the SFWMD sponsored a research project with the University of Florida for biological control of Brazilian pepper using insects from its native range. Two such insects—the Brazilian pepper sawfly (*Heteroperryia hubrichi*) and the Brazilian pepper thrips (*Liothrips ichini*)—are now in the Gainesville quarantine lab undergoing extensive host specificity testing designed to determine whether they will feed on other plants.

Many species of grass have become problematic throughout the state, especially along district rights-of-way. The key nonindigenous grass species are torpedo grass (*Panicum repens*) and cogon grass. Torpedo grass has become a major component of the Lake Okeechobee marsh community and, once established, can quickly become the dominant species. Between 1988 and 1992, hundreds of hectares of native sedge and rush communities were invaded and replaced by monospecific stands of torpedo grass. Torpedo grass now covers approximately 6100 ha of the lake's marsh and is rapidly expanding its range on lands where the hydrology has been restored. Funding shortfalls and the plant's general resistance to herbicide controls have thwarted management efforts. The SFWMD recently sponsored a research project on management options for torpedo grass.

Cogon grass is one of the most widespread invasive plants in the SJRWMD. It invades disturbed areas—roadsides, fence and fire lines, pastures, levees, power-line rights-of-way, and logged-over lands—and is beginning to invade relatively undisturbed habitats. In 1994, the district first attempted to control cogon grass on power-line rights-of-way, fence lines, and district levees. Small but widespread infestations of cogon grass occur throughout the SWFWMD and SFWMD. Because these populations are still small, intensive herbicide applications, along with mechanical controls, have been implemented. Anecdotal evidence to date suggests it is difficult to contain cogon grass with conventional control techniques.

A multitude of invasive vine species plagues southern Florida. Particularly alarming is the Old World climbing fern (*Lygodium microphyllum*), which appears to be rapidly expanding its range in southern Florida's wetlands (Austin 1978b). This fern threatens Everglades tree islands and the region's cypress forests. Little has been done to counter this threat, but the SFWMD intends to evaluate herbicide controls.

After Hurricane Andrew, the loss of tree canopy in Dade County's tropical hardwood hammocks and the consequent spread of alien vines such as sewer vine (*Paederia cruddasiana*), air potato, and wood rose (*Merremia tuberosa*) have posed the greatest threat to the reestablishment of native vegetation. Air potato and skunk vine have also invaded and modified habitat on SWFWMD lands. SWFWMD personnel have been evaluating the effectiveness and selectivity of several herbicides.

Recently a highly invasive vine characteristic of the traditional Deep South, kudzu (*Pueraria montana*), was identified in the Everglades. This species was established on the levee system by the Soil Conservation Service in the 1950s, when a trial planting was made to determine the suitability of kudzu as a vegetative cover for the limestone levees (Broward County Soil Conservation Service 1952). Despite repeated mowings for 40 years, small patches of kudzu persisted. Once identified, these small infestations were quickly treated with herbicides by SFWMD staff, and long-term monitoring should ensure eradication.

Control of Nonindigenous Vertebrates

Several nonindigenous vertebrate species inhabit district-owned lands. Feral pigs (*Sus scrofa*) are presently the only introduced animals known to cause serious environmental damage. They interfere with natural area management by competing with native wildlife species for available food, by eating or destroying native plant seedlings, and by disrupting wetland and upland plant populations with their year-round rooting. Severe rooting damage creates erosion and siltation.

SWFWMD staff, through routine aerial and ground surveys and reports from other staff, regularly assess the impact of feral pigs on all district-owned lands and classify them as major (widespread and nonseasonal) or minor (small-scale or seasonal). On major-impact sites, staff conduct a detailed evaluation by either baited transects or a rooting survey. After a baseline pig population index is established for a major-impact area, control options are assessed and an implementation plan is developed. This plan and the options assessment constitute supporting documents for a review and approval process. One recent option employed was a "special dog and gun hunt" for two weeks on approximately 12,000 ha on the SWFWMDs Green Swamp West property. The district's total cost to administer the hunt was $41,641. The SFWMD coordinates feral-pig hunts with the Florida Game and Fresh Water Fish Commission on some Save Our Rivers properties. Some 380 pigs were removed during the 50 designated hunt days in 1994. Since the initiation of these hunts in 1989, there has been a slight improvement in general conditions on these sites, but no population survey or rooting index is in place to measure their success.

Presently no SJRWMD funds are used to control pigs, but the district does license a trapper to control pigs in one parcel. The trapper is given an exclusive-use trapping contract but is not paid by the district. The district spends from $5000 to $8000 a year repairing and reseeding its flood control levees as a result of damage caused by feral pigs. Domestic cats (*Felis catus*) are occasionally seen hunting on district lands, and one can assume they are killing native animals.

Lessons

Throughout Florida, nonindigenous species problems have expanded at a much greater pace than the financial resources available for their management. Moreover, our basic understanding of each species' biology and subsequent development of sound control options have lagged far behind the identification of new nonindigenous species problems. Only the floating aquatic plant management program can be viewed as a "maintenance control" program within the state. Under the Save Our Rivers and Preservation 2000 programs, the state has bought a great deal of land critical to sustaining regional ecosystems. Maintenance management of nonindigenous species on such lands is essential to the restoration and preservation of these natural resources.

Careful coordination is needed to optimize the impact of the limited financial and human resources available to manage these pests. The state's water management districts work closely with the Florida Exotic Pest Plant Council, a nonprofit organization of natural resource managers from both the private and public sectors concerned with the problem of nonindigenous species and their impact on Florida's remaining native ecosystems. The council has been effective in identifying nonindigenous species issues and assisting land managers with specific problems, but its powers are limited. Clearly, nonindigenous species management in Florida must be coordinated at the state level. In 1971, the state created the Bureau of Aquatic Plant Management, administered by the Florida Department of Environmental Protection (formerly the Department of Natural Resources), to coordinate the management of Florida's aquatic plant problems. The control of other nonindigenous species in Florida needs the same kind of commitment.

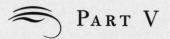

PART V

The Regulatory Framework

21 The Federal Government's Role

DON C. SCHMITZ AND
RANDY G. WESTBROOKS

 The federal government devotes significant resources to managing nonindigenous species that are agricultural pests and preventing introduction of new ones, but it deals inadequately with such species that threaten our nation's natural areas. This chapter focuses on federal programs that affect nonindigenous species in natural areas and the ways in which those efforts help, or fail, to protect Florida from invasion.

The Office of Technology Assessment's 1993 overview (U.S. Congress 1993) concluded that the present federal framework for dealing with harmful nonindigenous species is largely an uncoordinated patchwork of laws, regulations, policies, and programs and that, in general, current federal efforts do not match the problems at hand. Federal programs include those intended to restrict the entry of harmful species into the United States, to limit their movement among states, and to control or eradicate damaging species once they are introduced. But they are also designed to introduce nonindigenous species considered beneficial, to assist in their dissemination, to encourage their use by industry, and to conduct research on both harmful and beneficial species. Harmful nonindigenous species are either intentionally or accidentally introduced animals, diseases, and plants that cause significant harm to natural, managed, and agricultural ecosystems.

Efforts to keep such species out of the United States are among the largest federal programs. The U.S. Department of Agriculture's Animal and Plant Health Inspection Service spent at least $100 million in fiscal year 1992 for agricultural quarantine and port inspection (most of which was spent on agricultural diseases and disease vectors). The U.S. Fish and Wildlife Service requested $3 million for port inspections of fish and wildlife species in 1992. These funds were intended to enforce conservation laws and monitor commercial traffic, however, not to prevent entry of nonindigenous species into the United States. Until 1993, it was impossible to assess the effectiveness of these federal efforts because these agencies lacked performance standards for their port inspection activities or even routine evaluations of their programs (U.S. Congress 1993). Realistically, the Office of Technology Assessment report concluded that total exclusion of invasive nonindigenous species at U.S. ports is probably not feasible.

Moreover, present federal efforts have failed to prevent the introduction of many invasive nonindigenous species because further entry is denied only after the species have become established or have caused economic or environmental damage. The Federal Noxious Weed Act and the Lacey Act—the two major laws that restrict entry of nonindigenous species—permit a harmful species to be imported legally until it is declared, by listing, to be undesirable. Excluding a nonindigenous plant species from entry into the United States requires its addition to the Federal Noxious Weed list, a process time consuming and often impossible. It took more than five years, for example, to list the Australian melaleuca tree (*Melaleuca quinquenervia*) as a federal noxious weed species. Without the support of the entire Florida congressional delegation, it is doubtful that melaleuca would ever have been listed. Another example is tropical soda apple (*Solanum viarum*), a herbaceous perennial from Brazil that has recently become a serious problem in pastures and natural areas of Florida (Mullahey et al. 1993) and is spreading to other southern states (Bryson et al. 1995). Tropical soda apple was proposed for listing as a federal noxious weed in 1993. The process of listing was still going on in August 1995. At least 250 weeds meeting the Federal Noxious Weed Act's definition of a federal noxious weed remain unlisted (Singletary 1990).

At least 20 federal agencies have regulatory oversight concerning research, use, prevention, or control of nonindigenous species (U.S. Congress 1993). This patchwork of agency involvement has often prevented a prompt and timely reaction to the introduction of harmful nonindigenous species, thus allowing their establishment and subsequent spread (Singletary 1994).

Management of nonindigenous species, or lack thereof, on federal lands can have long-term impacts on surrounding state, local, and private lands in containing or spreading unwanted species. The federal government is a substantial landowner and land manager in Florida. A total of 909,895 ha is included in Everglades National Park, Big Cypress National Preserve, and the Loxahatchee National Wildlife Refuge (Palm Beach County) (one of the largest complexes of preserved federally managed land in the eastern United States). In addition,

370, 271 ha is preserved in another 27 national wildlife refuges throughout Florida, and 460,902 ha is managed in national forests. Military bases, too, account for substantial landholdings in Florida: approximately 210,000 ha is managed by the Department of Defense.

The following section examines the roles and responsibilities of major federal agencies and how their actions have affected or will affect the establishment and control of nonindigenous species in Florida.

U.S. Department of Agriculture (USDA)

A number of subagencies of the USDA have responsibilities related to nonindigenous species. The broadest assignment is that of the Animal and Plant Health Inspection Service (APHIS), which restricts movement of agricultural pests and pathogens into the United States by inspection, prohibition, and requirement of permits for the entry of agricultural products, seeds, live plants and animals, and other articles that may carry nonindigenous organisms (U.S. Congress 1993). Traditionally, the agency's primary concern has been species that threaten agricultural interests and not those capable of invading natural areas. But because of pressure from local and regional nonprofit organizations and national professional societies, APHIS has begun developing a new weed policy that would include those in natural areas as well.

Most of APHIS's pest exclusion occurs at ports of entry. In 1990, inspections of incoming passenger baggage and cargo at U.S. ports of entry found more than a million violations and assessed $723,345 in penalties, less than $1 per violation on average (U.S. Congress 1993). Sixteen APHIS ports of entry in the United States, including Miami and Orlando, are authorized to receive imported live propagated plant material. Approximately 90 percent of live plant material imported into the United States comes into Miami (T. Dobbs, pers. comm.). Generally, at least 2 percent of each plant shipment is inspected for insect pests, pathogens, and federally listed noxious weeds. Because it is impossible to inspect every plant in an imported shipment, some shipments are "precleared" by inspections or treatment to eliminate pests before they are imported into the United States. Such shipments can enter the country without further inspection.

APHIS restricts interstate movement of agricultural plant pests or pathogens by imposing domestic quarantines and regulations. Moreover, the agency regulates interstate shipment of livestock, animal products, and other items that could transport animal pathogens, as well as nursery stock, soil, and equipment that could spread plant pathogens listed in domestic quarantines (U.S. Congress 1993). The U.S. mail can be an important conduit of pest introduction (Figure 21.1). Presently APHIS can inspect first class mail only on issuance of a

*Shell Flower of Water Lettuce   *Water Hyacinth

FIGURE 21.1   Water lettuce (*Pistia stratiotes*) and water hyacinth (*Eichhornia crassipes*) advertised for sale in a mail-order catalog. Photo by Don Schmitz.

warrant. APHIS confiscated more than 4000 pounds of prohibited plant material in a trial inspection program between Hawaii and the mainland United States during five months in 1990 (U.S. Congress 1993).

APHIS regulates the interstate shipment of federally listed noxious weeds only when a quarantine is declared and an eradication program is implemented. To date, the agency has imposed only one domestic quarantine for an agricultural weed: witchweed (*Striga asiatica*). Because of this narrowly interpreted view of the Federal Noxious Weed Act, APHIS has allowed interstate commerce of approximately 10 listed species (Westbrooks 1990). For example, seeds of water spinach (*Ipomoea aquatica*), a federally listed noxious weed, are commonly sold from state to state as an agricultural crop, even though this non-indigenous aquatic plant can invade and disrupt the ecology of Florida's ponds, lakes, and canals. Since the early 1980s, numerous deliberately planted infestations of water spinach have been found and eradicated by the Florida Department of Environmental Protection. Because APHIS does not stop interstate sale of this species, it is only a matter of time before it becomes widespread throughout Florida.

APHIS eradicates and controls certain species that are either confined to a specific area or recently introduced. Once a nonindigenous species has become widespread, however, management responsibilities are often shifted to other federal or state agencies. Cat-claw mimosa (*Mimosa pigra*), for example, a federally listed noxious weed, is presently confined to 395 sparsely infested hectares in southern Florida. This Central American native was found to be ideally suited to the environmental conditions in southern Florida's Everglades wetland system (Sutton 1994). In 1985, APHIS conducted a joint survey with state agencies to determine the extent of cat-claw mimosa in Florida and subsequently provided limited funding to the state for its control and eradication. Since then, even though this federally listed noxious weed is confined to a small area, the responsibility for eradication has been completely shifted to the state of Florida. In 1992, APHIS contributed only 23 percent of the cost for the control program; in 1993, APHIS funds were discontinued (R. Kipker, pers. comm.).

The APHIS Animal Damage Control Program controls or eradicates native or nonindigenous wildlife that can affect agriculture. By default, it has become the lead agency for controlling the brown tree snake (*Bioga irregularis*), but to date all funds for control of this species have come from the U.S. Department of the Interior and the U.S. Department of Defense (Fritts 1995). For example, the program was awarded a $2 million contract from the U.S. Department of Defense for 1993–1994 to trap brown tree snakes in military transport facilities on Guam. From January through March 1994, dogs trained to detect brown tree snakes found only four snakes in or near such facilities on other islands (Hall 1994). Given the high population densities of this snake on Guam, however, current efforts may not suffice to prevent its dispersal from the island. Because of Florida's subtropical climate, "islandlike habitat," highly disturbed landscape, and abundance of military bases, a brown tree snake invasion is a serious possibility in Florida's future (Fritts 1994).

The Federal Seed Act authorizes the USDA to regulate the labeling and content of agricultural seeds imported into the United States. Presently nine agricultural weeds that are widespread in the United States (such as Johnson grass, *Sorghum halepense*) are regulated under the Federal Seed Act. APHIS is now in the process of listing all federal noxious weeds under the Federal Seed Act (with a zero tolerance for contamination of seed shipments), but the act does not regulate ornamental plant seeds and may leave Florida vulnerable to new invasive plant species. Of related interest, the initial spread of many former ornamental plants that are now invasive in Florida—such as Australian pine (*Casuarina equisetifolia*), Brazilian pepper (*Schinus terebinthifolius*), kudzu (*Pueraria montana*), and water hyacinth (*Eichhornia crassipes*)—is linked to the early commercial seed trade (Mack 1991).

The USDA Forest Service manages the 77-million-ha national forest system for timber production, outdoor recreation, rangeland grazing, watershed preservation, and fish and wildlife habitat. This agency actively manages noxious weeds and has a strong interest in biological control. It also fights invasive nonindigenous insect species like the European gypsy moth (*Lymantria dispar*), which is found throughout the Northeast and in several disjunct areas. The agency considers the management of fish and wildlife in national forests to be a state responsibility (U.S. Congress 1993). As general policy, the Forest Service endorses the introduction of native and "desirable" nonnative fish and wildlife species, and nonnative introductions desired by the public may be allowed. The agency does, however, consider the impact of a potential new introduction, especially if it is controversial (U.S. Congress 1993). Introduction of species at new sites involves agreements between the Forest Service and state fish and wildlife agencies and coordination with the U.S. Fish and Wildlife Service.

The USDA's Agricultural Research Service conducts research on the prevention, control (including biological control), and eradication of harmful nonindigenous species, sometimes in cooperation with APHIS. From 1986 through 1992, for example, federal, state, and county agencies provided more than $1.4 million to the Agricultural Research Service for research on biological control agents for the Australian melaleuca tree in Florida. The agency is also a repository of seeds and other plant materials (germ plasm) for agricultural use. An average of 8500 accessions per year were incorporated into the National Plant Germplasm System between 1985 and 1989, of which 90 percent were of foreign origin (U.S. Congress 1993). Extreme care is taken to ensure that imported germ plasm is not a vector of nonindigenous insects and diseases, but the potential invasiveness of plant species and varieties is not screened.

The USDA Natural Resource Conservation Service, formerly the U.S. Soil Conservation Service, was established in 1935 to protect land from soil erosion. This agency offers technical advice to public agencies and private landowners on grasses, forages, trees, and shrubs suitable for erosion control. During the 1930s, when it was known as the Soil Erosion Service, this agency distributed approximately 85 million kudzu seedlings to southern landowners for land revitalization and to reduce soil erosion (Everest et al. 1991). By the 1950s, kudzu had become a nuisance species. By 1962, the agency recommended kudzu only for sites far from developed areas (Everest et al. 1991). By 1991, kudzu had infested almost 3 million hectares of the Southeast, including extensive areas of northern Florida.

A number of plants introduced and recommended by the USDA Natural Resource Conservation Service are nonindigenous species. The Agricultural Research Service is the principal source of foreign plant materials in the United States. Moreover, there is no agencywide policy in the USDA Natural Resource Conservation Service that addresses the use of nonindigenous plant species. Although the agency does provide general guidance to its plant materials centers (there are 20 in the United States) on testing species for toxicity and for propensity to become an agricultural pest, present review processes fail to screen out potential pests to the environment. At least 7 of the 22 nonindigenous plant species released between 1980 and 1990 were reported to have the potential to invade natural areas (U.S. Congress 1993). And compounding the problem, this federal agency does not control or eradicate its introduced species once they become invasive.

U.S. Department of the Interior

The U.S. Fish and Wildlife Service (FWS) regulates the importation and interstate shipment of injurious fish and wildlife under the Lacey Act. Present regulations prohibit or restrict entry of a number of fishes, mammals, birds, reptiles, and shellfish and two fish pathogens. The FWS port-of-entry inspection program is much smaller, in budget and number of employees, than the USDA/APHIS program. Nationally, only 22 percent of imported wildlife shipments in 1990 were inspected (U.S. Congress 1993).

Under Executive Order 11987 (24 May 1977) by President Jimmy Carter, all federal agencies were instructed to restrict introductions of "exotic" species into U.S. ecosystems and to encourage state and local governments, along with private citizens, to prevent such introductions. The FWS was meant to take the lead in drafting federal regulations, but when attempts to implement this order met strong opposition from agriculture, the pet trade, and other special-interest groups (U.S. Congress 1993), the formal regulatory effort was largely abandoned. Although the FWS claims that regulations drafted under this executive order make it agency policy to discourage introductions of nonindigenous species, that policy has not been uniformly adopted throughout the agency (Edwards 1992). The FWS's Federal Aid Program, for example, allows states to recover up to 75 percent of costs associated with fish and wildlife "restoration" but historically the program has supported numerous introductions of nonindigenous fish and wildlife species. In Florida, the blue tilapia (*Oreochromis aureus*) was brought into the state as a potential sport fish with FWS support funds. The blue tilapia has since spread to at least 20 counties and is now the dominant species in many lakes in central Florida (Chapter 7 in this volume). Consequently, the program now discourages the introduction of species not yet established and requires states to assess the environmental impacts of any introductions they propose. The FWS's participation in the Aquatic Species Task Force has forced some internal evaluation of the agency's role in nonindigenous species introductions. Nonindigenous species of fish and wildlife are still being introduced, although now at a much lower rate (U.S. Congress 1993). The FWS has no comprehensive management or control program for nonindigenous species. For example, management and control operations for nonindigenous plants and animals in Florida's national wildlife refuges are generally paid for with funds designated for base operations and maintenance funding (Chapter 16 in this volume).

Much of the research on the distribution, biology, and control of aquatic nuisance species (primarily fish) is conducted at the National Fisheries Research Centers in Gainesville (Alachua County), Florida; Ann Arbor, Michigan; and LaCrosse, Wisconsin. The Gainesville lab (now under the auspices of the National Biological Service) monitors the spread of nonindigenous fishes in the United States. An inspection service operated by the FWS at the Warm Springs Regional Fisheries Center in Georgia certifies that individual nonnative grass carp (*Ctenopharyngodon idella*) intended for use in aquatic plant control are triploid and thus not able to reproduce (U.S. Congress 1993).

Less than 1 percent of the National Park Service's budget is allocated to research, management, and control of invasive nonindigenous species. Surveys conducted in the mid-1980s found that such control was a significant management concern (U.S. Congress 1993). In Florida, effective management of melaleuca and Brazilian pepper in Everglades National Park and the Big Cypress National Preserve is hampered by lack of dedicated funds (Pernas 1994; Chapter 17 in this volume).

Other U.S. Department of the Interior agencies, such as the Bureau of Land Management (BLM) and the Bureau of Indian Affairs, deal with nonindigenous species invasions on the lands they manage. Both of these agencies have modest management programs in Florida. The BLM is planning to remove Australian pine (*Casuarina* spp.), melaleuca, and Brazilian pepper from a 32-ha remnant scrub community within the city of Jupiter (Palm Beach County) (U.S. Department of the Interior 1995). Tropical soda apple is controlled under the direction of the Bureau of Indian Affairs on approximately 2023 ha of pastureland on the two Indian reservations in southern Florida (Harriman 1995).

Other Federal Agencies

The National Oceanic and Atmospheric Administration (NOAA) deals with nonindigenous species in the management of the Great Lakes and coastal resources. Much of the research on the zebra mussel (*Dreissena polymorpha*), both inside and outside the federal government, has been funded by this agency. Zebra mussels are expected to expand their southern range into northern Florida (Chapter 6 in this volume). Moreover, the National Marine Fisheries Service of NOAA inspects imported shellfish for nonindigenous parasites and pathogens.

Movement of military equipment by the Department of Defense (DOD) can transport harmful nonindigenous species as well. Although the DOD formerly employed its own military customs inspectors trained by APHIS (the program was discontinued in 1995), the brown tree snake has been discovered at military airports on uninfested Pacific islands (U.S. Congress 1993). The DOD is the fifth-largest land manager in the United States, and through cooperative agreements with the FWS and state agencies, management of fish and wildlife and new species introductions occur only when all parties agree. The DOD's efforts to control nonindigenous aquatic plants are primarily limited to those of the U.S. Army Corps of Engineers. The Corps of Engineers helps to control invasive aquatic weeds through its Aquatic Plant Control Program (Chapter 14 in this volume), but limited removal of invasive nonnative plants and animals is sometime undertaken at military bases throughout Florida (C. Bauer, pers. comm.).

Additional federal agencies play smaller roles. When the Asian tiger mosquito (*Aedes albopictus*) was introduced into Florida in 1986 and spread to all but one county (O'Meara 1994), for example, the Department of Health and Human Services Public Health Service promulgated regulations for fumigating used tire imports after they were found to be the major pathway of introduction of the Asian tiger mosquito into the United States. The U.S. Coast Guard issued mandatory ballast-management regulations to prevent further introductions of harmful nonindigenous species into the Great Lakes.

The U.S. government, then, plays a critical role in preventing the entry of invasive nonindigenous species—and, once they are here, in controlling the subsequent spread of those that become disruptive to agricultural and managed and natural ecosystems. Generally, federal agencies have been reluctant to address the issue of nonindigenous species substantively, especially as it relates to invasions in natural areas. Thus, Florida is forced to rely on its own governmental resources, laws, and expertise to protect its unique and fragile ecosystems from being overwhelmed by nonindigenous species.

22 The State's Role

TOM C. BROWN

 Today a largely uncoordinated group of laws, administrative rules, policies, and programs govern the introduction of nonindigenous species into Florida. Attempts to control harmful introductions, unintentional and intentional, have focused on controlling the pathways of introduction, and little emphasis has been placed on achieving maintenance control of existing infestations.

Most harmful nonindigenous species have been intentionally introduced into Florida as the result of commerce or the desire to modify the natural habitat—for example, for landscape planting or agricultural purposes. These introductions have often been well intentioned; some are even supported and encouraged by state agencies. Nonindigenous species have been introduced to provide additional fishing and hunting opportunities (Chapters 7 and 10 in this volume), for example, but many of these species have caused significant environmental damage and economic loss.

Many harmful introductions have come as by-products of agricultural interests such as forestry, aquaculture, horticulture, soil conservation efforts, and seed commerce, all of which have provided pathways for introduction of harmful plants (Austin 1978a; Chapter 3 in this volume). Most agricultural products grown in the state are nonindigenous. The vast majority have proved to pose no threat, but those few that have become problematic have devastated natural ecosystems, as can be illustrated by the impact of the melaleuca tree (*Melaleuca quinquenervia*) on southern Florida's wetlands (Chapter 3 in this volume).

Natural disasters have also played a role in introducing and spreading non-indigenous species in Florida. Hurricanes have certainly been important; examples include the release of many nonindigenous animals and the spread of nonindigenous plants by Hurricane Andrew (U.S. Congress 1993). Natural disasters such as fires and floods create disturbances that continue to provide pathways for introduction. Other dispersal agents for nonindigenous species include wind, water, and animals.

Such natural dispersal recognizes no state or jurisdictional boundary lines. Historically, government's response to natural dispersal has been immediate and thorough when species introductions threaten agriculture; otherwise the response has been virtually nil. The limited federal funding provided to deal with the threat by nonindigenous species to tropical hardwood hammocks in southern Florida after Hurricane Andrew destroyed their tree canopies was the exception rather than the rule. No federal response was forthcoming when Australian pine trees (*Casuarina* spp.) substantially increased their range in southern Florida after Hurricanes Donna and Betsy in the 1960s (Klukas and Truesdell 1969).

Because governmental entities focusing narrowly on specific programs and species have emphasized only pathways of introduction, the larger problem has been addressed only in general terms, and government efforts have failed in at least two respects. First, governments have not prevented further introductions of harmful species—because they have failed to curtail intentional introductions and because it is impossible to prevent introductions resulting from natural dispersal. Therefore introductions, both unintentional and intentional, will continue and will inevitably include some that are harmful. Although perfect detection and control will always be impossible, screening can be an important tool in detecting future problems. More could be done in this arena. Second, because the state has done little to control harmful nonindigenous species once they have become established, such species continue to spread and cause environmental and economic damage. With a few exceptions—for example, recent improvements in and around Everglades National Park—state management activities during the past 80 years have been generally uncoordinated and mostly ineffective.

Because state government has sought mainly to exclude agricultural pests, literally hundreds of such species are excluded each year—for example, it is unlawful to introduce, sell, or distribute any unwanted race of honeybee, such as the African honeybee (*Apis mellifera scutellata*), or any regulated article infested or infected with any honeybee pest declared by rule to be a threat to the state's apiary industry. But no screening at either the federal or state level addresses negative environmental impacts, so nonindigenous species continue to be allowed entry simply because they pose no danger to established agricultural products and enterprises and because they have economic value to some commercial interest. Although these species do not appear on either federal or state noxious weed lists, they have sometimes become among the worst invaders, destroying biological diversity and displacing native species. Classic examples in Florida are cogon grass (*Imperata cylindrica*) and torpedo grass (*Panicum repens*), imported as forage for cattle.

Florida statutes give the Florida Department of Agriculture and Consumer Services (FDACS) the authority to make rules under which plants and plant products, including nursery stock, may be brought into Florida from other states, territories, and foreign countries, but this authority is limited to actions necessary for the eradication, control, or prevention of the dissemination of plant pests or noxious weeds. A "noxious weed" is defined as "any living stage, including, but not limited to, seeds and productive parts, of a parasitic or other plant of a kind, or subdivision of a kind, which may be a serious agricultural threat in Florida." This definition excludes the authority to consider adverse environmental impacts. Moreover, these statutes define "plant pest" as an entity that can "directly or indirectly injure or cause disease or damage in any plants or plant parts or any processed, manufactured, or other plant products." Again, by definition, environmental concerns are excluded.

The U.S. Fish and Wildlife Service (FWS) and the Public Health Service restrict entry of nonindigenous fish, wildlife, and potential human disease vectors. Current FWS regulations prohibit or restrict entry into the United States of two families of fishes; 18 genera or species of mammals, birds, reptiles, and shellfish; and two fish pathogens. But the continued entry into Florida of harmful nonindigenous species even in regulated categories suggests these agencies are not entirely successful (U.S. Congress 1993).

Once invasive nonindigenous plants and animals arrive in the United States, the federal government and the state of Florida have no coordinated management strategy to prevent their spread. Although efforts are made to prevent interstate movement of agricultural pests, virtually nothing is done to prevent the interstate movement of introduced, invasive nonindigenous plants that do not affect agriculture but are harmful to the environment. Once introduced into the country, nonindigenous plants are free to move about through interstate commerce without control, except in rare cases. Although federal law makes interstate commerce in some species (such as water hyacinth, *Eichhornia crassipes*) illegal, virtually no enforcement action is taken at the federal level, and the law is all but ignored (Schwalbe 1992). Water hyacinth is openly cultivated in many states and shipped in interstate commerce. Many of these shipments are catalog sales made through the postal service.

Individual states have no meaningful power to regulate international or interstate commerce. States do establish geographical barriers, such as agricultural inspection stations, against the spread of agricultural pests, but they seem powerless against the spread of invasive nonindigenous species through mail-order sales. These sales are not monitored or reported by the postal system and, unless reported by some concerned individual, go undetected by enforcement officials. When they are discovered, states are virtually powerless to enforce their laws prohibiting such sales. Many species on Florida's prohibited aquatic plant list are openly advertised in magazines and newspapers within the state by mail-order companies. No federal law prevents the cultivation, promotion, or sale of countless nonindigenous species that have demonstrated environmentally destructive potential, so shipments continue throughout the states and even into foreign countries (see Chapter 21 in this volume).

Federal agencies manage about 30 percent of the nation's lands and play a major role in determining the distributions and population sizes of nonindigenous plants and animals. Agency policies range from strict to nonexistent. Even those, such as the U.S. National Park Service, with stringent policies to conserve indigenous species and exclude nonindigenous ones, do little to control harmful nonindigenous plants and animals once they have been introduced. Most federal land-management agencies favor the use of indigenous species in fish and wildlife stocking but have no similar policy regarding plant introductions. These policies harm Florida's environment. Moreover, management programs, if they exist at all, are small and inadequately funded (see Chapter 14 in this volume). Florida receives about $1 million each year through the U.S. Army Corps of Engineers to match with state funds to keep navigable waterways open—a problem created primarily by invasive nonindigenous aquatic plants in Florida's public waterways. The state receives no assistance with invasive nonindigenous species in wetland and upland forest areas, even when these species threaten the biological integrity of nearby national parks and reserves.

Millions of hectares of property in private ownership are affected by the Conservation Reserve Program of the Agricultural Stabilization and Conservation Service; there are no requirements for planting indigenous species or controlling nonindigenous species on lands enrolled in this program. In 1990, some 58 percent of Conservation Reserve Program lands were planted with grasses nonindigenous to the United States; only 24 percent were planted with indigenous grasses (U.S. Congress 1993).

Activities by the state of Florida mirror those of the federal government: because of limited statutory authority, priority is given to agricultural pests. As at the national level, agriculture in Florida has served as an important pathway for introductions of harmful nonindigenous species. Over the years the landscape nursery, pet, aquaculture, and similar industries have strongly advocated further introductions of nonindigenous plants and animals to broaden their markets and have opposed regulation of the import, culture, and sale of nonindigenous species, even if they are harmful to the environment (Gaskalla 1993). The Florida Game and Fresh Water Fish Commission (FGFC) has also imported fish and wildlife for recreational purposes that have sometimes harmed native species. The blue tilapia (*Oreochromis aureus*) was deliberately introduced and impedes the spawning of native fishes in many waters.

Some critics with ties to agriculture continue to argue that many nonindigenous species may have value and that screening for environmental impacts, which is sometimes costly, time consuming, and complex, should be suspended until adverse ecological outcomes have been demonstrated (Lugo 1990; see also Coblentz 1991). The problem with this view is that, once harm has been demonstrated, it is usually too late, or extremely expensive, to repair the damage. Florida and Hawaii are the two states with the most nonindigenous species—and they are suffering the greatest environmental impacts. The Florida experience has shown that benefits to humans can easily be exceeded by costs resulting from loss of biodiversity and management costs that run to millions of dollars annually. The ecological reality is that nonindigenous species, at least weedy ones, cannot be eradicated once they are established. Clearly reason and economic interest dictate that species be proved harmless to the environment, beyond reasonable doubt, *before* introduction.

Although the public is beginning to recognize the importance of biological diversity, problems created by invasive nonindigenous plants and animals are not commonly understood. Florida's regulations regarding nonindigenous plants and animals, like their federal counterparts, have developed piecemeal over many years and are confusing.

The Florida Department of Environmental Protection (FDEP) Bureau of Aquatic Plant Management, Division of Recreation and Parks, and Bureau of Coastal and Aquatic Managed Areas all engage in some management of nonindigenous species; more is being done on aquatic vegetation than on uplands. The water management districts engage in limited nonindigenous species management on lands of district responsibility (Chapter 20 in this volume). Rights-of-way are managed by the Florida Department of Transportation (FDOT) for a few nonindigenous species (Caster 1994). FDACS controls some nonindigenous pests (Hardin 1994) that pose a threat to agricultural interests. With the exception of the work done by the FDEP's Bureau of Aquatic Plant Management, there is little coordination of these activities by the various agencies.

The problems created by nonindigenous plants and animals on most public lands are simply ignored. As a result, the spread of these invaders on public lands is epidemic. The same can be said of many thousands of hectares of land in private ownership in Florida.

Table 22.1 lists Florida's statutory provisions dealing with nonindigenous species, the agencies affected, and the purposes of the provisions. Most of the

TABLE 22.1.
Florida Statutes References to Nonindigenous Species

Statute	Purpose					Kingdom	
	Funding	Prevention	Control	Research	Education	Plants	Animals
Department of Environmental Protection							
212.69	X					X	
253.7829		X	X				X
253.023	X		X			X	
327.28	X					X	
369.20–.251		X	X	X	X	X	
370.081		X					X
375.045	X		X			X	
861.04		X				X	
Florida Game and Fresh Water Fish Commission							
372.26		X					X
372.265		X					X
372.98		X					X
372.981		X					X
Water Management District							
369.251			X			X	
373.185		X				X	
373.59	X		X			X	
Department of Agriculture and Consumer Services							
581.083		X				X	X
581.031		X				X	
581.091		X				X	
581.101		X				X	
581.111		X				X	
581.011		X				X	X
581.145		X				X	

Note: This list does not include indirect references.

statutes focus on established nonindigenous species, with the aim of preventing their further spread by deliberate introductions, but the problems of natural dispersal and maintenance control are hardly addressed. In any event, these statutes have little meaning without the specific appropriation of funds by the legislature to engage in research, screening, control, and eradication.

Table 22.2 lists the rules in the Florida Administrative Code that address nonindigenous species. Again, aside from rules affecting the FDEP's Bureau of Aquatic Plant Management, little in these rules addresses environmental problems.

Table 22.2.

Florida Administrative Code References to Nonindigenous Species

Rule	Purpose					Kingdom	
	Funding	Prevention	Control	Research	Education	Plants	Animals
Department of Environmental Protection							
62-312			X			X	
62-321			X			X	
62-660		X					X
62C-20		X	X	X		X	
62C-52		X		X	X	X	
62C-54	X	X	X		X	X	
62D-2		X	X				X
62D-5		X	X		X		
62D-15		X	X			X	X
Florida Game and Fresh Water Fish Commission							
39-4		X					X
39-6		X					X
39-12		X					X
39-23		X					X
Water Management District							
40C-42			X			X	
40D-45		X	X			X	
40E-6		X				X	
Department of Agriculture and Consumer Services							
5A-1		X	X			X	X
5B-47		X				X	
5B-57		X		X		X	X
5E-5		X				X	
Administration Commission							
28-25		X				X	
28-27		X				X	
Department of Transportation							
14-13		X				X	
14-40		X				X	
Department of Management Services							
60D-14		X				X	

Note: This list does not include indirect references.

The Jurisdictional Problem

The problems associated with invasive nonindigenous species in Florida extend across administrative and political boundaries, over large landscapes, across land and water, for long periods of time. When dealing with nonindigenous species, scientists recognize the ecosystem connections among all biological systems and realize that it is not enough to focus on just one level of an ecosystem. Government's attempts to deal with the problem have not taken the interrelatedness of natural systems sufficiently into account. Because natural systems cannot be made to conform to political or administrative boundaries, a way must be found to manage nonindigenous species as though jurisdictional boundaries did not exist. Such management, to be effective, must be applied on a scale sufficiently large (in space and time) and must cover the affected landscape, both land and water. Such a centralized management scheme can be seen in the cooperative effort administered by the FDEP's Bureau of Aquatic Plant Management, in which federal, state, and local government interests are all coordinated on an ecosystem-wide basis.

Ideally the cooperation would also include private landowners. Scientists would integrate conflicting interests, legal mandates, and management goals within the confines of existing boundaries. Present jurisdictional realities, however, greatly complicate this effort. The more jurisdictions there are involved, the more difficult it is to organize a cooperative effort to combat nonindigenous species. The coordination of activities in response to cross-jurisdictional and administrative problems created by nonindigenous species is far more complex than most would imagine. Indeed it is a problem that will be a legislative nightmare to untangle.

Managing nonindigenous species requires greater emphasis on comprehensive, long-term efforts and the use of biophysical data. Scientists must improve their ability to collect data on nonindigenous species, synthesize data from many sources, and analyze that information. Important gaps in the data must be noted and ways devised to fill them—for example, ways to survey widespread nonindigenous species at lower cost and to identify management needs must be found. These data must be accessible to all agencies involved in nonindigenous-species-management programs and to private landowners interested in controlling such species on their lands. In Florida, some data are available from the Natural Areas Inventory program in the FDEP, the Bureau of Aquatic Plant Management in the same agency, and the FGFC, but they are not comprehensive and do not reflect the conditions on vast tracts of private property.

Scientists must also track the results of their actions so that success or failure can be assessed. Evaluations and audits create ongoing feedback of useful information (Hester 1991). A common fault of these management programs is insufficient follow-up evaluation to determine whether desired objectives are being achieved. The failure to collect and report data relating to methods and results—and to share this information with other agencies both public and private—results in repetition of mistakes. Currently the typical pattern is reactive rather than proactive management—that is, a crisis occurs, a solution (management activity, new policy, statute, or rule) is implemented, and all concerned return to business as usual until the next crisis. An ongoing evaluation would provide a continuing source of feedback that would validate success and allow for adjustment or abandonment of a failing course of action before a new crisis develops. This is the essence of adaptive management. The management of nonindigenous species should be a learning process in which scientists incorporate the results of previous actions and adjust future recommendations accordingly. The cycle of data collection, management, evaluation, and auditing provides for this adjustment.

The jurisdictional problem, meanwhile, remains enormous. The following brief review of the jurisdictions involved in the management process in Florida indicates the magnitude of the problem, which is shared in virtually every state in the nation.

Private Sector

Businesses and individuals who own property in Florida must often deal with many entities with jurisdictional and administrative responsibilities at both the federal and state level. Sufficiently large tracts of lands may be subject to multijurisdictional land-use and development ordinances enacted to implement differing local government comprehensive plans. Particularly in southern Florida, they may be included in the comprehensive plans of several different cities, counties, and regional planning councils. The inability of a private landowner or government agency to deal uniformly with a tract of land, because of jurisdictional differences, may lead to significant difficulties in addressing nonindigenous species problems and lead to less effective resource management. With the exception of certain laws affecting agriculture, no state laws require private landowners to remove nonindigenous species from their lands or prevent their introduction. Populations on these lands serve as "seed banks" to infest other private and public lands—and, of course, private lands are often infested with nonindigenous species from contiguous public lands.

From the point of view of the private sector, current efforts to address nonindigenous species are problematic at best. Probably the most troublesome problem is that there is no single entity "in charge" to whom the property owner may go for definitive answers. It seems clear that if management of nonindigenous species is to succeed, some agency must be authorized to guide the decision-making process to assure that all concerns are evaluated prior to action. The decision-making process is difficult when local, regional, state, and federal agencies are all involved, but no single entity has all the data needed to make an informed decision. Although coordination may be difficult to accomplish from the federal perspective, it is certainly possible at the state level.

Local Government

Each city or county government in Florida is required by state law to adopt a comprehensive plan for guiding land development and government programs. County authority is established by state law and each county's charter. City authority is established pursuant to Florida Statutes and each city's charter. Local plans must be adopted by local ordinance and must include many elements addressing or incorporating environmental issues or program aspects. For example, each plan must include a "conservation element" for the conservation, use, and protection of natural resources in the area. These conservation elements may or may not contain plans for dealing with noxious nonindigenous species.

All land development regulations and governmental development activities must be consistent with the adopted plan. The appropriate regional planning council, as we shall see, reviews these plans to determine regional impacts, including the effects on natural resources, and to ensure consistency with the regional plan. Plans cannot be adopted by the local government until they have completed review by the Florida Department of Community Affairs, another state agency. This review incorporates comments by other state and regional agencies. These plans are used to establish conservation areas that have sensitive environmental features. Special environmental ordinances may be used to require the removal of nonindigenous species and to limit disturbance of habitat sensitive to invasion by nonindigenous species.

Regional Planning Councils

Each of the 11 regional planning councils was established pursuant to Florida Statutes to assist local governments in resolving common problems, to engage in area-wide comprehensive and functional planning, to administer certain federal and state grants-in-aid, and to provide a regional focus for multiple programs undertaken on an area-wide basis. Each district must formulate a regional plan that identifies and addresses significant regional resources and facilities. The regional plans must list all natural resources of regional significance by specific geographic location. Regional plans are used for planning, not for permitting or regulatory purposes, and must be consistent with the state comprehensive plan.

State Comprehensive Plan

The Executive Office of the Governor is statutorily required to prepare a state comprehensive plan to provide long-range guidance for orderly social, economic, and physical growth. This plan includes goals, objectives, and policies related to many program areas, including natural resources and environmental management. Every state agency's strategic plan must be consistent with the state comprehensive plan. All state-agency budgets and programs must be consistent with the state comprehensive plan and the agency's strategic plan and, moreover, must support its goals and policies.

State-Agency Plans

Each state agency is required by law to prepare and submit a strategic plan to the governor. These plans state the directions an agency will take to carry out its mission within the context of the state comprehensive plan. Priority Issue One in the FDEP's strategic plan, for example, addresses ecosystem management: "The Department is committed to an ecosystems approach, a range of regulatory, acquisition and planning strategies to insure the long-term health, restoration, and growth of natural systems." An item of concern under this issue is "exotic species."

Department of Environmental Protection

Presently the only invasive nonindigenous plants being managed on a statewide basis in Florida are aquatic plants in public waters. The lead agency coordinating this program is the FDEP, which matches funds provided by the Corps of Engineers and conducts the largest program in the country aimed at controlling the harmful effects of widespread invasive nonindigenous plants on public lands. The program includes funding for aquatic plant management and is a cooperative effort among the FDEP, the Corps of Engineers, the FGFC, the water management districts, and local governments. All parties cooperate by helping to establish priorities based on surveys of needs, establishing work plans, and monitoring activities and results. Moreover, this program includes permitting and technical sections and maintains seven district offices throughout the state (see Chapter 14 in this volume). This program has also been given responsibility by the legislature for management of certain invasive wetland and upland nonindigenous plant species. Some nonindigenous species are also managed by the FDEP on a limited basis, primarily in state parks, with regular expense funds or Pollution Recovery Trust Fund money (penalties levied against violators of environmental laws and rules).

An estimated 607,000 ha of Florida's public lands has experienced significant invasions by nonindigenous plants, primarily in wetland and upland forest habitats. Although the organization and technical expertise to manage these species in an integrated manner exist under the FDEP's aquatic plant management program, the agency has not been statutorily designated to do so.

Other State Agencies

As noted earlier, state agencies other than the FDEP are involved to some degree in management of nonindigenous species. These include the Florida Game and Fresh Water Fish Commission (FGFC), the Florida Department of Transportation (FDOT), and the Florida Department of Agriculture and Consumer Services (FDACS).

THE FLORIDA GAME AND FRESH WATER FISH COMMISSION The FGFC derives its authority from both the Florida constitution and Florida Statutes. Its authority comprises the exercise of regulatory and executive powers with respect to all wild animal life and freshwater aquatic life, including the importation and dissemination of nonindigenous wild animal species and nonindigenous freshwater aquatic life.

Some of the nonindigenous species problems with which the FGFC must deal are a result of the agency's own activities. Its policy has in the past included authorizing the introduction of nonindigenous fish and animal species intended to enhance fishing and hunting, but some of these species have become serious threats to native species. The FGFC does, however, work closely with the FDEP's Bureau of Aquatic Plant Management to manage invasive nonindigenous aquatic plants. The FGFC has statutory authority for permitting the use of grass carp (*Ctenopharyngodon idella*) as a biological control agent for managing aquatic weeds. It also conducts extensive land and vegetation management practices for the benefit of both game and nongame animal-habitat preservation and improvement that entails removing certain nonindigenous species on lands managed by the agency. The FGFC deals with other state agencies, federal agencies, and local governments in many projects related to nonindigenous species.

THE FLORIDA DEPARTMENT OF TRANSPORTATION   The FDOT's involvement with nonindigenous species centers primarily on its rights-of-way or roadside management program. The statewide highway landscape-grant program promotes the use of native plants, native wild flowers, compost, and Xeriscape practices. In the past, planting certain nonindigenous species was actually promoted, but department rules now prohibit planting certain nonindigenous species on rights-of-way or any plants "which directly compete with and displace desirable native plants." One local government attempted to persuade the Florida legislature to pass a law in 1994, and again in 1995, to allow cultivation and planting of the invasive nonindigenous Australian pine on FDOT rights-of-way. This legislation passed in 1996. Moreover, the FDOT faces a major problem in controlling the reinfestation of rights-of-way by invasive nonindigenous species from adjoining private properties.

Although the FDOT is authorized to remove nonindigenous species from its rights-of-way, it lacks authority to enter into agreements with adjacent landowners to control nonindigenous species on private lands to prevent reinfestation of rights-of-way. This problem has been particularly persistent with cogon grass. Another problem for the FDOT is the dissemination of nonindigenous species through the movement of soil and sod on rights-of-way. This activity in highway construction has created massive problems by inadvertently spreading invasive nonindigenous species such as cogon grass, tropical soda apple (*Solanum viarum*), and cat-claw mimosa (*Mimosa pigra*).

THE FLORIDA DEPARTMENT OF AGRICULTURE AND CONSUMER SERVICES   The FDACS has the authority under state law to eradicate, to control, or to prevent the dissemination of plant pests and noxious weeds, but statutory definitions limit this authority to species that pose an "agricultural threat." Nonindigenous species are screened for this "threat" to agriculture, but the statute does not include authority to consider invasiveness elsewhere.

Many of the most severe nonindigenous plant problems in Florida have been caused by importation of nonindigenous species for agricultural purposes—for the aquarium and ornamental plant industries, for example. Examples include melaleuca, Brazilian pepper tree (*Schinus terebinthifolius*), and hydrilla (*Hydrilla verticillata*), three of the most invasive nonindigenous species in the state.

The FDACS has veto power over FDEP's prohibited aquatic plant species list and exercised that veto in 1993—thus preventing the addition of the invasive nonindigenous species torpedo grass to the list—because certain agricultural interests were selling this nonindigenous species, sometimes mixed with other grasses, for hay (Riherd 1992). Torpedo grass is presently threatening the western marsh area in Lake Okeechobee and is rapidly spreading statewide.

The FDACS also supported the cultivation and sale of water hyacinth—one of the most invasive aquatic weeds in the world—and sought legislative approval for its sale, even though Florida spends almost $2 million a year to keep this plant under maintenance control (Gaskalla 1993). The agency defended its action on the grounds that water hyacinth was to be sold to Canada, where it posed no real environmental threat because of the cold climate. The FDACS also failed to note that the sales to Canada were primarily to mail-order businesses that in turn shipped this species worldwide—including the United States and Florida. Water hyacinth poses no threat to agricultural interests in Florida or elsewhere, but it is a significant threat to aquatic ecosystems throughout the world, and its range continues to expand. This legislation also passed in the 1996 legislative session.

The FDACS has taken steps to assist in the removal of certain invasive nonindigenous plant species in environments where they have been demonstrated to pose a threat to agriculture. The agency took a role in removing wild red rice (*Oryza rufipogon*) from Everglades National Park, for example, and is active in management activities for tropical soda apple, primarily on rangelands.

Water Management Districts

Each of the five water management districts created under Florida law must develop a water management plan. These plans are not regulatory but result in significant interaction with local governments and state agencies. Their implementation is expected to be accomplished through a number of different management tools, including "land acquisition and management." Water management districts are authorized to use a portion of land management funding for the control of nonindigenous species.

Three of the five water management districts participate in the FDEP's aquatic-plant-management cooperative program and control invasive nonindigenous aquatic plant species as cooperators. The Northwest Florida and Suwannee River districts do not participate because of the paucity of aquatic plants and insufficient staff.

Gaps and Options

Successful conservation of biodiversity through the management of nonindigenous species will require a new level of coordination involving private and governmental entities. Most of the tools that have proved effective in dealing with nonindigenous species are available and, moreover, are applicable to public lands, but they are not being used because of lack of funding and lack of a lead agency to coordinate the effort.

Approximately 20 percent of Florida's uplands and nonsovereign wetlands (about 2.8 million hectares) are a part of public conservation lands. Perhaps 3 million additional hectares of submerged lands lie within boundaries of aquatic preserves and the Florida Keys National Marine Sanctuary, comprising about 40 percent of Florida's sovereign submerged lands. The removal and control of nonindigenous species from all public lands should be based on an organizational plan similar to the one presently effective in dealing with invasive nonindigenous aquatic plants. Wherever more than one jurisdictional or administrative entity exists within the same ecosystem, the FDEP should be allowed to coordinate activities so that removal of nonindigenous species will complement other restoration activities. Where upland restoration is to occur, for example, an emphasis on monitoring and removal during the "disturbed" phase of the restoration will help keep nonindigenous plant species out until restored native vegetation can be established. At the same time, the species that pose the greatest threat to an entire ecosystem (such as melaleuca in the Everglades) can be given first removal priority.

The fundamental legal question is whether present laws suffice to deal with the threat of nonindigenous species or whether Florida needs more stringent and comprehensive policies. Elected officials and the public alike seem indifferent to this growing problem. One reason is the management dilemma posed by the presence of nonindigenous species. The actions required to deal with invasive nonindigenous plants and animals, which often entail killing or removing them, are far less popular than are establishment of reserves and preserves, protecting species from exploitation, restoration planting of native species, or regulating obviously destructive development activities. Considerable public outcry usually meets efforts to control Australian pine in southern Florida, for example, where many consider it aesthetically pleasing and value the shade it provides.

The consensus among scientists is that Florida does need more stringent policies: there is no policy today that truly authorizes and funds a coordinated effort to protect natural areas from the damage caused by nonindigenous species. The state's ability to deal with the problem is limited, too, by inadequate data. Although the Florida Natural Areas Inventory (FNAI) provides valuable information in many instances, we need better species descriptions, better inventories of abundance and distribution, and better data on the impact on ecosystems. The FNAI was established in 1981 as a cooperative effort between the state and The Nature Conservancy (an international nonprofit conservation organization). Although the inventory gives priority to acquiring information on the rarest and most threatened plants and animals, it gathers other information on ecological diversity including data on nonindigenous species. The FNAI has been used primarily to identify significant natural areas and establish land-protection priorities in the state. It has listed more than 675 managed areas encompassing more than 2.4 million hectares. Much of this information has also been incorporated into a geographic information system (GIS).

More research is needed on best management practices, including the most viable means of restoring formerly infested areas. Many of the mechanisms whereby nonindigenous species affect native species are poorly understood. Recognizing the cause and extent of the problems created by these species is a first step toward responsible stewardship.

Obvious gaps in state policy stem from several sources, and legislative options must be proposed to correct them. First, no agency has broad authority over nonindigenous species in order to protect natural areas and ecosystems from these species, to preserve biological diversity, or to ensure that environmental assessments consider the impact of nonindigenous species on the ecosystem. The development of management solutions requires considering many options, including biological control, habitat alteration, physical removal, and genetic manipulation. Moreover, no effective management scheme can ignore human needs and political influence. Solutions to problems created by invasive nonindigenous species must address the conflict between a desire to restore and protect natural systems and people's need for food, water, transportation, and recreation. Clearly, state agencies and local governments cannot all become expert in these management options; only a lead agency can be expected to do so. As the regulation of wild animals and freshwater aquatic life is part of the constitutional authority granted to the FGFC and cannot be changed without a constitutional amendment resulting from a vote of the people, these responsibilities and those relating to nonindigenous species on public lands are unlikely ever to come fully under one agency. Interagency coordination is the best one can hope for.

Second, state agencies with statutory authority are reluctant to exercise that authority or to institute rules where statutes provide no criteria or are unclear about the scope or intent of the authority. Clear statutory mandates are necessary.

Third, authority for proactive action is necessary. Piecemeal, after-the-fact attention has been ineffective. With the possible exception of the FGFC constitutional authority over wildlife and FDACS authority to screen for agricultural pests, no state agency has the statutory authority to take meaningful steps to halt the introduction of harmful nonindigenous species into the state. Thus invasive nonindigenous species are introduced, become established, and come to present an expensive, interminable control problem (Coblentz 1991).

Fourth, few data are available on the origins, numbers, distribution, potential impacts, and methods of control for nonindigenous species—again because no lead agency has been designated to gather, compile, store, process, and distribute this information. A lead agency should be given the statutory mandate and funding to collect and maintain a database on the presence of invasive nonindigenous species in the natural environment and to coordinate with the activities of other agencies to avoid duplication of effort. These data should document the magnitude of the problem, assess the funding needed to deal with it, and serve as the basis for planning and the selection of management techniques and the coordination of effort. Improper control techniques and practices exacerbate the damage caused by invasive species—as when stress caused by management efforts causes melaleuca trees in southern Florida to release their seeds.

Finally, the Florida legislature has never provided adequate funding to undertake eradication and maintenance control of nonindigenous species on public lands. A logical source of such funds is the same one used to purchase these lands for preservation, conservation, and recreational purposes. Funds should be appropriated directly and specifically for maintenance control of nonindigenous species, both plants and animals. The value of land management policy has been recognized by the Nature Conservancy, which now insists that a substantial stewardship fund be established before it will purchase any land it will manage. In a state as highly developed and highly disturbed as Florida, the lack of commitment to managing natural resources to assure their preservation is the moral equivalent of a commitment to their destruction. It is poor public policy to purchase lands for their biological heritage and diversity, in an effort to preserve them for future generations, only to allow nonindigenous species to destroy them.

Maintaining the status quo—doing nothing—also comes with a risk. The example that looms largest is the crown jewel of ecosystems in Florida: the Everglades. Much attention has been given recently to water quality in the Everglades, and significant funds have been dedicated to its improvement. But this system may well be totally destroyed in just a few short years, no matter how improved its water quality, if we do not meet the threat from invasive nonindigenous plants. The thousands of hectares that have already been affected by nonindigenous species may never be fully restored as part of a naturally functioning ecosystem—and certainly not without vast amounts of money.

Lessons for Other States

Florida has made many mistakes in its efforts to deal with nonindigenous species, mistakes from which other states might learn and benefit. These mistakes fall into six categories. First, Florida has failed to integrate planning for nonindigenous species into the comprehensive planning process designed in part to protect the environment from the adverse effects associated with rapid population growth. But with a population of 14 million that adds 500,000 annually (Bouvier and Weller 1992), and with some 40 million visitors per year, most of the state has not remained pristine: native species are being increasingly confined to limited areas with artificial boundaries, surrounded by significantly disturbed lands. Nonindigenous species seem to become established more easily in these disturbed areas (Hobbs 1989; Panetta and Hopkins 1991). Although nonindigenous species continue to spread at an alarming rate, no mandatory provisions for managing them are included in the comprehensive planning process at any administrative or jurisdictional level.

Second, Florida's land management activities on public lands have failed to include the planning, funding, or implementation of components to address nonindigenous species adequately. Although Florida Statutes provide for conservation, preservation, and protection of Florida's biodiversity through acquisition and management of public lands, far greater commitment has been given to acquisition than to management. Usually the goal of conservation management is to maintain species diversity and ecosystems; priorities for management must now be established to accomplish this goal (Saunders et al. 1991). Although millions of hectares of land have been acquired by the state, funding for management activities, including the management of nonindigenous species, has not kept pace. Under present law, those with management responsibilities for public lands do not have the statutory authority, or necessary funding, to protect natural areas in public lands from invasions.

Third, Florida has failed to require all entities introducing nonindigenous species into the state to undertake prerelease evaluations for their potential to harm native species diversity and ecosystems, and no such screening protocol has been legislatively mandated. This lacuna, in conjunction with other contributing factors, such as the state's subtropical climate and its growing pet, aquarium, and ornamental plant industries, has resulted in some of the most severe nonindigenous species problems in the United States (U.S. Congress 1993).

Fourth, Florida has failed to expand its prohibited plant lists to include nonindigenous species that have already harmed species diversity and ecosystems in Florida or elsewhere or that may have adverse effects if introduced. The Florida legislature has created an internal inconsistency in the state's efforts to control nonindigenous plants by giving an agency, other than the agency charged with environmental protection, veto power over species added to the prohibited plant list. This agency's primary interests are in promoting agricultural enterprises, not in protecting the natural environment.

Fifth, Florida has failed to designate a lead agency to coordinate the efforts of state agencies to manage nonindigenous species in nonagricultural settings. Many state and local government agencies are involved in the limited management of nonindigenous species. But most often they operate independently, without adequate funding, without coordination, and without looking at entire ecosystems. Presently, only nonindigenous aquatic plants are managed on an ecosystem basis by a lead agency that provides matching funds and coordinates efforts by the federal government, state agencies, and local governments.

And sixth, Florida has failed to offer incentive programs or assistance to private landowners for the control of nonindigenous species on private lands. These lands, often contiguous to public lands, suffer from the same nonindigenous species problems and serve as sources of reinfestation when control activities are undertaken on public lands. Coordination and cooperation between public and private interests are the key to solving this problem.

 PART VI

Conclusion

23

Why We Should Care and What We Should Do

Daniel Simberloff, Don C. Schmitz, and Tom C. Brown

 Florida harbors a plethora of nonindigenous species. Why should we care about them? Staggering economic costs are associated with some invaders, but, except for control costs, this book has touched on economics largely in passing. Rather, we have focused on the ecological impact of these invaders on native species, communities, and ecosystems. Biologists, economists, and philosophers have advanced ethical, psychological, and economic reasons why we should concern ourselves with the conservation of native biodiversity. (See, for example, Norton 1987; Wilson 1992; Callicott 1994.) Cox et al. (1994) briefly discuss the staggering economic value of Florida's native species, communities, and ecosystems in terms of consumptive and nonconsumptive natural resource uses. If ethical considerations matter at all, Florida's burden is particularly great. Of 632 species and subspecies in the United States listed as endangered under the Endangered Species Act (Chadwick 1995), 80 are found in Florida (Anonymous 1995a)—one of the highest concentrations in the nation. Another 26 have the next highest classification: threatened. A recent Nature Conservancy study shows Florida to have the third highest total of highly threatened species and subspecies (522) in the United States, after California and Hawaii (B. Stein, pers. comm.). Many of Florida's species are endemic—about 17 percent of the 668 terrestrial and freshwater vertebrates and 8 percent of the 3500 or so vascular plants (Muller et al. 1989). Moreover, of the 81 native plant communities (Florida Natural Areas Inventory 1990), 13 are endemic (Muller et al. 1989). Rare and

FIGURE 23.1   Hydrilla (*Hydrilla verticillata*) filling Lake Okeechobee to the surface, September 1995. Photo by Jeffrey D. Schardt.

endemic communities include mangrove swamp, pine rockland, sandhill, and scrub. All of these communities are diminished and fragmented by activities associated with human population growth, and nonindigenous species threaten to damage these communities further by displacing native species and degrading ecosystem function. (Figure 23.1)

Regional Differences

Many chapters in this book point to an apparent disparity between northern and southern Florida in numbers and impact of nonindigenous species. For some taxa—such as freshwater fishes (Chapter 7) and especially amphibians and reptiles (Chapter 8)—the number of nonindigenous species in the south dramatically exceeds that in the north. For others—such as birds (Chapter 9)—the same pattern is clear but not so pronounced.

Despite numerous hypotheses (Chapter 3), the reason for this pattern is unclear. It is true that, for most taxa, the number of native species declines from north to south, encouraging some version of a "biotic resistance" hypothesis (Chapter 1) to explain the greater number of nonindigenous species in the south, but no real evidence shows that arriving species survive or fail in differential numbers because of differing resistance posed by the resident biota. In fact, there are few data on the numbers of failed introductions. If we wish to understand the reason for the gradient in number of surviving nonindigenous species, we will probably require at least a systematic consideration of failure rates.

Many workers have argued that southern Florida is more "disturbed" than northern Florida and that this disturbance has made it more prone to invasion by interlopers. Yet the many kinds of disturbance, both natural and anthropogenous, cannot all be ranked on the same scale as to severity. In any event, northern Florida has been greatly disturbed, although the results are perhaps not so apparent to the human eye (Chapter 3). Surely an understanding of the role of disturbance in establishment of nonindigenous species will require a careful consideration of the specific kind of disturbance and the specific ways in which both native and nonindigenous species respond to each kind (Chapters 1 and 4). The preliminary but detailed data on the effects of Hurricane Andrew in southern Florida in 1992 show that this disturbance benefited certain native species over certain nonindigenous potential invaders, while other nonindigenous species were favored over other natives. Further, the responses of the various species to anthropogenous disturbance did not adumbrate their responses to the hurricane (Chapter 4). With respect to the relative invasibility of north and south, it would be extremely enlightening to compare ongoing similar studies of the impacts wrought by Hurricane Kate in 1985 in northern Florida, but published research (such as Platt and Rathbun 1993) does not address nonindigenous species.

A simple "null" hypothesis for the greater number of established nonindigenous species in the south has been posed by C. Lippincott (pers. comm.) for plants: more propagules of more species arrive in southern Florida, either by human means or on their own. The data for some taxa—such as plants (Chapter 2), reptiles and amphibians (Chapter 8), and fishes (Chapter 7)—are at least superficially consistent with this hypothesis, and it would be informative to have systematically gathered data on numbers of arriving propagules for many species.

Differences among Taxa

Ecological effects of Florida's nonindigenous species vary enormously. Two types of invaders are particularly likely to have major ecosystem effects: species that constitute new habitats and species that modify habitats by altering ecosystem processes (Chapter 3). Because plants form the biological matrix for most communities, and several nonindigenous plant species in Florida have formed or affect habitats over substantial areas, plants currently have the greatest impact among all nonindigenous taxa in the state. For terrestrial vertebrates—birds (Chapter 9), mammals (Chapter 10), and reptiles and amphibians

(Chapter 8)—the documented effects on native ecosystems are few. Almost all of the contributors to this volume confess that most data on effects are anecdotal and sketchy, and, as noted in Chapter 1, many effects of nonindigenous species are subtle and can easily be overlooked. Nevertheless, relative to invertebrates and plants, terrestrial vertebrates are quite well studied, and it seems unlikely that many of their nonindigenous species are problematic in Florida without our knowing it. Perhaps feral pigs cause the most damage today (Chapters 12 and 18–20). Other terrestrial vertebrates could become great threats in the future. The brown tree snake, for example, should it reach Florida, might devastate many native species (Chapter 8), and an increase in monk parakeet populations could be greatly detrimental to native species, although opinions vary (Chapter 9).

Ecological impacts of nonindigenous freshwater fishes in Florida are probably even more poorly known than are those of the terrestrial vertebrates, but there are compelling arguments (Chapter 7) that, even though no extinctions of native species can be attributed to nonindigenous fishes, many of the latter are substantial threats. In other parts of the United States, nonindigenous fishes have often played a key role in recent cases in which native fishes have been endangered or extinguished (Chapter 7).

Insect species, of course, greatly outnumber all vertebrate species combined. Their impacts, too, are usually poorly understood, so it is not surprising that the impacts of nonindigenous insects, except those that affect agriculture, are barely studied (Chapter 5). It seems possible, however, that nonindigenous insects already affect native systems more strongly than do vertebrates. Part of this effect is beneficial—for example, there are introduced insects that control alligator-weed (Chapter 3, 5, and 15). In contrast, the cactus moth threatens two native cactus species (Chapters 1, 5, and 15), a nonindigenous leaf beetle threatens two native endangered morning glory species (Chapter 5), a Mexican weevil devastates native bromeliad populations (Chapter 5), and the gypsy moth, should it become established in Florida (Chapter 5), could devastate deciduous trees and communities based on them. Although the imported fire ant affects primarily disturbed habitats in Florida (Chapter 5), it potentially threatens many native species as a predator and competitor (Vinson 1994)—and the advent of a polygynous form exacerbates the threat (Porter et al. 1988; Vinson 1994).

Among nonindigenous freshwater invertebrates in Florida, the Asiatic clam may already affect entire native ecosystems, and the zebra mussel, which may arrive soon, would probably have far-reaching impacts (Chapter 6). Several nonindigenous snails already present may affect native species by either competition or herbivory (Chapter 6). If terrestrial and freshwater invertebrates are poorly understood, marine invertebrates are even more so; in many cases their geographic range is unknown and their status as native or nonindigenous is a mystery (Chapter 11). One cannot even hazard a guess as to whether nonindigenous marine invertebrates are affecting native ecosystems, although major demonstrated impacts in well-studied areas (Carlton 1989) suggest that nonindigenous marine invertebrates may already seriously affect some Florida systems. Carlton and Ruckelshaus (Chapter 11), for example, contend that the isopod *Sphaeroma terebrans* is nonindigenous; a similar congener affects an entire ecosystem in Costa Rica by boring red mangrove roots (Perry 1988). The extent of invasion of Florida's coastal waters by nonindigenous marine plants, and their impact, is virtually unknown, but major effects elsewhere are well documented (Chapter 11).

Predicting Ecological Effects

As noted in Chapter 1, nonindigenous species can have many types of effects (direct, indirect, and synergistic) on native species, communities, and ecosystems. One could reasonably argue that if a potential nonindigenous species can form a new habitat or greatly alter ecological processes, its importation should be forbidden. Among plants, for example, if a potential invader can shade out other species or modify the fire regime, we have reason to expect problems (Chapter 3). The great majority of nonindigenous plant species, however, are not invasive or environmentally harmful. The fraction that have substantial effects has not been estimated in Florida, but other estimates center on about 10 percent (references in Simberloff 1991; Chapter 3).

Most ornamentals and agricultural plants stay where they are planted. Do they then need no regulation? Even an environmental purist would be hard pressed to object to certain nonindigenous species. The Venus's-flytrap (*Dionaea muscipula*), for example, is native to a small, boggy region in North Carolina. Because of habitat destruction and poaching for the ornamental trade, its range has dwindled from 18 to 11 counties, and it has become a "species of special concern" in North Carolina (Culotta 1994). In 1992, it was added to the Convention on International Trade in Endangered Species of Wild Fauna and Flora list of species requiring a permit for export, but its range continues to contract. It has been planted in northern Florida, survives as small, self-sustaining populations at a few sites, and shows no evidence of even slight invasive tendencies (S. Hermann, pers. comm.). Should this introduction be regulated?

Well, . . . yes! Though most introduced species do not become invasive, scientists have not been very good at predicting which introductions will be scourges and which innocuous (Hobbs and Humphries 1995). Effects are often subtle and surprising (Chapter 1). Further, a previously noninvasive species can become invasive because of subsequent environmental change or because of another introduction: ornamental Asian fig trees in southern Florida were not invasive until their pollinator wasps arrived separately (McKey and Kaufmann 1991). And sometimes a species becomes a major problem only after a long, inexplicable lag (Chapter 3).

Because of the inherent dispersal abilities of living organisms (Chapter 1), there is no guarantee that they will not, on their own, disperse from an area of release where they are unproblematic to sites in which they are pests. Three biological control agents (a wasp, a nematode, and a fly) were introduced into Florida in the 1980s, for example, to control pestiferous nonindigenous mole crickets. Although in Florida native mole crickets are not attacked (J. H. Frank, pers. comm.), no apparent consideration was given to the possibility that the biological control species might spread. An American mole cricket found further north (*Gryllotalpa major*) is a candidate endangered species. Might a biocontrol agent spread to within its range? Moreover, a nonindigenous species that is not a pest in areas in which it was intended to stay can be carried by humans to areas where it can be troublesome. One hypothesis for the spread of the cactus moth (Chapters 1 and 5) to Florida is that it was carried in the international trade in cut flowers.

The effects of nonindigenous species are inherently unpredictable in another way that is different from, for example, the effects of a chemical pollutant. Species can evolve. In North America, the Dutch elm disease fungus evolved more pathogenic strains (von Broembsen 1989), and numerous initially virulent diseases have evolved benign strains (Ewald 1983). Such evolution is one possible reason for the long time lags occasionally seen before invasiveness is manifested (Ewel 1986). Although we know of no demonstration in Florida of increased ecological damage caused by evolution of a nonindigenous species—no one has even attempted to quantify the probability of such an event—the potential harm is enormous.

In sum, then, the inherent unpredictability of nonindigenous species means that their entry must be much more tightly regulated than it is. They should be assumed guilty until proven innocent (Ruesink et al. 1995). The first and least costly line of defense is keeping them out of Florida. Although political and economic considerations make complete exclusion of all nonindigenous species an unrealistic goal, every single proposed introduction should at least receive thoughtful consideration. The "blacklist" approach—only species determined a priori to be potentially harmful are subject to regulation—has never worked well (Ruesink et al. 1995; Wade 1995). All nonindigenous species are potentially harmful. Thus, a "whitelist" approach is needed: every potential import should be assumed harmful until shown to pose a low risk, after which point it can be placed on a whitelist. No blanket exceptions should be allowed. Further, at the first observation of unexpected invasiveness, a species should be removed from the whitelist and further importation forbidden.

If an invasive species gets into Florida because of inadvertent transport, on its own, or because it was erroneously put on the whitelist, an effort should be made, where feasible, to eradicate the species quickly (Chapter 13)—for once it has spread, the cost of either eradication or continuing control will be much greater. If eradication fails, a comprehensive plan for maintenance control must be established and consistently followed (Chapters 14 and 15).

Controlling Nonindigenous Species

Myers and Ewel (1990b) list two recurrent themes in *Ecosystems of Florida* (Myers and Ewel 1990a): each of Florida's terrestrial ecosystems is but a fraction as large as it was at the beginning of the century and nature can no longer maintain what is left without human management. Similarly, many of Florida's watersheds have experienced increased nutrient loadings because of burgeoning urbanization and agricultural runoff (Canfield et al. 1983b; Schmitz et al. 1993), and intensive management is needed to preserve even a vestige of our native aquatic ecosystems. The present volume has shown that nonindigenous species are a key component of the degradation of Florida ecosystems, and better management of them is crucial, but it has also shown that our understanding of Florida's invasion biology is incomplete—in fact, for some taxa it is rudimentary. Similarly, ecological restoration of sites infested by nonindigenous species often fails because of a lack of understanding of invasion biology (Chapter 12). Improved knowledge will aid management, but management must be improved even as research into the biology of nonindigenous species continues. In light of the facts adduced here, what should be done now?

Preventing Entry

Presently there is virtually no screening of the vast majority of imported plants and animals for their potential invasiveness (Chapter 22). Federal efforts are fragmentary at best and primarily aimed at preventing new agricultural pests (Wade 1995; Chapter 21). Activities by the state of Florida mirror those of the federal government: priority is given mainly to agricultural pests (Chapter 22). Gordon and Thomas (Chapter 2) argue that because the federal response to biological invasions is inadequate, Florida should independently develop a comprehensive Florida Noxious Weeds list that includes species which invade nonagricultural lands. We would go further: no species on the noxious weeds list could be admitted to a whitelist, but neither could any other species be admitted without a detailed consideration of its potential invasiveness. Similarly, Courtenay (Chapter 7) pleads for more stringent regulation of importation of nonindigenous fishes combined with a better effort to educate the public about the dangers of releasing nonindigenous fishes. From the standpoint of state government, Brown (Chapter 22) observes a failure to establish a screening protocol for whether a nonindigenous species of any taxon poses a threat to Florida's ecosystems. Similar failures have beset other blacklist approaches (Wade 1995). Brown also notes a failure to expand blacklists of known invasive species and points to the internal inconsistency of allowing agencies that introduce nonindigenous species to veto the addition of invasive species to blacklists designed to protect public lands and waterways.

As noted earlier, the inherent unpredictability of living entities dictates a philosophy of guilty unless proven innocent—or at least until substantial research buttresses a claim of low probability of harmful effect.

Eradication and Maintenance Control

Management of nonindigenous species that arrive in Florida has usually been poorly funded and fragmentary at best. A success story is the maintenance control of water hyacinth (Chapter 14), which has been reduced to a minor component of the aquatic flora at an approximate annual cost of $2.7 million. Similarly, the total eradication of the snail *Achatina fulica* cost less than $2 million (Chapter 13). These campaigns contrast with many others in which management funding is insufficient and inconsistent. In 1985, for example, $2.5 million was needed to manage hydrilla in Florida's public waterways. Because funding did not keep pace with hydrilla's growth and expansion, the plant's coverage doubled between 1992 and 1994. Thus $14 million was needed to manage hydrilla adequately in 1995; but only $5 million was appropriated (Chapter 14). Early investment may allow eradication that becomes impossible once a species has spread (Chapter 13), and the cost of implementing maintenance control can quickly exceed that of early eradication. The basic problem is that living organisms do not sit around waiting for us to find funding to deal with them: they grow, move, and evolve.

Certain potentially invasive nonindigenous species are not managed at all—for example, many plants (Chapter 17), most vertebrates (Chapter 7–10), almost all insects that do not affect agriculture (Chapter 5), and freshwater and marine invertebrates (Chapters 6 and 11). Even when there is a management effort, it is usually piecemeal. No lead agency is designated to coordinate efforts of state agencies to manage nonindigenous species in nonagricultural situations. Nor do substantial incentive programs or assistance to private landowners encourage their integration into the control process (Chapter 22).

Prospects

Florida is besieged by nonindigenous species. Virtually every ecosystem is already heavily affected or a potential target. Nevertheless, there is little scientific research on most aspects of invasion biology in Florida; for some taxa, virtually nothing is known. Several factors contribute to this state of affairs: inadequate publicity about the problems, the idiosyncrasies of researchers' taste, and failure of governmental and private sources to provide funding for research. To some extent, these same problems beset all aspects of conservation. But they are particularly acute for nonindigenous species because of their importance to conservation and their continuing spread.

Efforts to exclude nonindigenous species are inadequate, and the dearth of knowledge about many aspects of invasion biology argues for much more rigorous exclusion. Once exclusion has failed, technologies often exist to eradicate or control invasive newcomers, but governmental structure and inadequate funding usually hinder the implementation of these methods.

In sum, problems of invasive nonindigenous species *are* largely soluble. Exclusionary regulations would help immensely, and methods can be developed to deal with established species. The key is our willingness to try hard enough.

REFERENCES

Abrahamson, W. G., and D. C. Hartnett. 1990. Pine flatwoods and dry prairies. In R.L. Myers and J. J. Ewel (eds.), *Ecosystems of Florida*. Orlando: University of Central Florida Press.

Adams, F. H. 1996. First bat colonies in the Keys. *Florida Wildlife* 50:2–5.

Albin, C. L. 1994. Non-indigenous plant species find a home in mined lands. In D. C. Schmitz and T. C. Brown. (eds.), *An Assessment of Invasive Non-Indigenous Species in Florida's Public Lands*. Technical Report TSS-94-100. Tallahassee: Florida Department of Environmental Protection.

Aldrich, J. W., and J. S. Weske. 1978. Origin and evolution of the eastern house finch populations. *Auk* 95:528–536.

Alexander, T. R. 1967. A tropical hammock on the Miami (Florida) limestone—a twenty-five-year study. *Ecology* 48:863–867.

———. 1971. Sawgrass biology related to the future of the Everglades ecosystem. *Proceedings of the Soil Science Society of Florida* 31:72–74.

Alien Species Action Plan Working Group. 1994. Hawaii alien species action plan: A multi-agency commitment. Working document.

Allen, E. R., and W. T. Neill. 1954. The Florida deer. *Florida Wildlife* 7:21, 37.

Allen, F. E. 1953. Distribution of marine invertebrates by ships. *Australian Journal of Marine and Freshwater Research* 4:307–316.

Allen, J. A. 1871. On the mammals and winter birds of east Florida, with an examination of certain assumed specific characters in birds, and a sketch of the bird-fauna of eastern North America. *Bulletin of the Museum of Comparative Zoology* 2:161–540.

Allen, J. C., J. L. Foltz, W. N. Dixon, A. M. Liebhold, J. J. Colbert, J. Regniere, D. R. Gray, J. W. Wilder, and I. Christie. 1993. Will the gypsy moth become a pest in Florida? *Florida Entomologist* 76:102–113.

American Ornithologists' Union. 1983. *Check-List of North American Birds*. Lawrence, Kans.: Allen Press.

Andres, L. A. 1977. The economics of biological control of weeds. *Aquatic Botany* 3:111–123.

———. 1981a. Conflicting interests and the biological control of weeds. In E. S. Delfosse (ed.), *Proceedings of the 5th International Symposium on Biological Control of Weeds*. Brisbane: Commonwealth Scientific and Industrial Research Organization, Australia.

———. 1981b. Biological control of naturalized and native plants: Conflicting interests. In G. C. Papavizas (ed.), *Biological Control in Crop Production*. Beltsville Agricultural Research Center Symposium 5. Totowa, N. J.: Allanheld, Osmun.

———. 1981c. Insects in the biological control of weeds. In D. Pimentel (ed.), *CRC Handbook of Pest Management in Agriculture*, vol. 2. Boca Raton: CRC Press.

Andres, L. A., C. J. Davis, P. Harris, and W. J. Wapshere. 1976. Biological control of weeds. In C. B. Huffaker and P. S. Messenger (eds.), *Theory and Practice of Biological Control*. New York: Academic Press.

Anonymous. 1942. With the park ranger. *The Bulletin* (Journal of Highlands Hammock), August.

———. 1955. Roving rodent. *Florida Wildlife* 8:10.

———. 1963. Sambar deer bagged near Port St. Joe. *Florida Wildlife* 16:5.

———. 1967. Wanted: A game species of aquatic-type deer. *Florida Wildlife* 21:31.

———. 1975. Wildlife research report: Grass carp. *Florida Wildlife* 28:6.

———. 1988. Element stewardship abstract for *Schinus terebinthifolius*, Brazilian peppertree. Arlington, Va.: The Nature Conservancy.

———. 1992. *Genetic Manipulation: The Threat or the Glory?* Canberra: Parliament of the Commonwealth of Australia.

———. 1994. Element stewardship abstract for *Paederia foetida*. Arlington, Va.: The Nature Conservancy.

———. 1995a. *Endangered Species Fact Sheet for the State of Florida*. Washington, D.C.: Endangered Species Coalition.

———. 1995b. EPA approves limited use of bromoxynil. *Gene Exchange* 5(4), 6(1):1, 12.

Armentano, T. V., R. F. Doren, W. J. Platt, and T. Mullins. 1995. Effects of Hurricane Andrew on coastal and interior forests of southern Florida: Overview and synthesis. *Journal of Coastal Research* 21:111–144.

Ashton, R. E., Jr., and P. S. Ashton. 1981. *Handbook of Reptiles and Amphibians of Florida*. Part 1: *The Snakes*. Miami: Windward.

———. 1985. *Handbook of Reptiles and Amphibians of Florida*. Part 2: *Lizards, Turtles, and Crocodilians*. Miami: Windward.

———. 1988. *Handbook of Reptiles and Amphibians of Florida*. Part 3: *The Amphibians*. Miami: Windward.

Atkinson, I. 1989. Introduced animals and extinction. In D. Western and M. Pearl (eds.), *Conservation for the Twenty-First Century*. New York: Oxford University Press.

Atkinson, T. H., and S. B. Peck. 1994. Annotated checklist of the bark and ambrosia beetles (Coleoptera: Platypodidae and Scolytidae) of southern Florida. *Florida Entomologist* 77:313–329.

Austin, D. F. 1978a. Exotic plants and their effects in southeastern Florida. *Environmental Conservation* 5:25–34.

———. 1978b. Spread of the exotic fern *Lygodium microphyllum* in Florida. *American Fern Journal* 68:65–66.

———. 1993. E.P.P.C. 1993 list of invasive exotics. *Resource Management Notes* 5:27–31.

Austin, D. F., K. Coleman-Marois, and D. R. Richardson. 1977. Vegetation of southeastern Florida—II–V. *Florida Scientist* 40:331–361.

Ayers, E. L. 1957. The two medfly eradication programs in Florida. *Proceedings of the Florida Horticultural Society* 70:67–69.

Bailey, H. H. 1924a. The armadillo in Florida and how it reached there. *Journal of Mammalogy* 5:264–265.

Bailey, L. H. 1924b. *Manual of Cultivated Plants.* New York: Macmillan.

———. 1971. *Manual of Cultivated Plants.* New York: Macmillan.

Baird, K. 1989. High quality restoration of riparian ecosystems. *Restoration and Management Notes* 7(2):60–65.

Baker, H. G. 1965. Characteristics and modes of origins of weeds. In H. G. Baker and G. L. Stebbins (eds.), *The Genetics of Colonizing Species.* New York: Academic Press.

———. 1974. The evolution of weeds. In R. F. Johnston, P. W. Frank, and C. D. Michener (eds.), *Annual Review of Ecology and Systematics.* Palo Alto: Annual Reviews.

Balciunas, J. K., and D. W. Burrows. 1993. The rapid suppression of the growth of *Melaleuca quinquenervia* saplings in Australia by insects. *Journal of Aquatic Plant Management* 31:265–270.

Balciunas, J. K., D. W. Burrows, and M. F. Purcell. 1994. Field and laboratory host ranges of the Australian weevil, *Oxyops vitiosa* (Coleoptera: Curculionidae), a potential biological control agent for the paperbark tree, *Melaleuca quinquenervia.* *Biological Control* 4:351–360.

Baloch, G. M., A. G. Khan, and M. A. Ghani. 1972. Phenology, biology, and host-specificity of some stenophagous insects attacking *Myriophyllum* spp. in Pakistan. *Hyacinth Control Journal* 10:13–16.

Bancroft, G. T., J. S. Godley, D. T. Gross, N. N. Rojas, D. A. Sutphen, and R. W. McDiarmid. 1983. Large scale operations management test of the use of the white amur for control of problem plants; the herpetofauna of Lake Conway, Florida: Species accounts. Vicksburg: Miss.: U.S. Army Corps of Engineers Water Experiment Station CE.

Bancroft, G. T., W. Hoffman, R. J. Sawicki, and J. C. Ogden. 1992. The importance of the water conservation areas in the Everglades to the endangered wood stork (*Mycteria americana*). *Conservational Biology* 6:392–398.

Barbour, T. 1910. *Eleutherodactylus ricordii* in Florida. *Proceedings of the Biological Society of Washington* 23:100.

Bard, A. M. 1994. Memorandum on information for 1993 legislative report. April 11, 1994. Parks District 3, Florida Park Service, Orlando.

Barkley, F. A. 1944. *Schinus* L. *Brittonia* 5:160–198.

Bartlett, R. D. 1988. *In Search of Reptiles and Amphibians.* New York: E. J. Brill.

———. 1994. Florida's alien herps. *Reptile and Amphibian Magazine* (March–April): 56–73, 103–109.

———. 1995a. The teids of the southeastern U.S. *Tropical Fish Hobbyist* (March): 112–126.

———. 1995b. The anoles of the United States. *Reptiles* (March): 48–65.

Bartodziej, W. 1992. Amphipod contribution to waterhyacinth (*Eichhornia crassipes* [Mart.] Solms) decay. *Florida Scientist* 55:103–111.

Bass, D. G., Jr. 1993. North Florida Streams Research Project. Study I. Fish Community Analysis. Completion Report, Wallop-Breaux Project F-36. Tallahassee: Florida Game and Fresh Water Fish Commission.

Bass, D. G., Jr., and V. G. Hitt. 1974. Ecological distribution of the introduced Asiatic clam, *Corbicula manilensis,* in Florida. Report to the Florida Game and Fresh Water Fish Commission, Tallahassee.

Bate, C. S. 1866. Carcinological gleanings No. 2. *Annals and Magazine of Natural History* (3)17:24–31.

Baumhover, A. H., A. J. Graham, B. A. Bitter, D. E. Hopkins, W. D. New, F. H. Dudley, and R. C. Bushland. 1955. Screw-worm control through release of sterilized flies. *Journal of Economic Entomology* 48:462–466.

Beatley, T. 1991. Protecting biodiversity in coastal environments: Introduction and overview. *Coastal Management* 19:1–19.

Beck, W. M., J. L. Hulbert, and T. Pearce. 1970. Unpublished report on Lake Okeechobee bottom fauna surveys, 1969–1970. Florida State Board of Health, Tallahassee.

Beckner, J. 1968. *Lygodium microphyllum,* another fern escaped in Florida. *American Fern Journal* 58:93–94.

Beetle, D. E. 1972. A note on land snails associated with kudzu vine. *Nautilus* 86:18–19.

Beever, J. W., III. 1994. Mangroves and Brazilian pepper invasion. In D. C. Schmitz and T. C. Brown (eds.), *An Assessment of Invasive Non-indigenous Species in Florida's Public Lands.* Technical Report TSS-94-100. Tallahassee: Florida Department of Environmental Protection.

Beissinger, S. R., and N. F. R. Snyder. 1992. *New World Parrots in Crisis: Solutions from Conservation Biology.* Washington, D.C.: Smithsonian Institution Press.

Belden, R., and W. Frankenberger. 1977. Management of feral hogs in Florida—past, present, and future. In G. W. Wood (ed.), *Research and Management of Wild Hog Populations: Proceedings of a Symposium.* Georgetown, S.C.: Belle W. Baruch Forest Science Institute of Clemson University.

Bell, M., III. 1966. *Some Notes and Reflections upon a Letter from Benjamin Franklin to Noble Wimberly Jones, October 7, 1772.* Darien, Ga.: Ashantily Press.

Belleville, B. 1994. Critter patrol. *Orlando Sentinel,* 29 May, Florida Section.

Belshe, J. F. 1961. Observation of an introduced tropical fish (*Belonesox belizanus*) in southern Florida. M.S. thesis, University of Miami, Coral Gables, Florida.

Bennett, C. A. 1993. Quarantine biological control operations. In *Proceedings of the 27th Annual Meeting, Aquatic Plant Control Research Program.* Miscellaneous Paper A-93-2. Vicksburg, Miss.: U.S. Army Engineer Waterways Experiment Station.

Bennett, F. D. 1993. Do introduced parasitoids displace native ones? *Florida Entomologist* 76:54–63.

———. 1994. Rhodesgrass mealybug. In D. Rosen, F. D. Bennett, and J. L. Capinera (eds.), *Pest Management in the Subtropics: Biological Control—a Florida Perspective.* Andover, U.K.: Intercept.

Bennett, F. D., and D. H. Habeck. 1991. Brazilian peppertree—prospects for biological control in Florida. In T. D. Center et al. (eds.), *Proceedings of the Symposium on Exotic Pest Plants.* Technical Report NPS/NREVER/ NRTR-91/06. Denver: U.S. Department of the Interior, National Park Service.

Berger, J. L. 1993. Ecological restoration and non-indigenous plant species: A review. *Restoration Ecology* 1:74–82.

Bergquist, G. T., A. M. Pable, and J. Jerrigan. 1994. *Strategic Assessment of Florida's Environment: SAFE.* Tallahassee: Florida Center for Public Management, Florida State University.

Berner, L., and M. L. Pescador. 1988. *The Mayflies of Florida.* Rev. ed. Tallahassee and Gainesville: University Presses of Florida.

Bertness, M. D. 1984. Habitat and community modification by an introduced herbivorous snail. *Ecology* 65:370–381.

Bhatkar, A. P. 1988. Confrontation behavior between *Solenoposis invicta* and *S. geminata,* and competitiveness of certain Florida ant species against *S. invicta.* In J. C. Trager (ed.), *Advances in Myrmecology.* Boulder: Westview Press.

Bigler, W. J., E. Lassing, E. Buff, A. L. Lewis, and G. L. Hoff. 1975. Arbovirus surveillance in Florida: Wild vertebrate studies 1965–1974. *Journal of Wildlife Diseases* 11:348–356.

Bingham, B. L. 1992. Life histories in an epifaunal community coupling of adult and larval processes. *Ecology* 73:2244–2259.

Bird, C. J., M. J. Dadswell, and D. W. Grund. 1993. First record of the potential nuisance alga *Codium fragile* ssp. *tomentosoides* (Chlorophyta, Caulerpales) in Atlantic Canada. *Proceedings of the Nova Scotia Institute of Science* 40:11–17.

Blackburn, R. D., K. K. Steward, and L. W. Weldon. 1967. *USDA Agricultural Research Service Weed Investigations Aquatic and Non-Crop Areas Annual Report for 1967.* Fort Lauderdale, Fla.: USDA, Plantation Field Laboratory.

Blair, R. M., and M. J. Langlinais. 1960. Nutria and swamp rabbits damage baldcypress seedlings. *Journal of Forestry* 58:388–389.

Blake, N. M. 1980. *Land into Water—Water into Land: A History of Water Management in Florida.* Tallahassee: University Presses of Florida.

Bodle, M. J., A. P. Ferriter, and D. D. Thayer. 1994. The biology, distribution and ecological consequences of *Melaleuca quinquenervia* in the Everglades. In S. M. Davis and J. C. Ogden (eds.), *Everglades: The Ecosystem and Its Restoration.* Delray Beach, Fla.: St. Lucie Press.

Boender, R. 1995. The first commercial butterfly farm and public exhibition in the United States. *Florida Entomologist* 78:36–38.

Boudouresque, C. F., A. Meinesz, M. Verlaque, and M. Knoepffler-Peguy. 1992. The expansion of the tropical alga *Caulipera taxifolia* (Chlorophyta) in the Mediterranean. *Cryptogamie, Algologie* 13:144–145.

Boudouresque, C. F., A. Meinesz, and V. Gravez. 1994. *First International Workshop on* Caulerpa taxifolia. Marseille: GIS Posidonie.

Bousfield, E. L. 1973. *Shallow-Water Gammaridean Amphipoda of New England.* Ithaca: Cornell University Press.

Bouvier, L. F., and B. Weller. 1992. *Florida in the 21st Century.* Washington, D.C.: Center for Immigration Studies.

Bowen, W. W. 1968. Variation and evolution of Gulf Coast populations of beach mice, *Peromyscus polionotus. Bulletin of the Florida State Museum, Biological Sciences* 12:1–91.

Bowers, M. D., N. E. Stamp, and S. K. Collinge. 1992. Early stage of host range expansion by a specialist herbivore, *Euphydryas phaeton* (Nymphalidae). *Ecology* 73:526–536.

Boyter, H. H., Jr. 1994. Non-indigenous species on lands within Florida's wildlife management areas. In D. C. Schmitz and T. C. Brown (eds.), *An Assessment of Invasive Non-indigenous Species in Florida's Public Lands*. Technical Report TSS-94-100. Tallahassee: Florida Department of Environmental Protection.

Brady, J. R., and II. W. Campell. 1983. Distribution of coyotes in Florida. *Florida Field Naturalist* 11:40–41.

Braithwaite, R. W., W. M. Lonsdale, and J. A. Estbergs. 1989. Alien vegetation and native biota in tropical Australia: The impact of *Mimosa pigra*. *Biological Conservation* 48:189–210.

Brakhage, D. 1987. Comments on the symposium. In *Proceedings of the Grass Carp Symposium, Gainesville, Florida*. Vicksburg, Miss.: U.S. Army Corps of Engineers.

Bratton, S. P. 1975. The effect of the European wild boar, *Sus scrofa*, on grey beech forest in the Great Smokey Mountains. *Ecology* 56:1356–1366.

Breeman, A. M. 1988. Relative importance of temperature and other factors in determining geographic boundaries of seaweeds: Experimental and phenological evidence. *Helgolander Meeresuntersuchungen* 42:199–241.

Breeman, A. M., and M. D. Guiry. 1989. Tidal influences on the photoperiodic induction of tetrasporogenesis in *Bonnemaisonia hamifera* (Rhodophyta). *Marine Biology* 102:5–14.

Breise, L., and M. Smith. 1973. Competition between *Mus musculus* and *Peromyscus polionotus*. *Journal of Mammalogy* 54:968–969.

Brisbin, I. L., Jr. 1977. The pariah. *Pure-Bred Dogs American Kennel Gazette* 94:22–29.

———. 1986. Unusual "Carolina dog" under study in Aiken area. *South Carolina Out-of-Doors* 13:4.

Brokaw, N. V. L., and L. R. Walker. 1991. Summary of the effects of Caribbean hurricanes on vegetation. *Biotropica* 23:442–447.

Bronstein, J. L., and M. Hossaert-McKey. 1995. Hurricane Andrew and a Florida fig pollination mutualism: Resilience of an obligate interaction. *Biotropica* 27: 373–381.

Brothers, T. S., and A. Spingarn. 1992. Forest fragmentation and alien plant invasion of central Indiana old-growth forests. *Conservation Biology* 6:91–100.

Broward County Soil Conservation Service. 1952. Broward County Soil Conservation Service Annual Report.

Brown, J. H. 1995. *Macroecology*. Chicago: University of Chicago Press.

Brown, J. L., and N. R. Spencer. 1973. *Vogtia malloi*, a newly introduced phycitine moth (Lepidoptera: Pyralidae) to control alligatorweed. *Environmental Entomology* 2:519–523.

Brown, L. N. 1969. An exotic squirrel. *Florida Wildlife* 23:4–5.

———. 1974. Breeding biology of the nutria (*Myocastor coypus*) in the Tampa Bay area, Florida. Unpublished report, Abstract papers of the 54th annual meeting of the American Society of Mammalogists.

———. 1975. Ecological relationships and breeding biology of the nutria (*Myocastor coypus*) in the Tampa, Florida, area. *Journal of Mammalogy* 56:928–930.

Brown, L. N., and R. J. McGuire. 1969. Status of the red-bellied squirrel (*Sciurus aureogaster*) in the Florida Keys. *American Midland Naturalist* 82:629–630.

———. 1975. Field ecology of the exotic Mexican red-bellied squirrel in Florida. *Journal of Mammalogy* 56:405–419.

Brown, R. B., E. L. Stone, and V. W. Carlisle. 1990. Soils. In R. L. Myers and J. J. Ewel (eds.), *Ecosystems of Florida*. Orlando: University of Central Florida Press.

Brown, T. C. 1994. Existing state authorities and responsibilities. In D. C. Schmitz and T. C. Brown (eds.), *An Assessment of Invasive Non-indigenous Species in Florida's Public Lands*. Technical Report TSS–94–100. Tallahassee: Florida Department of Environmental Protection.

Brunner, M. C., and T. R. Batterson. 1984. The effect of three sediment types on tuber production in hydrilla (*Hydrilla verticillata* L. F. Royle). *Journal of Aquatic Plant Management* 22:95–97.

Bryson, C., J. Byrd, and R. Westbrooks. 1995. Tropical soda apple (*Solanum viarum* Dunal) in the United States. Fact Sheet. Jackson: Mississippi Department of Agriculture and Commerce, Bureau of Plant Industry.

Buchanan, C. D., and R. V. Talmage. 1954. The geographical distribution of the armadillo in the United States. *Texas Journal of Science* 6:142–150.

Bucher, E. H. 1992. Neotropical parrots as agricultural pests. In S. R. Beissinger and N. F. R. Snyder (eds.), *New World Parrots in Crisis: Solutions from Conservation Biology*. Washington, D.C.: Smithsonian Institution Press.

Buck, H. 1979. Optimism swells with the possibility of a sterile hybrid grass carp. *Fisheries* 4(5):31.

Buckingham, G. R. 1990. Quarantine laboratory: Research for hydrilla control. In *Proceedings of the 24th Annual Meeting, Aquatic Plant Control Research Program*. Miscellaneous Paper A-90-3. Vicksburg, Miss.: U.S. Army Engineer Waterways Experiment Station.

———. 1992. Temperate biological control insects for Eurasian watermilfoil and hydrilla. In *Proceedings of the 26th Annual Meeting, Aquatic Plant Control Research Program*. Miscellaneous Paper A-92-2. Vicksburg, Miss.: U.S. Army Engineer Waterways Experiment Station.

———. 1993. Foreign research on insect biological control agents. In *Proceedings of the 27th Annual Meeting, Aquatic Plant Control Research Program*. Miscellaneous Paper A-93-2. Vicksburg, Miss.: U.S. Army Engineer Waterways Experiment Station.

———. 1994. Biological control of aquatic weeds. In D. Rosen, F. D. Bennett, and J. L. Capinera (eds.), *Pest Management in the Subtropics: Biological Control—a Florida Perspective*. Andover, U.K.: Intercept.

Buker, G. E. 1982. Engineers vs. Florida's green menace. *Florida Historical Quarterly* (April):413–427.

Bump, G. 1951. Game introductions—when, where, and how. *Transactions of the North American Association of Fish and Wildlife Agencies* 16:316–325.

────. 1968. *Foreign Game Investigations: A Federal–State Cooperative Program.* Washington, D.C.: U.S. Department of the Interior.

Bump, G., and W. B. Bohl. 1964. *Summary of Foreign Game Bird Propagation and Liberations.* U.S. Fish and Wildlife Service Special Scientific Report 80. Washington, D.C.: U.S. Department of the Interior.

Buscemi, P. A. 1958. Littoral oxygen depletion produced by a cover of *Elodea canadensis. Oikos* 9:239–245.

Buschman, L. L. 1976. Invasion of Florida by the "lovebug" *Plecia nearctica* (Diptera: Bibionidae). *Florida Entomologist* 59:191–194.

Buschman, L. L., W. H. Whitcomb, T. M. Neal, and D. L. Mays. 1977. Winter survival and hosts of the velvetbean caterpillar in Florida. *Florida Entomologist* 60:267–273.

Butterfield, B. P., J. B. Hauge, and W. E. Meshaka, Jr. 1993. The occurrence of *Hemidactylus mabouia* on the United States mainland. *Herpetological Review* 24:111–112.

Butterfield, B. P., W. E. Meshaka, Jr., and R. L. Kilhefner. 1994. Two anoles new to Broward County, Florida. *Herpetological Review* 25:77–78.

Cairns, J., Jr. 1986. Restoration, reclamation, and regeneration of degraded or destroyed ecosystems. In M. E. Soulé (ed.), *Conservation Biology: The Science of Scarcity and Diversity.* Sunderland, Mass.: Sinauer.

Callicott, J. B. 1994. A brief history of American conservation philosophy. In W. W. Covington and L. F. DeBano (eds.), *Sustainable Ecological Systems: Implementing an Ecological Approach to Land Management.* Fort Collins, Colo.: USDA Forest Service.

Caltagirone, L. E., and R. L. Doutt. 1989. The history of the vedalia beetle importation to California and its impact on the development of biological control. *Annual Review of Entomology* 34:1–16.

Cameron, G. N., and S. R. Spencer. 1989. Rapid leaf decay and nutrient release in a Chinese tallow forest. *Oecologia* 80:222–228.

Campbell, C. W., S. E. Malo, and J. Popenoe. 1987. "Tikal," an early-maturing sapodilla cultivar. *Proceedings of the Florida State Horticultural Society* 100:281–283.

Campbell, G. 1985. Part 1: The rich and varied mammal fauna of Sanibel. *Sanibel-Captiva Islander,* 5 November.

Canfield, D. E., Jr., M. J. Maceina, and J. V. Shireman. 1983a. Effects of hydrilla and grass carp on water quality in a Florida lake. *Water Research Bulletin* 19:773–778.

────. 1983b. Trophic state classifications of lakes with aquatic macrophytes. *Canadian Journal of Fisheries and Aquatic Science* 40:1713–1718.

Canfield, D. E., Jr., K. A. Langeland, S. B. Linda, and W. T. Haller. 1985. Relations between water transparency and maximum depth of macrophyte colonization in lakes. *Journal of Aquatic Plant Management* 23:25–28.

Carey, J. 1982. Mangroves—swamps nobody likes. *International Wildlife* 12(5):19–28.

Carey, J. R. 1992a. The Mediterranean fruit fly in California: Taking stock. *California Agriculture* 46(1):12–17.

———. 1992b. The medfly in California. Response to letters to the editor. *Science* 255:514–516.

Carl, K. P. 1982. Biological control of native pests by introduced natural enemies. *Biological Control News and Information* 3:191–200.

Carleton, A. R. 1971. Studies on a population of the red-whiskered bulbul, *Pycnonotus jocosus* (Linnaeus), in Dade County, Florida. M.S. thesis, University of Miami, Coral Gables, Florida.

Carleton, A. R., and O. T. Owre. 1975. The red-whiskered bulbul in Florida: 1960–71. *Auk* 92:40–57.

Carlton, J. T. 1979a. History, biogeography, and ecology of the introduced marine and estuarine invertebrates of the Pacific coast of North America. Ph.D. dissertation, University of California, Davis.

———. 1979b. Introduced invertebrates of San Francisco Bay. In T. J. Conomos (ed.), *San Francisco Bay: The Urbanized Estuary.* San Francisco: American Association for the Advancement of Science, Pacific Division.

———. 1985. Transoceanic and interoceanic dispersal of coastal marine organisms: The biology of ballast water. *Oceanography and Marine Biology Annual Review* 23:313–371.

———. 1987. Patterns of transoceanic marine biological invasions in the Pacific Ocean. *Bulletin of Marine Science* 41:452–465.

———. 1989. Man's role in changing the face of the ocean: Biological invasions and implications for conservation of near-shore environments. *Conservation Biology* 3:265–273.

———. 1992a. Dispersal of living organisms into aquatic ecosystems as mediated by aquaculture and fisheries activities. In A. Rosenfield and R. Mann (eds.), *Dispersal of Living Organisms into Aquatic Ecosystems.* College Park: Maryland Sea Grant Program.

———. 1992b. Introduced marine and estuarine mollusks of North America: An end-of-the-20th-century perspective. *Journal of Shellfish Research* 11:489–505.

———. 1992c. Blue immigrants: The marine biology of maritime history. *The Log* (Mystic Seaport Museum) 44:31–36.

———. 1992d. Overview of the issues concerning marine species introductions and transfers. In R. DeVoe (ed.), *Introductions and Transfers of Marine Species: Achieving a Balance between Economic Development and Resource Protection.* Proceedings of the Conference and Workshop, 30 October–2 November, 1991, Hilton Head Island, South Carolina. Charleston: South Carolina Sea Grant Consortium.

———. 1993. Dispersal mechanisms of the zebra mussel (*Dreissena polymorpha*). In T. F. Nalepa and D. W. Schloesser (eds.), *Zebra Mussels: Biology, Impacts, and Control.* Boca Raton: CRC Press.

———. 1996. Biological invasions and cryptogenic species. *Ecology* 77:1653–1655.

Carlton, J. T., and J. Geller. 1993. Ecological roulette: The global transport and invasion of nonindigenous marine organisms. *Science* 261:78–82.

Carlton, J. T., and J. Hodder. 1995. Biogeography and dispersal of coastal marine organisms: Experimental studies on a replica of a sixteenth century sailing vessel. *Marine Biology* 121:721–730.

Carlton, J. T., and E. W. Iverson. 1981. Biogeography and natural history of *Sphaeroma walkeri* Stebbing (Crustacea: Isopoda) and its introduction to San Diego Bay, California. *Journal of Natural History* 15:31–48.

Carlton, J. T., and J. A. Scanlon. 1985. Progression and dispersal of an introduced alga: *Codium fragile* ssp. *tomentosoides* (Chlorophyta) on the Atlantic coast of North America. *Botanica Marina* 28:155–165.

Carlton, J. T., D. M. Reid, and H. van Leeuwen. 1995. *Shipping Study: The Role of Shipping in the Introduction of Non-indigenous Aquatic Organisms to the Coastal Waters of the United States (Other Than the Great Lakes) and an Analysis of Control Options.* National Sea Grant College Program/Connecticut Sea Grant Project R/ES-6. Department of Transportation, U.S. Coast Guard, Washington, D.C., and Groton, Connecticut. Report Number CG-D-11-95. Government Accession Number AD-A294809.

Carr, A. F. 1940. A contribution to the herpetology of Florida. University of Florida Publication, Biological Science Series, vol. 3, no. 1.

———. 1982. Armadillo dilemma. *Animal Kingdom* 85:40–43.

Cassani, J. R. 1986. Arthopods on Brazilian peppertree, *Schinus terebinthifolius* (Anacardiaceae) in South Florida. *Florida Entomologist* 69:184–196.

Cassani, J. R., D. R. Maloney, D. H. Habeck, and F. D. Bennett. 1989. New insect records on Brazilian peppertree, *Schinus terebinthifolius* (Anacardiaceae) in South Florida. *Florida Entomologist* 72:714–716.

Caster, J. 1994. Invasive alien species on D.O.T. right-of-way managed lands. In D. C. Schmitz and T. C. Brown (eds.), *An Assessment of Invasive Non-indigenous Species in Florida's Public Lands.* Technical Report TSS-94-100. Tallahassee: Florida Department of Environmental Protection.

Caughley, G., and A. Gunn. 1996. *Conservation Biology in Theory and Practice.* Cambridge, Mass.: Blackwell Scientific.

Causey, M. K., and C. A. Cude. 1978. Feral dog predation of gopher tortoise, *Gopherus polyphemus* (Reptilia, Testudines, Testudinidae), in southeast Alabama. *Herpetological Review* 9:94–95.

Center, T. D. 1982. The waterhyacinth weevils *Neochetina eichhorniae* and *N. bruchi.* *Aquatics* 4(2):8–19.

———. 1987. Insects, mites, and pathogens as agents of waterhyacinth (*Eichhornia crassipes* [Mart.] Solms) leaf and ramet mortality. *Journal of Lake and Reservoir Management* 3:285–293.

———. 1992a. Biological control of weeds in waterways and on public lands in the south-eastern United States. In J. H. Combellack, K. J. Levick, J. Parsons, and R. G. Richardson (eds.), *Proceedings of the 1st International Weed Control Congress.* Melbourne: Weed Science Society of Victoria.

———. 1992b. Release and establishment of new biological control agents of *Hydrilla verticillata.* In *Proceedings of the 26th Annual Meeting, Aquatic Plant Control Research Program.* Miscellaneous Paper A-92-2. Vicksburg, Miss.: U.S. Army Engineer Waterways Experiment Station.

Center, T. D., and W. C. Durden. 1981. Release and establishment of *Sameodes al-*

biguttalis for the biological control of waterhyacinth. *Environmental Entomology* 10:75–80.

Center, T. D., and N. R. Spencer. 1981. The phenology and growth of water hyacinth (*Eichhornia crassipes* [Mart.] Solms) in a eutrophic North-Central Florida lake. *Aquatic Botany* 10:1–32.

Center, T. D., and T. K. Van. 1989. Alteration of waterhyacinth (*Eichhornia crassipes* [Mart.] Solms) leaf dynamics and photo-chemistry by insect damage and plant density. *Aquatic Botany* 35:181–195.

Center, T. D., A. F. Cofrancesco, and J. K. Balciunas. 1990. Biological control of aquatic and wetland weeds in the southeastern United States. In E. S. Delfosse (ed.), *Proceedings of the 7th International Symposium on Biological Control of Weeds.* Rome: Istituto Sperimentale per la Patologia Vegetale, Ministerio dell'Agricoltura e dell Foreste.

Chadwick, D. H. 1995. Dead or alive: The Endangered Species Act. *National Geographic* 187(3):4–41.

Chapman, J. W., and J. T. Carlton. 1991. A test of criteria for introduced species: The global invasion by the isopod *Synidotea laevidorsalis* (Miers, 1881). *Journal of Crustacean Biology* 11:386–400.

———. 1995. Predicted discoveries of the introduced isopod *Synidotea laevidorsalis* (Miers, 1881). *Journal of Crustacean Biology* 14:700–714.

Chapman, L. 1994. Established non-indigenous monkey populations and the simian herpes B virus. In D. C. Schmitz and T. C. Brown (eds.), *An Assessment of Invasive Non-indigenous Species in Florida's Public Lands.* Technical Report TSS-94-100. Tallahassee: Florida Department of Environmental Protection.

Charudattan, R., and H. L. Walker. 1982. *Biological Control of Weeds with Plant Pathogens.* New York: Wiley.

Chen, E., and J. F. Gerber. 1990. Climate. In R. L. Myers and J. J. Ewel (eds.), *Ecosystems of Florida.* Orlando: University of Central Florida Press.

Chervinsky, J. 1967. *Tilapia nilotica* (Linné) from Lake Rudolf, Kenya and its hybrids resulting from a cross with *T. aurea* (Steindachner) (Pisces, Cichlidae). *Bamidgeh* 19:81–96.

Chesler, J. 1994. Not just bilge water. *American Conchologist* 22:13.

Choate, P. M. 1995. Ground beetles (Coleoptera: Carabidae). In J. H. Frank and E. D. McCoy, Precinctive insect species in Florida. *Florida Entomologist* 78:21–35.

Civeyrel, L., and D. Simberloff. 1996. A tale of two snails: Is the cure worse than the disease? *Biodiversity and Conservation* 5:1231–1252.

Clark, R. A., and H. V. Weems, Jr. 1989. Detection, quarantine, and eradication of fruitflies invading Florida. *Proceedings of the Florida State Horticultural Society* 102:159–164.

Clark, R. S. 1994. The present "piecemeal" approach to melaleuca management. In D. C. Schmitz and T. C. Brown (eds.), *An Assessment of Invasive Non-indigenous Species in Florida's Public Lands.* Technical Report TSS-94-100. Tallahassee: Florida Department of Environmental Protection.

Clarke, A. H. 1986. Competitive exclusion of *Canthyria* (Unionidae) by *Corbicula fluminea* (Müller). Malacology Data Net, Ecoresearch Series 1:3-10.

Clausen, C. P., B. R. Bartlett, E. C. Bay, P. DeBach, R. D. Goeden, E. F. Legner, J. A.

McMurtry, E. R. Oatman, and D. Rosen. 1978. *Introduced Parasites and Predators of Arthropod Pests and Weeds: A World Review.* Agricultural Handbook 480. Washington D.C.: U.S. Department of Agriculture.

Clench, W. J. 1970. *Corbicula manilensis* (Philippi) in lower Florida. *Nautilus* 84(1):36.

Coblentz, B. 1991. A response to Temple and Lugo. *Conservation Biology* 5:5–6.

Cofrancesco, A. F. 1991. A history and overview of biological control technology. In *Proceedings of the 25th Annual Meeting, Aquatic Plant Control Research Program.* Miscellaneous Paper A-91-3. Vicksburg, Miss.: U.S. Army Engineer Waterways Experiment Station.

Cohen, R. R. H., P. V. Dresler, E. J. P. Phillips, and R. L. Cory. 1984. The effect of the Asiatic clam, *Corbicula fluminea,* on phytoplankton of the Potomac River, Maryland. *Limnology and Oceanography* 29:170–180.

Coile, N. C. 1993. *Tropical Soda Apple,* Solanum viarum *Dunal: The Plant from Hell.* Botany Circular 27. Florida Department of Agriculture and Consumer Services, Division of Plant Industry.

Coleman, J. S., and S. A. Temple. 1993. Rural residents' free-ranging domestic cats: A survey. *Wildlife Society Bulletin* 21:381–390.

Colle, D. E. 1982. Hydrilla—miracle or migraine for Florida's sportfish. *Aquatics* 4:6.

Colle, D. E., and J. V. Shireman. 1980. Coefficients of condition for largemouth bass, bluegill, and redear sunfish in hydrilla-infested lakes. *Transactions of the American Fisheries Society* 109:521–531.

Colle, D. E., J. V. Shireman, W. T. Haller, D. E. Canfield, and J. C. Joyce. 1985. Influence of hydrilla on angler utilization, harvest and monetary expenditure at Orange Lake, Florida. University of Florida, Gainesville.

Collette, B. B. 1961. Correlations between ecology and morphology in anoline lizards from Havana, Cuba and southern Florida. *Bulletin of the Museum of Comparative Zoology* 125:137–162.

———. 1990. Problems with gray literature in fishery science. In J. Hunter (ed.), *Writing for Fishery Journals.* Bethesda, Md.: American Fisheries Society.

Collins, J. T. 1993. *Amphibians and Reptiles in Kansas.* 3rd ed. Lawrence: University of Kansas Museum of Natural History.

Colman, P. H. 1978. An invading giant. *Wildlife in Australia* 15(2):46–47.

Colvin, D. L., J. Gaffney, and D. G. Shilling. 1994. *Cogongrass: Biology, Ecology and Control in Florida.* Weeds in the Sunshine. SS-AGR-52. Florida Cooperative Extension Service, Institute of Food and Agricultural Sciences, University of Florida.

Conant, R., and J. T. Collins. 1991. *A Field Guide to Reptiles and Amphibians of Eastern and Central North America.* Boston: Houghton Mifflin.

Connell, J. H. 1978. Diversity in tropical rainforests and coral reefs. *Science* 199: 1302–1310.

Conner, W. H., and J. W. Day, Jr. 1989. Response of coastal wetland forests to human and natural changes in the environment with emphasis on hydrology. In D. D. Hook and R. Lea (eds.), *Proceedings of the Symposium: The Forested Wetlands in the Southern United States.* General Technical Report SE-50. USDA Forest Service.

Conner, W. H., and K. Flynn. 1989. Growth and survival of baldcypress (*Taxodium distichum* [L.] Rich.) planted across a flooding gradient in a Louisiana bottomland forest. *Wetlands* 9:207–217.

Conner, W. H., and J. R. Toliver. 1987. Vexar seedling protectors did not reduce nutria damage to planted baldcypress seedlings. *USDA Tree Planter's Note* 38:26–29.

——. 1988. The problem of planting Louisiana swampland when nutria (*Myocastor coypus*) are present. In N. R. Holler (ed.), *Proceedings of the Third Eastern Wildlife Damage Control Conference*, University of Alabama, Auburn, Alabama.

——. 1990. Long-term trends in the bald-cypress (*Taxodium distichum*) resource in Louisiana (U.S.A.). *Forest Ecology and Management* 33/34:543–557.

Conover, M. R. 1979. Effect of gastropod shell characteristics and hermit crabs on shell epifauna. *Journal of Experimental Marine Biology and Ecology* 40:81–94.

Cookson, L. J. 1991. Australasian species of Limnoriidae (Crustacea: Isopoda). *Memoirs of the Museum of Victoria* 52:137–262.

Cope, E. D. 1875. Check-list of North American batrachia and reptilia. *Bulletin of the U.S. National Museum* 1:1–104.

——. 1889. The batrachia of North America. *Bulletin of the U.S. National Museum* 34:1–525.

Coppel, H. C., and J. W. Mertins. 1977. *Biological Insect Pest Suppression*. New York: Springer-Verlag.

Cordo, H. A., and C. J. DeLoach. 1978. Host specificity of *Sameodes albiguttalis* in Argentina, a biological control agent for waterhyacinth. *Environmental Entomology* 7:322–328.

Cornwell, G. 1970. Exotic waterfowl program. *Florida Wildlife* 24(5):18–22.

Coulson, J. R. 1977. *Biological Control of Alligatorweed, 1959–1972: A Review and Evaluation*. Technical Bulletin 1547. Washington, D.C.: USDA.

Coulson, J. R., and C. J. DeLoach. In press. Epilogue: Accomplishments and current status of ARS research on classical biological control of arthropods and weeds. In J. R. Coulson, P. V. Vail, M. E. Dix, D. A. Nordlund, and W. C. Kauffman (eds.), *110 Years of Biological Control Research and Development in the United States Department of Agriculture*. Washington, D.C.: USDA.

Courtenay, W. R., Jr. 1989. Exotic fishes in the National Park system. In L. K. Thomas (ed.), *Management of Exotic Species in Natural Communities*. Proceedings of the 4th Triennial Conference on Research in the National Parks and Equivalent Reserves, vol. 5. Washington, D.C.: George Wright Society and National Park Service.

——. 1990. Fish conservation and the enigma of introduced species. *Proceedings of the Bureau of Rural Research* (Canberra) 8:11–20.

——. 1993. Species pollution through fish introductions. In B. N. McKnight (ed.), *Biological Pollution: The Control and Impact of Invasive Exotic Species*. Indianapolis: Indiana Academy of Sciences.

——. 1994. Non-indigenous fishes in Florida. In D. C. Schmitz and T. C. Brown (eds.), *An Assessment of Invasive Non-Indigenous Species in Florida's Public Lands*. Technical Report TSS-94-100. Tallahassee: Florida Department of Environmental Protection.

———. 1995. The case for caution with fish introductions. In H. L. Schramm, Jr., and R. G. Piper (eds.), *Uses and Effects of Cultured Fishes in Aquatic Ecosystems*. American Fisheries Society Symposium 15. Bethesda, Md.: American Fisheries Society.

Courtenay, W. R., Jr., and D. A. Hensley. 1979. Range expansion in southern Florida of the introduced spotted tilapia, with comments on biological impress. *Biological Conservation* 6:149–151.

Courtenay, W. R., Jr., and W. W. Miley II. 1975. Range expansion and environmental impress of the introduced walking catfish in the United States. *Environmental Conservation* 2:145–148.

Courtenay, W. R., Jr., and C. R. Robins. 1973. Exotic aquatic organisms in Florida with emphasis on fishes. *Transactions of the American Fisheries Society* 102:1–12.

———. 1975. Exotic organisms: An unsolved, complex problem. *BioScience* 25:306–313.

———. 1989. Fish introductions: Good management, mismanagement, or no management? *Reviews in Aquatic Sciences* 1:159–172.

Courtenay, W. R., Jr., and J. R. Stauffer, Jr. 1990. The introduced fish problem and the aquarium fish industry. *Journal of the World Aquaculture Society* 21:145–159.

Courtenay, W. R., Jr., and J. D. Williams. 1992. Dispersal of exotic species from aquaculture sources, with emphasis on freshwater fishes. In A. Rosenfield and R. Mann (eds.), *Dispersal of Living Organisms into Aquatic Ecosystems*. College Park: Maryland Sea Grant Program.

Courtenay, W. R., Jr., H. F. Sahlman, W. W. Miley II, and D. J. Herrema. 1974. Exotic fishes in fresh and brackish waters of Florida. *Biological Conservation* 6:292–302.

Courtenay, W. R., Jr., D. A. Hensley, J. N. Taylor, and J. A. McCann. 1984. Distribution of exotic fishes in the continental United States. In W. R. Courtenay, Jr., and J. R. Stauffer, Jr. (eds.), *Distribution, Biology, and Management of Exotic Fishes*. Baltimore: Johns Hopkins University Press.

———. 1986. Distribution of exotic fishes in North America. In C. H. Hocutt and E. O. Wiley (eds.), *The Zoogeography of North American Freshwater Fishes*. New York: Wiley.

Courtenay, W. R., Jr., D. P. Jennings, and J. D. Williams. 1991. Appendix 2: Exotic fishes of the United States and Canada. In C. R. Robins et al. (eds.), *A List of Common and Scientific Names of Fishes from the United States and Canada*. 5th ed. Special Publication 20. Bethesda, Md.: American Fisheries Society.

Cox, J., R. Kautz, M. MacLaughlin, and T. Gilbert. 1994. *Closing the Gaps in Florida's Wildlife Habitat Conservation System*. Tallahassee: Florida Game and Fresh Water Fish Commission.

Craig, R. J. 1993. Regeneration of native Mariana Island forest in disturbed habitats. *Micronesica* 26:99–108.

Craighead, F. C., Jr., and R. F. Dasmann. 1974. Exotic big game on public lands. In J. A. Bailey, W. Elder, and R. D. McKinney (eds.), *Readings in Wildlife Conservation*. Bethesda, Md.: Wildlife Society.

Crawley, M. J. 1987. What makes a community invasible? In A. J. Gray, M. J. Crawley, and P. J. Edwards (eds.), *Colonization, Succession, and Stability*. Oxford: Blackwell Scientific.

————. 1989. Chance and timing in biological invasions. In J. A. Drake et al. (eds.), *Biological Invasions: A Global Perspective.* Chichester: Wiley.

Creed, R. P., and S. P. Sheldon. 1991. The potential for biological control of Eurasian watermilfoil (*Myriophyllum spicatum*): Results of Brownington Pond, Vermont, survey and multi-lake state survey. In *Proceedings of the 25th Annual Meeting, Aquatic Plant Control Research Program.* Miscellaneous Paper A-91-3. Vicksburg, Miss.: U.S. Army Engineer Waterways Experiment Station.

————. 1992. Further investigations into the effect of herbivores on Eurasian watermilfoil (*Myriophyllum spicatum*). In *Proceedings of the 26th Annual Meeting, Aquatic Plant Control Research Program.* Miscellaneous Paper A-92-2. Vicksburg, Miss.: U.S. Army Engineer Waterways Experiment Station.

————. 1994. The effect of two herbivorous insect larvae on Eurasian watermilfoil. *Journal of Aquatic Plant Management* 32:21–26.

Cromack, K., Jr., and C. D. Monk. 1975. Litter production, decomposition, and nutrient cycling in a mixed hardwood watershed and a white pine watershed. In F. G. Howell, J. B. Gentry, and M. H. Smith (eds.), *Mineral Cycling in Southeastern Ecosystems.* Washington, D.C.: Energy Research and Development Administration.

Crosby, A. W. 1986. *Ecological Imperialism: The Biological Expansion of Europe, 900–1900.* Cambridge: Cambridge University Press.

Crowder, J. P. 1974. Exotic pest plants of south Florida. South Florida Environmental Project. Ecological Report DI-SFEP-74-23.

Culotta, E. 1994. Vanishing fly-traps. *Audubon* 96(2):16–18.

Cunningham, V. D., and R. D. Dunford. 1970. Recent coyote record from Florida. *Quarterly Journal of the Florida Academy of Sciences* 33:279–280.

Curnutt, J. L. 1989. Breeding bird use of a mature stand of Brazilian pepper. *Florida Field Naturalist* 17(3):53–60.

Czaran, T., and S. Bartha. 1992. Spatiotemporal dynamic models of plant populations and communities. *Trends in Ecology and Evolution* 7:38–42.

Dahlsten, D. L. 1986. Control of invaders. In H. A. Mooney and J. A. Drake (eds.), *Ecology of Biological Invasions of North America and Hawaii.* New York: Springer-Verlag.

Dahm, C. N., K. W. Cummins, H. M. Valett, and R. L. Coleman. 1995. An ecosystem view of the restoration of the Kissimmee River. *Restoration Ecology* 3:225–238.

Dalrymple, G. H. 1988. The herpetofauna of Long Pine Key, Everglades National Park, in relation to vegetation and hydrology. In R. C. Szaro, K. E. Severson, and D. R. Patten (eds.), *Proceedings of the Symposium on Management of Amphibians, Reptiles, and Small Mammals in North America.* General Technical Report RM-166. Washington, D.C.: USDA Forest Service.

————. 1994. Non-indigenous amphibians and reptiles in Florida. In D. C. Schmitz and T. C. Brown (eds.), *An Assessment of Invasive Non-indigenous Species in Florida's Public Lands.* Technical Report TSS-94-100. Tallahassee: Florida Department of Environmental Protection.

Dalrymple, N. K., G. H. Dalrymple, and K. A. Fanning. 1993. Vegetation of restored rock-plowed wetlands in the East Everglades. *Restoration Ecology* 1:220–225.

Daniels, P. 1983. Prowlers on the Mexican border. *National Wildlife* 21:14–17.

D'Antonio, C. M., and P. M. Vitousek. 1992. Biological invasions by exotic grasses, the grass/fire cycle, and global change. *Annual Review of Ecology and Systematics* 23:63–87.

Dasmann, R. F. 1971. *No Further Retreat.* New York: Macmillan.

Daughtrey, M. 1993. Dogwood anthracnose disease: Native fungus or exotic invader? In B. N. McKnight (ed.), *Biological Pollution: The Control and Impact of Invasive Exotic Species.* Indianapolis: Indiana Academy of Sciences.

Davidson, N. A., and N. D. Stone. 1989. Imported fire ants. In D. L. Dahlsten and R. Garcia (eds.), *Eradication of Exotic Pests.* New Haven: Yale University Press.

Davis, D. E., K. Myers, and J. B. Hoy. 1976. Biological control among vertebrates. In C. B. Huffaker and P. S. Messenger (eds.), *Theory and Practice of Biological Control.* New York: Academic Press.

Davis, D. R., and J. E. Peña. 1990. Biology and morphology of the banana moth, *Opogona sacchari* (Bojer) (Lepidoptera: Tineidae), and its introduction into Florida. *Proceedings of the Entomological Society of Washington* 92:593–618.

Davis, F. E., and J. S. Reel. 1983. Water quality management strategy for Lake Okeechobee, Florida. In J. Taggart and L. Moore (eds.), *Lake Restoration, Protection and Management.* Proceedings of the second annual conference of the North American Lake Management Society. EPA 440/5-83-001. Washington, D.C.: U.S. Environmental Protection Agency.

Davis, J. H., Jr. 1967. General map of the natural vegetation of Florida. Circular S-178. Gainesville: Institute of Food and Agricultural Sciences, Agricultural Experiment Station, University of Florida.

Davis, R. 1995. Southern Atlantic coast region. *National Audubon Society Field Notes* 49:34–37.

Dawes, C. J. 1974. *Marine Algae of the West Coast of Florida.* Coral Gables: University of Miami Press.

Day, J. F., E. E. Storrs, L. M. Stark, A. L. Lewis, and S. Williams. 1995. Antibodies to St. Louis encephalitis virus in armadillos from southern Florida. *Journal of Wildlife Diseases* 31:10–14.

Dean, H. A., M. J. Schuster, J. C. Boling, and P. T. Riherd. 1979. Complete biological control of *Antonina graminis* in Texas with *Neodusmetia sangwani* (a classic example). *Bulletin of the Entomological Society of America* 25:262–267.

DeBach, P. 1964. The scope of biological control. In P. DeBach and E. I. Schlinger (eds.), *Biological Control of Insect Pests and Weeds.* New York: Reinhold.

———. 1974. *Biological Control by Natural Enemies.* London: Cambridge University Press.

DeBach, P., and D. Rosen. 1991. *Biological Control by Natural Enemies.* 2nd ed. London: Cambridge University Press.

Degner, R. F., L. W. Rodan, W. K. Mathis, and E. P. J. Gibbs. 1983. The recreational and commercial importance of feral swine in Florida. *Preventive Veterinary Medicine* 1:371–381.

Delfosse, E. S., and J. M. Cullen. 1981. New activities in biological control of weeds in Australia. II. *Echium plantagineum:* curse or salvation? In E. S. Delfosse (ed.),

Proceedings of the 5th International Symposium on Biological Control of Weeds.
Brisbane: Commonwealth Scientific and Industrial Research Organization,
Australia.

DeLoach, C. J. 1976. *Neochetina bruchi,* a biological control agent of waterhyacinth:
Host specificity in Argentina. *Annals of the Entomological Society of America*
60:635–642.

DeLoach, C. J., and H. A. Cordo. 1976. Life cycle and biology of *Neochetina bruchi,* a
weevil attacking waterhyacinth in Argentina. *Annals of the Entomological Society of
America* 60:643–652.

———. 1978. Life history and ecology of the moth, *Sameodes albiguttalis,* a candidate
for biological control of waterhyacinth. *Environmental Entomology* 7:309–321.

Dennill, G. B., and D. Donnelly. 1991. Biological control of *Acacia longifolia* and related weed species (Fabaceae) in South Africa. *Agricultural Ecosystems and Environment* 37:115–135.

Denslow, J. S. 1980. Patterns of plant species diversity during succession under different disturbance regimes. *Oecologia* (Berlin) 46:18–21.

Denslow, J. S., and A. E. Gomez Diaz. 1990. Seed rain to treefall gaps in a neotropical
rain forest. *Canadian Journal of Forest Restoration* 20:642–648.

DePourtales, L. F. 1877. Hints on the origin of the flora and fauna of the Florida Keys.
American Naturalist 11:137–144.

de Villele, X., and M. Verlaque. 1995. Changes and degradation in a *Posidonia oceanica*
bed invaded by the introduced tropical alga *Caulerpa taxifolia* in the north western
Mediterranean. *Botanica Marina* 38:79–87.

de Vos, A. R., R. H. Manville, and R. G. Van Gelder. 1956. Introduced mammals and
their influence on native biota. *Zoologica* 41:163–194.

DeVries, D. M. 1995. *East Everglades Exotic Plant Control Annual Report, 1995.*
Homestead: Everglades National Park.

Deyrup, M. 1995. Ants (Hymenoptera: Formicidae). In J. H. Frank and E. D. McCoy,
Precinctive insect species in Florida. *Florida Entomologist* 78:21–35.

Diamond, J. M. 1984. Island population biology: Possible effects of unrestricted pesticide use on tropical birds. *Nature* 310:452.

Diamond, J. M., and T. J. Case. 1986. Overview: Introductions, extinctions, extermi-
nations, invasions. In J. M. Diamond and T. J. Case (eds.), *Community Ecology.*
New York: Harper & Row.

Dickey, C. 1976. Are hunting preserves for you? *Florida Wildlife* 29(9):14–18.

Dickson, J. D., III, R. O. Woodbury, and T. R. Alexander. 1953. Check list of flora of
Big Pine Key, Florida, and surrounding Keys. *Quarterly Journal of the Florida
Academy of Sciences* 16:181–200.

Directorate General for the Environment. 1994. Second international workshop on
Caulerpa taxifolia. Summary of the results presented. Press release.

Di Stefano, J. F., and R. F. Fisher. 1983. Invasion potential of *Melaleuca quinquenervia*
in southern Florida. *Forest Ecology and Management* 7:133–141.

Dixon, J. R. 1987. *Amphibians and Reptiles of Texas.* College Station: Texas A&M University Press.

Dixon, P. S., and L. M. Irvine. 1977. *Seaweeds of the British Isles*. Pt. I: *Introduction, Nemaliales, Gigartinales*. London: British Museum (Natural History).

Dodd, C. K., Jr., and R. A. Seigel. 1991. Relocation, repatriation, and translocation of amphibians and reptiles: Are they conservation strategies that work? *Herpetologica* 47:336–350.

Dolbeer, R. A. 1988. Current status and potential of lethal means of reducing bird damage in agriculture. In *Acta XIX Congressus Internationalis Ornithologici*, vol. 1. National Museum of Natural Sciences. Ottawa: University of Ottawa Press.

Doren, R. F., and D. T. Jones. 1994. Non-native species management in Everglades National Park. In D. C. Schmitz and T. C. Brown (eds.), *An Assessment of Invasive Non-indigenous Species in Florida's Public Lands*. Technical Report TSS-94-100. Tallahassee: Florida Department of Environmental Protection.

Doren, R. F., and G. Rochefort. 1983. *Exotic Plant Control Handbook*. Homestead: Everglades National Park.

Doren, R. F., and L. D. Whiteaker. 1990a. Comparison of economic feasibility of chemical control strategies on differing age and density classes of *Schinus terebinthifolius*. *Natural Areas Journal* 10:28–34.

———. 1990b. Effects of fire on different size individuals of *Schinus terebinthifolius*. *Natural Areas Journal* 10:107–113.

Doren, R. F., L. D. Whiteaker, G. Molnar, and D. Sylvia. 1990. Restoration of former wetlands within the Hole-in-the-Donut in Everglades National Park. In F. J. Webb, Jr. (ed.), *Proceedings of the 7th Annual Conference on Wetlands Restoration and Creation*. Tampa: Hillsborough Community College, Institute of Florida Studies.

Doren, R. F., L. D. Whiteaker, and A. M. LaRosa. 1991a. Evaluation of fire as a management tool for controlling *Schinus terebinthifolius* as secondary growth on abandoned agricultural land. *Environmental Management* 15:121–129.

Doren, R. F., L. D. Whiteaker, and R. Rochefort. 1991b. Seasonal effects on herbicide basal bark treatment of *Schinus terebinthifolius*. In T. D. Center et al. (eds.), *Proceedings of the Symposium on Exotic Pest Plants*. NPS/NREVER/NRTR-91/06 Technical Report. Denver: U.S. Department of the Interior, National Park Service.

Douglass, J. F., and C. E. Winegarner. 1977. Predation of eggs and young of the gopher tortoise, *Gopherus polyphemus* (Reptilia, Testudines, Testudinidae), in southern Florida. *Journal of Herpetology* 11:236–238.

Doutt, R. L. 1964. The historical development of biological control. In P. DeBach and E. I. Schlinger (eds.), *Biological Control of Insect Pests and Weeds*. New York: Reinhold.

Dowell, R. V., G. E. Fitzpatrick, and J. A. Reinert. 1979. Biological control of citrus blackfly in southern Florida. *Environmental Entomology* 8:595–597.

Drake, J. A., H. A. Mooney, F. di Castri, R. H. Groves, F. J. Kruger, M. Rejmánek, and M. Williamson (eds.). 1989. *Biological Invasions: A Global Perspective*. Chichester: Wiley.

Dray, F. A., Jr., and T. D. Center. 1992. Biological control of *Pistia stratiotes* L. (waterlettuce) using *Neohydronomus affinis* Hustache (Coleoptera: Curculionidae). Technical Report A-92-1. Vicksburg, Miss.: U.S. Army Engineer Waterways Experiment Station.

Dray, F. A., Jr., T. D. Center, D. H. Habeck, C. R. Thompson, A. F. Cofrancesco, and J. K. Balciunas. 1990. Release and establishment in the southeastern U.S. of *Neohydronomus affinis* (Coleoptera: Curculionidae), an herbivore of waterlettuce (*Pistia stratiotes*). *Environmental Entomology* 19:799–803.

Dray, F. A., T. D. Center, and D. H. Habeck. 1993. Phytophagous insects associated with *Pistia stratiotes* in Florida. *Environmental Entomology* 22:1146–1155.

Duellman, W. E., and A. Schwartz. 1958. Amphibians and reptiles of southern Florida. *Bulletin of the Florida State Museum* 3:181–324.

Duever, M. J., J. E. Carlson, J. F. Meeder, L. C. Duever, L. H. Gunderson, L. A. Riopelle, T. R. Alexander, R. F. Myers, and D. P. Spangler. 1979. *Resource Inventory and Analysis of the Big Cypress National Preserve.* Gainesville: Center for Wetlands, University of Florida, and Ecological Research Unit, National Audubon Society.

DuMond, F. V. 1967. Semi-free-ranging colonies of monkeys at Goulds Monkey Jungle. In C. Jarvis (ed.), *International Zoo Yearbook* 7. London: Zoological Society of London.

———. 1968. The squirrel monkey in a seminatural environment. In L. A. Rosenblum and R. W. Cooper (eds.), *The Squirrel Monkey.* New York: Academic Press.

Dunkle, S. W. 1995. Dragonflies and damselflies (Odonata). In J. H. Frank and E. D. McCoy, Precinctive insect species in Florida. *Florida Entomologist* 78:21–35.

Duquesnel, J. G. 1994. Supplemental report on exotic removals for District 5, March 5, 1993. Key Largo: Parks District 5, Florida Park Service.

Durrett, R. 1988. *Lecture Notes on Particle Systems and Percolation.* Pacific Grove, Calif.: Wadsworth.

Dye, R. L., II. 1975. Zoogeography and economic impact of the nutria (*Myocastor coypus*) in the United States. M.S. thesis, University of Florida, Gainesville.

Ebenhard, T. 1988. Introduced birds and mammals and their ecological effects. *Swedish Wildlife Research* (Viltrevy) 13(4):1–107.

Edscorn, J. B. 1976. Avian highlights of the year (August 1975–July 1976). In *Lake Region Naturalist 1976–1977 Yearbook.* Lakeland, Fla.: Lake Region Audubon Society.

Edwards, G. B. 1992. Letter from assistant director, Fisheries Division, Fish and Wildlife Service, U.S. Department of the Interior, Washington, D.C., to Phyllis Windle, U.S. Congress, Office of Technology Assessment, March 16, 1992.

Ehler, L. E., and L. A. Andres. 1983. Biological control: Exotic natural enemies to control exotic pests. In C. L. Wilson and C. L. Graham (eds.), *Exotic Plant Pests and North American Agriculture.* New York: Academic Press.

Ehrlich, P. R. 1986. Which animal will invade? In H. A. Mooney and J. A. Drake (eds.), *Ecology of Biological Invasions of North America and Hawaii.* New York: Springer-Verlag.

Eldredge, L. G., and S. E. Miller. 1995. How many species are there in Hawaii? *Bishop Museum Occasional Papers* 41:3–18.

Elliot, C. 1967. Welcome to the royal duck. *Outdoor Life* 140(3):44–47.

El-Sayed, S. Z., W. M. Sackett, L. M. Jeffrey, A. D. Fredericks, R. P. Saunders, P. S. Conger, G. A. Fryxell, K. A. Steidinger, and S. A. Earle. 1972. Chemistry, primary productivity and benthic algae of the Gulf of Mexico. Folio 22 in V. C. Bushnell (ed.), *Serial Atlas of the Marine Environment.* New York: American Geographical Society.

Elton, C. 1958. *The Ecology of Invasions by Animals and Plants.* London: Methuen.

Embree, D. G. 1966. The role of introduced parasites in the control of winter moth in Nova Scotia. *Canadian Entomology* 98:1159–1168.

Emerson, S. 1994. Wildlife vs. growers: Who's winning the battle? *Citrus & Vegetable Magazine* 58:13, 16–17.

Emmel, T. C. 1995. Butterflies (Lepidoptera: Papilionoidea, Hesperioidea). In J. H. Frank and E. D. McCoy, Precinctive insect species in Florida. *Florida Entomologist* 78:21–35.

Espinoza, J. 1990. The southern limit of *Sargassum muticum* (Yendo) Fensholt (Phaeophyta, Fucales) in the Mexican Pacific. *Botanica Marina* 33:193–196.

Estevez, E. D. 1994. Inhabitation of tidal salt marshes by the estuarine wood-boring isopod *Sphaeroma terebrans* in Florida. In M.-F. Thompson, R. Nagabhushanam, R. Sarojini, and M. Fingerman (eds.), *Recent Developments in Biofouling Control.* Rotterdam: Balkema.

Estevez, E. D., and J. L. Simon. 1975. Systematics and ecology of *Sphaeroma* (Crustacea: Isopoda) in the mangrove habitats of Florida. *Proceedings of the International Symposium on Biological Management of Mangroves* 1:286–304.

Evans, J. 1970. About nutria and their control. U.S. Department of the Interior, Bureau of Sport Fisheries and Wildlife, Resource Publication 86:1–65.

Everest, J. W., J. H. Miller, D. M. Ball, and M. G. Patterson. 1991. *Kudzu in Alabama.* Circular ANR-65. Auburn University: Alabama Cooperative Extension Service.

Everglades National Park. 1982. *Everglades National Park Resource Management Plan and Environmental Assessment.* Washington, D.C.: National Park Service.

Ewald, P. W. 1983. Host–parasite relations, vectors, and the evolution of disease severity. *Annual Review of Ecology and Systematics* 14:465–485.

Ewel, J. J. 1986. Invasibility: Lessons from south Florida. In H. A. Mooney and J. A. Drake (eds.), *Ecology of Biological Invasions of North America and Hawaii.* New York: Springer-Verlag.

Ewel, J. J., R. Meador, R. Myers, L. Conde, and B. Sedlik. 1976. *Studies of Vegetation Changes in South Florida.* Report to U.S. Forest Service on Research Agreement 18-492, University of Florida, Gainesville.

Ewel, J. J., D. S. Ojima, D. A. Karl, and W. F. DeBusk. 1982. Schinus *in Successional Ecosystems of Everglades National Park.* South Florida Research Center Report T-676. Homestead: National Park Service.

Ewel, K. C., H. T. Davis, and J. E. Smith. 1989. Recovery of Florida cypress swamps from clearcutting. *Southern Journal of Applied Forestry* 13:123–126.

Exotic Pest Plant Council. 1991. Skunk vine on the mid-west coast of Florida. *Florida Exotic Pest Plant Council Newsletter* 1(4):3.

———. 1992. Hurricane Andrew one, Dade County zero. *Florida Exotic Pest Plant Council Newsletter* 2(3):2–3.

———. 1993. Status of Burma reed in Dade County pine rocklands. *Florida Exotic Pest Plant Council Newsletter* 3(1):3.

————. 1995. Florida Exotic Pest Plant Council's 1995 list of Florida's most invasive species. *Florida Exotic Pest Plant Council Newsletter* 5(1):5.

Fairchild, D. 1938. *The World Was My Garden.* New York: Scribner's.

Feder, J. L. 1995. The effects of parasitoids on sympatric host races of *Rhagoletis pomonella* (Diptera: Tephritidae). *Ecology* 76:801–813.

Fernandez, C., L. M. Gutierrez, and J. M. Rico. 1990. Ecology of *Sargassum muticum* on the north coast of Spain: Preliminary observations. *Botanica Marina* 33:423–428.

Fernandez, D. S., and N. Fetcher. 1991. Changes in light availability following Hurricane Hugo in a subtropical montane forest in Puerto Rico. *Biotropica* 23:393–399.

Fichter, E., and A. D. Linder. 1964. *The Amphibians of Idaho.* Pocatello: Idaho State University Museum.

Fitch, H. S., P. Goodrum, and C. Newman. 1952. The armadillo in the southeastern United States. *Journal of Mammalogy* 33:21–37.

Fitzpatrick, G. E., and N. S. Carter. 1984. Relative competition and weed potential of *Leucaena leucocephala. Proceedings of the Florida State Horticultural Society* 97:240–241.

Fix, J. 1966. Perky's low batting average. *Florida Wildlife* 20:32–33, 44.

Fletcher, R. L., G. Blunden, B. E. Smith, D. J. Rogers, and B. C. Fish. 1989. Occurrence of a fouling, juvenile stage of *Codium fragile* ssp. *tomentosoides* (Goor) Silva (Chlorophyceae, Codiales). *Journal of Applied Phycology* 1:227–237.

Floch, J. Y., R. Pajot, and I. Wallentinus. 1991. The Japanese brown alga *Undaria pinnatifida* on the coast of France and its possible establishment in European waters. *Journal du Conseil* 47:379–390.

Florida Department of Agriculture and Consumer Services. 1993. *Florida Aquatic Plant Locator.* 2nd ed. Tallahassee: Bureau of Seafood and Aquaculture.

Florida Department of Environmental Protection (FDEP). 1995. Status of the Aquatic Plant Maintenance Program in Florida Public Waters. Annual report, fiscal year 1993–1994, to the governor. Tallahassee: Florida Department of Environmental Protection, Bureau of Aquatic Plant Management.

Florida Exotic Pest Plant Council. 1995. List of Florida's most invasive species. *Florida EPPC Newsletter* 5(1).

Florida Game and Fresh Water Fish Commission (FGFC). 1992. Memorandum. Free-roaming macaques. Specific nuisance and safety hazards: 1976–1984. Tallahassee: Division of Law Enforcement.

————. 1994a. *Agency Strategic Plan.* Tallahassee.

————. 1994b. *Everglades Regional Fishing Guide.* Tallahassee.

————. 1995. *1993–94 Annual Report.* Tallahassee.

Florida Natural Areas Inventory. 1990. *Guide to the Natural Communities of Florida.* Tallahassee: Florida Natural Areas Inventory.

Florida Park Service. 1994a. *Florida Park Service Operations Procedures Manual.* Tallahassee: Florida Department of Environmental Protection, Division of Recreation and Parks.

————. 1994b. *Florida Park Service 1994 Resource Management Annual Report.*

Tallahassee: Florida Division of Recreation and Parks, Bureau of Natural and Cultural Resources.

Flowers, J. D. 1991. Subtropical fire suppression in *Melaleuca quinquenervia*. In T. D. Center et al. (eds.), *Proceedings of the Symposium on Exotic Pest Plants, November 1988*. Technical Report NPS/NREVER/NRTR–91/06. Denver: U.S. Department of the Interior, National Park Service.

Floyd, J. 1971. Florida takes new look at waterfowl. *Outdoor Life* 147(6):49.

Flynn, L. B., S. M. Shea, J. C. Lewis, and R. L. Marchinton. 1990. Part 3. Population statistics, health, and habitat use. In *Ecology of Sambar Deer on St. Vincent National Wildlife Refuge, Florida*. Bulletin of the Tall Timbers Research Station 25.

Forcella, F., J. T. Wood, and S. P. Dillon. 1986. Characteristics distinguishing invasive weeds within *Echium* (bugloss). *Weed Research* 26:351–364.

Forrester, D. J. 1991. *Parasites and Diseases of Wild Mammals in Florida*. Gainesville: University Presses of Florida.

Forrester, D. J., J. H. Porter, R. C. Belden, and W. B. Frankenberger. 1981. Lungworms of feral swine in Florida. *Journal of American Veterinary Medical Association* 181:1278–1280.

Fowler, H. W. 1915. Cold-blooded vertebrates from Florida, the West Indies, Costa Rica, and eastern Brazil. *Proceedings of the Academy of Natural Sciences of Philadelphia* 67:244–269.

Francis, M. A., E. R. Suh, and B. S. Dudock. 1989. The nucleotide sequence and characterization of four chloroplast tRNAs from the alga *Codium fragile*. *Journal of Biological Chemistry* 264:243–249.

Frank, J. H. 1994. Inoculative biological control of mole crickets. In A. R. Leslie (ed.), *Handbook of Integrated Pest Management for Turf and Ornamentals*. Boca Raton: Lewis Publishers.

———. 1995. Rove beetles (Coleoptera: Staphylinidae, *sensu stricto*). In J. H. Frank and E. D. McCoy, Precinctive insect species in Florida. *Florida Entomologist* 78:21–35.

Frank, J. H., and F. D. Bennett. 1990. Conclusion on classical biological control. In D. H. Habeck, F. D. Bennett, and J. H. Frank (eds.), *Classical Biological Control in the Southern United States. Southern Cooperative Series Bulletin* 355:i–viii, 1–197.

Frank, J. H., and E. D. McCoy. 1990. Endemics and epidemics of shibboleths and other things causing chaos. *Florida Entomologist* 73:1–9.

———. 1991. Medieval insect behavioral ecology, and chaos. *Florida Entomologist* 74:1–9.

———. 1992. The immigration of insects to Florida, with a tabulation of records published since 1970. *Florida Entomologist* 75:1–28.

———. 1993. The introduction of insects into Florida. *Florida Entomologist* 76:1–53.

———. 1994. Commercial importation into Florida of invertebrate animals as biological control agents. *Florida Entomologist* 77:1–20.

———. 1995a. Invasive adventive insects and other organisms in Florida. *Florida Entomologist* 78:1–15.

———. 1995b. Precinctive insect species in Florida. *Florida Entomologist* 78:21–35.

Frank, J. H., and M. C. Thomas. 1994. *Metamasius callizona* (Chevrolat) (Coleoptera:

Curculionidae), an immigrant pest, destroys bromeliads in Florida. *Canadian Entomologist* 128:673–682.

Frank, J. H., J. P. Parkman, and F. D. Bennett. 1995. *Larra bicolor* (Hymenoptera: Sphecidae), a biological control agent of *Scapteriscus* mole crickets (Orthoptera: Gryllotalpidae), established in northern Florida. *Florida Entomologist* 78:619–623.

Frank, P. A. 1992. Conservation and ecology of the Anastasia Island beach mouse. *Endangered Species Update* 9:9.

Frank, P. A., and S. Humphrey. 1992. *Anastasia Beach Mouse Study.* Nongame Technical Report 12. Tallahassee: Florida Game and Fresh Water Fish Commission.

Fritts, T. H. 1994. Does the brown tree snake pose a threat to Florida and its tourist industry? In D. C. Schmitz and T. C. Brown (eds.), *An Assessment of Invasive Non-Indigenous Species in Florida's Public Lands.* Technical Report TSS-94-100. Tallahassee: Florida Department of Environmental Protection.

———. 1995. Letter from Chief, Biological Survey Project, National Biological Service, National Museum of Natural History, Washington D.C., to Dr. Daniel Simberloff, Florida State University, Department of Biological Science, Tallahassee, dated 15 June.

Frodge, J. D., G. L. Thomas, and J. B. Pauley. 1990. Effects of canopy formation by floating and submersed aquatic macrophytes on water quality of two shallow Pacific Northwest lakes. *Aquatic Botany* 38:231–248.

Fuller, S. L. H. 1974. Clams and mussels (Mollusca: Bivalvia). In C. W. Hart, Jr., and S. L. H. Fuller (eds.), *Pollution Ecology of Freshwater Invertebrates.* New York: Academic Press.

Funasaki, G. Y., P. Y. Lai, L. M. Nakahara, J. W. Beardsley, and A. K. Ota. 1988. A review of biological control introductions in Hawaii: 1890 to 1985. *Proceedings of the Hawaiian Entomological Society* 28:105–160.

Futch, C. R., and S. A. Willis. 1992. Characteristics of the procedures for marine species introductions in Florida. In R. DeVoe (ed.), *Introductions and Transfers of Marine Species: Achieving a Balance between Economic Development and Resource Protection.* Proceedings of the Conference and Workshop, 30 October–2 November, 1991, Hilton Head Island, South Carolina. Charleston: South Carolina Sea Grant Consortium.

Gann, G., and D. Gordon. 1994. *Paederia foetida* (skunk vine) and *P. cruddasiana* (sewer vine): Threats and management strategies. Element Stewardship Abstract. Arlington, Va.: The Nature Conservancy.

Garman, S. 1887. On West Indian reptiles. Iguanidae. *Bulletin of the Essex Institute* 19:1–26.

Gaskalla, R. 1993. Memorandum from the director, Division of Plant Industry, Florida Department of Agriculture and Consumer Services, Gainesville, to Dr. Martha Roberts, deputy commissioner for food safety, Florida Department of Agriculture and Consumer Services, dated April 8.

Gayol, P., C. Falconetti, J. R. M. Chisholm, and J. M. Jaubert. 1995. Metabolic responses of low-temperature-acclimated *Caulerpa taxifolia* (Chlorophyta) to rapidly-elevated temperature. *Botanica Marina* 38:61–67.

Geldenhuys, C. J., P. J. le Roux, and K. H. Cooper. 1986. Alien invasions in indigenous evergreen forest. In I. A. W. Macdonald, F. J. Kruger, and A. A. Ferrar (eds.), *The Ecology and Management of Biological Invasions in Southern Africa.* Proceedings of the National Synthesis Symposium on the Ecology of Biological Invasions. Cape Town: Oxford University Press.

Geller, J. B., J. T. Carlton, and D. A. Powers. 1994. PCR-based detection of mtDNA haplotypes of native and invading mussels on the northeastern Pacific coast: Latitudinal pattern of invasion. *Marine Biology* 119:243–249.

Gipson, P. S., I. K. Gipson, and J. A. Sealander. 1975. Reproductive biology of wild *Canis* (Canidae) in Arkansas. *Journal of Mammalogy* 56:605–612.

Girardin, D. 1994. Non-indigenous species on St. Johns River Water Management District lands. In D. C. Schmitz and T. C. Brown (eds.), *An Assessment of Invasive Non-indigenous Species in Florida's Public Lands.* Technical Report TSS-94-100. Tallahassee: Florida Department of Environmental Protection.

Givenaud, T., J. Cosson, A. Givernaud-Mouradi, M. Elliott, and J.-P. Ducrotoy. 1989. Study of populations of *Sargassum muticum* (Yendo) Fensholt on the coasts of the lower Normandie (France). In Estuarine and Coastal Sciences Association (ed.), *Estuaries and Coasts: Spatial and Temporal Intercomparisons.*

Givenaud, T., J. Cosson, and A. Givernaud-Mouradi. 1990. Regeneration of the brown seaweed *Sargassum muticum* (Phaeophyceae, Fucales). *Cryptogamie, Algologie* 11:293–304.

Glass, C. M., R. G. McLean, J. B. Katz, D. S. Maehr, C. B. Cropp, L. J. Kirk, A. J. McKeirnan, and J. F. Evermann. 1994. Isolation of pseudorabies (Aujeszky's disease) virus from a Florida panther. *Journal of Wildlife Diseases* 30:180–184.

Godfrey, R. K., and J. W. Wooten. 1981. *Aquatic and Wetland Plants of Southeastern United States: Dicotyledons.* Athens: University of Georgia Press.

Goeden, R. D., and L. T. Kok. 1986. Comments on a proposed "new" approach for selecting agents for the biological control of weeds. *Canadian Entomologist* 118:51–58.

Goff, L. J., L. Liddle, P. C. Silva, M. Voytek, and A. W. Coleman. 1992. Tracing species invasion in *Codium,* a siphonous green alga, using molecular tools. *American Journal of Botany* 79:1279–1285.

Goodyear, N. C. 1992. Spatial overlap and dietary selection of native rice rats and exotic black rats. *Journal of Mammalogy* 73:186–200.

Gopal, B. 1987. *Water Hyacinth.* Amsterdam: Elsevier.

Gore, J., and T. Schaefer. 1993. *Santa Rosa Beach Mouse Survey.* Nongame Wildlife Program Final Performance Report. Tallahassee: Florida Game and Fresh Water Fish Commission.

Gosling, M. 1989. Extinction to order. *New Scientist* 121:44–49.

Goyer, R. A., and J. D. Stark. 1981. Suppressing waterhyacinth with an imported weevil. *Ornamental South* 3(6):21–22.

———. 1984. The impact of *Neochetina eichhorniae* on waterhyacinth in southern Louisiana. *Journal of Aquatic Plant Management* 22:57–61.

Greathead, D. J. 1986. Parasitoids in classical biological control. In J. Waage and D. Greathead (eds.), *Insect Parasitoids: 13th Symposium of the Royal Entomological Society of London.* New York: Academic Press.

REFERENCES 393

Greenway, J. C. 1967. *Extinct and Vanishing Birds of the World.* New York: Dover.

Griffo, J. V., Jr. 1957. The status of the nutria in Florida. *Quarterly Journal of the Florida Academy of Sciences* 20:209–215.

Grodowitz, M. J. 1991. Biological control of waterlettuce using insects: Past, present, and future. In *Proceedings of the 25th Annual Meeting, Aquatic Plant Control Research Program.* Miscellaneous Paper A-91-3. Vicksburg, Miss.: U.S. Army Engineer Waterways Experiment Station.

Grodowitz, M. J., R. M. Stewart, and A. F. Cofrancesco. 1991. Population dynamics of waterhyacinth and the biological control agent *Neochetina eichhorniae* (Coleoptera: Curculionidae) at a southeast Texas location. *Environmental Entomology* 20:652–660.

Grodowitz, M. J., W. Johnson, and L. D. Nelson. 1992. Status of biological control of waterlettuce in Louisiana and Texas using insects. Miscellaneous Paper A-92-3. Vicksburg, Miss.: U.S. Army Engineer Waterways Experiment Station.

Grodowitz, M. J., T. Center, E. Snoddy, and E. Rives. 1993. Status of the release and establishment of insect biological control agents of hydrilla. In *Proceedings of the 27th Annual Meeting, Aquatic Plant Control Research Program.* Miscellaneous Paper A-93-2. Vicksburg, Miss.: U.S. Army Engineer Waterways Experiment Station.

Grow, G. 1984. New threats to the Florida panther. *ENFO Newsletter* (October):1–2, 4–8, 10–12.

Grue, C. E. 1985. Pesticides and the decline of Guam's native birds. *Nature* 316:301.

Gunner, H. B. 1983. Biological control technology development: Microbiological control of Eurasian watermilfoil. In *Proceedings of the 17th Annual Meeting, Aquatic Plant Control Research Program.* Miscellaneous Paper A-83-3. Vicksburg, Miss.: U.S. Army Engineer Waterways Experiment Station.

Habeck, D. H. 1976. The case for biological control of lantana in Florida citrus groves. *Proceedings of the Florida State Horticultural Society* 89:17–18.

Habeck, D. H., and F. D. Bennett. 1990. *Cactoblastis cactorum* Berg (Lepidoptera: Pyralidae), a phycitine new to Florida. Entomology Circular 333. Tallahassee: Florida Department of Agriculture and Consumer Services, Division of Plant Industry.

Habeck, D. H., F. D. Bennett, and E. E. Grissell. 1989. First record of a phytophagous seed chalcid from Brazilian peppertree in Florida. *Florida Entomologist* 72:378–379.

Habeck, M. H., S. B. Lovejoy, and J. G. Lee. 1993. When does investing in classical biological control make economic sense? *Florida Entomologist* 76:96–101.

Habeck, D. H., F. D. Bennett, and J. K. Balciunas. 1994. Biological control of terrestrial and wetland weeds. In D. Rosen, F. D. Bennett, and J. L. Capinera (eds.), *Pest Management in the Subtropics: Biological Control—a Florida Perspective.* Andover, U.K.: Intercept.

Hadfield, M. G., and B. S. Mountain. 1980. A field study of a vanishing species, *Achatinella mustelina* (Gastropoda; Pulmonata), in the Waianae Mountains of Oahu. *Pacific Science* 34:345–358.

Hagen, K. S., and J. M. Franz. 1973. A history of biological control. In R. F. Smith, T. E. Mittler, and C. N. Smith (eds.), *History of Entomology.* Palo Alto: Annual Reviews.

Hale, M. M., J. E. Crumpton, and R. J. Schuler, Jr. 1995. In H. L. Schramm, Jr., and R. G. Piper (eds.), *Uses and Effects of Cultured Fishes in Aquatic Ecosystems.* American Fisheries Society Symposium 15. Bethesda, Md.: American Fisheries Society.

Hall, D. G. 1991. Sugarcane lace bug *Leptodictya tabida,* an insect pest new to Florida. *Florida Entomologist* 74:148–149.

———. 1992. Japanese climbing fern. *Resource Management Notes* 4(4):43.

Hall, R. W., L. E. Ehler, and B. Bisrabi-Ershadi. 1980. Rate of success in classical biological control of arthropods. *Bulletin of the Entomological Society of America* 26:111–114.

Hall, T. 1994. Progress report—brown tree snake control for the government of Guam and the Department of Defense. Hyattsville, Md.: USDA/APHIS–ADC.

Hallegraeff, G. M., and C. J. Bolch. 1991. Transport of toxic dinoflagellate cysts via ships' ballast water. *Marine Pollution Bulletin* 22:27–30.

———. 1992. Transport of diatom and dinoflagellate resting spores in ships' ballast water: Implications for plankton biogeography and aquaculture. *Journal of Plankton Research* 14:1067–1084.

Haller, W. T. 1978. *Hydrilla: A New and Rapidly Spreading Aquatic Weed Problem.* Circular S–245. Gainesville: Institute of Food and Agricultural Services, University of Florida.

Haller, W. T., and D. L. Sutton. 1975. Community structure and competition between hydrilla and vallisneria. *Hyacinth Control Journal* 13:48–50.

Hamer, J. L. (ed.). 1985. Southeastern Branch, Insect Detection, Evaluation and Prediction Report, 1984. Entomological Society of America, Vol. 9.

Hammer, R. L. 1995a. Air-potato times three. *Resource Management Notes* 7(3):5.

———. 1995b. Kudzu: Further south than you think. *Resource Management Notes* 7(3):7–8.

Hamon, A. B. 1995. Soft scales (Homoptera: Coccidae). In J. H. Frank and E. D. McCoy, Precinctive insect species in Florida. *Florida Entomologist* 78:21–35.

Hansen, K. L., E. G. Ruby, and R. L. Thompson. 1971. Trophic relationships in the water hyacinth community. *Quarterly Journal of the Florida Academy of Sciences* 34:107–113.

Hanson, R. P., and L. Karstad. 1959. Feral swine in the southeastern United States. *Journal of Wildlife Management* 23:64–75.

Hardin, D. E. 1994. Non-indigenous species on forestry-managed lands. In D. C. Schmitz and T. C. Brown (eds.), *An Assessment of Invasive Non-Indigenous Species in Florida's Public Lands.* Technical Report TSS-94-100. Tallahassee: Florida Department of Environmental Protection.

Hardt, R. A. 1986. Japanese honeysuckle: From "one of the best" to ruthless pest. *Arnoldia* (Boston) 46(2):27–34.

Harley, K.L.S. 1990. The role of biological control in the management of waterhyacinth, *Eichhornia crassipes. Biocontrol News and Information* 11(1):11–22.

Harris, P. 1973. The selection of effective agents for the biological control of weeds. *Canadian Entomologist* 105:1495–1503.

———. 1979. Costs of biological control of weeds by insects in Canada. *Weed Science* 27:242–250.

————. 1988. Environmental impacts of weed control insects. *BioScience* 38:542–548.

————. 1989. Practical considerations in a biological control of weeds program. In *International Symposium on Biological Control Implementation Proceedings and Abstracts.* Bulletin 6. McAllen, Texas: North American Plant Protection Organization.

————. 1991. Classical biological control of weeds: Its definition, selection of effective agents, and administrative-political problems. *Canadian Entomologist* 123:827–849.

————. 1993. Effects, constraints and the future of weed biological control. *Agricultural Ecosystems and Environment* 46:289–303.

Harris, P., and M. Maw. 1984. *Hypericum perforatum* L., St. Johns wort (Hypericaceae). In J. S. Kelleher and M. A. Hume (eds.), *Biological Programmes against Insects and Weeds in Canada 1969–1980.* Farnham Royal, U.K.: Commonwealth Agricultural Bureaux.

Harris, S. C., T. H. Martin, and K. W. Cummins. 1995. A model for aquatic invertebrate response to Kissimmee River restoration. *Restoration Ecology* 3:181–194.

Harrison, K., and D. M. Holdich. 1984. Hemibranchiate sphaeromatids (Crustacea: Isopoda) from Queensland, Australia, with a world-wide review of the genera discussed. *Zoological Journal of the Linnean Society* 81:275–387.

Hastings, A. 1995a. Models of spatial spread: A synthesis. *Biological Conservation,* 6/78:143–148.

————. 1995b. Models of spatial spread: Is the theory complete? *Ecology,* 6/77:1675:1679.

Hay, C. H. 1990. The dispersal of sporophytes of *Undaria pinnatifida* by coastal shipping in New Zealand, and implications for further dispersal of *Undaria* in France. *British Phycological Journal* 25:301–313.

Heard, W. H. 1964. *Corbicula fluminea* in Florida. *Nautilus* 77(3):105–107.

————. 1966. Further records of *Corbicula fluminea* (Müller) in the southern United States. *Nautilus* 79(4):142–143.

————. 1979. Identification manual of the freshwater clams of Florida. *Florida Department of Environmental Regulation Technical Series* 4(2):1–83.

Hebert, P.D.N., B. W. Muncaster, and G. L. Mackie. 1989. Ecological and genetic studies on *Dreissena polymorpha* (Pallas): A new mollusc in the Great Lakes. *Canadian Journal of Fisheries and Aquatic Sciences* 46:1857–1951.

Heinsohn, G. E. 1955. Life history and ecology of the freshwater clam, *Corbicula fluminea.* M.S. thesis, University of California, Santa Barbara.

Hengeveld, R. 1989. *Dynamics of Biological Invasions.* London: Chapman & Hall.

————. 1992. Potential and limitations of predicting invasion rates. *Florida Entomologist* 75:60–72.

Henry, D. A., and P. A. McLaughlin. 1975. The barnacles of the *Balanus amphitrite* complex (Cirripedia, Thoracica). *Zoologische Verhandlingen* (Leiden) 141.

Hersh, S. L. 1981. Ecology of the Key Largo wood rat (*Neotoma floridana smalli*). *Journal of Mammalogy* 62:201–206.

Hester, F. E. 1991. The U.S. National Park Service experience with exotic species. *Natural Areas Journal* 11(3):127–128.

Heywood, V. H. 1989. Patterns, extents, and modes of invasions by terrestrial plants. In

J. A. Drake et al. (eds.), *Biological Invasions: A Global Perspective.* Chichester: Wiley.

Hicks, D. W., and J. W. Tunnell. 1993. Invasion of the south Texas coast by the edible brown mussel, *Perna perna* (Linnaeus, 1758). *Veliger* 36:92–94.

Hill, E. P., P. W. Sumner, and J. B. Wooding. 1987. Human influences on range expansion of coyotes in the southeast. *Wildlife Society Bulletin* 15:521–524.

Hinkle, M. K. 1992. Environmental issues of biological control regulation. In R. Charudattan and H. W. Browning (eds.), *Regulations and Guidelines: Critical Issues in Biological Control.* Proceedings of a USDA/CSRS National Workshop, June 1991, Vienna, Virginia. Gainesville: Institute of Food and Agricultural Sciences, University of Florida.

Hoagland, D. B., G. R. Horst, and C. W. Kilpatrick. 1989. Biogeography and population biology of the mongoose in the West Indies. In C. A. Woods and H. H. Genways (eds.), *Biogeography of the West Indies.* Gainesville: Sandhill Crane Press.

Hoagland, K. E. 1986. Genetic variation in seven wood-boring teredinid and pholadid bivalves with different patterns of life history and dispersal. *Malacologia* 27:323–339.

Hoagland, K. E., and R. D. Turner. 1986. Evolution and adaptive radiation of wood-boring bivalves (Pholadacea). *Malacologia* 21:111–148.

Hobbs, R. J. 1989. The nature and effects of disturbance relative to invasions. In J. A. Drake et al. (eds.), *Biological Invasions: A Global Perspective.* Chichester: Wiley.

———. 1991. Disturbance as a precursor to weed invasion in native vegetation. *Plant Protection Quarterly* 6:99–104.

Hobbs, R. J., and L. F. Huenneke. 1992. Disturbance, diversity and invasion: Implications for conservation. *Conservation Biology* 6:324–337.

Hobbs, R. J., and S. E. Humphries. 1995. An integrated approach to the ecology and management of plant invasions. *Conservation Biology* 9:761–770.

Hodges, E. M., and D. W. Jones. 1950. *Torpedo Grass.* Circular S–14. University of Florida Agricultural Experiment Station.

Hoeck, H. N. 1984. Introduced fauna. In R. Perry (ed.), *Key Environments: Galapagos.* Oxford: Pergamon Press.

Hoelmer, K. A., and J. K. Grace. 1989. Citrus blackfly. In D. L. Dahlsten and R. Garcia (eds.), *Eradication of Exotic Pests.* New Haven: Yale University Press.

Hoffman, J. R. 1991. Biological control of weeds in South Africa. *Agricultural Ecosystems and Environment* 37:1–225.

Hoffmeister, D. F. 1975. Bat tower on Sugarloaf Key, southern Florida. *Bat Research News* 16:18.

Hofstetter, R. H. 1975. Effects of fire in the ecosystem. Pt. II, app. K. South Florida Environmental Project. Coral Gables: University of Miami.

———. 1991. The current status of *Melaleuca quinquenervia* in southern Florida. In T. D. Center et al. (eds.), *Proceedings of the Symposium on Exotic Pest Plants, November 1988.* Technical Report NPS/NREVER/NRTR-91/06. Denver: U.S. Department of the Interior, National Park Service.

Hofstetter, R. H., and R. S. Sonenshein. 1990. Vegetative changes in a wetland in the vicinity of a well field, Dade County, Florida. Water Resources Investigations Report 89–4155. U.S. Geological Survey.

Hokkanen, H. 1985. Success in classical biological control. *CRC Critical Reviews in Plant Sciences* 3:35–72.

Hokkanen, H., and D. Pimentel. 1984. New approach for selecting biological control agents. *Canadian Entomologist* 116:1109–1121.

Holdridge, L. R. 1940. Some notes on the mangrove swamps of Puerto Rico. *Caribbean Forester* 1:19–29.

Holler, N. 1992. Perdido Key beach mouse. In S. R. Humphrey (ed.), *Rare and Endangered Biota of Florida.* Vol. 1: *Mammals.* Gainesville: University Presses of Florida.

Honnell, D. R., J. D. Madsen, and R. M. Smart. 1993. Effects of selected exotic and native aquatic plant communities on water temperature and dissolved oxygen. Vol. A-93-2. U.S. Army Corps of Engineers Aquatic Plant Control Research Program.

Horvitz, C. C. 1994. Hammocks and hurricanes: A surprisingly diverse array of nonindigenous plants threaten the natural regeneration of hardwood hammocks after hurricanes. Box 3.7a in D. C. Schmitz and T. C. Brown (eds.), *An Assessment of Invasive Non-Indigenous Species in Florida's Public Lands.* Technical Report TSS-94-100. Tallahassee: Florida Department of Environmental Protection.

Horvitz, C. C., S. McMann, and A. Freedman. 1995. Exotics and hurricane damage in three hardwood hammocks in Dade County Parks, Florida. *Journal of Coastal Research* (Special Hurricane Andrew Issue) 21:145–158.

Horvitz, C. C., J. B. Pascarella, S. McMann, A. Freedman, and R. H. Hofstetter. In press. Regeneration guilds of invasive nonindigenous plants in hurricane-affected subtropical hardwood forests. *Ecological Applications* (Special Issue for the 1994 Ecological Society of America Symposium on Use of Ecological Concepts in Conservation Biology: Lessons from the Southeast).

Howard, R. A., and L. Schockman. 1995. Recovery responses of tropical trees after Hurricane Andrew. *Harvard Papers in Botany* 6:37–74.

Howarth, F. G. 1983. Classical biological control: Panacea or Pandora's box? *Proceedings of the Hawaiian Entomological Society* 24:239–244.

Howell, A. H. 1932. *Florida Bird Life.* New York: Coward-McCann.

Huenneke, L. 1988. SCOPE program on biological invasions: A status report. *Conservation Biology* 2:8–10.

Huffaker, C. B., and P. S. Messenger (eds.). 1976. *Theory and Practice of Biological Control.* New York: Academic Press.

Huffaker, C. B., P. S. Messenger, and P. DeBach. 1971. The natural enemy component in natural control and the theory of biological control. In C. B. Huffaker (ed.), *Biological Control.* New York: Plenum Press.

Huffaker, C. B., F. J. Simmonds, and J. E. Laing. 1976. The theoretical and empirical basis of biological control. In C. B. Huffaker and P. S. Messenger (eds.), *Theory and Practice of Biological Control.* New York: Academic Press.

Huffstodt, J. 1989. Get ready for peacock bass. *Florida Wildlife* 43(4):14–18.

Hughes, C. E., and B. T. Styles. 1987. The benefits and potential risks of woody legume introductions. *Tree Crops Journal* 4:209–248.

Humphrey, S. R. 1974. Zoogeography of the nine-banded armadillo (*Dasypus novem-cinctus*) in the United States. *BioScience* 24:457–462.

Humphrey, S. R., and D. B. Barbour. 1981. Status and habitat of three subspecies of *Peromyscus polionotus* in Florida. *Journal of Mammalogy* 62:840–844.

Hunt, K. W. 1947. The Charleston woody flora. *American Midland Naturalist* 37:670–756.

Hutchison, A. M. 1992. A reproducing population of *Trachemys scripta elegans* in southern Pinellas County, Florida. *Herpetological Review* 23:74–75.

Hutt, A. 1967. New ducks for Florida. *Florida Wildlife* 21(2):14–17.

Huttenmeyer, B. 1994. Butterfly peacock bass: Fishing for the record. *Florida Wildlife* 48(3):32–35.

Hyler, W. R. 1995. Vervet monkeys in the mangrove ecosystems of southeastern Florida: Preliminary census and ecological data. *Florida Scientist* 58:38–43.

Jackson, C. G., Jr., C. M. Holcomb, and M. M. Jackson. 1972. Strontium-90 in the ex-oskeletal ossicles of *Dasypus novemcinctus*. *Journal of Mammalogy* 53:921–922.

James, F. C. 1990. The selling of wild birds: Out of control? *Living Bird* 9:8–15.

———. 1994. An assessment of non-indigenous birds in Florida's public lands. In D. C. Schmitz and T. C. Brown (eds.), *An Assessment of Invasive Non-indigenous Species in Florida's Public Lands*. Technical Report TSS-94-100. Tallahassee: Florida Department of Environmental Protection.

Jamieson, G. S., and R. S. McKinney. 1937. *Stillingia* oil. *Oil and Soap* 15:295–296.

Janzen, D. H. 1986. The eternal external threat. In M. E. Soulé (ed.), *Conservation Biology: The Science of Scarcity and Diversity*. Sunderland, Mass.: Sinauer.

Jennings, W. L. 1958. The ecological distribution of bats in Florida. Ph.D. thesis, University of Florida, Gainesville.

Jennings, W. L., N. J. Schneider, A. L. Lewis, and J. E. Scatterday. 1960. Fox rabies in Florida. *Journal of Wildlife Management* 24:171–179.

Jensen, D. B., and D. J. Vosick. 1994. Introduction. In D. C. Schmitz and T. C. Brown (eds.), *An Assessment of Invasive Non-indigenous Species in Florida's Public Lands*. Technical Report TSS-94-100. Tallahassee: Florida Department of Environmental Protection.

Jensen, J. B. 1994. Geographic distribution: *Phrynosoma cornutum*. *Herpetological Review* 25:165.

Johnson, A. F. 1994. Coastal impacts of non-indigenous species. In D. C. Schmitz and T. C. Brown (eds.), *An Assessment of Invasive Non-indigenous Species in Florida's Public Lands*. Technical Report TSS-94-100. Tallahassee: Florida Department of Environmental Protection.

Johnson, A. F., and M. G. Barbour. 1990. Dunes and maritime forests. In R. L. Myers and J. J. Ewel (eds.), *Ecosystems of Florida*. Orlando: University of Central Florida Press.

Johnson, A. F., and J. W. Muller. 1993. *An Assessment of Florida's Remaining Coastal Upland Natural Communities: Southwest Florida*. Tallahassee: Florida Natural Areas Inventory Report.

Johnson, A. F., J. W. Muller, and K. A. Bettinger. 1993. *An Assessment of Florida's Remaining Coastal Upland Natural Communities: Southeast Florida.* Tallahassee: Florida Natural Areas Inventory Report.

Johnson, D. M., and P. D. Stiling. 1996. Host specificity of *Cactoblastis cactorum* (Lepidoptera: Pyralidae), an exotic *Opuntia*-feeding moth, in Florida. *Environmental Entomology* 25:743–748.

Johnson, F. A. 1987. Lake Okeechobee's waterfowl habitat: Problems and possibilities. *Aquatics* 9(2):20–21.

Johnson, F. A., and F. Montalbano III. 1984. Selection of plant communities by wintering waterfowl on Lake Okeechobee, Florida. *Journal of Wildlife Management* 48:174–178.

Johnston, R. F., and M. Janiga. 1995. *Feral Pigeons.* New York: Oxford University Press.

Jones, R. H., and K. W. McLeod. 1989. Shade tolerance in seedlings of Chinese tallow tree, American sycamore, and cherry bark oak. *Bulletin of the Torrey Botanical Club* 116:371–377.

Jones, S. R., and S. A. Phillips. 1987. Aggressive and defensive propensities of *Solenopsis invicta* (Hymenoptera: Formicidae) and three indigenous ant species in Texas. *Texas Journal of Science* 39:107–115.

Jordan, J. 1990. Memorandum to Greg Jubinsky, regional biologist, Bureau of Aquatic Plants, West Palm Beach, Florida, dated 28 June 1990.

Joyce, J. C. 1985. Benefits of maintenance control of water hyacinth. *Aquatics* 7:11–13.

Jubinsky, G. 1993. *A Review of the Literature:* Sapium sebiferum. Florida Department of Natural Resources Technical Report, TSS 93–03.

Julien, M. (ed.). 1982. *Biological Control of Weeds: A World Catalogue of Agents and Their Target Weeds.* 1st ed. Slough, U.K.: Commonwealth Institute of Biological Control.

———. 1992. *Biological Control of Weeds: A World Catalogue of Agents and Their Target Weeds.* 3rd ed. Slough, U.K.: Commonwealth Institute of Biological Control.

Kapraun, D. F., and R. B. Searles. 1990. Planktonic bloom of an introduced species of *Polysiphonia* (Ceramiales, Rhodophyta) along the coast of North Carolina, U.S.A. *Hydrobiologia* 204/205:269–274.

Kareiva, P., I. M. Parker, and M. Pascual. 1996. Can we use experiments and models in predicting the invasiveness of genetically engineered organisms? *Ecology* 77:1670–1675.

Katahira, L. K., P. Finnegan, and C. P. Stone. 1993. Eradicating feral pigs in montane mesic habitat at Hawaii Volcanoes National Park. *Wildlife Society Bulletin* 21: 269–274.

Kautz, R. S. 1993. Trends in Florida wildlife habitat 1936–1987. *Florida Scientist* 56(1):7–24.

Keller, J. E. 1963. Status of the red junglefowl in the southeastern States. *Proceedings of the Southeastern Association of Fish and Wildlife Agencies* 17:107–108.

Kellert, S. 1979. American attitudes toward and knowledge of animals: An update. *International Journal for the Study of Animal Problems* 1:87–119.

Kellog, F. E., T. H. Eleazer, and T. R. Colvin. 1978. Transmission of blackhead from jungle fowl to turkey. *Proceedings of the Southeastern Association of Fish and Wildlife Agencies* 32:378–379.

Kensley, B., and M. Schotte. 1989. *Guide to the Marine Isopod Crustaceans of the Caribbean.* Washington, D.C.: Smithsonian Institution Press.

Kensley, B., W. G. Nelson, and M. Schotte. 1995. Marine isopod diversity of the Indian River lagoon, Florida. *Bulletin of Marine Science* 57:136–142.

Killinger, G. B., G. E. Ritchey, C. B. Blickensdcrfcr, and W. Jackson. 1951. *Argentine Bahia Grass.* University of Florida Agricultural Experiment Station S-31.

Kimball, F. 1993. Elvis is alive and well in Arcadia! *Heartland Farmer & Rancher,* 2 August.

King, W., and T. Krakauer. 1966. The exotic herpetofauna of southeastern Florida. *Quarterly Journal of the Florida Academy of Sciences* 29:144–154.

Kingsolver, J. M. 1995. Seed beetles (Coleoptera: Bruchidae). In J. H. Frank and E. D. McCoy, Precinctive insect species in Florida. *Florida Entomologist* 78:21–35.

Kluge, A. G., and M. J. Eckardt. 1969. *Hemidactylus garnotii* Dumeril and Bibron, a triploid all-female species of geckonid lizard. *Copeia* 1969:651–664.

Klukas, R. W. 1969. The Australia pine problem in Everglades National Park. Management Report. Homestead, Fla.: Everglades National Park.

Klukas, R. W., and W. G. Truesdell. 1969. The Australian pine problem in Everglades National Park. Pt. 1: The problem and some possible solutions. Unpublished report for the South Florida Research Center, Everglades National Park, Homestead, Fla.

Knott, J. R. 1976. Letter to Mrs. Alfred G. Kay.

Koebel, J. W., Jr. 1995. An historical perspective on the Kissimmee River restoration project. *Restoration Ecology* 3:149–159.

Koehler, P. G., and R. J. Brenner. 1995. Cockroaches (Blattodea). In J. H. Frank and E. D. McCoy, Precinctive insect species in Florida. *Florida Entomologist* 78:21–35.

Koepp, W. P. 1979. The status of *Schinus* manipulation in Everglades National Park. In *Technical Proceedings of Techniques for Control of* Schinus *in South Florida.* Sanibel, Fla.: Sanibel-Captiva Conservation Foundation.

Kravitz, M. J. 1987. First record of *Boccardiella ligerica* (Ferronniere) (Polychaeta: Spionidae) from the east coast of North America. *Northeast Gulf Science* 9:39–42.

Krebs, C. J. 1985. *Ecology: The Experimental Analysis of Distribution and Abundance.* 3rd ed. New York: Harper & Row.

———. 1994. *Ecology: The Experimental Analysis of Distribution and Abundance.* 4th ed. New York: HarperCollins.

Krebs, J. W., M. L. Wilson, and J. E. Childs. 1995. Rabies—epidemiology, prevention, and future research. *Journal of Mammalogy* 76:681–694.

Kretchman, D. W. 1962. *Torpedograss and Citrus Groves.* Circular S-136. University of Florida Agriculture Experiment Station.

Kushlan, J. A. 1990. Freshwater marshes. In R. L. Myers and J. J. Ewel (eds.), *Ecosystems of Florida.* Orlando: University of Central Florida Press.

Lai, P. -Y. 1988. Biological control: A positive point of view. *Proceedings of the Hawaiian Entomological Society* 28:179–190.

Lampton, B. 1994. Pennies for the peacocks—adding these fish to the water has created jobs and enhanced the south Florida lifestyle. *Florida Sportsman* 26:198–205.

Lande, R. 1988. Genetics and demography in biological conservation. *Science* 241:1455–1460.

Langeland, K. A. 1990. Hydrilla (*Hydrilla verticillata* [L.F.] Royle), a continuing problem in Florida waters. Circular 884. Gainesville: Institute of Food and Agricultural Services, University of Florida.

Langford, F. H., F. J. Ware, and R. D. Gasaway. 1978. Status and harvest of introduced *Tilapia aurea* in Florida lakes. In R. O. Smitherman, W. L. Shelton, and J. H. Grover (eds.), *Culture of Exotic Fishes Symposium.* Auburn, Ala.: Fish Culture Section, American Fisheries Society.

Laroche, F. B. 1994. Melaleuca *Management Plan for Florida.* 2nd ed. Recommendations from the Melaleuca Task Force. West Palm Beach, Fla.: Exotic Pest Plant Council.

Laroche, F. B., and A. P. Ferriter. 1992. Estimating expansion rates of melaleuca in south Florida. *Journal of Aquatic Plant Management* 30:62–65.

LaRosa, A. M., R. F. Doren, and L. Gunderrson. 1992. Alien plant management in Everglades National Park: An historical perspective. In C. P. Stone, C. W. Smith, and J. T. Tunison (eds.), *Alien Plant Invasions in Native Ecosystems of Hawaii: Management and Research.* Honolulu: University of Hawaii, Cooperative National Park Studies Unit.

Lassuy, D. R. 1994. Aquatic nuisance organisms: Setting national policy. *Fisheries* (Bethesda) 19(4):14–17.

———. 1995. Introduced species as a factor in extinction and endangerment of native fish species. In H. L. Schramm, Jr., and R. G. Piper (eds.), *Uses and Effects of Cultured Fishes in Aquatic Ecosystems.* American Fisheries Society Symposium 15. Bethesda, Md.: American Fisheries Society.

Lauritsen, D. D. 1986. Filter-feeding in *Corbicula fluminea* and its effect on seston removal. *Journal of the North American Benthological Society* 5:165–172.

Lawson, R., P. G. Frank, and D. L. Martin. 1991. A gecko new to the United States herpetofauna, with notes on geckoes of the Florida Keys. *Herpetological Review* 22:11–12.

Layne, J. N. 1965. Occurrence of black-tailed jackrabbits in Florida. *Journal of Mammalogy* 46:502.

———. 1969. Strange animals in Florida. *Florida Naturalist* 42:50.

———. 1971. Fleas (Siphonaptera) of Florida. *Florida Entomologist* 54:35–51.

———. 1974. The land mammals of south Florida. In P. J. Gleason (ed.), *Environments of South Florida: Present and Past.* Memoir 2. Miami: Miami Geological Society.

———. 1976. The armadillo—one of Florida's oddest animals. *Florida Naturalist* 49:8–12.

———. 1978. Florida red wolf. In J. N. Layne (ed.), *Rare and Endangered Biota of Florida.* Vol. 1: *Mammals.* Gainesville: University Presses of Florida.

———. 1992. Sherman's short-tailed shrew. In S. R. Humphrey (ed.), *Rare and Endangered Biota of Florida.* Vol. 1: *Mammals.* Gainesville: University Presses of Florida.

———. 1993a. History of an introduction of elk in Florida. *Florida Field Naturalist* 21:77–80.

———. 1993b. Long-tailed weasel observations in south-central Florida. *Florida Field Naturalist* 21:108–114.

Layne, J. N., and D. Glover. 1977. Home range of the armadillo in Florida. *Journal of Mammalogy* 58:411–413.

Lazell, J. D., Jr. 1989. *Wildlife of the Florida Keys*. Washington, D.C.: Island Press.

Lazell, J. D., Jr., and K. F. Koopman. 1985. Notes on bats of Florida's lower keys. *Florida Scientist* 48:37–41.

Lee, D. S., and E. Bostelman. 1969. The red fox in central Florida. *Journal of Mammalogy* 50:161.

Lee, H. G. 1987. Immigrant mussel settles in Northside generator. *Shell-O-Gram* (Jacksonville Shell Club) 28:7–9.

Lekic, M., and L. Mihajlovic. 1971. Entomofauna of *Myriophyllum spicatum* L. (Halorrhagidaceae), an aquatic weed in Yugoslav territory. *Arhiv za Poljoprivredne Nauke* 23(82):63–76.

Lemaitre, R. 1995. *Charybdis helleri* (Milne Edwards, 1867), a nonindigenous portunid crab (Crustacea, Decapoda, Brachyura) discovered in the Indian River lagoon system of Florida. *Proceedings of the Biological Society of Washington.* 108:643–648.

Leslie, A. J. 1994. Submersed plants, wildlife and water quality: Don't be fooled by invasive aliens. In D. C. Schmitz and T. C. Brown (eds.), *An Assessment of Invasive Non-indigenous Species in Florida's Public Lands*. Technical Report TSS-94-100. Tallahassee: Florida Department of Environmental Protection.

Leslie, D., J. Van Dyke, R. Hestand, and B. Thompson. 1987. Management of aquatic plants in multi-use lakes with crass carp (*Ctenopharyngodon idella*). *Lake and Reservoir Management* 3:110–125.

Lever, C. 1987. *Naturalized Birds of the World*. Avon, U.K.: Bath Press.

Levins, R., and H. Heatwole. 1973. Biogeography of the Puerto Rican Bank: Introduction of species onto Palominitos Island. *Ecology* 54:1056–1064.

Lewis, J. C., L. B. Flynn, R. L. Marchinton, S. M. Shea, and E. M. Marchinton. 1990. Pt. 1: Introduction, study area description, and literature review. In *Ecology of Sambar Deer on St. Vincent National Wildlife Refuge, Florida*. Bulletin of the Tall Timbers Research Station 25.

Lewis, R. 1990. Wetlands restoration/creation/enhancement terminology: Suggestions for standardization. In J. A. Kusler and M. E. Kentula (eds.), *Wetland Creation and Restoration: The Status of the Science*. Washington, D.C.: Island Press.

———. 1994. Ecological restoration of lands infested with non-indigenous plant species. In D. C. Schmitz and T. C. Brown (eds.), *An Assessment of Invasive Non-Indigenous Species in Florida's Public Lands*. Technical Report TSS-94-100. Tallahassee: Florida Department of Environmental Protection.

Lidicker, W. Z., Jr. 1991. Introduced mammals in California. In R. H. Groves and F. Di Castri (eds.), *Biogeography of Mediterranean Invasions*. Cambridge: Cambridge University Press.

Lind, A. O. 1969. Coastal landforms of Cat Island, Bahamas. Research Paper 122. Department of Geology, University of Chicago.

Linder, A. D., and E. Fichter. 1970. *The Reptiles of Idaho.* Pocatello: Idaho State University Press.

Lipscomb, D. J. 1989. Impacts of feral hogs on longleaf pine regeneration. *Southern Journal of Applied Forestry* 13:177–181.

Lloyd, J. E. 1995. Fireflies (Coleoptera: Lampyridae). In J. H. Frank and E. D. McCoy, Precinctive insect species in Florida. *Florida Entomologist* 78:21–35.

Lockwood, J. A. 1993. Environmental issues involved in biological control of rangeland grasshoppers (Orthoptera: Acrididae) with exotic agents. *Environmental Entomology* 22:503–518.

Lodge, T. E. 1994. *The Everglades Handbook: Understanding the Ecosystem.* Delray Beach: St. Lucie Press.

Lofgren, C. S. 1986. History of imported fire ants in the United States. In C. S. Lofgren and R. K. Vander Meer (eds.), *Fire Ants and Leafcutting Ants, Biology and Control.* Boulder: Westview Press.

Loftus, W. F. 1989. Distribution and ecology of exotic fishes in Everglades National Park. In L. K. Thomas (ed.), *Management of Exotic Species in Natural Communities.* Proceedings of the 4th Triennial Conference on Research in the National Parks and Equivalent Reserves, vol. 5. Washington, D.C.: George Wright Society and National Park Service.

Long, J. L. 1981. *Introduced Birds of the World.* New York: Universe Books.

Long, R. W. 1974a. Origin of the vascular flora of southern Florida. In P. J. Gleason (ed.), *Environments of South Florida: Present and Past.* Memoir 2. Miami: Miami Geological Society.

———. 1974b. The vegetation of southern Florida. *Florida Scientist* 37(1):33–45.

Long, R. W., and O. Lakela. 1976. *A Flora of Tropical Florida.* Miami: Banyan Books.

Lonsdale, W. M. 1992a. The biology of *Mimosa pigra.* In K. L. S. Harley (ed.), *A Guide to the Management of* Mimosa pigra. Canberra: CSIRO.

———. 1992b. The impact of weeds in national parks. In J. H. Combellack, K. J. Levick, J. Parsons, and R. G. Richardson (eds.), *Proceedings of the 1st International Weed Control Congress.* Melbourne: Weed Science Society of Victoria.

———. 1994. Inviting trouble: Introduced pasture species in northern Australia. *Australian Journal of Ecology* 19:345–354.

Lonsdale, W. M., I. L. Miller, and I. W. Forno. 1989. The biology of Australian weeds 20. *Mimosa pigra* L. *Plant Protection Quarterly* 4:119–131.

Loope, L. L. 1992. An overview of problems with introduced plant species in national parks and biosphere reserves of the United States. In C. P. Stone, C. W. Smith, and J. T. Tunison (eds.), *Alien Plant Invasions in Native Ecosystems of Hawaii: Management and Research.* Honolulu: University of Hawaii, Cooperative National Park Studies Unit.

Loope, L. L., and V. L. Dunevitz. 1981a. Impact of fire exclusion and invasion of *Schinus terebinthifolius* on limestone rockland pine forests of southeastern Florida. Technical Report T-645. Everglades National Park, South Florida Research Center.

———. 1981b. Investigations of early plant succession on abandoned farmland in Everglades National Park. Technical Report T-644. Everglades National Park, South Florida Research Center.

Loope, L. L., and P. G. Scowcroft. 1985. Vegetation response within exclosures in Hawaii: A review. In C. P. Stone and J. M. Scott (eds.), *Hawaii's Terrestrial Ecosystems: Preservation and Management.* Honolulu: University of Hawaii, Cooperative National Park Resources Studies Unit.

Loope, L., M. Duever, A. Herndon, J. Snyder, and D. Jansen. 1994. Hurricane impacts on uplands and freshwater swamp forest. *BioScience* 44:238–246.

Lorence, D. H., and R. W. Sussman. 1986. Exotic species invasion into Mauritius wet forest remnants. *Journal of Tropical Ecology* 2:147–162.

Losos, J. B. 1992a. A critical comparison of the taxon-cycle and character-displacement models for size evolution of *Anolis* lizards in the Lesser Antilles. *Copeia* 1992:279–288.

———. 1992b. The evolution of convergent structure in Caribbean *Anolis* communities. *Systematic Biology* 41:403–420.

———. 1994. Integrative approaches to evolutionary ecology: *Anolis* lizards as model systems. *Annual Review of Ecology and Systematics* 25:467–493.

Losos, J. B., J. C. Marks, and T. W. Schoener. 1993. Habitat use and ecological interactions of an introduced and a native species of *Anolis* lizard on Grand Cayman, with a review of the outcomes of anole introductions. *Oecologia* 95:525–532.

Lowery, G. H., Jr. 1974. *The Mammals of Louisiana and Its Adjacent Waters.* Baton Rouge: Louisiana State University Press.

Ludlow, M. E. 1994. Memorandum on panhandle impacts of exotic species for 1993 annual report. Florida Caverns State Park, Florida Park Service.

Lugo, A. E. 1990. Letter. *Conservation Biology* 4:345.

Lund, T. A. 1980. *American Wildlife Law.* Berkeley: University of California Press.

Macdonald, I.A.W., L. L. Loope, M. B. Usher, and O. Hamann. 1989. Wildlife conservation and the invasion of nature reserves by introduced species: A global perspective. In J. A. Drake et al. (eds.), *Biological Invasions: A Global Perspective.* Chichester: Wiley.

Macdonald, I.A.W., C. Thebaud, W. A. Strahm, and D. Strasberg. 1991. Effects of alien plant invasions on native vegetation remnants on La Réunion (Mascarene islands, Indian Ocean). *Environmental Conservation* 18:51–61.

McHargue, L. 1994. After Andrew: Vines threaten Dade County hammocks. *Resource Management Notes* 6(1):2.

McInnes, D. 1994. Comparative ecology and factors affecting the distribution of north Florida fire ants. Ph.D. dissertation, Florida State University, Tallahassee.

Mack, R. N. 1986. Alien plant invasions into the intermountain West: A case history. In H. A. Mooney and J. A. Drake (eds.), *Ecology of Biological Invasions of North America and Hawaii.* New York: Springer-Verlag.

———. 1989. Temperate grasslands vulnerable to plant invasions: Characteristics and consequences. In J. A. Drake et al. (eds.), *Biological Invasions: A Global Perspective.* Chichester: Wiley.

———. 1991. The commercial seed trade: An early dispenser of weeds in the United States. *Economic Botany* 45:257–273.

McKey, D. B., and S. C. Kaufmann. 1991. Naturalization of exotic *Ficus* species (Moraceae) in south Florida. In T. D. Center et al. (eds.), *Proceedings of the Symposium on Exotic Pest Plants, November 1988.* Technical Report NPS/NR-EVER/NRTR-91/06. Denver: U.S. Department of the Interior, National Park Service.

McMahon, R. F. 1983. Ecology of an invasive pest bivalve, *Corbicula.* In W. D. Russell-Hunter (ed.), *The Mollusca.* Vol. 6: *Ecology.* New York: Academic Press.

McPherson, B. F., W. H. Sonntag, and M. Sabanskas. 1984. Fouling community of the Loxahatchee River estuary, Florida, 1980–81. *Estuaries* 7:149–157.

McVea, C., and C. E. Boyd. 1975. Effects of waterhyacinth cover on water chemistry, phytoplankton, and fish in ponds. *Journal of Environmental Quality* 4(3): 375–378.

Maddox, D. M. 1968. Bionomics of an alligatorweed flea beetle, *Agasicles* sp. in Argentina. *Annals of the Entomological Society of America* 61:1299–1305.

———. 1970. Bionomics of a stem borer, *Vogtia malloi* (Lepidoptera: Pyralidae) on alligatorweed in Argentina. *Annals of the Entomological Society of America* 63: 1267–1273.

Maddox, D. M., L. A. Andres, R. D. Hennessey, R. D. Blackburn, and N. R. Spencer. 1971. Insects to control alligatorweed: An invader of aquatic ecosystems in the United States. *BioScience* 21:985–991.

Maehr, D., and G. B. Caddick. 1995. Demographics and genetic introgression in the Florida panther. *Conservation Biology* 9:1295–1298.

Maehr, D., J. Roof, D. Land, J. McCown, R. Belden, and W. Frankenberger. 1989. Fates of wild hogs released into occupied Florida panther home ranges. *Florida Field Naturalist* 17:42–43.

Maehr, D. S., R. C. Belden, E. D. Land, and L. Wilkins. 1990. Food habits of panthers in southwest Florida. *Journal of Wildlife Management* 54:420–423.

Maffei, M. D. 1991. Melaleuca control on Arthur R. Marshall Loxahatchee National Wildlife Refuge. In T. D. Center et al. (eds.), *Proceedings of the Symposium on Exotic Pest Plants, November 1988.* Technical Report NPS/NREVER/NRTR-91/06. Denver: U.S. Department of the Interior, National Park Service.

Maguire, J. 1990. Letter to H. Zebuth, biologist, Biological Resources Section, Environmental Resources Management, Dade County, Miami, Florida.

Manville, R. H. 1963. *The Nutria in the United States.* Wildlife Leaflet 445. U.S. Fish and Wildlife Service, Bureau of Sports Fisheries and Wildlife.

Maples, W. R. 1979. Summary of report on rhesus monkeys (*M. mulatta*). Exotic and Nonnative Species Conference, Florida Atlantic University, Boca Raton, 20 April 1979.

Maples, W. R., A. B. Brown, and P. M. Hutchins. 1976. Introduced monkey populations at Silver Springs, Florida. *Florida Anthropologist* 29:133–136.

Marchand, L. J. 1946. The saber crab, *Platychirograpsus typicus* Rathbun, in Florida: A case of accidental dispersal. *Quarterly Journal of the Florida Academy of Sciences* 9:93–100.

Markham, R. H. 1986. Biological control agents of *Hydrilla verticillata:* Final report on surveys in East Africa, 1981–1984. Miscellaneous Paper A-84-4. Vicksburg, Miss.: U.S. Army Engineer Waterways Experiment Station.

Marlatt, R. B. 1975. Growth of *Sansevieria trifasciata* "Laurentii" in south Florida. *Proceedings of the Florida State Horticultural Society* 88:596–598.

Martin, R. G. 1976. Exotic fish problems and opportunities in the southeast. *Proceedings of the Southeastern Association of Fish and Wildlife Agencies* 30:15–17.

Martin, S. W. 1949. *Florida's Flagler.* Athens: University of Georgia Press.

Mayer, J. J., and I. L. Brisbin, Jr. 1991. *Wild Pigs in the United States.* Athens: University of Georgia Press.

Mayfield, M., L. Avila, and N. Rappaport. 1994. Atlantic Hurricane Season of 1992. Preliminary report (updated 2 March 1993) on Hurricane Andrew 16–28 August 1992. *Monthly Weather Review* 122:517–538.

Mazourek, J. C., and P. N. Gray. 1994. The Florida duck or the mallard. *Florida Wildlife* 48(3):29–31.

Mazzotti, F. J., W. Ostrenko, and A. T. Smith. 1981. Effects of the exotic plants *Melaleuca quinquenervia* and *Casuarina equisetifolia* on small mammal populations in the eastern Florida Everglades. *Florida Scientist* 44:65–71.

Mead, A. R. 1979. *Ecological Malacology: With Particular Reference to* Achatina fulica. Vol. 2b of V. Fretter, J. Fretter, and J. Peake (eds.), *Pulmonates.* London: Academic Press.

Mead, F. W., and F. D. Bennett. 1987. Casuarina spittlebug, *Clastoptera undulata* Uhlar (Homoptera: Cercopidae). Entomology Circular 294. Gainesville: Florida Department of Agriculture and Consumer Services, Division of Plant Industry.

Meador, R. E. 1977. The role of mycorrhizae in influencing succession on abandoned Everglades farmland. M.S. thesis, University of Florida, Gainesville.

Means, D. B., and D. Simberloff. 1987. The peninsula effect: Habitat-correlated species decline in Florida's herpetofauna. *Journal of Biogeography* 14:551–568.

Meier, R. E. 1994. Coexisting patterns and foraging behavior of introduced and native ants (Hymenoptera Formicidae) in the Galapagos Islands (Ecuador). In D. F. Williams (ed.), *Exotic Ants: Biology, Impact, and Control of Introduced Species.* Boulder: Westview Press.

Meinesz, A., and B. Hesse. 1991. Introduction of the tropical green alga *Caulerpa taxifolia* and its invasion of the northwestern Mediterranean. *Oceanologica Acta* 14:415–426.

Meinesz, A., J. de Vaugelas, B. Hesse, and X. Mari. 1993. Spread of the introduced tropical green alga *Caulerpa taxifolia* in Mediterranean waters. *Journal of Applied Phycology* 5:141–147.

Meinesz, A., J. de Vaugelas, J. -M. Cottalorda, G. Caye, S. Charrier, T. Commeau, L. Delahaye, M. Febvre, F. Jaffrennou, R. Lemee, H. Molenaar, and D. Pietkiewicz. 1994. Suivi de l'invasion de l'algue tropicale *Caulerpa taxifolia* devant les côtes françaises de la Méditerranée: Situation au 31 décembre 1994. Laboratoire Environnement Marin Littoral, Université de Nice–Sophia Antipolis.

Menzies, R. J. 1957. The marine borer family Limnoriidae (Crustacea, Isopoda). *Bulletin of Marine Science* 7:101–200.

Mericas, D., P. Gremillion, and E. Terczak. 1990. Aquatic weed removal as a nutrient export mechanism in Lake Okeechobee, Florida. In J. J. Berger (ed.), *Environmental Restoration: Science and Strategies for Restoring the Earth.* Washington, D.C.: Island Press.

Merrill, L. D. 1989. Citrus canker. In D. L. Dahlsten and R. Garcia (eds.), *Eradication of Exotic Pests.* New Haven: Yale University Press.

Meshaka, W. E., Jr. 1994. Ecological correlates of successful colonization in the life history of the Cuban treefrog, *Osteopilus septentrionalis* (anuran; Hylidae). Ph.D. dissertation, Florida International University, Miami.

———. 1996. Theft or cooperative foraging in the barred owl? *Florida Field Naturalist* 24:15.

Meshaka, W. E., Jr., and B. Ferster. 1995. Two species of snakes prey on the Cuban treefrogs in southern Florida. *Florida Field Naturalist* 23:97–99.

Meshaka, W. E., Jr., and J. Lewis. 1994. *Cosymbotus platyurus* in Florida: Ten years of stasis. *Herpetological Review* 25:127.

Meshaka, W. E., Jr., B. P. Butterfield, and B. Hauge. 1994. *Hemidactylus frenatus* established on the lower Florida Keys. *Herpetological Review* 25:127–128.

Metzger, R. J., and P. L. Shafland. 1986. Use of detonating cord for sampling fish. *North American Journal of Fisheries Management* 6:113–118.

Mikkelsen, P. M., P. S. Mikkelsen, and D. J. Karlen. 1995. Molluscan biodiversity in the Indian River lagoon, Florida. *Bulletin of Marine Science* 57:94–127.

Miley, W. W., II. 1978. Ecological impact of the pike killifish, *Belonesox belizanus* Kner (Poeciliidae) in southern Florida. M.S. thesis, Florida Atlantic University, Boca Raton.

Miller, J. H., and B. Edwards. 1983. Kudzu: Where did it come from? And how can we stop it? *Southern Journal of Applied Forestry* 7:165–169.

Miller, M., and G. Aplet. 1993. Biological control: A little knowledge is a dangerous thing. *Rutgers Law Review* 45(2):285–334.

Miller, M. A. 1968. Isopoda and Tanaidacea from buoys in coastal waters of the continental United States, Hawaii, and the Bahamas (Crustacea). *Proceedings of the U.S. National Museum* 125.

Miller, N. G. 1990. The genera of Meliaceae in the southeastern United States. *Journal of the Arnold Arboretum* 71:453–486.

Miller, R. C., and M. McClure. 1931. The freshwater clam industry of the Pearl River. *Lingnan Science Journal* 10:307–323.

Milon, J. W., J. Yingling, and J. E. Reynolds. 1986. *An Economic Analysis of the Benefits of Aquatic Weed Control in North-Central Florida with Special Reference to Orange and Lochloosa Lakes.* Economics Report 113. Institute of Food and Agricultural Science, University of Florida.

Minckley, W. L, and J. E. Deacon (eds.). 1991. *Battle against Extinction: Native Fish Management in the American West.* Tucson: University of Arizona Press.

Mitchell, D. J., F. N. Martin, and R. Charudattan. 1994. Biological control of plant pathogens and weeds in Florida. In D. Rosen, F. D. Bennett, and J. L. Capinera (eds.), *Pest Management in the Subtropics: Biological Control—a Florida Perspective.* Andover, U.K.: Intercept.

Mitchell, D. S. 1976. The growth and management of *Eichhornia crassipes* and *Salvinia* spp. in their native environment and in alien situations. In C. K. Varshney and J. Rzoska (eds.), *Aquatic Weeds of South East Asia.* The Hague: Junk.

Mitchell, J. C. 1994. *The Reptiles of Virginia.* Washington, D.C.: Smithsonian Institution Press.

Miyamoto, M. M., M. P. Hayes, and M. R. Tennant. 1986. Biochemical and morphological variation in Floridian populations of the bark anole (*Anolis distichus*). *Copeia* 1986:76–86.

Moler, P. E. 1992. Introduction. In P. E. Moler (ed.), *Rare and Endangered Biota of Florida.* Vol. 3: *Amphibians and Reptiles.* Gainesville: University Presses of Florida.

Montalbano, F., III, S. Hardin, and W. M. Hetrick. 1979. Utilization of hydrilla by ducks and coots in central Florida. *Proceedings of the Annual Conference of the Southeastern Association of Fish and Wildlife Agencies* 33:36–42.

Montegut, R. S., R. D. Gasaway, D. F. DuRant, and J. L. Atterson. 1976. An ecological evaluation of the effects of grass carp (*Ctenopharyngodon idella*) introduction in Lake Wales, Florida. Florida Game and Fresh Water Fish Commission Report, Tallahassee.

Moody, M. E., and R. N. Mack. 1988. Controlling the spread of plant invasions: The importance of nascent foci. *Journal of Applied Ecology* 25:1009–1021.

Moody, P. 1986. Peacock bass come to Florida. *Florida Wildlife* 40(6):12–15.

Mook, D. 1983. Indian River fouling organisms: A review. *Florida Scientist* 46: 162–167.

Mooney, H. A., and J. A. Drake (eds.), 1986. *Ecology of Biological Invasions of North America and Hawaii.* New York: Springer-Verlag.

Mooney, H. A., S. P. Hamburg, and J. A. Drake. 1986. The invasions of plants and animals into California. In H. A. Mooney and J. A. Drake (eds.), *Ecology of Biological Invasions of North America and Hawaii.* New York: Springer-Verlag.

Morton, B. 1987. Recent marine introductions into Hong Kong. *Bulletin of Marine Science* 41:503–513.

Morton, J. F. 1976. Pestiferous spread of many ornamental and fruit species in South Florida. *Proceedings of the Florida State Horticultural Society* 89:348–353.

———. 1978. Brazilian pepper—its impact on people, animals and environment. *Economic Botany* 32:353–359.

———. 1980. The Australian pine or beefwood (*Casuarina equisetifolia* L.), an invasive "weed" tree in Florida. *Proceedings of the Florida State Horticultural Society* 93:87–95.

———. 1983. Woman's tongue, or cha-cha (*Albizia lebbeck* Benth.), a fast-growing weed tree in Florida, is prized for timber, fuel, and forage elsewhere. *Proceedings of the Florida State Horticultural Society* 96:173–178.

———. 1985. The earleaf acacia, a fast-growing, brittle, exotic "weed" tree in Florida. *Proceedings of the Florida State Horticultural Society* 98:309–314.

Motie, A., D. M. Myers, and E. E. Storrs. 1986. A serologic survey for leptospires in nine-banded armadillos (*Dasypus novemcinctus* L.) in Florida. *Journal of Wildlife Diseases* 22:423–424.

Mount, R. H. 1981. The red imported fire ant, *Solenopsis invicta* (Hymenoptera: Formicidae), as a possible serious predator on some native southeastern vertebrates: Direct observations and subjective impressions. *Journal of the Alabama Academy of Science* 52:71–78.

——— (ed.). 1984. *Vertebrate Wildlife of Alabama.* Auburn: Alabama Agricultural Experiment Station, Auburn University.

Moxley, D. J., and F. H. Langford. 1982. Beneficial effects of hydrilla on two eutrophic lakes in central Florida. *Proceedings of the Annual Conference of the Southeastern Association of Fish and Wildlife Agencies* 36:280–286.

Moyle, P. 1986. Fish introductions into North America: Patterns and ecological impact. In H. A. Mooney and J. A. Drake (eds.), *Ecology of Biological Invasions of North America and Hawaii.* New York: Springer-Verlag.

Mullahey, J. J., and D. L. Colvin. 1993. Tropical soda apple: A new noxious weed in Florida. Fact Sheet WRS-7. University of Florida Cooperative Extension Service.

Mullahey, J., M. Nee, R. Wunderlin, and K. Delaney. 1993. Tropical soda apple (*Solanum viarum*): A new weed threat in subtropical regions. *Weed Technology* 7:783–786.

Muller, J. W., E. D. Hardin, D. R. Jackson, S. E. Gatewood, and N. Caire. 1989. *Summary Report on the Vascular Plants, Animals, and Plant Communities Endemic to Florida.* Tallahassee: Florida Game and Fresh Water Fish Commission.

Mulliken, T., and J. B. Thomsen. 1990. U.S. bird trade: The controversy continues although imports decline. *Traffic* (U.S.A.) 10(3):1–11.

Murdoch, W. W., J. Chesson, and P. L. Chesson. 1985. Biological control in theory and practice. *American Naturalist* 125:344–366.

Murry, R. E. 1963. The black francolin. *Proceedings of the Southeastern Association of Fish and Wildlife Agencies* 17:117–120.

Myers, R. L. 1975. The relationship of site conditions to the invading capability of *Melaleuca quinquenervia* in southwest Florida. M.S. thesis, University of Florida, Gainesville.

———. 1976. *Melaleuca* field studies. In J. Ewel et al. (eds.), *Studies of Vegetation Changes in South Florida.* U.S. Forest Service Research Agreement 18–492.

———. 1983. Site susceptibility to invasion by the exotic tree *Melaleuca quinquenervia* in southern Florida. *Journal of Applied Ecology* 20:645–658.

———. 1984. Ecological compression of *Taxodium distichum* var. *nutans* by *Melaleuca quinquenervia* in southern Florida. In K. C. Ewel and H. T. Odum (eds.), *Cypress Swamps.* Gainesville: University Presses of Florida.

———. 1990. Scrub and high pine. In R. L. Myers and J. J. Ewel (eds.), *Ecosystems of Florida.* Orlando: University of Central Florida Press.

Myers, R. L., and J. J. Ewel (eds.). 1990a. *Ecosystems of Florida.* Orlando: University of Central Florida Press.

———. 1990b. Problems, prospects, and strategies for conservation. In R. L. Myers and J. J. Ewel (eds.), *Ecosystems of Florida.* Orlando: University of Central Florida Press.

Nadel, H., J. H. Frank, and R. J. Knight, Jr. 1992. Escapees and accomplices: The naturalization of exotic *Ficus* and their associated faunas in Florida. *Florida Entomologist* 75:29–38.

National Academy of Sciences. 1987. *Research Briefings 1987: Report of the Research Briefing Panel on Biological Control in Managed Ecosystems.* Washington, D.C.: National Academy Press.

National Park Service. 1988. *Management Policies.* Washington, D.C.: U.S. Department of the Interior.

———. 1991. *Natural Resources Management Guideline.* NPS-77. Washington, D.C.: U.S. Department of the Interior.

National Research Council (NRC). 1984. Casuarinas: *Nitrogen-fixing Trees for Adverse Sites.* Advisory Committee on Technology Innovation, Board on Science and Technology for International Development, Office of International Affairs. Washington, D.C.: National Academy Press.

———. 1992. *Restoration of Aquatic Ecosystems.* Washington, D.C.: National Academy Press.

Natkiel, R., and A. Preston. 1986. *Atlas of Maritime History.* New York: Facts on File.

Nature Conservancy and Natural Resources Defense Council. 1992. *The Alien Pest Species Invasion in Hawaii: Background Study and Recommendations for Interagency Planning.* Honolulu: Nature Conservancy of Hawaii.

Nauman, C. E., and D. F. Austin. 1978. Spread of the exotic fern *Lygodium microphyllum* in Florida. *American Fern Journal* 68:65–66.

Nechols, J. R., W. C. Kauffman, and P. W. Schaeffer. 1992. Significance of host specificity in classical biological control. In W. C. Kauffman and J. R. Nechols (eds.), *Selection Criteria and Ecological Consequences of Importing Natural Enemies.* Lanham, Md.: Entomological Society of America.

Nechols, J. R., L. A. Andres, J. W. Beardsley, R. D. Goeden, and C. G. Jackson. 1995. *Biological Control in the Western United States.* Publication 3361. Oakland: University of California, Division of Agriculture and Natural Resources.

Nehrling, H. 1944. *My Garden in Florida.* Vol. 1. Estero, Fla.: American Eagle.

Neill, W. T. 1952. The spread of the armadillo in Florida. *Ecology* 33:282–284.

———. 1961. On the trail of the jaguarondi. *Florida Wildlife* 15:10–13.

———. 1977. Another look at the jaguarondi. *Florida Wildlife* 30:2–5.

Nellis, D. W., N. F. Eichholz, T. W. Regan, and C. Feinstein. 1978. Mongoose in Florida. *Wildlife Society Bulletin* 6:249–250.

Nelson, B. V., and T. F. Richards. 1994. Non-indigenous species management operations within the Southwest Florida Water Management District. In D. C. Schmitz, and T. C. Brown (eds.), *An Assessment of Invasive Non-indigenous Species in Florida's Public Lands.* Technical Report TSS-94-100. Tallahassee: Florida Department of Environmental Protection.

Nelson, L. K. 1963. Introductions of the blackneck pheasant group and crosses into the southeastern States. *Proceedings of the Southeastern Association of Fish and Wildlife Agencies* 17:111–121.

Nesbitt, S. A. 1978. Whooping crane. In H. W. Kale II (ed.), *Rare and Endangered Biota of Florida,* vol. 2. *Birds.* Gainesville: University Presses of Florida.

Neushul, M., C. D. Amsler, D. C. Reed, and R. J. Lewis. 1992. Introduction of marine plants for aquacultural purposes. In A. Rosenfield and R. Mann (eds.), *Dispersal of Living Organisms into Aquatic Ecosystems*. College Park: Maryland Sea Grant Program.

Newman, C. 1948. Florida's big game. *Florida Wildlife* 1:4–5, 18.

———. 1949. Florida's armored invasion. *Florida Wildlife* 3:3–5.

Newsom, L. D. 1978. Eradication of plant pests—con. *Bulletin of the Entomological Society of America* 24:35–40.

Newsome, A. E., and I. R. Noble. 1986. Ecological and physiological characters of invading species. In R. H. Groves and J. J. Burdon (eds.), *Ecology of Biological Invasions*. Cambridge: Cambridge University Press.

Nilsson, G. 1981. *The Bird Business: A Study of the Commercial Cage Bird Trade*. Washington, D.C.: Animal Welfare Institute.

Noble, R. L., and R. D. Germany. 1986. Changes in fish populations of Trinidad Lake, Texas, in response to abundance of blue tilapia. In R. H. Stroud (ed.), *Fish Culture in Fisheries Management*. Bethesda, Md.: American Fisheries Society.

Norton, B. G. 1987. *Why Preserve Natural Variety?* Princeton: Princeton University Press.

Norton, T. A., and L. E. Deysher. 1989. The reproductive ecology of *Sargassum muticum* at different latitudes. In J. S. Ryland and P. A. Tyler (eds.), *Reproduction, Genetics and Distributions of Marine Organisms*. Fredersburg, Denmark: Olsen and Olsen, International Symposium Series.

O'Brien, C. W. 1995. Weevils (Coleoptera: Curculionidae *sensu lato*). In J. H. Frank and E. D. McCoy, Precinctive insect species in Florida. *Florida Entomologist* 78:21–35.

O'Brien, L. B. 1995. Planthoppers (Homoptera: Fulgoroidea). In J. H. Frank and E. D. McCoy, Precinctive insect species in Florida. *Florida Entomologist* 78:21–35.

O'Brien, S. J., M. E. Roelke, N. Yuhki, K. W. Richards, W. E. Johnson, W. L. Franklin, A. E. Anderson, O. L. Bass, Jr., R. C. Belden, and J. S. Martenson. 1990. Genetic introgression within the Florida panther *Felis concolor coryi*. *National Geographic Research* 6:485–494.

Office of Environmental Services. 1995. *Conservation and Recreation Lands (CARL) Annual Report 1995*. Tallahassee: Florida Department of Environmental Protection, Division of State Lands.

Office of Park Planning. 1995. Jurisdictional Report. Tallahassee: Florida Department of Environmental Protection, Division of Recreation and Parks.

Ogawa, H., R. Hirano, and I. Hanyu. 1990. Effects of temperature of the maturation and early development of *Sargassum muticum* (Yendo) Fensholt, Phaeophyta. In *Proceedings of the Second Asian Fisheries Forum, Tokyo*.

Ogilvie, V. E. 1966. Report on the peacock bass project including Venezuelan trip report and a description of five *Cichla* species. Unpublished report, Florida Game and Freshwater Fish Commission, Boca Raton, Florida.

O'Hara, J. 1967. Invertebrates found in water hyacinth mats. *Florida Scientist* 30:73–80.

Oliver, J. D. 1992. Carrotwood: An invasive plant new to Florida. *Aquatics* 14(3):4–9.

Olmstead, I. C., L. L. Loope, and R. P. Russell. 1981. Vegetation of Southern Coastal Region of Everglades National Park between Flamingo and Joe Bay. Report T-620. National Park Service, Everglades National Park, South Florida Research Center.

O'Meara, G. F. 1994. Immigrant mosquito species in Florida. In D. C. Schmitz and T. C. Brown (eds.), *An Assessment of Invasive Non-indigenous Species in Florida's Public Lands.* Technical Report TSS-94-100. Tallahassee: Florida Department of Environmental Protection.

Opler, P. A. 1978. Insects of American chestnut: Possible importance and conservation concern. In W. McDonald (ed.), *The American Chestnut Symposium.* Morgantown, W.V.: American Chestnut Association.

Orth, P. G., and R. A. Conover. 1975. Changes in nutrients resulting from farming the Hole-in-the-Donut, Everglades National Park. *Proceedings of the Florida State Horticultural Society* 88:221–225.

Ostrenko, W., and F. Mazzotti. 1981. Small mammal populations in *Melaleuca quinquenervia* communities in the eastern Florida Everglades. In R. K. Geiger (ed.), *Proceedings of Melaleuca Symposium.* Tallahassee: Florida Division of Forestry.

Owre, O. T. 1973. A consideration of the exotic avifauna of southeastern Florida. *Wilson Bulletin* 85:491–500.

Panetta, F. D. 1993. A system of assessing proposed plant introductions for weed potential. *Plant Protection Quarterly* 8:10–14.

Panetta, F. D., and A.J.M. Hopkins. 1991. Weeds in corridors: Invasion and management. In D. A. Saunders and R. J. Hobbs (eds.), *Nature Conservation: The Role of Corridors.* Chipping Norton: Surrey Beatty & Sons.

Panetta, F. D., P. Pheloung, M. Lonsdale, S. Jacobs, M. Mulvaney, and W. Wright. 1994. Screening plants for weediness: A procedure for assessing species proposed for importation into Australia. Australian Weeds Committee.

Paradiso, J. L. 1966. Recent records of coyotes, *Canis latrans,* from southeastern United States. *Southwestern Naturalist* 11:500–501.

Parks, P. 1973. About bats. *Key West Citizen,* 30 December.

Parsons, P. A. 1982. Adaptive strategies of colonizing animal species. *Geological Review* 57:117–148.

Pascarella, J. B. 1995. The effects of Hurricane Andrew on the population dynamics and the mating system of the tropical understory shrub *Ardisia escallonioides* (Myrsinaceae). Ph.D. dissertation, University of Miami, Coral Gables, Florida.

Pascarella, J. B., and C. C. Horvitz. In press. The importance of hurricanes in the population dynamics of a tropical understory shrub: Megamatrix elasticity analysis. *Ecology.*

Pearson, O. P. 1964. Carnivore–mouse predation: An example of its intensity and bioenergetics. *Journal of Mammalogy* 45:177–188.

Pemberton, R. W. 1980. Exploration for natural enemies of *Hydrilla verticillata* in eastern Africa. Miscellaneous Paper A-80-1. Vicksburg, Miss.: U.S. Army Engineer Waterways Experiment Station.

———. 1985a. Native plant considerations in the biological control of leafy spurge. In E. S. Delfosse (ed.), *Proceedings of the 6th International Symposium on Biological Control of Weeds.* Ottawa: Agriculture Canada.

———. 1985b. Native weeds as candidates for biological control research. In E. S. Delfosse (ed.), *Proceedings of the 6th International Symposium on Biological Control of Weeds.* Ottawa: Agriculture Canada.

———. 1995. *Cactoblastis cactorum* in the United States: An immigrant biological control agent or an introduction of the nursery industry? *American Entomologist* 41:230–232.

Penfound, W. T., and T. T. Earle. 1948. The biology of the water hyacinth. *Ecological Monographs* 18:448–472.

Perkins, J. H. 1989. Eradication: Scientific and social questions. In D. L. Dahlsten and R. Garcia (eds.), *Eradication of Exotic Pests.* New Haven: Yale University Press.

Pernas, T. J. 1994. Non-native species management in Big Cypress National Preserve. In D. C. Schmitz and T. C. Brown (eds.), *An Assessment of Invasive Non-indigenous Species in Florida's Public Lands.* Technical Report TSS-94-100. Tallahassee: Florida Department of Environmental Protection.

Peroni, P. A., and W. G. Abrahamson. 1986. Succession in Florida Sandridge vegetation: A retrospective study. *Florida Scientist* 49:176–191.

Perrins, J., M. Williamson, and A. Fitter. 1992a. A survey of differing views of weed classification: Implications for regulation of introductions. *Biological Conservation* 60:47–56.

———. 1992b. Do annual weeds have predictable characters? *Acta Oecologia* 13:517–533.

Perry, D. M. 1988. Effects of associated fauna on growth and productivity in red mangrove. *Ecology* 69:1064–1075.

Perry, D. M., and R. C. Brusca. 1989. Effects of the root-boring isopod *Sphaeroma terebrans* on red mangrove forests. *Marine Ecology Progress Series* 57:287–292.

Peschken, D. P. 1972. *Chrysolina quadrigemina* (Coleoptera: Chrysomelidae) introduced from California to British Columbia against the weed *Hypericum perforatum:* Comparison of behavior, physiology, and colour in association with post-colonization adaptation. *Canadian Entomologist* 104:1689–1698.

Peters, R. H. 1991. *A Critique for Ecology.* Cambridge: Cambridge University Press.

Phillips, J. C. 1928. *Wild Birds Introduced or Transplanted in North America.* Technical Bulletin 61. Washington D.C.: U.S. Department of Agriculture.

Pielou, E. C. 1979. *Biogeography.* New York: Wiley.

Pimentel, D. 1963. Introducing parasites and predators to control native pests. *Canadian Entomologist* 95:785–792.

Pimm, S. L., G. E. Davis, L. E. Loope, C. T. Roman, T. J. Smith III, and J. T. Tilmant. 1994. Hurricane Andrew. *BioScience* 44:224–229.

PlantFinder. 1994. *PlantFinder with Interior Scan Section.* Fort Lauderdale: Betrock Publishing.

Platt, W. J., and S. L. Rathbun. 1993. Dynamics of an old-growth longleaf pine population. In S. M. Hermann (ed.), *The Longleaf Pine Ecosystem: Ecology, Restoration and Management.* Proceedings of the 18th Tall Timbers Fire Ecology Conference. Tallahassee: Tall Timbers Research Station.

Porak, W. A., S. Crawford, D. Renfro, R. L. Cailteau, and J. Chadwick. 1990. *Large*

Mouth Bass Population Responses to Aquatic Plant Management Strategies. Wallop-Breaux project F-24-R. Study XIII.

Porter, S. D., B. Van Eimeren, and L. E. Gilbert. 1988. Invasion of red imported fire ants (Hymenoptera: Formicidae): Microgeography and competitive replacement. *Annals of the Entomological Society of America* 81:913–918.

Pullen, R.S.V., and R. H. Lowe-McConnell (eds.). 1982. *The Biology and Culture of Tilapias.* Manila: International Centre for Living Aquatic Resources Management.

Putz, F. E. 1991. Silvicultural effects of lianas. In F. E. Putz and H. A. Mooney (eds.), *The Biology of Vines.* Cambridge: Cambridge University Press.

Quimby, P. C. 1982. Impact of diseases on plant populations. In R. Charudattan and H. L. Walker (eds.), *Biological Control of Weeds with Plant Pathogens.* New York: Wiley.

Quinn, T. G. 1994a. Are there imported crocodiles in Florida's future? In D. C. Schmitz and T. C. Brown (eds.), *An Assessment of Invasive Non-indigenous Species in Florida's Public Lands.* Technical Report TSS-94-100. Tallahassee: Florida Department of Environmental Protection.

———. 1994b. Non-indigenous animal regulatory program. In D. C. Schmitz and T. C. Brown (eds.), *An Assessment of Invasive Non-indigenous Species in Florida's Public Lands.* Technical Report TSS-94-100. Tallahassee: Florida Department of Environmental Protection.

Rackleff, R. B. 1972. *Close to Crisis: Florida's Environmental Problems.* Tallahassee: New Issues Press.

Ramakrishnan, P. S., and P. M. Vitousek. 1989. Ecosystem-level processes and the consequences of biological invasions. In J. A. Drake et al. (eds.), *Biological Invasions: A Global Perspective.* Chichester: Wiley.

Randall, J. M. 1993. Exotic weeds in North American and Hawaiian natural areas: The Nature Conservancy's plan of attack. In B. N. McKnight (ed.), *Biological Pollution: The Control and Impact of Invasive Exotic Species.* Indianapolis: Indiana Academy of Sciences.

Read, R. W. 1962. *Jasminum* species in cultivation in Florida and their correct names. *Proceedings of the Florida State Horticultural Society* 75:430–437.

Reasoner Bros. 1900. Royal Palm Nurseries Annual Mail Order Catalog. Oneco, Fla.

Reddy, K. R., and P. D. Sacco. 1981. Decomposition of waterhyacinth in agricultural drainage water. *Journal of Environmental Quality* 10:228–234.

Rehm, A. E. 1976. The effects of the wood-boring isopod *Sphaeroma terebrans* on the mangrove communities of Florida. *Environmental Conservation* 3:47–57.

Rehm, A. E., and H. J. Humm. 1973. *Sphaeroma terebrans:* A threat to the mangroves of southwestern Florida. *Science* 182:173–174.

Rejmánek, M. 1994. What makes a species invasive? In P. Pyšek, K. Prach, M. Rejmánek, and P. M. Wade (eds.), *Plant Invasions.* The Hague: SPB Academic Publishing.

Rhoads, S. N. 1894. Contributions to the mammalogy of Florida. *Proceedings of the Academy of Natural Sciences of Philadelphia* 46:152–161.

Rhymer, J., and D. Simberloff. 1996. Extinction by hybridization and introgression. *Annual Review of Ecology and Systematics* 27:83–109.

Ribi, G. 1981. Does the wood boring isopod *Sphaeroma terebrans* benefit red mangroves (*Rhizophora mangle*)? *Bulletin of Marine Science* 31:925–928.

———. 1982. Differential colonization of roots of *Rhizophora mangle* by the wood boring isopod *Sphaeroma terebrans* as a mechanism to increase root density. *Publicazione di Stazione Zoologica di Napoli: Marine Ecology* 3:13–19.

Richard, D. I., J. W. Small, and J. A. Osborne. 1985. Response of zooplankton to the reduction and elimination of submerged vegetation by grass carp and herbicide in four Florida lakes. *Hydrobiologia* 123:97–108.

Richardson, D. M., R. M. Cowling, and D. C. Le Maitre. 1990. Assessing the risk of invasive success in *Pinus* and *Banksia* in South African mountain fynbos. *Journal of Vegetation Science* 1:629–642.

Richardson, D. R. 1977. Vegetation of the Atlantic Coastal Ridge of Palm Beach County, Florida. *Florida Scientist* 40:281–330.

Richardson, H. 1897. Description of a new species of *Sphaeroma*. *Proceedings of the Biological Society of Washington* 11:105–107.

Rickard, E. R., and C. B. Worth. 1951. Complement-fixation tests for murine typhus on the sera of wild-caught cotton rats in Florida. *American Journal of Hygiene* 53:332–336.

Ridings, W. H., D. J. Mitchell, C. L. Schoulties, and N. E. El-Gholl. 1977. Biological control of milkweed vine in Florida citrus groves with a pathotype of *Phytophthora citrophthora*. In T. E. Freeman (ed.), *Proceedings of the 4th International Symposium on Biological Control of Weeds*. Gainesville: Institute of Food and Agricultural Sciences, University of Florida.

Riherd, C. 1992. Letter from assistant director, Division of Plant Industry, Florida Department of Agriculture and Consumer Services, Gainesville, to the Florida Department of Natural Resources, 10 November.

Riser, N. W. 1970. Biological studies on *Taenioplana teredini* Hyman, 1944. *American Zoologist* 10:553 (abstract).

———. 1974. Epilogue. In N. W. Riser and M. P. Morse (eds.), *Biology of the Turbellaria*. New York: McGraw-Hill.

Roberts, R., and D. Richardson. 1994. Climbing fern, *Lygodium microphyllum*, a pest species in south Florida wetlands. *Palmetto* 10:16 (abstract).

Robertson, W. B., Jr. 1955. An analysis of breeding bird populations of tropical Florida in relation to the vegetation. Ph.D. dissertation, University of Illinois, Urbana.

Robertson, W. B., Jr., and P. C. Frederick. 1994. The faunal chapters: Contexts, synthesis, and departures. In S. M. Davis and J. C. Ogden (eds.), *Everglades: The Ecosystem and Its Restoration*. Delray Beach, Fla.: St. Lucie Press.

Robertson, W. B., Jr., and J. A. Kushlan. 1974. The southern Florida avifauna. In *Environments of South Florida: Present and Past*. Memoir 2. Miami: Miami Geological Society.

Robertson, W. B., Jr., and G. E. Woolfenden. 1992. *Florida Bird Species: An Annotated List*. Special Publication 6. Gainesville: Florida Ornithological Society.

Rodda, G. H., and T. H. Fritts. 1992. The impact of the introduction of the colubrid snake *Boiga irregularis* on Guam's lizards. *Journal of Herpetology* 26:166–174.

Roelke, M. E., J. S. Martenson, and S. J. O'Brien. 1993. The consequences of demographic reduction and genetic depletion in the endangered Florida panther. *Current Biology* 3:340–350.

Rogers, H. H., and D. E. Davis. 1972. Nutrient removal by water-hyacinth. *Weed Science* 20:423–428.

Rohwer, G. G. 1958. The Mediterranean fruit fly in Florida—past, present, and future. *Florida Entomologist* 41:23–25.

Room, P. M., D.P.A. Sands, I. W. Forno, M.F.J. Taylor, and M. H. Julien. 1985. A summary of research into biological control of salvinia in Australia. In E. S. Delfosse (ed.), *Proceedings of the 6th International Symposium on Biological Control of Weeds.* Ottawa: Agriculture Canada.

Ross, M. S., J. F. Meeder, G. Telesnicki, and C. Weekley. 1995. *Terrestrial Ecosystems of the Crocodile Lakes National Wildlife Refuge: The Effects of Hurricane Andrew.* Final Report to the Department of the Interior, U.S. Fish and Wildlife Service, Contract 1448-0004-93-035 (May 1995).

Ross, S. T. 1991. Mechanisms structuring stream fish assemblages: Are there lessons from introduced species? *Environmental Biology of Fishes* 30:359–368.

Roughgarden, J. 1992. Comments on the paper by Losos: Character displacement versus taxon loop. *Copeia* 1992:288–295.

———. 1995. Anolis *Lizards of the Caribbean: Ecology, Evolution, and Plate Tectonics.* Oxford: Oxford University Press.

Rueness, J. 1989. *Sargassum muticum* and other introduced Japanese macroalgae: Biological pollution of European coasts. *Marine Pollution Bulletin* 20:173–176.

Ruesink, J. L., I. M. Parker, M. J. Groom, and P. K. Kareiva. 1995. Reducing the risks of nonindigenous species introductions. *BioScience* 45:465–477.

Ryerson, K. A. 1967. The history of plant exploration and introduction in the United States Department of Agriculture. In *Proceedings of the International Symposium on Plant Introduction.* Tegucigalpa, Honduras: Escuela Agricola Panamericana.

Sailer, R. I. 1978. Our immigrant insect fauna. *Bulletin of the Entomological Society of America* 24:1–11.

———. 1981. Progress report on importation of natural enemies of insect pests in the U.S.A. In J. R. Coulson (ed.), *Proceedings of the Joint American-Soviet Conference on Use of Beneficial Organisms in the Control of Crop Pests.* College Park, Md.: Entomological Society of America.

———. 1983. History of insect introductions. In C. L. Wilson and C. L. Graham (eds.), *Exotic Plant Pests and North American Agriculture.* New York: Academic Press.

Sailer, R. I., J. A. Reinert, D. Boucias, P. Busey, R. L. Kepner, T. G. Forrest, W. G. Hudson, and T. J. Walker. 1994. *Mole Crickets in Florida.* Bulletin 846. Gainesville: Florida Agricultural Experiment Station, University of Florida.

Salisbury, E. J. 1961. *Weeds and Aliens.* New York: Macmillan.

Salzburg, M. A. 1984. *Anolis sagrei* and *Anolis cristatellus* in southern Florida: A case study in interspecific competition. *Ecology* 65:14–19.

Samol, H. H. 1972. Rat damage and control in the Florida sugarcane industry. In M. T. Henderson (ed.), *Proceedings of the International Society for Sugarcane Technology, 14th Congress.* New Orleans. Baton Rouge: Franklin Press.

Samuel, D. E. 1975. The kiskadee in Bermuda. *Newsletter of the Bermuda Biological Station* 4.

Sanders, R. W. 1987. Identity of *Lantana depressa* and *L. ovatifolia* (Verbenaceae) of Florida and the Bahamas. *Systematic Botany* 12:44–60.

Sanderson, J. C. 1990. A preliminary survey of the distribution of the introduced macroalga, *Undaria pinnatifida* (Harvey) Suringer on the east coast of Tasmania, Australia. *Botanica Marina* 33:153–157.

Sasek, T. W., and B. R. Strain. 1991. Effects of CO_2 enrichment on the growth and morphology of a native and an introduced honeysuckle vine. *American Journal of Botany* 78:69–75.

Saul, S. 1992. Letter: The medfly in California. *Science* 255:515.

Saunders, D. A., R. J. Hobbs, and C. R. Margules. 1991. Biological consequences of ecosystem fragmentation: A review. *Conservation Biology* 5:18–32.

Savage, J. M. 1982. The enigma of the Central American herpetofauna: Dispersals or vicariance? *Annals of the Missouri Botanical Garden* 69:464–547.

Savidge, J. 1984. Guam: Paradise lost for wildlife. *Biological Conservation* 30:305–317.

———. 1987. Extinction of an island forest avifauna by an introduced snake. *Ecology* 68:660–668.

Scanlon, G. 1976. Concerns regarding the introduction of exotic species in wildlife management programs. *Proceedings of the Southeastern Association of Fish and Wildlife Agencies* 29:674-679.

Schardt, J. D. 1992. 1992 Florida aquatic plant survey report. Technical Report 942-CGA. Tallahassee: Florida Department of Environmental Protection, Bureau of Aquatic Plant Management.

———. 1994. History and development of Florida's aquatic plant management program. In D. C. Schmitz and T. C. Brown (eds.), *An Assessment of Invasive Non-indigenous Species in Florida's Public Lands.* Technical Report TSS-94-100. Tallahassee: Florida Department of Environmental Protection.

Schardt, J. D., and J. A. Ludlow. 1993. 1993 Florida aquatic plant survey. Technical Report 952-CGA. Tallahassee: Florida Department of Environmental Protection, Bureau of Aquatic Plant Management.

Scheffrahn, R. H. 1995. Termites (Isoptera). In J. H. Frank and E. D. McCoy, Precinctive insect species in Florida. *Florida Entomologist* 78:21–35.

Schick, J. M., and A. N. Lamb. 1977. Asexual reproduction and genetic population structure in the colonizing sea anemone *Haliplanella luciae. Biological Bulletin* 153:604–617.

Schmitz, D. C., and J. A. Osborne. 1984. Zooplankton densities in a hydrilla infested lake. *Hydrobiologia* 111:127–132.

Schmitz, D. C., B. V. Nelson, L. E. Nall, and J. D. Schardt. 1991. Exotic aquatic plants in Florida: A historical perspective and review of the present aquatic plant regulation program. In T. D. Center et al. (eds.), *Proceedings of the Symposium on Exotic Pest Plants.* Technical Report NPS/NREVER/NRTR-91/06. Denver: U.S. Department of the Interior, National Park Service.

Schmitz, D. C., J. D. Schardt, A. J. Leslie, F. A. Dray, Jr., J. A. Osborne, and B. V. Nelson. 1993. The ecological impact and management history of three invasive alien aquatic plant species in Florida. In B. N. McKnight (ed.), *Biological Pollution: The Control and Impact of Invasive Exotic Species.* Indianapolis: Indiana Academy of Sciences.

Schneider, R. F. 1967. Range of the Asiatic clam in Florida. *Nautilus* 81(2):68–69.

Schoener, T. W. 1975. Presence and absence of habitat shift in some widespread lizard species. *Ecological Monographs* 45:233–258.

Schomer, N. S., and R. D. Drew. 1982. An ecological characterization of the lower Everglades, Florida Bay, and the Florida Keys. FWS/OBS-82/58.1. Washington, D.C.: U.S. Fish and Wildlife Service.

Schortemeyer, J. L., R. E. Johnson, and J. D. West. 1981. A preliminary report on wildlife occurrence in melaleuca heads in the Everglades Wildlife Management Area. In R. K. Geiger (ed.), *Proceedings of Melaleuca Symposium.* Tallahassee: Florida Division of Forestry.

Schotte, M. 1989. Two new species of wood-boring *Limnoria* (Crustacea: Isopoda) from New Zealand, *L. hicksi* and *L. reniculus. Proceedings of the Biological Society of Washington* 102:716–725.

Schramm, H. L., Jr., and K. J. Jirka. 1989. Effects of aquatic macrophytes on benthic macroinvertebrates in two Florida lakes. *Journal of Freshwater Ecology* 5:1–12.

Schramm, H. L., Jr., K. J. Jirka, and M. V. Hoyer. 1987. Epiphytic macroinvertebrates on dominant macrophytes in two central Florida lakes. *Journal of Freshwater Ecology* 4:151–161.

Schwalbe, C. P. 1992. Letter to Florida Department of Natural Resources from U.S. Department of Agriculture, Animal and Plant Health Inspection Service, Hyattsville, Md.

Schwartz, A. 1952. The land mammals of southern Florida and the Upper Keys. Ph.D. dissertation, University of Michigan, Ann Arbor.

———. 1971. *Anolis distichus. Catalogue of American Amphibians and Reptiles* 108.1–108.4.

Schwartz, A., and R. W. Henderson. 1991. *Amphibians and Reptiles of the West Indies: Descriptions, Distributions, and Natural History.* Gainesville: University Presses of Florida.

Schwartz, M. W. 1988. Species diversity patterns in woody flora on three North American peninsulas. *Journal of Biogeography* 15:759–774.

Schwartz, M. W., and J. M. Randall. 1995. Valuing natural areas and controlling non-indigenous plants. *Natural Areas Journal* 15:98–100.

Scott, J. K., and F. D. Panetta. 1993. Predicting the Australian weed status of southern African plants. *Journal of Biogeography* 20:87–93.

Scott, J., S. Mountainspring, F. Ramsey, and C. Kepler. 1986. Forest bird communities of the Hawaiian Islands: Their dynamics, ecology, and conservation. *Studies in Avian Biology* 9. Los Angeles: Cooper Ornithological Society, University of California.

Seal, U. S., and R. C. Lacy. 1989. *Florida Panther Viability Analysis and Species Survival Plan.* Cooperative Agreement 14-16-004-90-902. Gainesville: Captive Breeding Specialist Group, U.S. Fish and Wildlife Service.

Searles, R. B., M. H. Hommersand, and C. D. Amsler. 1984. The occurrence of

Codium fragile subsp. *tomentosoides* and *C. taylori* (Chlorophyta) in North Carolina. *Botanica Marina* 27:185–187.

Seavey, R., and J. Seavey. 1994. *Ardisia elliptica* in Everglades National Park: An overview through 1993. Unpublished report. Homestead: Everglades National Park.

Shaffer, M. L. 1981. Minimum population sizes for species conservation. *BioScience* 31:131–134.

Shafland, P. 1979. Lab at Boca Raton. *Florida Wildlife* 33(4):17–19.

———. 1984. A proposal for introducing peacock bass (*Cichla* spp.) in southeast Florida canals. Unpublished report. Boca Raton: Florida Game and Fresh Water Fish Commission.

———. 1986. A review of Florida's efforts to regulate, assess, and manage exotic fishes. *Fisheries* (Bethesda) 11(2):20–25.

———. 1993. An overview of Florida's introduced butterfly peacock bass (*Cichla ocellaris*) sportfishery. *Natura* 96:26–29.

———. 1995. Introduction considerations and establishment of a successful butterfly peacock fishery in Florida. In H. L. Schramm, Jr., and R. G. Piper (eds.), *Uses and Effects of Cultured Fishes in Aquatic Ecosystems.* American Fisheries Society Symposium 15. Bethesda, Md.: American Fisheries Society.

———. In press. Exotic fish management in Florida. *Fisheries* 21.

———. 1996a. Exotic fishes in Florida—1994. *Reviews in Fisheries Science* 4:101–122.

———. 1996b. Exotic fish assessments: An alternative view. *Reviews in Fisheries Science* 4:123–132.

Shafland, P., and K. J. Foote. 1979. A reproducing population of *Serrasalmus humeralis* Valenciennes in southern Florida. *Florida Scientist* 42:206–214.

Shafland, P., and W. M. Lewis, Sr. 1984. Terminology associated with introduced organisms. *Fisheries* 9(4):17–18.

Shaw, J. H. 1985. *Introduction to Wildlife Management.* New York: McGraw-Hill.

Shearer, J. F. 1992. Pathogen biological control studies of hydrilla and Eurasian watermilfoil. In *Proceedings of the 26th Annual Meeting, Aquatic Plant Control Research Program.* Miscellaneous Paper A-92-2. Vicksburg, Miss.: U.S. Army Engineer Waterways Experiment Station.

———. 1993. Biological control of hydrilla and milfoil using plant pathogens. In *Proceedings of the 27th Annual Meeting, Aquatic Plant Control Research Program.* Miscellaneous Paper A-93-2. Vicksburg, Miss.: U.S. Army Engineer Waterways Experiment Station.

Shelton, W. L., and R. O. Smitherman. 1984. Exotic fishes in warmwater aquaculture. In W. R. Courtenay, Jr., and J. R. Stauffer, Jr. (eds.), *Distribution, Biology, and Management of Exotic Fishes.* Baltimore, Md.: Johns Hopkins University Press.

Sherman, H. B. 1943. The armadillo in Florida. *Florida Entomologist* 26:54–59.

———. 1952. A list and bibliography of the mammals of Florida, living and extinct. *Quarterly Journal of the Florida Academy of Sciences* 15:86–126.

Shrader-Frechette, K. S., and E. D. McCoy. 1993. *Method in Ecology.* Cambridge: Cambridge University Press.

Sieg, J., and R. N. Winn. 1981. The Tanaidae (Crustacea: Tanaidae) of California, with a key to the world genera. *Proceedings of the Biological Society of Washington* 94:315–343.

Silver, H., and W. T. Silver. 1969. Growth and behavior of the coyote-like canid of northern New England with observations of canid hybrids. *Wildlife Monographs* 17:1–41.

Silver, J. 1927. The introduction and spread of house rats in the United States. *Journal of Mammalogy* 8:58–60.

Silverstein, A., and V. Silverstein. 1974. *Animal Invaders: The Story of Imported Wildlife.* New York: Atheneum.

Silvertown, J. W. 1993. *Introduction to Plant Population Biology.* London: Blackwell Scientific.

Simberloff, D. 1981. Community effects of introduced species. In M. H. Nitecki (ed.), *Biotic Crises in Ecological and Evolutionary Time.* New York: Academic Press.

———. 1985. Predicting ecological effects of novel entities: Evidence from higher organisms. In H. O. Halvorson, D. Pramer, and M. Rogul (eds.), *Engineered Organisms in the Environment: Scientific Issues.* Washington, D.C.: American Society for Microbiology.

———. 1986. Introduced insects: A biogeographic and systematic perspective. In H. A. Mooney and J. A. Drake (eds.), *Ecology of Biological Invasions of North America and Hawaii.* New York: Springer-Verlag.

———. 1989. Which insect introductions succeed and which fail? In J. A. Drake et al. (eds.), *Biological Invasions: A Global Perspective.* Chichester: Wiley.

———. 1990. Reconstructing the ambiguous: Can island ecosystems be restored? In D. R. Towns, C. H. Daugherty, and I. A. E. Atkinson (eds.), *Ecological Restoration of New Zealand Islands.* Wellington, N.Z.: Department of Conservation.

———. 1991. Keystone species and community effects of biological introductions. In L. R. Ginzburg (ed.), *Assessing Ecological Risks of Biotechnology.* Boston: Butterworth-Heinemann.

———. 1992. Conservation of pristine habitats and unintended effects of biological control. In W. C. Kauffman and J. E. Nechols (eds.), *Selection Criteria and Ecological Consequences of Importing Natural Enemies.* Lanham, Md.: Entomological Society of America.

———. 1995a. Introduced species. In *Encyclopedia of Environmental Biology,* vol. 2. San Diego: Academic Press.

———. 1995b. Why do introduced species appear to devastate islands more than mainland areas? *Pacific Science* 49:87–97.

Simberloff, D., B. J. Brown, and S. Lowrie. 1978. Isopod and insect root borers may benefit Florida mangroves. *Science* 201:630–632.

Simmonds, F. J., and F. D. Bennett. 1966. Biological control of *Opuntia* spp. by *Cactoblastis cactorum* in the Leeward Islands (West Indies). *Entomophaga* 11:183–189.

Simmonds, F. J., J. M. Franz, and R. I. Sailer. 1976. History of biological control. In C. B. Huffaker and P. S. Messenger (eds.), *Theory and Practice of Biological Control.* New York: Academic Press.

Simpson, C. T. 1932. *Florida Wildlife: Observations on the Flora and Fauna of the State and the Influence of Climate and Environment on Their Development.* New York: Macmillan.

Simpson, R. H., and M. B. Lawrence. 1971. Atlantic hurricane frequencies along the U.S. coastline. NOAA Technical Memorandum NWS SR-58.

Sinclair, R. M. 1971. Annotated bibliography on the exotic bivalve *Corbicula* in North America, 1900–1971. *Sterkiana* 43:11–18.

Sinclair, R. M., and B. G. Isom. 1963. *Further Studies on the Introduced Asiatic Clam (*Corbicula*) in Tennessee.* Nashville: Tennessee Stream Pollution Board, Tennessee Department of Public Health.

Singer, F. J., W. T. Swank, and E.E.C. Clebsch. 1984. Effects of wild pig rooting in a deciduous forest. *Journal of Wildlife Management* 48:466–473.

Singletary, H. M. 1990. Statement with testimony submitted at hearings before the Senate Subcommittee on Agricultural Research and General Legislation, Committee on Agriculture, Nutrition, and Forestry, 28 March 1990, pp. 354–356, 380–382.

———. 1994. Statement with testimony submitted at hearings before the Committee on Governmental Affairs, U.S. Senate, 11 March 1994, pp. 8–11, 74–76.

Slater, H. H., W. J. Platt, D. B. Baker, and H. A. Johnson. 1995. Effects of Hurricane Andrew on damage and mortality of trees in subtropical hardwood hammocks of Long Pine Key, Everglades National Park, Florida, U.S.A. *Journal of Coastal Research* (Special Hurricane Andrew issue) 21:197–207.

Small, J. K. 1927. Among flora aborigines: A record of exploration of Florida in the winter of 1922. *Journal of the New York Botanical Garden* 28:1–20, 25–40.

———. 1933. *Manual of the Southeastern Flora.* Published by the author, New York.

Smallwood, K. S. 1990. Turbulence and ecology of invading species. Ph.D. dissertation, University of California, Davis.

Smith, G. 1969. St. Vincent Island. *Florida Wildlife* 23:12–17.

———. 1970. Tilapia in review. *Florida Wildlife* 23:18–23.

Smith, H. M., and R. H. McCauley. 1948. Another new anole from south Florida. *Proceedings of the Biological Society of Washington* 61:159–166.

Smith, P. W. 1987. The Eurasian collared dove arrives in the Americas. *American Birds* 41:1370–1379.

Smith, P. W., and S. A. Smith. 1993. An exotic dilemma for birders: The canary-winged parakeet. *Birding* 25(6):426–430.

Snyder, G. H., J. F. Morton, and W. G. Genung. 1981. Trials of *Ipomoea aquatica,* nutritious vegetable with high protein- and nitrate-extraction potential. *Proceedings of the Florida State Horticultural Society* 94:230–235.

Snyder, J. R., A. Herndon, and W. B. Robertson, Jr. 1990. South Florida rockland. In R. L. Myers and J. J. Ewel (eds.), *Ecosystems of Florida.* Orlando: University of Central Florida Press.

Soil Conservation Service. 1982. *National List of Scientific Plant Names.* Vol. 1: *List of Plant Names.* Publication SCS-TP-159. U.S. Department of Agriculture, Soil Conservation Service.

Sosa, O. 1985. The sugarcane delphacid, *Perkinsiella saccharicida* (Homoptera: Delphacidae), a sugarcane pest new to North America detected in Florida. *Florida Entomologist* 68:357–360.

South, G. R. 1984. A checklist of marine algae of eastern Canada, second revision. *Canadian Journal of Botany* 62:680–704.

Sowder, A., and S. Woodall. 1985. Small mammals of melaleuca stands and adjacent environments in southwestern Florida. *Florida Scientist* 48(1):41–44.

Spatz, G., and D. Mueller-Dumbois. 1975. Succession after pig digging in grassland communities on Mauna Loa, Hawaii. *Phytocoecologica* 3:346–373.

Spivey, H. R. 1979. First records of *Trypetesa* and *Megalasma* (Crustacea: Cirripedia) in the Gulf of Mexico. *Bulletin of Marine Science* 29:497–508.

Steck, G. J. 1995. Fruit flies (Diptera: Tephritidae). In J. H. Frank and E. D. McCoy, Precinctive insect species in Florida. *Florida Entomologist* 78:21–35.

Stejneger, L. 1922. Two geckos new to the fauna of the United States. *Copeia* 1922:56.

Stern, W. L. 1988. The long-distance dispersal of *Oeceoclades maculata*. *American Orchid Society Bulletin* 57:960–971.

Stevens, L. M., A. L. Steinhauer, and J. R. Coulson. 1975. Suppression of Mexican bean beetle on soybeans with annual inoculative releases of *Pediobius foveolatus*. *Environmental Entomology* 4:947–952.

Stevenson, H. M. 1976. *Vertebrates of Florida*. Gainesville: University Presses of Florida.

Stevenson, H. M., and B. H. Anderson. 1994. *The Birdlife of Florida*. Gainesville: University Presses of Florida.

Stevenson, H. M., and R. L. Crawford. 1974. Spread of the armadillo into the Tallahassee-Thomasville area. *Florida Field Naturalist* 2:8–10.

Stevenson, J. A. 1990. Memorandum to Don Schmitz, 2 March 1990, Florida Department of Natural Resources, Division of Recreation and Parks, Tallahassee.

Steward, K. K. 1993. Seed production in monoecious and dioecious populations of *Hydrilla*. *Aquatic Botany* 46:169–183.

Steward, K. K., T. K. Van, V. Carter, and A. H. Pieterse. 1984. *Hydrilla* invades Washington, D.C., and the Potomac. *American Journal of Botany* 71:162–163.

Stiling, P. D. 1996. *Ecology: Theories and Applications* 2nd ed. Upper Saddle River, New Jersey: Prentice Hall.

Stone, C. P. 1985. Alien animals in Hawaii's native ecosystems: Toward controlling the effects of introduced vertebrates. In C. P. Stone and J. M. Scott (eds.), *Hawaii's Terrestrial Ecosystems: Preservation and Management*. Honolulu: Cooperative National Park Resources Studies Unit, University of Hawaii.

Stone, C. P., L. W. Cuddihy, and J. T. Tunison. 1992. Responses of Hawaiian ecosystems to removal of feral pigs and goats. In C. P. Stone, C. W. Smith, and J. T. Tunison (eds.), *Alien Plant Invasions in Native Ecosystems of Hawaii: Management and Research*. Honolulu: Cooperative National Park Studies Unit, University of Hawaii.

Storrs, E. E., and H. P. Burchfield. 1985. Epidemiology of leprosy in wild armadillos. *Florida Scientist* 48 (Suppl. 1):50.

Storrs, E. E., H. P. Burchfield, and R.J.W. Rees. 1988. Superdelayed parturition in armadillos: A new mammalian survival strategy. *Leprosy Review* 59:11–15.

Streng, D. R., J. S. Glitzenstein, and W. J. Platt. 1993. Evaluating effects of season of burn in longleaf pine forests: A critical literature review and some results from an ongoing long-term study. In S. Hermann (ed.), *The Longleaf Pine Ecosystem: Ecology, Restoration and Management.* Proceedings of the 18th Tall Timbers Fire Ecology Conference. Tallahassee: Tall Timbers Research Station.

Suasa-Ard, W., and B. Napompeth. 1982. Investigations on *Episammia pectinicornis* (Hampson) (Lepidoptera: Noctuidae) for biological control of the waterlettuce in Thailand. Technical Bulletin 3. Bangkok: National Biological Control Research Center.

Sutton, D. L. 1994. Does catclaw mimosa have the potential to invade South Florida's wetlands? In D. C. Schmitz and T. C. Brown (eds.), *An Assessment of Invasive Non-Indigenous Species in Florida's Public Lands.* Technical Report TSS-94-100. Tallahassee: Florida Department of Environmental Protection.

Sutton, D. L., and K. A. Langeland. 1993. Can *Mimosa pigra* be eradicated in Florida? *Proceedings of the Southern Weed Control Conference* 46:239–243.

Sutton, D. L., and V. V. Vandiver. 1986. Grass carp, a fish for biological management of hydrilla and other aquatic weeds in Florida. Bulletin 867. Institute of Food and Agricultural Science, University of Florida, Florida Cooperative Extension Service.

Swift, C. C., C. R. Gilbert, S. A. Bortone, G. H. Burgess, and R. W. Yerger. 1986. Biogeography of the freshwater fishes of the southeastern United States: Savannah River to Lake Pontchartrain. In C. H. Hocutt and E. O. Wiley (eds.), *The Zoogeography of North American Freshwater Fishes.* New York: Wiley.

SWIM. 1990. FY 1989–90 Major Emphasis. Unpublished photocopy.

———. 1992. Tampa Bay Surface Improvement and Management Plan. Draft (February). Photocopy.

Tabita, A., and J. W. Woods. 1962. History of waterhyacinth control in Florida. *Hyacinth Control Journal* 1:19–23.

Tallahassee Democrat. 1975. Hydrilla: Even experts aren't sure if the mysterious plant is really dangerous. State agencies disagree on ways to control weed. 31 August, Section C1.

Tanner, G. W., and M. R. Werner. 1986. Cogongrass in Florida . . . an Encroaching Problem. Florida Cooperative Extension Service Fact Sheet WRS-5.

Tanner, R. D., S. S. Hussain, L. A. Hamilton, and F. T. Wolf. 1980. Kudzu (*Pueraria lobata*): Potential agricultural and industrial resource. *Economic Botany* 33:400–412.

Tardieu, V. 1995. Alerte virale en Australie. *Le Monde,* 1 November, p. 19.

Tarver, D. P., J. A. Rodgers, M. J. Mahler, and R. L. Lazor. 1985. *Aquatic and Wetland Plants of Florida.* 3rd ed. Tallahassee: Bureau of Aquatic Plant Management, Florida Department of Natural Resources.

Taylor, J. N., D. B. Snyder, and W. R. Courtenay, Jr. 1986. Hybridization between two introduced substrate-spawning tilapias (Pisces: Cichlidae) in Florida. *Copeia* 1986:903–909.

Taylor, L. L., R. G. Lessnau, and S. M. Lehman. 1994. Prevalence of whipworm

(*Trichuris*) ova in two free-ranging populations of rhesus macaques (*Macaca mulatta*) in the Florida Keys. *Florida Scientist* 57:102–107.

Taylor, W. R. 1960. Notes on three Bermudian marine algae. *Contribution from the Bermuda Biological Station for Research* 294:277–283.

———. 1967. A *Caulerpa* newly recorded for the West Indies. Extrait des Travaux de Biologie Végétale dédiés au Professeur P. Dangeard. *Le Botaniste* série L:467–471.

Teagarden, E. J. 1954. Letter to editor of "Sticks and Stones." *Florida Wildlife* 7:4.

Temple, S. 1990. The nasty necessity: Eradicating exotics. *Conservation Biology* 4:113–115.

———. 1992. Exotic birds: A growing problem with no easy solution. *Auk* 109:395–397.

Thakur, N. K. 1978. On the food of the air-breathing catfish, *Clarias batrachus* (Linn.), occurring in wild waters. *Internationale Revue der gesamten Hydrobiologie* 63:421–431.

Thayer, D. D., and A. Ferriter. 1994. Invasive non-indigenous vegetation management on South Florida Water Management District lands. In D. C. Schmitz and T. C. Brown (eds.), *An Assessment of Invasive Non-indigenous Species in Florida's Public Lands.* Technical Report TSS-94-100. Tallahassee: Florida Department of Environmental Protection.

Thayer, D. D., and V. Ramey. 1986. *Mechanical Harvesting of Aquatic Weeds: Literature Review.* Technical Report. Tallahassee: Florida Department of Natural Resources.

Theriot, E. A., A. F. Cofrancesco, and J. F. Shearer. 1993. Pathogen biological control research for aquatic plant management. In *Proceedings of the 27th Annual Meeting, Aquatic Plant Control Research Program.* Miscellaneous Paper A-93-2. Vicksburg, Miss.: U.S. Army Engineer Waterways Experiment Station.

Thomas, M. C. 1993. The flat bark beetles of Florida (Coleoptera: Silvanidae, Passandridae, Laemophloeidae). *Arthropods of Florida and Neighboring Land Areas* 15:i–vii, 1–93.

———. 1994a. *Chelymorpha cribraria* (Fabricius), a tortoise beetle new to Florida. Florida Department of Agriculture and Consumer Services, Division of Plant Industry, *Entomological Circular* 363:1–2.

———. 1994b. Insect immigrants indirectly attributable to Hurricane Andrew. Box 2.4a. in D. C. Schmitz and T. C. Brown (eds.), *An Assessment of Invasive Non-indigenous Species in Florida's Public Lands.* Technical Report TSS-94-100. Tallahassee: Florida Department of Environmental Protection.

———. 1995. Invertebrate pets and the Florida Department of Agriculture and Consumer Services. *Florida Entomologist* 78:39–44.

Thomas, R. A. 1994. Geographic distribution: *Ramphotyphlops braminus* (Braminy blind snake). *Herpetological Review* 25:34.

Thompson, C. R., and D. H. Habeck. 1989. Host specificity and biology of the weevil *Neohydronomus pulchellus*, biological control agent of waterlettuce (*Pistia stratiotes*). *Entomophaga* 34:299–306.

Thompson, F. G. 1984. *The Freshwater Snails of Florida*. Gainesville: University of Florida Press.

Thompson, J. B. 1919. Napier and merker grasses: Two new forage crops for Florida. *University of Florida Agricultural Experiment Station Bulletin* 153:237-249.

Thompson, J. D. 1991. The biology of an invasive plant. *BioScience* 41:393-401.

Thomsen, J. B., and T. A. Mulliken. 1992. Trade in neotropical psittacines and its conservation implications. In S. R. Beissinger and N.F.R. Snyder (eds.), *New World Parrots in Crisis, Solutions from Conservation Biology*. Washington, D.C.: Smithsonian Institution Press.

Thurston County Department of Water and Waste Management. 1995. *Long Lake Eurasian Watermilfoil Eradication Project*. Olympia, Wash.: Thurston County Department of Water and Waste Management.

Tietz, H. M. 1972. *An Index to the Described Life Histories, Early Stages and Hosts of the Macrolepidoptera of the Continental United States and Canada*. Vol. 1. Sarasota: A. C. Allyn.

Tilmant, J. T. 1980. Investigations of rodent damage to the thatch palms *Thrinax morrisii* and *Thrinax radiata* on Elliott Key, Biscayne National Park, Florida. Report M-589. Everglades National Park, South Florida Research Center.

Timmer, C. E., and L. W. Weldon. 1967. Evapotranspiration and pollution of water by water hyacinth. *Hyacinth Control Journal* 6:34-37.

Tisdell, C. A., B. A. Auld, and K. M. Menz. 1984. On assessing the value of biological control of weeds. *Protection Ecology* 6:169-179.

Tomlinson, P. B. 1980. *The Biology of Trees Native to Tropical Florida*. Allston, Mass.: P. B. Tomlinson.

Toth, L., D. A. Arrington, M. A. Brady, and D. A. Muszick. 1995. Conceptual evaluation of factors potentially affecting restoration of habitat structure within the channelized Kissimmee River ecosystem. *Restoration Ecology* 3:160-180.

Towne, C. W., and E. N. Wentworth. 1950. *Pigs: From Cave to Corn Belt*. Norman: University of Oklahoma Press.

Traffic (U.S.A.). 1990. *U.S. Imports of Wildlife*. Washington, D.C.: World Wildlife Fund.

———. 1991. *U.S. Fish and Wildlife Service Designated Port Activity*. Washington, D.C.: World Wildlife Fund.

———. 1993. *U.S. Fish and Wildlife Service Designated Port Activity*. Washington, D.C.: World Wildlife Fund.

Trewavas, E. 1983. *Tilapiine Fishes of the Genera* Sarotherodon, Oreochromis *and* Danakilia. Ithaca: Cornell University Press.

Trexler, J. C. 1995. Restoration of the Kissimmee River: A conceptual model of past and present fish communities and its consequences for evaluating restoration success. *Restoration Ecology* 3:195-210.

Trost, C. H., and J. H. Hutchinson. 1963. Foods of the barn owl in Florida. *Quarterly Journal of the Florida Academy of Sciences* 26:382-384.

Tschinkel, W. R. 1988. Distribution of fire ants *Solenopsis invicta* and *S. geminata* in north Florida in relation to habitat and disturbance. *Annals of the Entomological Society of America* 81:76–81.

———. 1993. The fire ant (*Solenopsis invicta*): Still unvanquished. In B. N. McKnight (ed.), *Biological Pollution: The Control and Impact of Invasive Exotic Species.* Indianapolis: Indiana Academy of Sciences.

Turner, R. D. 1955. The family Pholadidae in the Western Atlantic and the Eastern Pacific. Part II—Martesiinae, Jouannetiinae and Xylophaginae. *Johnsonia* 3:65–160.

———. 1966. *A Survey and Illustrated Catalogue of the Teredinidae (Mollusca: Bivalvia).* Cambridge: Museum of Comparative Zoology, Harvard University.

———. 1971. Identification of marine wood-boring molluscs. In E.B.G. Jones and S. K. Eltringham (eds.), *Marine Borers, Fungi and Fouling Organisms of Wood.* Paris: Organization for Economic Cooperation and Development.

Turner, R. D., and A. C. Johnson. 1971. Biology of marine wood-boring molluscs. In E.B.G. Jones and S. K. Eltringham (eds.), *Marine Borers, Fungi and Fouling Organisms of Wood.* Organization for Economic Cooperation and Development.

U.S. Army Corps of Engineers, Jacksonville District. 1973. Final Environmental Impact Statement. Aquatic Plant Control Program, State of Florida.

U.S. Congress. 1957. Water hyacinth obstructions in the waters of the Gulf and south Atlantic states: Letter from the Secretary of the Army. 35th Congress. Document 37.

———. 1965. Expanded project for aquatic plant control: Letter from the Secretary of the Army. 89th Congress. Document 251.

———. 1993. *Harmful Non-indigenous Species in the United States.* U.S. Congress/OTA-F-565. Washington, D.C.: U.S. Congress, Office of Technology Assessment.

U.S. Department of the Interior. 1995. Florida: Resource management plan and record of decision. Bureau of Land Management, Jackson District, Eastern States, June 1995.

U.S. Fish and Wildlife Service. 1993. *Annual Report of Lands under Control of the U.S. Fish and Wildlife Service as of September 30, 1993.* Washington, D.C.: U.S. Department of the Interior.

———. 1994. *An Ecosystem Approach to Fish and Wildlife Conservation: An Approach to More Effectively Conserve the Nation's Biodiversity.* Washington, D.C.: U.S. Department of the Interior.

Van, T. K., W. T. Haller, and G. Bowes. 1976. Comparison of photosynthetic characteristics of three submersed aquatic macrophytes. *Plant Physiology* 58:761–768.

Vance, D. R., and R. L. Westemeier. 1979. Interactions of pheasants and prairie chickens in Illinois. *Wildlife Society Bulletin* 7:221–225.

van den Bosch, F., R. Hengeveld, and J.A.J. Metz. 1992. Analysing the velocity of animal range expansion. *Journal of Biogeography* 19:135–150.

van den Bosch, R. 1968. Comments on population dynamics of exotic insects. *Bulletin of the Entomological Society of America* 14:112–115.

van den Bosch, R., E. I. Schlinger, E. J. Dietrich, J. C. Hall, and B. Puttler. 1964. Studies on succession, distribution, and phenology of imported parasites of *Therioaphis trifolii* (Monell) in southern California. *Ecology* 45:602–621.

van der Leek, M. L., H. N. Becker, P. Humphrey, C. L. Adams, R. C. Belden, W. B. Frankenberger, and P. L. Nicoletti. 1993. Prevalence of *Brucella* sp. antibodies in feral swine in Florida. *Journal of Wildlife Diseases* 29:410–415.

Van der Werff, H. 1982. Effects of feral pigs and donkeys on the distribution of selected food plants. *Noticias de Galápagos* 36:17–18.

van Dijk, G. 1985. *Vallisneria* and its interactions with other species. *Aquatics* 7(3):6–10.

van Driesche, R. G. 1994. Classical biological control of environmental pests. *Florida Entomologist* 77:20–33.

Van Gelder, R. G. 1979. Mongooses on mainland of North America. *Wildlife Society Bulletin* 7:197–198.

Van Name, W. G. 1936. *The American Land and Fresh-Water Isopod Crustacea.* Bulletin of the American Museum of Natural History 71.

Van Oppen, M.J.H., S.G.A. Draisma, J. L. Olsen, and W. T. Stam. 1995. Multiple trans-Arctic passages in the red alga *Phycodrys rubens:* Evidence from nuclear rDNA ITS sequences. *Marine Biology* 23:179–188.

van Riper, C., III, S. G. van Riper, M. L. Goff, and M. Laird. 1986. The epizootiology and ecological significance of malaria in Hawaiian land birds. *Ecological Monographs* 56:327–344.

Vaugelas, J. de, J. Blachier, J. -M. Cottalorda, T. Komatsu, R. Lemee, A. Meinesz, H. Molenaar, and D. Pietkiewicz. 1994. Premiers résultats de la campagne de sensibilisation européenne sur l'invasion de *Caulerpa taxifolia.* Situation sur les côtes françaises de la Méditerranée à la fin de 1993. In C. F. Boudouresque, A. Meinesz, and V. Gravez (eds.), *First International Workshop on* Caulerpa taxifolia. Marseille: GIS Posidonie.

Veitch, C. R., and B. D. Bell. 1990. Eradication of introduced animals from the islands of New Zealand. In D. R. Towns, C. H. Daugherty, and I.A.E. Atkinson (eds.), *Ecological Restoration of New Zealand Islands.* Wellington, N.Z.: Department of Conservation.

Verlaque, M. 1994. Inventaire des plantes introduites en Méditerranée: Origines et répercussions sur l'environnement et les activités humaines. *Oceanologica Acta* 17:1–23.

Verlaque, M., and R. Riouall. 1989. Introduction de *Polysiphonia nigrescens* et d'*Antithamnion nipponicum* (Rhodophyta, Ceramiales) sur le littoral Méditerranéen français. *Cryptogamie, Algologie* 10:313–323.

Verrill, A. H. 1942. The dog-killer. *Florida Game and Fish* 3:6–9.

Villadolid, D. V., and F. G. Del Rosario. 1930. Some studies on the biology of tulla (*Corbicula manilensis* Philippi), a common food clam of Laguna de Bay and its tributaries. *Philippine Agriculturalist* 19:355–382.

Vinson, S. B. 1994. Impact of the invasion of *Solenopsis invicta* (Buren) on native food webs. In D. F. Williams (ed.), *Exotic Ants: Biology, Impact, and Control of Introduced Species.* Boulder: Westview Press.

Vitousek, P. M. 1986. Biological invasions and ecosystem properties: Can species make a difference? In H. A. Mooney and J. A. Drake (eds.), *Ecology of Biological Invasions of North America and Hawaii.* New York: Springer-Verlag.

Vitousek, P. M., L. L. Loope, and C. P. Stone. 1987. Introduced species in Hawaii: Biological effects and opportunities for ecological research. *Trends in Ecology and Evolution* 2:224–227.

Vogt, G. B. 1960. Exploration for natural enemies of alligator weed and related plants in South America. Special Report PI-4. Washington, D.C.: U.S. Department of Agriculture, Agricultural Research Service.

———. 1961. Exploration for natural enemies of alligator weed and related plants in South America. Special Report PI-5. Washington, D.C.: U.S. Department of Agriculture, Agricultural Research Service.

von Broembsen, S. L. 1989. Invasions of natural ecosystems by plant pathogens. In J. A. Drake et al. (eds.), *Biological Invasions: A Global Perspective.* Chichester: Wiley.

Voss, H. J. 1992. Letters: The medfly in California. *Science* 255:514–515.

Voss, N. A. 1959. Studies on the pulmonate gastropod *Siphonaria pectinata* (Linnaeus) from the southeast coast of Florida. *Bulletin of Marine Science of the Gulf Caribbean* 9:84–99.

Vtorov, I. P. 1993. Feral pig removal: Effects on soil microarthropods in a Hawaiian rain forest. *Journal of Wildlife Management* 57:875–880.

Waage, J. K. 1990. Ecological theory and the selection of biological control agents. In M. Mackauer, L. E. Ehler, and J. Roland (eds.), *Critical Issues in Biological Control.* Andover, U.K.: Intercept.

Wade, D. D. 1981. Some melaleuca-fire relationships including recommendations for homesite protection. In R. K. Geiger (ed.), *Proceedings of Melaleuca Symposium.* Tallahassee: Florida Division of Forestry.

Wade, D., J. Ewel, and R. Hofstetter. 1980. Fire in South Florida ecosystems. U.S. Forest Service General Technical Report SE17. Asheville, N.C.: Southeast Forest Experiment Station.

Wade, S. A. 1995. Stemming the tide: A plea for new exotic species legislation. *Journal of Land Use and Environmental Law* 10:343–370.

Walford, L. A., and R. I. Wicklund. 1968. Monthly sea temperature structure from the Florida Keys to Cape Cod. Folio 15 in W. Webster (ed.), *Serial Atlas of the Marine Environment.* New York: American Geographic Society.

Walker, T. J. 1995. Grasshoppers and crickets (Orthoptera). In J. H. Frank and E. D. McCoy, Precinctive insect species in Florida. *Florida Entomologist* 78:21–35.

Wallace, H. E. 1963. 1963 nutria survey. Florida Game and Fresh Water Fish Commission, 11 February (mimeographed).

Walsh, G. E. 1967. An ecological study of a Hawaiian mangrove swamp. In G. H. Lauff (ed.), *Estuaries.* Washington, D.C.: American Association for the Advancement of Science.

Wamer, N. O. 1978. Avian diversity and habitat in Florida: Analysis of a peninsular diversity gradient. M.S. thesis, Florida State University, Tallahassee.

Wapshere, A. J. 1975. A protocol for programmes for biological control of weeds. *Pest Articles and News Summaries* 21(3):295–303.

———. 1981. Recent thoughts on exploration and discovery for biological control of weeds. In E. G. Delfosse (ed.), *Proceedings of the 5th International Symposium on*

Biological Control of Weeds. Brisbane: Commonwealth Scientific and Industrial Research Organization, Australia.

Ward, D. B. 1989. How many plant species are native to Florida? *Palmetto* 9(4):3–5.

Ware, F. 1974. Progress with *Morone* hybrids in fresh water. *Proceedings of the Southeastern Association of Fish and Wildlife Agencies* 28:48–54.

———. 1995. *Florida Striped Bass.* Educational Bulletin 1. Tallahassee: Division of Fisheries, Florida Game and Fresh Water Fish Commission.

Ware, F., R. Gasaway, R. Martz, and T. Drda. 1975. Investigations of herbivorous fishes in Florida. In P. Brezonik and J. Fox (eds.), *Water Quality Management through Biological Control.* Gainesville: Department of Environmental Engineering Sciences, University of Florida.

Warner, R. E. 1968. The role of introduced species in the extinction of the endemic Hawaiian avifauna. *Condor* 70:101–120.

Warren, G. L., and M. J. Vogel. 1991. Aquatic invertebrate communities of Lake Okeechobee. In Lake Okeechobee–Kissimmee River–Everglades Resource Evaluation. Completion Report to U.S. Department of the Interior, Wallop-Breaux Project No. F-52. Tallahassee: Florida Game and Fresh Water Fish Commission.

Warren, M. L., and B. M. Burr. 1994. Status of freshwater fishes of the United States: Overview of an imperiled fauna. *Fisheries* (Bethesda) 19(1):6–18.

Wassmer, D. A., D. D. Guenther, and J. N. Layne. 1988. Ecology of the bobcat in south-central Florida. *Bulletin of the Florida State Museum, Biological Sciences* 33:159–228.

Watkins, C. E., II, J. V. Shireman, and W. T. Haller. 1983. The influence of aquatic vegetation upon zooplankton and benthic macroinvertebrates in Orange Lake, Florida. *Journal of Aquatic Plant Management* 21:78–83.

Weatherhead, P. J., S. Tinker, and H. Greenwood. 1982. Indirect assessment of avian damage to agriculture. *Journal of Applied Ecology* 19:773–782.

Webb, S. D. 1974. Chronology of Florida Pleistocene mammals. In S. D. Webb (ed.), *Pleistocene Mammals of Florida.* Gainesville: University Presses of Florida.

Weimer, J. E. 1994. Supplemental report on exotic species impacts at Paynes Prairie. Paynes Prairie State Preserve, Florida Park Service.

Weller, M. W. 1969. Potential dangers of exotic waterfowl introductions. *Wildfowl* 20:55–58.

Wenner, A. S., and D. H. Hirth. 1984. Status of the feral budgerigar in Florida. *Journal of Field Ornithology* 55:214–219.

Wenner, E. L., and D. M. Knott. 1992. Occurrence of Pacific white shrimp, *Penaeus vannamei,* in coastal waters of South Carolina. In R. DeVoe (ed.), *Introductions and Transfers of Marine Species: Achieving a Balance between Economic Development and Resource Protection.* Proceedings of the Conference and Workshop, 30 October–2 November 1991, Hilton Head Island, South Carolina. Charleston: South Carolina Sea Grant Consortium.

Westbrooks, R. 1990. Interstate sale of aquatic Federal Noxious Weeds as ornamentals in the United States. *Aquatics* 12:16–18, 24.

———. 1993. Exclusion and eradication of foreign weeds from the United States by

U.S.D.A. A.P.H.I.S. In B. N. McKnight (ed.), *Biological Pollution: The Control and Impact of Invasive Exotic Species*. Indianapolis: Indiana Academy of Sciences.

Westbrooks, R. G., and R. E. Eplee. 1989. Federal noxious weeds in Florida. *Proceedings of the Southern Weed Science Society* 42:316–321.

Westervelt, K. A. 1995. Hurricane Andrew sets the stage for ecological restoration at Cape Florida Park. *Florida Naturalist* 68:14–15.

Westman, W. E. 1990. Park management of exotic plant species: Problems and issues. *Conservation Biology* 4:251–260.

Wheeler, R. J. 1990. Behavioral characteristics of squirrel monkeys at the Bartlett Estate, Ft. Lauderdale. *Florida Scientist* 53:312–316.

Whiteaker, L. D., and R. F. Doren. 1989. Exotic plant species management strategies and list of exotic species in prioritized categories for Everglades National Park. NPS-Research Manual Report Series 89/04. Atlanta: National Park Service.

Whitehead, D. R., and A. G. Wheeler. 1990. What is an immigrant arthropod? *Annals of the Entomological Society of America* 83:9–14.

Whitemore, T. C. 1989. Canopy gaps and two major groups of forest trees. *Ecology* 70:536–537.

Wiens, J. A. 1989. *The Ecology of Bird Communities*. Vol. 2: *Processes and Variations*. Cambridge: Cambridge University Press.

Wilbur, H. M. 1984. Complex life cycles and community organization in amphibians. In P. W. Price, C. N. Slobodchikoff, and W. S. Gaud (eds.), *A New Ecology: Novel Approaches to Interactive Systems*. New York: Wiley.

Williams, D. F. (ed.). 1994. *Exotic Ants: Biology, Impact, and Control of Introduced Species*. Boulder: Westview Press.

Williams, E. E. 1969. The ecology of colonization as seen in the zoogeography of anoline lizards on small islands. *Quarterly Review of Biology* 44:345–389.

———. 1983. Ecomorphs, faunas, island size, and diverse end points in island radiations of *Anolis*. In R. B. Huey, E. R. Pianka, and T. W. Schoener (eds.), *Lizard Ecology: Studies of a Model Organism*. Cambridge: Harvard University Press.

Williams, J. E., J. E. Johnson, D. A. Hendrickson, S. Contreras-Balderas, J. D. Williams, M. Navarro-Mendoza, D. E. McAllister, and J. E. Deacon. 1989. Fishes of North America endangered, threatened, or of special concern: 1989. *Fisheries* (Bethesda) 14(6):2–20.

Williams, M. C. 1980. Purposefully introduced plants that have become noxious or poisonous weeds. *Weed Science* 28:300–305.

Williamson, M., and A. Fitter. 1996. The varying success of invaders. *Ecology* 77:1661–1666.

Wilson, D. E., and D. M. Reeder. 1993. *Mammal Species of the World*. 2nd ed. Washington, D.C.: Smithsonian Institution Press.

Wilson, E. O. 1992. *The Diversity of Life*. Cambridge, Mass.: Harvard University Press.

Wilson, F. 1965. Biological control and the genetics of colonizing species. In H. G. Baker and G. L. Stebbins (eds.), *The Genetics of Colonizing Species*. New York: Academic Press.

Wilson, F., and C. B. Huffaker. 1976. The philosophy, scope, and importance of biological control. In C. B. Huffaker and P. S. Messenger (eds.), *Theory and Practice of Biological Control.* New York: Academic Press.

Wilson, L. D., and L. Porras. 1983. The ecological impact of man on the South Florida herpetofauna. *University of Kansas Museum of Natural History Special Publication* 9:1–89.

Winberry, J. J., and D. M. Jones. 1974. Rise and decline of the "miracle vine": Kudzu in the southern landscape. *Southeastern Geography* 13(2):61–70.

Winston, J. E. 1982. Marine bryozoans (Ectoprocta) of the Indian River area (Florida). *Bulletin of the American Museum of Natural History* 173: 99–176.

———. 1995. Ectoproct diversity of the Indian River coastal lagoon. *Bulletin of Marine Science* 57:84–93.

Winters, H. F. 1967. The mechanics of plant introduction. In *Proceedings of the International Symposium on Plant Introduction.* Tegucigalpa, Honduras: Escuela Agricola Panamericana.

Wirth, W. W. 1979. *Siolimyia amazonica* Fittkau, an aquatic midge new to Florida with nuisance potential. *Florida Entomologist* 62:134–135.

Wojcik, D. P. 1994. Impact of red imported fire ant on native ant species in Florida. In D. F. Williams (ed.), *Exotic Ants: Biology, Impact, and Control of Introduced Species.* Boulder: Westview Press.

Wolfe, J. L. 1968. Armadillo distribution in Alabama and northwest Florida. *Quarterly Journal of the Florida Academy of Sciences* 31:209–212.

Wolfe, L. D., and E. H. Peters. 1987. History of the freeranging rhesus monkeys (*Macaca mulatta*) of Silver Springs. *Florida Scientist* 50:234–245.

Wood, D. 1995. *Official Lists of the Endangered and Potentially Endangered Fauna and Flora in Florida.* Tallahassee: Florida Game and Fresh Water Fish Commission.

Wood, G. W., and R. H. Barrett. 1977. Status of wild pigs in the United States. *Wildlife Society Bulletin* 7(4):237–246.

Wood, J. E. 1959. Relative estimates of fox population levels. *Journal of Wildlife Management* 23:53–63.

Woodall, S. L. 1980. Site requirements for melaleuca seedling establishment. In R. K. Geiger (ed.), *Proceedings of Melaleuca Symposium.* Tallahassee: Florida Division of Forestry.

Wooding, J. B., and T. S. Hardisky. 1990. Coyote distribution in Florida. *Florida Field Naturalist* 18:12–14.

Woodruff, R. E. 1995. Scarab beetles (Coleoptera: Scarabaeidae). In J. H. Frank and E. D. McCoy, Precinctive insect species in Florida. *Florida Entomologist* 78:21–35.

Woods, J. W. 1963. Aerial application of herbicides. *Hyacinth Control Journal* 2:20.

Woods, K. D. 1993. Effects of invasion by *Lonicera tatarica* L. on herbs and tree seedlings in four New England forests. *American Midland Naturalist* 130:62–74.

Woods Hole Oceanographic Institution. 1952. *Marine Fouling and Its Prevention.* Annapolis, Md.: United States Naval Institute.

Workman, R. 1979. *Schinus.* Technical proceedings of techniques for control of

Schinus in South Florida: A workshop for natural area managers. Sanibel: Sanibel-Captiva Conservation Foundation.

Wunderlich, W. E. 1967. The use of machinery in the control of aquatic vegetation. *Hyacinth Control Journal* 6:22-24.

Wunderlin, R. P. 1982. *Guide to the Vascular Plants of Central Florida.* Gainesville: University Presses of Florida.

Wurtz, C. B., and S. S. Roback. 1955. The invertebrate fauna of some Gulf Coast rivers. *Proceedings of the Academy of Natural Sciences of Philadelphia* 107:167-206.

Wynne, M. J. 1993. The recognition of a new species of *Prionitis* (Halymeniaceae, Rhodophyta) from the Texas coast. *Journal of Phycology* 29, suppl. 8.

Yih, K., D. H. Boucher, J. H. Vandermeer, and N. Zamora. 1991. Recovery of the rain forest of southeastern Nicaragua after destruction by Hurricane Joan. *Biotropica* 23:106-111.

Yotsui, T., and S. Migita. 1989. Cultivation of a green alga *Codium fragile* by regeneration of medullary threads. *Bulletin of the Japanese Society of Science and Fisheries* 55:41-44.

Young, S. P., and H.H.T. Jackson. 1951. *The Clever Coyote.* Washington, D.C.: Wildlife Management Institution.

Zaneski, C. T. 1995. Once scorned tree finds new champions. *Miami Herald,* 11 April, pp. 1A, 7A.

Zeiger, C. F. 1962. Hyacinth-obstruction to navigation. *Hyacinth Control Journal* 1:16-17.

Zibrowius, H. 1971. Les espèces Méditerranéenes du genre *Hydroides* (Polychaeta, Serpulidae) remarques sur le pretendu polymorphisme de *Hydroides uncinata.* *Tethys* 2:691-746.

Zullo, V. A. 1992. *Balanus trigonus* Darwin (Cirripedia, Balaninae) in the Atlantic basin—an introduced species. *Bulletin of Marine Science* 50:66-74.

CONTRIBUTORS

H. Hugh Boyter, Jr.
Florida Game and Fresh Water Fish
    Commission
620 South Meridian St.
Tallahassee, FL 32399-1600

Tom C. Brown
Bureau of Aquatic Plant Management
Florida Department of Environmental
    Protection
Tallahassee, FL 32310

Brian P. Butterfield
Department of Zoology and Wildlife Science
Auburn University
Auburn University, AL 36849-5414

James T. Carlton
Maritime Studies Program
Williams College–Mystic Seaport
P.O. Box 6000, 75 Greenmanville Ave.
Mystic, CT 06355

Ted D. Center
U.S. Department of Agriculture
Agricultural Research Service
Fort Lauderdale, FL 33314

Walter R. Courtenay, Jr.
Department of Biological Sciences
Florida Atlantic University
Boca Raton, FL 33431-0991

James A. Cox
Florida Game and Fresh Water Fish
    Commission
620 South Meridian St.
Tallahassee, FL 32399-1600

Robert F. Doren
South Florida Natural Resources Center
Everglades National Park
40001 State Road 9336
Homestead, FL 33034

F. Allen Dray, Jr.
University of Florida
Institute of Food and Agriculture Science
Fort Lauderdale Research Education Center
Fort Lauderdale, FL 33314

Amy Ferriter
South Florida Water Management District
Vegetation Management Division
West Palm Beach, FL 33416

J. Howard Frank
Entomology and Nematology Department
University of Florida
Gainesville, FL 32611-0620

David Girardin
Operations Department
St. Johns River Water Management District
Palatka, FL 32177

Mark W. Glisson
Bureau of Natural and Cultural Resources
Florida Department of Environmental
    Protection
3900 Commonwealth Blvd.
Tallahassee, FL 32399-3000

Doria R. Gordon
The Nature Conservancy
Department of Botany
P.O. Box 118526
University of Florida
Gainesville, FL 32611

Craig Guyer
Department of Zoology and Wildlife Science
Auburn University
Auburn University, AL 36849-5414

H. Glenn Hall
Entomology and Nematology Department
University of Florida
Gainesville, FL 32611-0620

William Haller
IFAS Center for Aquatic Plants
University of Florida
Gainesville, FL 32606

Ronald H. Hofstetter
Department of Biology
University of Miami
P.O. Box 249118
Coral Gables, FL 33124

Carol C. Horvitz
Department of Biology
University of Miami
P.O. Box 249118
Coral Gables, FL 33124

Frances C. James
Department of Biological Science
Florida State University
Tallahassee, FL 32306-2043

Deborah B. Jensen
International Home Office
The Nature Conservancy
Arlington, VA 22209

David T. Jones
South Florida Natural Resources Center
Everglades National Park
40001 State Road 9336
Homestead, FL 33034

James N. Layne
Archbold Biological Station
Lake Placid, FL 33852

Roy R. Lewis III
Lewis Environmental Services, Inc.
Tampa, FL 33622

Earl D. McCoy
Biology Department
University of South Florida
Tampa, FL 33620-5150

Mark D. Maffei
U.S. Fish and Wildlife Service
Division of Habitat Conservation
Arlington, VA 22203
Current address: Androck Hardware
   Corporation
711 19th St.
Rockford, IL 61104

Walter E. Meshaka, Jr.
Museum, Everglades National Park
40001 SR-9226
Homestead, FL 33034-6733

Brian Nelson
Operations Department
Southwest Florida Water Management
   District
Brooksville, FL 34609

George F. O'Meara
Florida Medical Entomology Laboratory
200 9th St. SE
Vero Beach, FL 32962

Lt. Thomas G. Quinn
Florida Game and Fresh Water Fish
   Commission
620 South Meridian St.
Tallahassee, FL 32399-1600

John M. Randall
The Nature Conservancy
Wildland Weeds Management and Research
Section of Plant Biology
University of California
Davis, CA 95616

Tony Richards
Land Resources Department
Southwest Florida Water Management
   District
Brooksville, FL 34609

Mary H. Ruckelshaus
Department of Biological Science
Florida State University
Tallahassee, FL 32306-2043

Jeffrey D. Schardt
Florida Department of Environmental
   Protection
2051 East Dirac Drive
Tallahassee, FL 32310

Don C. Schmitz
Florida Department of Environmental
   Protection
Bureau of Aquatic Plant Management
Tallahassee, FL 32310

Daniel Simberloff
Department of Biological Science
Florida State University
Tallahassee, FL 32306-2043

David Sutton
IFAS Research and Education Center
University of Florida
Fort Lauderdale, FL 33314

Dan Thayer
South Florida Water Management District
Vegetation Management Division
West Palm Beach, FL 33416

Kevin P. Thomas
The Nature Conservancy
Department of Botany
P.O. Box 118526
University of Florida
Gainesville, FL 32611

Walter R. Tschinkel
Department of Biological Science
Florida State University
Tallahassee, FL 32306-3050

Gary L. Warren
Florida Game and Fresh Water Fish
    Commission
Division of Fisheries
Okeechobee, FL 34974

Randy G. Westbrooks
Weed scientist
Whiteville, NC 28472

INDEX

1,000 Points of Light Foundation 212
2,4-D (phenoxy herbicide) 232, 234–235,
    259–260
Abrus precatorius, rosary pea 24, 68
acacia, earleaf, *Acacia auriculiformis* 24, 269
Acacia 33, 69
    auriculiformis, earleaf acacia 24, 269
    choriophylla, tamarindillo 293
Acanthophthalmus kuhli, kuhli loach 299
acara
    black, *Cichlasoma bimaculatum* 111, 114, 270
    blue, *Aequidens pulcher* 113
Acarapis woodi, honeybee tracheal mite 80, 91
Achatina fulica, giant African snail 223–225,
    253–254, 366
Achatinella mustelina 254
Acheta domesticus 98
Acipenser oxyrynchus desotoi, Gulf sturgeon 116
Acridotheres tristis, common myna 140–141, 147,
    153–154
Acyrthosiphon pisum, pea aphid 83, 89
Adams Key 167
Adenanthera pavonina, red sandalwood 26,
    67–68
adventive, definition of 245
Aedes
    aegypti, yellow-fever mosquito 86
    albopictus, Asian tiger mosquito, forest day
        mosquito 86, 337
Aequidens pulcher, blue acara 113
aesthetic value 257
Agaonidae, fig wasps (*see also* individual taxa)
    90–91
agaonid wasps (*see also* individual taxa) 77
Agasicles hygrophila, alligatorweed flea beetle 82,
    99, 248–249, 260
Agave sisalana, sisal hemp 26
Agelaius phoeniceus, red-winged blackbird 145
agoutis, *Dasyprocta* spp. 301
Agricultural Research Service 334
Agricultural Stabilization and Conservation
    Service 342
Agrilus hyperici 248
Agrypon flaveolatum 249
Aimophila aestivalis, Bachman's sparrow 147
air potato, *Dioscorea bulbifera* 24, 56, 269
    effects of in Florida 41–42, 51, 67, 215–216,
        292
    management of 216, 290, 322–323
    noninvasive congeners of 60–61
Alabama 97, 105, 127, 165, 170, 173, 261
Alachua County
    biological control in 87, 92
    hydrilla in 54, 239
    National Fisheries Research Center in 336
    nonindigenous birds in 150, 306

nonindigenous mammals in 168–172, 185
Albizia 69
    lebbeck, woman's tongue 26
Alectoris chukar, chukar 306
Aleochara 85
Aleurocanthus woglumi, citrus blackfly 88, 97, 99,
    222, 249, 259
Aleyrodidae, whiteflies 80, 88, 89
algae (*see also* individual taxa) 187, 196–198
alkalinity 46
Allee effect 6
allelopathy 10–11
alligator, *Alligator mississipiensis* 180, 303–304,
    316
Alligator Alley (Interstate 75) 311, 320
alligator farms 298–299
Alligator mississipiensis, see alligator
alligatorweed, *Alternanthera philoxeroides*
    biological control of 58, 82, 99, 248–249, 260,
        319, 362
    introduction and extent of in Florida 26, 269
Alopochen aegyptiaca, Egyptian goose 306
Alsophis vudii 137
Alternanthera philoxeroides, see alligatorweed
Amazona
    amazonica, orange-winged parrot 141, 154
    viridigenalis, red-crowned parrot 141
Ameiurus
    brunneus, snail bullhead 115
    serracanthus, spotted bullhead 115
Ameiva
    ameiva 124, 128, 131, 135
    auberi 137
Amelia Island 152
Amitus hesperidum 88, 99, 249
Ammotragus levuia, Barbary sheep 305
amphibians (*see also* individual taxa) 44,
    123–138, 263, 272, 316, 360–362
Amphipoda (*see also* individual taxa) 50, 101,
    188, 190–191, 193
amur, white, *see* carp, grass
Amynothrips andersoni, alligatorweed thrips 82,
    96, 248, 260
Anabantidae (*see also* gourami(es)) 111, 113
Anabas testudineus, climbing perch 113
Anaea troglodyta, Florida leafwing 78
Anagyrus antoninae 90
Ananas, pineapple 84
Anas
    fulvigula fulvigula, Florida mottled duck
        11–12, 147–148
    platyrhynchos, mallard 11–12, 147–148
Anastrepha ludens, Mexican fruit fly 87
Ancylostoma pluridentatum, a hookworm 181
Andros 136

angelfish
    Arabian, *Pomacanthus asfur* 116
    freshwater, *Pterophyllum scalare* 113
    yellow-mask, *Pomacanthus xanthometopon* 116
Animal and Plant Health Inspection Service
    (APHIS) 226, 329, 331–334, 336
Animal Damage Control Program 333
animals, nonindigenous (*see also* individual taxa)
    216–218, 270–272
Anisoplia austriaca, wheat cockchafer 247
Ann Arbor, Michigan 336
Annelida, nonindigenous (*see also* individual taxa)
    191
anole(s) (*see also* individual taxa) 7, 124–126, 134
    brown, *see Anolis sagrei*
    green, *see Anolis carolinensis*
Anolis
    angusticeps 137
    carolinensis, green anole 132–134, 272
    chlorocyanus 124–125
    cristatellus 6, 124, 128, 131, 134
    cybotes 124, 131
    distichus 124, 131–132, 134, 136
    equestris 124, 128, 131–132, 134
    garmani 124, 131
    pulchellus 6
    sagrei, brown anole ix, 44, 124, 126–127,
        131–135, 270, 272
    smaragdinus 137
Anoplura (*see also* individual taxa) 81
ant, Argentine, *Iridomyrmex humilis* 92
ant, Argentine, *Linepithema humile* 10
ant, big-headed, *Pheidole megacephala* 10
ant, crazy, *Paratrechina longicornis* 92
ant, little fire, *Wasmannia auropunctata* 71–72
ant, red imported fire, *Solenopsis invicta*
    beneficial effects of in Florida 92
    damage caused by in Florida 92
    effects of on natural areas 218, 272
    effects of on other biota 10, 92, 135, 137, 184,
        312
    factors affecting spread of 15   , 61, 221–222
ant, pharaoh, *Monomorium pharaonis* 92
anteater, lesser 158
ant lions, Neuroptera 79
Anthonomus grandis, boll weevil 83–84, 92, 277
Anthozoa (*see also* individual taxa) 10
Anthribidae, weevils (*see also* individual taxa) 79
Antigonon leptopus, coral vine 26
Antigua 94
Antilocapra americana, pronghorn 159, 174,
    305–306, 315
Antithamnion nipponicum 197, 199
Antonina graminis, Rhodes-grass mealybug 90
ants, Formicidae (*see also* individual taxa) ix, 79,
    92, 247, 362
anurans (*see also* individual taxa) 125
Aonidiella aurantii, California red scale 82, 89
Apalachee Bay 169
Apalachicola, Florida 174
Apalachicola River 104, 110, 115
apes, Celebes black 316

Aphelinidae, aphelinid wasps (*see also* individual
    taxa) 79, 91
Aphelinus semiflavus 249
Aphelocoma coerulescens, Florida scrub jay 147
aphid
    brown citrus, *Toxoptera citricida* 89
    green peach, *Myzus persicae* 83
    pea, *Acyrthosiphon pisum* 83, 89
    spirea, *Aphis spiraecola* 83, 89
    spotted alfalfa, *Therioaphis maculata* 83, 89
    yellow clover, *Therioaphis trifolii* 249
    yellow sugarcane, *Sipha flava* 83
Aphididae, aphids (*see also* individual taxa) 89
APHIS, *see* Animal and Plant Health Inspection
    Service
Aphis spiraecola, spirea aphid 83, 89
Aphytis
    holoxanthus 10, 89, 91
    lepidosaphes 89
    lingnanensis 91
Apidae, bees (*see also* individual taxa) 91–92
Apis mellifera 85, 91–92, 98, 340
Apopka, Lake 86
apple 256
aquaculture (*see also* mariculture) 120, 188, 199,
    339, 342
Aquarium du Musée Océanographique de
    Monaco 198
aquarium industry
    introduction of nonindigenous species by 102,
        107–108, 227, 237
    regulation of 200, 299
    role of in species introductions 114–115, 120,
        188, 227, 351, 356
Aquatic Plant Trust Fund 233
Aquatic Species Task Force 335
Aquilegia canadensis var. *australis,* columbine 292
Ara severa, chestnut-fronted macaw 141, 154
Aratinga
    acuticaudata, blue-crowned parakeet 141, 154
    erythrogenys, red-masked parakeet 141, 154
    weddellii, dusky-headed parakeet 141, 154
arboviruses 183
Arcadia, Florida 192
Archbold Biological Station 168
Archie Carr National Wildlife Refuge 268, 271
Ardisia 61
    crenulata 52–53, 292
    elliptica, shoebutton ardisia 24–25, 65–68,
        70–71, 73, 277–278, 285
    escallonioides, marlberry 69–70
    humilis, see Ardisia elliptica
    polycephala, see Ardisia elliptica
ardisia, shoebutton, *see Ardisia elliptica*
Argentina 82, 146, 260
Argusia gnaphalodes, sea lavender 293
Aripeka, Florida 169
Aristichthys nobilis, see carp, bighead
Arizona 127
armadillo, nine-banded, *Dasypus novemcinctus*
    135, 157–159, 164–165, 177–180,
    182–185, 270, 272, 293

aroma, *Dichrostachys cinerea* 27
Arrhenophagus albitibiae 97
arrowhead vine, *Syngonium podophyllum* 29
Arsenal 283
Arthur R. Marshall Loxahatchee National Wildlife
   Refuge 268, 270–273, 330
Artiodactyla (*see also* individual taxa) 159,
   173–176
Ascidiacea 192
ascidians (*see also* individual taxa) 188–189
Asparagus densiflorus, asparagus fern 26
asparagus fern, *Asparagus densiflorus* 26
Aspidiotus destructor, coconut scale 83, 89
Astronotus ocellatus, oscar 111, 114, 270, 272
Asystasia gangetica, Ganges primrose 26
atala, Florida, *Eumaeus atala* 78
Aucilla River 165
Australia
   animals from 82–83, 90, 102, 150, 193–194,
      247, 261
   biological control in 218, 247–248, 257, 263
   nonindigenous animals in 152, 154, 224, 255
   nonindigenous plants in 23, 34–35, 58, 68–69,
      197
   plants from 239, 251–252
Australian Biological Control Act of 1984
   255
Australian Commonwealth Scientific and
   Industrial Research 255
Australian pine, *Casuarina* spp. (*see also*
      individual species)
   distribution of in Florida ix, 40–42, 66, 163,
      269, 279, 321
   interaction of with other species 45, 56, 163,
      211, 284
   introduction and spread of 14, 24, 26, 333
   interaction of with habitat disturbance 51,
      54–55, 66–67, 277
   management of 210–211, 263, 273, 277–280,
      282–284, 286, 290, 294, 310, 320–321,
      336, 240
   popular opposition to removal of 228, 251,
      288, 294, 350, 353
Australian pine, suckering (*see Casuarina glauca*)
Australian Weeds Committee 34
Avicennia germinans, black mangrove 162, 164
aviculture industry 143–144, 155
avocado 97
Avon Park Bombing Range 174
Axis axis, axis deer 159, 173–174
Azadirachta spp. 33

baboons 158, 316
Bacardi wildlife research facility 169
bacteria 198
badgers 316
Bagous
   affinis 261
   hydrillae 261
Bahamas
   amphibians and reptiles from 128, 136–137,
      312

birds from 142, 148–149
fishes from 116
nonindigenous invertebrates from 97, 190
nonindigenous invertebrates in 94, 190
Bahia Honda Key 161
Baker County 268
Balanus
   amphitrite 190
   reticulatus 190
   trigonus 190
ballast 188, 191–192, 199–200
banana 95
banana poka 14–15
Bankia
   caribbea 192
   carinata 192
   fimbriatula 190, 192
Banksia spp. 32–33
banyan tree, *Ficus altissima* 27, 91
barb
   dwarf, *Puntius gelius* 112
   rosy, *Puntius conchonius* 112
   tiger, *Puntius tetrazona* 112
   tinfoil, *Barbodes schwanefeldi* 112
Barbodes schwanefeldi, tinfoil barb 112
barnacles 189–190, 194
barramundi cod, *Cromileptes altivelis* 112, 115
barrier islands 290–291
Bartlett Estate 161
Basiliscus
   plumifrons 124–125, 131
   vittatus 124, 128, 131
bass
   largemouth, *Micropterus salmoides floridanus*
      46, 114, 117, 119
   peacock, *see* cichlid, peacock
   sea, Serranidae 112
   shoal, *Micropterus* n. sp. 116
   "sunshine," *Morone chrysops* x *M. saxatilis*
      hybrid 307–308
   temperate, Moronidae 112, 120
   white, *Morone chrysops* 112
bat
   Brazilian free-tailed, *Tadarida brasiliensis*
      160–162
   velvety free-tailed, *Molossus molossus*
      tropidorhynchus 159–161
Bauhinia spp. 33
   variegata, orchid tree 26
Bay County 167
bayonet, Spanish, *Yucca aloifolia* 269
beach star, *Remirea maritima* 293
bear(s), Ursidae 180, 182, 302, 316
   black, *Ursus americanus* 180
bee(s), Apidae (*see also* individual taxa) 91–92
   African honey, *Apis mellifera scutellata* 91–92,
      340
   European honey, *Apis mellifera mellifera* 85,
      91, 98
   halictid 69
bee louse (lice), Braulidae 85, 91
   Braula caeca 85, 91

beefwood, scaly-bark, *Casuarina glauca* (*see also* Australian pine) 24, 269
beetle(s), Coleoptera (*see also* individual taxa) 79, 81–85, 247–248, 262, 362
  alligatorweed flea, *Agasicles hygrophila* 82, 99, 248–249, 260
  ambrosia, Platypodidae 79, 84
  bark, Scolytidae 79, 84
  coconut leaf-mining 12
  Colorado potato, *Leptinotarsa decemlineata* 9, 81
  dung, Scarabaeidae 79, 84, 291
  European elm bark, *Scolytus multistriatus* 252
  flat bark, Cucujidae 79, 83–84
  flat bark, Laemophloeidae 79, 83
  flat bark, Passandridae 79, 83
  flat bark, Silvanidae 79, 83
  flea 99, 257
  grain 83
  ground, Carabidae 79, 81
  Japanese 8
  ladybird, Coccinellidae 82–83
  leaf, Chrysomelidae 73, 81–82
  merchant grain, *Oryzaephilus mercator* 83
  Mexican bean, *Epilachna varivestis* 82
  rove, Staphylinidae 79, 85
  sap, Nitidulidae 79
  saw-toothed grain, *Oryzaephilus surinamensis* 83
  seed, Bruchidae 79, 81
  vedalia, *Rodolia cardinalis* 83, 90, 247
  white-fringed, *Graphognathus* spp. 83
Belize 190
Belle Glade, Florida 147, 170
bellflower, *Campanula americana* 292
Belonesox belizanus, pike killifish 111, 120, 270
Bemisia
  argentifolii, silverleaf whitefly 80, 88
  tabaci, sweet-potato whitefly 80, 88
Bermuda 197
Betta splendens, Siamese fighting fish 113
Bibionidae, march flies 85
Big Cypress National Preserve
  location and management of 230, 330, 336, 240
  nonindigenous species in 48, 114, 117, 165, 178, 240
Big Pine Key 48, 218
Bill Baggs State Park 66–67
Bimini 136
binturong 158, 316
biodiversity 73, 343, 352, 359
biological control 77, 88, 98–99, 218, 245–263, 350
  definition of 246
  economic factors related to 80, 90, 251, 258–260, 98
  factors hindering 90, 95, 222, 251–253
  of insects 5–7, 82, 96
  of plants 168, 218, 227, 236–237, 241, 319–320, 322

risks and benefits of 99, 250, 257–262, 334, 354, 364
  by vertebrates 153–154, 168, 350
Biosphere Reserves 275
biotic resistance 3, 51, 360
bird trade 139, 301
birdpox 146
birds (*see also* individual taxa) 143–156
  effects of nonindigenous invertebrates on 92, 106, 256, 272
  effects of nonindigenous plants on 44, 46, 50, 54, 58, 238
  effects of nonindigenous vertebrates on 178–179, 182, 217, 254, 272, 309
  effects of on other species 43, 120, 144–147, 281, 361–362
  number of species of in Florida 76, 140, 263, 360
  parasites of 10, 13
Biscayne Bay 115, 141, 158, 166
Biscayne National Park 179
Biscayne River Canal 224
bischofia, *Bischofia javanica* 26, 67–68
Bischofia javanica, bischofia, Bishopwood 26, 67–68
Bishopwood, *Bischofia javanica* 26, 67–68
bison, *Bison bison* 302
Bison bison, bison 302
bivalves, Pelecypoda 101
Bivalvia 191–192
blackbird, red-winged, *Agelaius phoeniceus* 145
blackfly, citrus, *Aleurocanthus woglumi* 88, 99, 222, 249, 259
Blackwater Forest State Park 172
bladderpod, *Sesbania vesicaria* 269
Blarina carolinensis shermani, Sherman's short-tailed shrew 179
Blattaria, cockroaches 81
Blattella asahinai, Asian cockroach 81
Blattodea, cockroaches 79
blind mosquitoes, *see* Chironomidae
blowflies, Calliphoridae 85
Blowing Rocks Preserve 211–212
bluebird, eastern, *Sialia sialis* 10, 147
bluegill sunfish, *Lepomis macrochirus* 46
Boa constrictor 7, 124, 131–132
boars, Eurasian wild 175–176
boas, Boidae 7, 124, 131–132, 304
bobcat, *Lynx rufus* 180–182, 316
Boca Chica Key 160
Boca Raton 166, 191, 311
Boccardiella ligerica 189, 191
Boidae, boas 7, 124, 131–132, 304
Boiga irregularis, brown tree snake 5, 137, 304, 314, 333, 336, 362
Boisgiraud 247
Bolivia 92
Bombyx mori, oriental silkworm 76, 98
Bonnemaisonia hamifera 197
Boreioglycaspis melaleucae 262

borer, alligatorweed stem, *Vogtia malloi* 82, 95, 260, 248
botfly (botflies), Oestridae 87
   horse, *Gasterophilus intestinalis* 87
   nose, *Gasterophilus haemorrhoidalis* 87
   sheep, *Oestrus ovis* 87
   throat, *Gasterophilus nasalis* 87
Botrylloides nigrum, see Botryllus niger
Botryllus
   niger 192
   schlosseri 192
bowstring hemp, *Sansevieria hyacinthoides* 29
Boynton Beach 191
Brachiaria mutica, para grass 269
Braconidae 91
Bradford County 169
Branchiura sowerbyi 102
Branta canadensis, Canada goose 307
Brassaia actinophylla 25
Braula caeca, bee louse 85, 91
Braulidae, bee lice 85, 91
Brazil 120, 193–194, 330
Brazilian pepper, *Schinus terebinthifolius*
   biological control of 260, 262–263
   distribution of in Florida ix, 40–41, 43–44, 66, 213–214, 269, 271, 279–282, 321
   effects of 14, 43–44, 55–57, 144, 209–211, 268–271, 291
   interaction of with habitat disturbance 51, 66, 68, 210, 284–285
   interaction of with other species 56, 71, 93, 97, 163, 281, 284–285
   introduction and spread of 15, 25, 30, 52, 59, 67, 71, 208–209, 268–271, 291
   management of 211, 229, 260, 262–263, 273, 277, 283–284, 286, 291, 310, 320–322, 336
Brazilian satintail, *Imperata brasiliensis* 27, 33, 45
Brentidae, weevils 79
Brevard County 141, 152, 161, 163–165, 169, 268, 291
British Columbia 154
British Ministry of Agriculture 226–227
bromeliads 84, 362
Brooker, Florida 169
Brooksville Plant Introduction Garden (USDA) 30
Brotogeris versicolurus, canary-winged parakeet 141, 143, 151
Broward County
   nonindigenous birds in 150, 154
   nonindigenous fishes in 114, 117, 119
   nonindigenous insects in 72–73, 82, 84, 92, 222
   nonindigenous mammals in 161, 163, 165–166, 171
   nonindigenous molluscs in 107, 224
   nonindigenous plants in 46, 48, 58, 232, 288
   nonindigenous reptiles in 136
brucellosis 183, 292

Bruchidae, seed beetles 79, 81
bryozoans (*see also* individual taxa) 188–189, 192
Bubulcus ibis, cattle egret 140
Buck Island Ranch 174
budgerigar, *Melopsittacus undulatus* 141, 143, 150
buffalo (*see also* bison) 7, 158, 316
Bufo 126
   marinus 124, 131–132, 247, 270, 272, 313
   terrestris 135
bug(s), Hemiptera (*see also* individual taxa) 79, 88, 247, 262
   southern green stink, *Nezara viridula* 88
   sugarcane lace, *Leptodictya tabida* 88
bulbul(s) 15, 141, 151–152
   red-whiskered, *Pycnonotus jocosus* 7, 141–144, 146, 151–152,
bullhead
   snail, *Ameiurus brunneus* 115
   spotted, *Ameiurus serracanthus* 115
bull's-eye model 206
Bumblebee Island 176, 179
bunting, indigo, *Passerina cyanea* 147
Bureau of Aquatic Plant Management 325, 343–344, 346, 350
Bureau of Coastal and Aquatic Managed Areas 343
Bureau of Indian Affairs 336
Bureau of Land Management 336
Burma reed, *Neyraudia reynaudiana* 25, 44, 56
burning, controlled or prescribed, *see* fire regime
burr, noogoora, *Xanthium strumarium* 249
butterflies and moths, Lepidoptera (*see also* individual taxa) 75, 78–79, 93–95, 102, 256
butterfly, Baltimore checkerspot, *Euphydryas phaeton* 256
butterfly zoos 94
buttonwood 280

cabbageworm, imported, *Pieris rapae* 94
Cactaceae, cacti (*see also* individual taxa) 94, 254, 362
Cactoblastis cactorum, see moth, cactus
cactus, semaphore, *Opuntia corallicola* 94, 256
Caesars Creek 167
Caiman crocodilus, caiman 124, 131–132, 303
caimans 7
   black 316
   Caiman crocodilus 124, 131–132, 303
Cairina moschata, Muscovy duck 141, 143, 148, 306
Calhoun County 168
California
   biological control in 89–90, 247, 262
   eradication campaigns in 154, 225, 227
   nonindigenous animals in 127, 142, 146, 150, 152–154, 160, 225
   nonindigenous plants in 222
   vulnerability of to invasion 14, 359
Callichthyidae, plated catfishes 112
Callichthys, cascarudo 112

Calliphoridae, blowflies 85
Callisia fragrans, inch plant, spironema 26
Caloosahatchee National Wildlife Refuge 268
Caloosahatchee River 184, 231
Calophyllum
    calaba mast wood, Alexandrian laurel 26
    inophyllum 26
Calosoma 81
    sycophanta 247
Calotes versicolor 124–126
camelias 89
Campanula americana, bellflower 292
camphor tree, *Cinnamomum camphora* 24
Canada 263, 351
canaries 316
canids (*see also* individual taxa)
    wild 181–182
Canis familiaris, see dog
Canis latrans, see coyote
Canis lupus, see dog
Canis rufus, red wolf 181, 272, 316
canistel, *Pouteria campechiana* 29
canker, citrus, *Xanthomonas campestris* pv. *citri*
    224
Cape Florida State Recreation Area 294
Cape Sable 276, 278
Capra hircus, goat 159, 176, 179, 226
Capra ibex, ibex 305
Captiva Island 165
capybara, *Hydrochaeris hydrochaeris* 159,
    169–170
Carabidae, ground beetles 79, 81
caracals 316
caracara, crested, *Caracara plancus* 180
Caracara plancus, crested caracara 180
Carassius auratus, goldfish 112
carbaryl 224
Card Sound 278
cardinal, red-crested, *Paroaria coronata* 141, 154
Caretta caretta, loggerhead sea turtle 211, 278
Carica papaya, papaya 64–65, 67–68, 72–73
Carnivora (*see also* individual taxa) 159–160,
    170–173, 175
Carolinas 226
carp(s), Cyprinidae 111–112
    bighead, *Aristichthys nobilis* 300
    bighead, *Hypophthalmichthys nobilis* 112
    common, *Cyprinus carpio* 110–111, 118, 214
    grass, *Ctenopharyngodon idella* 112, 214, 227,
        248, 253–254, 300, 308, 319, 336, 350
    silver, *Hypophthalmichthys molitrix* 300
    snail or black, *Mylopharyngodon piceus* 300
Carpodacus mexicanus, house finch 141–142,
    152–153
carrotwood, *Cupaniopsis anacardioides* 24,
    44–45, 51, 263, 269
Carter, President Jimmy 335
cascarudo, *Callichthys* 112
Cassia spp. 33
cassowaries 316
Castanea dentata, American chestnut 13
Castellow Hammock 65–74, 215

castor bean, *Ricinus communis* 65, 213
Casuarina
    cunninghamiana (*see also* Australian pine) 26,
        42
    equisetifolia (*see also* Australian pine) 24, 42, 66,
        211, 269, 278–280, 288, 290
    glauca, suckering Australian pine (*see also*
        Australian pine) 24, 42, 269, 278–280
    litorea (*see Casuarina equisetifolia*)
cat, African golden 316
cat, domestic, *Felis catus*
    biological control by 246–247, 253
    effects of on native biota 10, 179–183, 312
    origin and status of in Florida 143, 159, 175,
        185, 270, 293–294, 324
    parasites of 86
cat, fishing 316
cat, Temminck's golden 316
catbird, gray, *Dumetella carolinensis* 70–71
catfish(es)
    African electric, Malapteruridae 300
    air-breathing, Clariidae 111, 300
    air-sac, Heteropneustidae 300
    armored, *Hypostomus* sp. 111
    candiru, Trichomycteridae 299–300
    electric 301
    flathead, *Pylodictus olivaris* 111, 115–116,
        118, 121
    freshwater, Ictaluridae 111
    granulated, *Pterodoras granulosus* 112
    long-whiskered, Pimelodidae 112
    plated, Callichthyidae 112
    Raphael, *Platydoras costatus* 112
    red-tail, *Phractocephalus hemioliopterus* 112
    ripsaw, *Pseudodoras niger* 112
    sailfin, *Liposarcus multiradiatus* 111
    suckermouth, Loricariidae 111
    thorny, Doradidae 112
    vermiculated sailfin, *Liposarcus disjunctivus*
        111
    walking, *Clarias batrachus* 7, 111, 114, 116,
        135, 214, 270, 300
Cathartes aura, turkey vulture 180
cat's claw, *Macfadyena unguis-cati* 28
cattle 87, 183, 221
cattle grub
    common, *Hypoderma lineatum* 87
    northern, *Hypoderma bovis* 87
Caulerpa 200
    scapelliformis 197
    taxifolia 197–199
cedar, bay, *Suriana maritima* 293
Cedar Keys National Wildlife Refuge 268
Celithemis elisa 95
cellular automata 8
Center for Plant Conservation 218
Central and Southern Flood Control District 232
Centrarchidae, sunfishes (*see also* individual taxa)
    111–112, 120
Ceratitis capitata, Mediterranean fruit fly 87–88,
    223, 225, 252
Ceratocystis fagacearum, oak wilt disease 13

Ceratopogonidae, midges (*see also* individual taxa) 101
Cercopithecus aethiops, vervet monkey 159, 161–162, 183
Cercospora rodmanii 260
cereus, night-blooming, *Cereus undatus* 26
Cereus undatus, night-blooming cereus 26
Cervidae, deer (*see also* individual taxa) 280, 302
Cervus
	elaphus, elk 159, 174, 306
	unicolor, sambar deer 159, 174, 270, 272, 306
Cestrum diurnum, day jasmine 26
Cetonia 247
Chagas disease 183
channelization 214
Channidae, snakeheads 300
Chaoboridae, midges (*see also* individual taxa) 101
Chapman Field Station 30
Characidae, characins 112–113
characins, Characidae 112–113
character displacement 133
Charles Deering estate 215
Charles River Laboratory 163
Charlotte County 84, 150, 171, 184, 213, 268
Charybdis 201
	helleri 191
Chassahowitzka National Wildlife Refuge 268, 272
cheetahs 316
Cheilostomata, bryozoans 192
Chelone glabra, turtlehead 256
Chelonia mydas, green sea turtle 211, 278
Chelura terebrans 191
Chelymorpha cribraria 72–73, 77, 81–82
Cherax quadricarinatus, Australian red-claw crayfish 102–103
chestnut, American, *Castanea dentata* 13
chestnut blight fungus, *Cryphonectria parasitica* 13
Chiefland, Florida 172
chimpanzee, *Pan troglodytes* 302, 316
China 89, 93, 144, 261
chinaberry, *Melia azedarach* 25
chinchilla 158, 316
Chiococca alba, snowberry 71
chipmunks 316
Chironomidae, midges, "blind mosquitoes" 86, 101, 105
Chiroptera (*see also* individual taxa) 159–161
chlordane 259
Chloris gayana, Rhodes grass 90
Chlorocebus aethiops, see monkey, vervet
Chlorophyta, green algae (*see also* individual taxa) 197
Chlorostrymon maesites, maesites hairstreak 78
Choctawhatchee Bay 110
Choctawhatchee River 110
cholera virus, hog 183
Chordata, nonindigenous (*see also* individual taxa) 192
Chrysolina
	hyperici 248, 257

quadrigemina 248, 257
Chrysomelidae, leaf beetles 73, 81–82
Chrysomphalus aonidum, Florida red scale 89
Chrysomyia 85
chub, southern bluehead, *Nocomis leptocephalus bellicus* 111
chukar, *Alectoris chukar* 306
Cichla
	ocellaris, peacock cichlid, peacock bass 111, 114, 118–120, 307, 314
	temensis, speckled pavon 113–114, 118, 120, 307
Cichlasoma
	bimaculatum, black acara 111, 114, 270
	citrinellum, Midas cichlid 111
	cyanoguttatum, Rio Grande cichlid 111
	managuense, jaguar guapote 111, 115
	meeki, firemouth 111
	octofasciatum, Jack Dempsey 111
	salvini, yellow-belly cichlid 111
	trimaculatum, three-spot cichlid 113, 226
	urophthalmus, Mayan cichlid 111, 114–115
cichlid(s), Cichlidae 111, 113
	banded, *Heros severus* 113
	bass, *Cichla ocellaris* 111, 307
	Mayan, *Cichlasoma urophthalmus* 111, 114–115
	Midas, *Cichlasoma citrinellum* 111
	peacock, *Cichla ocellaris* 111, 114, 118–120, 307, 314
	Rio Grande, *Cichlasoma cyanoguttatum* 111
	three-spot, *Cichlasoma trimaculatum* 113, 226
	yellow-belly, *Cichlasoma salvini* 111
Cichlidae, cichlids (*see also* individual taxa) 111, 113–114, 122
Cinnamomum camphora, camphor tree 24
Cirripedia, nonindigenous (*see also* individual taxa) 190
CITES 155, 301, 363
citrus 182, 224, 247
Citrus County 150, 268, 291
citrus groves 171–173, 178
civets, *Viverra* spp. 301
Cladium jamaicense, saw grass 48, 53–55, 191, 278–281
Cladocera 101
clams, nonindigenous 191
	Asiatic, *Corbicula fluminea* 102–105, 362
Clarias batrachus, walking catfish 7, 111, 114, 116, 135, 214, 270, 300
Clariidae, air-breathing catfishes 111, 300
Clear Lake, California 227
Clemmys muhlenbergi, bog turtle 303
Clerodendrum speciosissimum, Java glorybower 269
Cleveland, Ohio 153
clustervine, beach, *Jacquemontia reclinata* 293
Cnemidophorus 126
	lemniscatus 124–125, 131
	motaguae 124, 131, 134
	sexlineatus 134

Cnidaria, nonindigenous (*see also* individual taxa) 190
Coachella Valley Preserve, California 222
coati 179
  white-nosed, *Nasua narica* 159, 170
Cobitidae, loaches (*see also* individual taxa) 111
Coccidae, soft scales (*see also* individual taxa) 79, 89
Coccinella septempunctata 83
Coccinellidae, ladybird beetles (*see also* individual taxa) 82–83
Coccus
  capparidis, brown soft scale 82, 89
  hesperidum, brown soft scale 82
Cochliomyia hominivorax, screwworm fly 180, 222
cockatiels 316
cockchafer, wheat, *Anisoplia austriaca* 247
cocklebur, common, *Xanthium strumarium* 249
cockroach(es)
  Asian, *Blattella asahinai* 81
  Blattaria 81
  Blattodea 79
  giant Madagascan hissing, *Gromphadorhina* sp. 7, 81, 98
Cocoa, Florida 164
Cocoa Beach, Florida 154
Coconut Grove, Florida 141
Codium fragile tomentosoides 196–197
Coelophora inaequalis 83
coffee, wild, *Psychotria nervosa* 70
Coleoptera, *see* beetles
Colisa, see gourami(es)
Collier County 268, 311
  nonindigenous animals in 84, 114, 120, 141, 150,   154, 170–171
  nonindigenous plants in 41, 44, 48, 54, 291
Colocasia esculenta, taro, elephant ear 27, 269
Colorado 149, 306
Colossoma, pacu 112, 299
  macropomum, tambaqui 112
Coluber constrictor, black racer 135
Colubrina asiatica, lather leaf 24, 48, 51, 212, 269, 271, 277, 285
Columba livia, rock dove 141, 143, 148–149, 154, 270
Columbia County 147
Columbia River, Washington 103
columbine, *Aquilegia canadensis* var. *australis* 292
Commelina gigas, climbing dayflower 179
Commonwealth Institute of Biological Control, Australia 255–256
competition by nonindigenous species 10, 156, 189, 272, 281
  with native birds 147
  with native fishes 272
  with native herpetofauna 133–134
  with native mammals 182, 271
conductivity 46
Congo River 194

Congressional Office of Technology Assessment 35–37
Conopeum "seurati" 192
Conservation Reserve Program 342
Convention on International Trade in Endangered Species of Wild Fauna and Flora (CITES) 155, 301, 363
Cooperative Aquatic Plant Management Program 319
Coos Bay, Oregon 189
Copepoda, copepods 101, 188
copra 83
Copsychus sechellarum, Seychelles magpie-robin 156
Coptotermes formosanus 93
Coragyps atratus, black vulture 180
Coral Gables, Florida 146
coral reefs 189, 191
coral vine, *Antigonon leptopus* 26
Corbicula fluminea, Asiatic clam 102–105, 362
Cordia sesbestena, Geiger tree 293
Cordylophora caspia 190
Corkscrew Swamp Sanctuary 114
corn 89
Corvus brachyrhynchos, American crow 135
corydoras, *Corydoras* sp. 112
Corydoras sp., corydoras 112
cost-benefit analyses 258–260
Costa Rica 195, 363
Cosymbotus platyurus 124–126
cotton 227
  wild, *Gossypium hirsutum* 277
cougar, *Felis concolor* 301–302, 309, 316
cowbird
  brown-headed, *Molothrus ater* 145
  shiny, *Molothrus bonariensis* 270
coyote, *Canis latrans* 159, 170–171, 177, 179, 180–185, 316
crab(s) (*see also* individual taxa) 191–192
  saber, *Platychirograpsus spectabilis* 192
Crandon Park Zoo, Miami 142
crane
  Florida sandhill, *Grus canadensis floridanus* 309
  whooping, *Grus americana* 309
crappie
  black, *Pomoxis nigromaculatus* 46, 105
  white, *Pomoxis annularis* 112
crayfish 106
  Australian red-claw, *Cherax quadricarinatus* 102–103
crickets, Orthoptera (*see also* individual taxa) 76, 79, 95–96, 98
crickets, mole 92, 95–97, 259, 364
  northern mole, *Neocurtilla hexadactyla* 92
  short-winged mole, *Scapteriscus abbreviatus* 95
  southern mole, *Scapteriscus borellii* 87, 95
  tawny mole, *Scapteriscus vicinus* 87, 95
Cricotopus 261
crocodile(s) 304, 316

American, *Crocodylus acutus* 42, 126, 284
Crocodile Lake National Wildlife Refuge 268, 270–271, 273
crocodilians 125
Crocothemis servilia 95
Cromileptes altivelis, barramundi cod, panther grouper 112, 115
crow, American, *Corvus brachyrhynchos* 135
Crustacea (*see also* individual taxa) 190, 194
Cryphonectria parasitica, chestnut blight fungus 13
Cryptochetum iceryae 90
cryptogenic species 188, 195
Cryptognatha nodiceps 83, 89
Cryptolaemus montrouzieri 83
Cryptostegia grandiflora, Palay rubber vine 27
Cryptosula pallasiana 192
Crystal River 230
Crystal River National Wildlife Refuge 268
Ctenocephalides felis, cat flea 96
Ctenopharyngodon idella, grass carp, white amur, see carp, grass
ctenopoma, two-spot, *Ctenopoma nigropannosum* 113
Ctenopoma nigropannosum, two-spot ctenopoma 113
Ctenosaura pectinata 124, 128, 131–132
Ctenostomata 192
Cuba 78, 87, 132–133, 137, 148, 159–160
Cucujidae, flat bark beetles 79, 83–84
Culex quinquefasciatus 9–10
Culicidae, *see* mosquitoes
Cupaniopsis anacardioides, carrotwood 24, 44–45, 51, 263, 269
Curaçao 222
Curculionidae, weevils (*see also* individual taxa) 79, 83–84
Cutthroat Seep, Tiger Creek Preserve, Florida 176
Cyclas formicarius, sweet-potato weevil 83–84
Cynomys ludovicianus, black-tailed prairie dog 159, 167
cypress, bald, *Taxodium distichum* 217–218
Cyprinidae, carps and minnows, *see* carps, Cyprinidae
Cyprinus carpio, common carp 110–111, 118, 214
Cyrtobagous salviniae 248
Cystoseira fimbriata 197
Cyzenis albicans 249
Czechoslovakia 7

Dactyloctenium aegyptium, crowfoot grass 290
Dactylopius ceylonicus 247
Dacus dorsalis, oriental fruit fly 252
Dade County
    birds in 141, 146, 148–152, 154, 307
    citrus canker in 224
    effects of Hurricane Andrew on 64–74
    fishes in 114–115, 117–120, 307

introduction of nonindigenous species through 218–299
invertebrates in 72–73, 78, 82, 84, 107
mammals in 166, 170
management of nonindigenous species in 87, 209, 215, 232
nonindigenous plants in 41–42, 44, 46–49, 61, 215, 232, 290–292, 294
reptiles and amphibians in 128, 131, 134, 136
dahlbergia, Indian, *Dalbergia sissoo* 27
Dalbergia sissoo, Indian dalbergia 27
dams 183
damselflies, Odonata 79, 95
Dania, Florida 161, 163, 183
Danio
    malabaricus, Malabar danio 112
    rerio, zebra danio 112
danio
    Malabar, *Danio malabaricus* 112
    zebra, *Danio rerio* 112
Dasyprocta spp., agoutis 301
Dasypus novemcinctus, see armadillo, nine-banded
dayflower, climbing, *Commelina gigas* 179
Decapoda (*see also* individual taxa) 101, 191
deer, Cervidae 280, 302
    axis, *Axis axis* 159, 173–174
    sambar, *Cervus unicolor* 159, 174, 270, 272, 306
    white-tailed, *Odocoileus virginianus* 182, 271, 306, 310
Deering Hammock County Park 65–74
delphacid, sugarcane, *Perkinsiella saccharicida* 89, 247
Delray Beach 154
Dendrocygna
    autumnalis, black-bellied whistling duck 142
    bicolor, fulvous whistling duck 140
dengue fever 86
density dependence 246
Dermochelys coriacea, leatherback sea turtle 211
DeSoto County 170–172, 174, 185
dholes, Indian 316
Diadophis punctatus 135
Dialeurodes
    citri, citrus whitefly 88
    citrifolii, cloudy-winged whitefly 88
Diaprepes abbreviatus, Apopka weevil 83
Diaspididae, armored scales 89
Diatraea saccharalis, sugarcane moth borer 87
Dichrostachys cinerea, aroma 27
diffusion 7–9
Dingell-Johnson Act 121–122
dingo, Australian 175
dinoflagellates 188, 196, 199
Dionaea muscipula, Venus's-flytrap 363
Dioscorea 65
    alata 61
    bulbifera, see air potato
    sansibarensis 61
Diptera, flies (*see also* fly (flies) and individual taxa) 79, 85–88, 102

446 INDEX

Dirofilaria immitis, heartworm 175
disease 57, 135–136, 146, 155, 180–181,
183–184
dispersal 7–8, 59, 66, 70–71, 91, 102, 156, 166,
177, 200
jump 7–9, 129
disturbance
effect of on animals 71–72, 116, 168, 182
effect of on invasibility 3, 16, 36, 51, 53,
63–74, 122, 214
effect of on plants 41–49, 54, 280, 291, 352
in fresh water 121
need for management as a result of 355
varieties and degrees of 52, 361
Division of Recreation and Parks 343
Dixie County 150, 154, 169, 268
Djalmabatista pulcher 102
Dodge Island, Port of Miami 186
dog, *Canis familiaris*
"Carolina" 175
effects of on native biota 10, 135, 179
free-ranging populations of 175, 143, 183,
185, 293–294
hybridization of with coyote 181
origin of and status of in Florida 159, 175
"pariah," definition of 175
regulations governing 316
Dominican Republic 94, 218
Doradidae, thorny catfishes 112
dorados, *Salminus* 300
Dorosoma, shads 308
dove(s) 316
African collared, *Streptopelia roseogrisea*
141–142, 149–150, 154, 270
Barbary, *Streptopelia roseogrisea* 141–142,
149–150, 154, 270
Eurasian collared, *Streptopelia decaocto*
141–142, 147, 149–150
mourning, *Zenaidura macroura* 147
rock, *Columba livia* 141, 143, 148–149, 154,
270
white-winged, *Zenaida asiatica* 142, 307
dragonflies, Odonata (*see also* individual taxa) 79,
95
dragons, Komodo 316
Dreissena polymorpha, zebra mussel 102–106,
336, 362
drills and mandrills 316
Dry Tortugas 168
Dry Tortugas National Park 284
Drymarchon 126
corais couperi, eastern indigo snake 180, 272,
291
duck(s)
black-bellied whistling, *Dendrocygna autumnalis* 142
Florida mottled, *Anas fulvigula fulvigula*
11–12, 147–148
fulvous whistling, *Dendrocygna bicolor* 140
mallard, *Anas platyrhynchos* 11–12, 147–148
Muscovy 141, 143, 148, 306

South American 306
Dumetella carolinensis, gray catbird 70–71
DuPuis Reserve State Forest 291
Duval County 150, 167

ear-pod tree, *Enterolobium contortislilquum* 27
eartheater
pearl, *Geophagus brasiliensis* 113
red-striped, *Geophagus surinamensis* 111
East Cape 276, 281
East Everglades Acquisition Area 276, 278–280,
282, 284–285
East Everglades Exotic Plant Control Project 282
East Everglades Resource Planning and
Management Committee 282
Echidnophaga gallinacea, sticktight flea 96
Echium 32
plantagineum 255
ecological impact 39–61, 103–107, 178–182, 201
ecological imperialism 13–14
ecotones 48, 54–55, 280–281, 286
Ectoprocta 192
eels, electric, Electrophoridae 7, 299–300
Egeria densa, Brazilian elodea 237, 269, 290
eggplant 252
Eglin Air Force Base 176
Egmont Key National Wildlife Refuge 268, 271
egret, cattle, *Bubulcus ibis* 140
Eichhornia crassipes, see water hyacinth
Elaeophora schneideri 178
Elaphe 136–137
guttata 135
obsoleta 135
Electrophoridae, electric eels 7, 299–300
elephants 316
Indian 158
elephant ear, *Colocasia esculenta* 27, 269
Eleutherodactylus
coqui 125
planirostris 124, 129, 131–132, 135, 270
Elimia (Goniobasis) 107
elk, *Cervus elaphus* 159, 174, 306
Elliott Key 158, 166–167, 179
elm disease fungus, Dutch, *Ophiosoma ulmi* 252,
364
elm, American 252
El Niño Southern Oscillation 72
elodea, Brazilian, *Egeria densa* 237, 269,
290
Enallagma
basidens 95
civile 95
Encarsia
formosa 88, 249
lahorensis 88, 91
opulenta 88, 91, 99
sankarani 91
smithi 91
encephalitis 86, 149, 183
Encyclia tampensis, butterfly orchid 293
Encyrtidae (*see also* individual taxa) 91
Endangered Species Act 122, 359

endemic communities in Florida 360
endemicity 359
England 12, 154, 225–227
Ennomos subsignarius, snow-white linden moth
    153
Enterolobium contortislilquum, ear-pod tree 27
Ephemeroptera, mayflies 79, 101
Ephydridae (*see also* individual taxa) 102
Epicrates striatus 137
Epilachna varivestis, Mexican bean beetle 82
Epilachninae 82
Epimecis detexta 97
Epipremnum 65
    pinnatum cv. Aureum, pothos 27
Episimus 262
eradication campaigns 221–243, 366
    against birds 154
    against citrus canker 224
    cost of 221, 224–227, 311
    distinguished from biological control 246
    against insects 87–88, 221
    factors affecting 108, 224–228, 232, 237, 251,
        365
    against fishes 110, 117, 311–312
    against mammals 225–227, 311, 314
    against plants 226–228, 232, 237, 277, 286,
        323
Ercolania fuscovittata 191
Erinaceus algirus, see hedgehogs
Eriobotrya japonica, loquat 144
erosion 40
Erythrinidae, trahiras or tigerfishes (*see also* indi-
    vidual taxa) 300
Escambia County 141, 152–153, 169
Escambia River 104, 110, 115
establishment rate
    difficulty of predicting 4–7, 155, 314–315,
        360–361
    factors affecting 5–7, 31, 101, 116, 156
    of insects 4–5
    of reptiles and amphibians 125–126
Estero, Florida 48
ethology 245
Eubrychius 261
Eucalyptus spp. 33
Eucerocoris suspectus 262
Eucoilidae 91
Eugenia
    axillaris, white stopper 71
    foetida, Spanish stopper 71
    uniflora, Surinam cherry 27, 65
Euglandia rosea 253–254
Euhrychiopsis lecontei 261
Eulophidae (*see also* individual taxa) 91
Eumaeus atala, Florida atala 78
Eunica tatila, Florida purplewing 78
Eupatorium spp., dog fennel 210
Euphydryas phaeton, Baltimore checkerspot but-
    terfly 256
Eupristina
    altissima 91
    masoni 91

euryphagy 245, 253
Eustis, Lake 162
eutrophic lakes 86, 214
Everglades (*see also* Everglades Conservation
    Areas and Everglades National Park) 143,
    355
    animals in 165, 178, 305–306
    plants in 48–49, 59, 323, 332–333, 353
Everglades City 276, 281
Everglades Conservation Areas 47, 53–54, 230,
    242, 320
Everglades National Park (*see also* Everglades)
    230, 275–276
    animals in 44, 70, 114–115, 117, 132, 165,
        171–173, 183
    effects of Hurricane Andrew in 65
    invasibility of 130, 132
    management of 330, 336, 340, 275–286
    plants in 42–44, 226, 240, 242, 275–286
    restoration in 207–209, 283
evotransporation rates 55
Exotic Fish Laboratory 311
Exotic Pest Plant Council 22–23, 40, 229,
    240–241, 263, 282, 286, 325
Expanded Project for Aquatic Plant Control 232

Fabaceae (*see also* individual taxa) 32
Fairchild, David 22
Fairchild Tropical Garden 22, 218
Falco sparverius, American kestrel 180
falconry 298, 304
federal government, management by 329–337
Federal Noxious Weed Act 30, 35–36, 237, 240,
    330, 332
Federal Noxious Weed List 330
Federal Seed Act 30, 35–36, 333
Felis
    catus, see cat, domestic
    concolor, cougar 301–302, 309, 316
    concolor coryi, Florida panther 180–181,
        271–272, 293, 309, 311
    pardalis, ocelot 158, 185, 316
    silvestris, see cat, domestic
    yagouaroundi, jaguarundi 157, 159, 172–173,
        179, 270
Felsmere, Florida 158
fennel, dog, *Eupatorium* spp. 210
feral 143
Fergusonina sp. 262
fern(s)
    Asian sword, *Nephrolepis multiflora* 28
    climbing (*see also Lygodium*) 41, 46, 310, 321
    giant water, *Salvinia molesta* 248
    incised halberd, *Tectaria incisa* 29
    Japanese climbing, *Lygodium japonicum* 28,
        263, 269, 271
    Old World climbing, *Lygodium microphyllum* 25,
        40–41, 51, 53, 56, 216, 323
    spangles, water, *Salvinia rotundifolia* 269
ferret, *Mustela putorius* 158, 185, 254, 316
Ficopomatus enigmaticus 193
Ficus 16, 30, 33, 91, 363–364

altissima, banyan tree, lofty fig 27, 91
aurea, strangler fig 70–71
benghalensis, banyan 91
benjamina, weeping fig 27
elastica, Indian rubber tree 27
microcarpa, laurel fig 24, 30, 59, 67, 91, 93
nitida see Ficus microcarpa
retusa var. *nitida, see Ficus microcarpa*
fig(s) 16, 30, 33, 91, 363–364
    laurel, *Ficus microcarpa* 24, 30, 59, 67, 91, 93
    lofty, *Ficus altissima* 27, 91
    weeping, *Ficus benjamina* 27
fighting fish, Siamese, *Betta splendens* 113
Fiji 12
finch(es) 316
    house, *Carpodacus mexicanus* 141–142,
        152–153
Fiorinia theae, tea scale 89
fire regime
    effects of on invasibility 15–16, 51–52, 54, 73,
        206
    effects of on nonindigenous species 43, 45, 48,
        206, 208, 210, 215–217, 275, 278,
        280–281, 283, 310
    effects of nonindigenous species on 11, 40,
        44–48, 56, 216–217, 281, 363
fireflies, Lampyridae 79, 84
firemouth, *Cichlasoma meeki* 111
fish(es) (*see also* individual taxa)
    biological control by 249–250
    bony tongue, Osteoglossidae (*see also*
        individual taxa) 300
    effects of habitat modification on 214
    effects of nonindigenous species on 50, 57,
        106, 197–198
    effects of on other species 110, 114–16, 135
    eradication of 226
    farms 298–299
    internet address of national database on non-
        indigenous 118
    introduction and frequency of nonindigenous
        109–122, 263, 360–361, 365
    parasites of 104–105
Flacourtia indica, governor's plum 27
flagfish, *Jordanella floridae* 117
Flagler County 164
Flamingo, Florida 132, 276–277, 285
flamingo, greater, *Phoenicopterus ruber* 147–148
flatwoods 45, 52, 172–173
flatworms (*see also* individual taxa) 188, 190
flea(s), Siphonaptera 96
    cat, *Ctenocephalides felis* 96
    Hoplopsyllus glacialis affinis 181
    human, *Pulex irritans* 96
    oriental rat, *Xenopsylla cheopis* 96
    predation on 92
    rabbit 253
    sticktight, *Echidnophaga gallinacea* 96
flies, *see* fly (flies)
Flint River, Georgia 115
flood control 40, 232, 237

flooding 109, 120
Florida Administrative Commission 345
Florida Aquatic Weed Control Act 233
Florida Bay 276–278, 284–285, 320–321
Florida Department of Agriculture and Consumer
    Services
    authority of 35, 344–345, 354
    dispersal of nonindigenous plants by 200
    management by 81, 341, 343–345, 351–352
    mission of 289
    research by 295
Florida Department of Community Affairs 348
Florida Department of Environmental Protection
    authority and strategic plan of 344–345, 349
    control of hydrilla by 237–239, 318–319
    control of melaleuca by 241
    control of water hyacinth by 233, 235–237,
        317–318
    management by 287–288, 302, 325, 332, 343,
        346, 349–350, 352
    regulation of marine species by 199
    vegetation surveys by 41
Florida Department of Management Services 345
Florida Department of Natural Resources, *see*
        Florida Department of Environmental
        Protection
Florida Department of Transportation 287,
        289–292, 294–295, 343, 345, 350–351
Florida Division of Forestry 287, 289–290, 294
Florida Division of Plant Industry 98, 223–224,
        295
Florida Game and Fresh Water Fish Commission
    eradication campaigns by 117
    introductions by 114–115, 144, 304, 306–308,
        343
    management by 232, 287–289, 294, 297–316,
        324, 349–350
    mission and authority of 288, 297, 344–345,
        354
    monitoring and inspection by 117–118, 299,
        301, 312
    regulation and enforcement by 298, 303
    research by 292–293, 346
    restoration by 210
Florida Institute of Food and Agricultural
        Sciences 22
Florida Keys
    other animals on 94, 110, 141, 256, 312
    mammals on 160–161, 163–165, 168, 181
    plants on 256, 290
    restoration on 218
Florida Keys National Marine Sanctuary 352
Florida Keys National Wildlife Refuge 271
Florida Natural Areas Inventory 41, 346, 353
Florida Noxious Weed list 36–37, 365
Florida Panther National Wildlife Refuge 268,
        271–272
Florida Park Service 287–288, 290, 292, 294
Florida Power & Light 320
Florida Water Resources Act 317
Floridichthys carpio, goldspotted killifish 117

Flueggea virosa, flueggea 27
flueggea, *Flueggea virosa* 27
flukes, parasitic 104, 107
fly (flies), Diptera (*see also* individual taxa) 79,
    85–88, 102
    agromyzid 90
    apple maggot, *Rhagoletis pomonella* 256
    Asian hydrilla, *Hydrellia pakistanae* 261
    Australian hydrilla, *Hydrellia balciunasi* 102,
        261
    biological control by 259, 261
    blow, Calliphoridae 85
    bot, *see* botfly
    Caribbean fruit 87–88
    fruit, Tephritidae 79, 87–88
    gall, *Fergusonina* sp. 262
    horn, *Haematobia irritans* 87, 92
    horse, Tabanidae 79
    house, *Musca domestica* 87
    louse, Hippoboscidae 86
    march, Bibionidae 85
    Mediterranean fruit, *Ceratitis capitata* 87–88,
        223, 225, 252
    Mexican fruit, *Anastrepha ludens* 87
    muscid, Muscidae 87
    oriental fruit, *Dacus dorsalis* 252
    papaya fruit 87–88
    saw, *see* sawfly
    screwworm, *Cochliomyia hominivorax* 180, 222
    stable, *Stomoxys calcitrans* 87
    tachinid, Tachinidae 87
    warble, Oestridae 87
Foley, Alabama 165
forensic entomology 85
forestry 339
Forficula 247
Formicidae, ants (*see also* individual taxa) ix, 79,
    92, 247, 362
Fort Lauderdale 92, 150–152, 161, 163, 222,
    318, 320
Fort Lauderdale-Hollywood International Airport
    163
Fort Lonesome, Florida 170
Fort Myers 136, 173, 318
Fort Pierce 150–151, 196
fouling communities 187, 189–192
four-o'clock, burrowing, *Okenia hypogaea* 293
fox
    black, *see* coyote
    gray, *Urocyon cinereoargenteus* 172, 182
    red, *Vulpes vulpes* 159, 171–172, 177,
        179–180, 182–185
France 193, 247
francolin, black, *Francolinus francolinus* 141,
    144, 306
Francolinus francolinus, black francolin 141, 144,
    306
Franklin County 169, 176, 268
Frankliniella occidentalis, western flower thrips
    96
frog(s) 179, 302–303

Cuban tree, *Osteopilus septentrionalis* 44, 124,
    131–132, 134–135, 270
    gopher, *Rana aesopus* 291
greenhouse, *Eleutherodactylus planirostris* 124,
    129, 131–132, 135, 270
fruitflies, *see* flies, fruit
Fulgoroidea, plant hoppers (*see also* individual
    taxa) 79, 89
Fundulus
    chrysotus, golden topminnow 117
    seminolis, Seminole killifish 117
fungus (fungi) (*see also* individual taxa) 260, 262,
    247, 249, 261
fur farms 226

Gainesville 92, 150, 322, 336
Galápagos 72
Gallus gallus, red jungle fowl 306
Gambusia 249–250
    holbrooki, mosquitofish 120
Gammarus tigrinus 193
Ganges primrose, *Asystasia gangetica* 26
gar, Florida, *Lepisosteus platyrhincus* 114
Garlon 4 283, 285
Gasterophilus
    haemorrhoidalis, nose botfly 87
    intestinalis, horse botfly 87
    nasalis, throat botfly 87
Gastrophryne 126
    Gastrophryne carolinensis 135
Gastropoda (*see also* individual taxa) 101, 191
gavials 316
gecko(s) 124, 137
    Gekko gecko 124, 126, 131
    Hemidactylus frenatus 124–125, 131
    Hemidactylus garnotii 124, 131–132
    Hemidactylus mabouia 124–125, 131–132
    Hemidactylus turcicus, Mediterranean gecko
        124, 131, 270
    Sphaerodactylus argus 124, 128, 131–132
    Sphaerodactylus cinerus, Indo-Pacific gecko
        125, 270
    Sphaerodactylus elegans 124, 128–129,
        131–132
    Sphaerodactylus nigropunctatus 137
    Sphaerodactylus notatus 132
Geiger tree, *Cordia sesbestena* 293
Gekko, see gecko(s)
Gelechiidae, gelechiid moths 93–94
genetic(s)
    diversity 6, 184
    effects of nonindigenous mammals 181–182
    importance of to biological control 245
    manipulation 227, 354
Geochelone spp., Galápagos tortoise 304
Geographic Information System 353
Geometridae, measuring worms (*see also*
    individual taxa) 79
Geophagus
    brasiliensis, pearl eartheater 113
    surinamensis, red-striped eartheater 111

Georgia 97, 170, 175, 261, 336
gerbils 316
Germany 154
gibbons 7, 316
Gila monster, *Heloderma suspectum* 303
Gilchrist County 169
Glades County 170
Global Positioning System 283
glorybower, Java, *Clerodendrum speciosissimum* 269
goat, *Capra hircus* 159, 176, 179, 226
Goeldichironomus amazonicus 86
Goethe State Forest 292
Golden Gate State Forest 292
goldfish, *Carassius auratus* 112
Gonatodes albogularis 124, 128, 131–132
Goniobasis, see Elimia
goose
     Canada, *Branta canadensis* 307
     Egyptian, *Alopochen aegyptiaca* 306
Gopherus
     agassizi, desert tortoise 135–136
     polyphemus, gopher tortoise 44, 136, 179, 272, 281, 291, 293
gorillas 316
Gossypium hirsutum, wild cotton 277
Goulds Monkey Jungle 161
gourami(es), Anabantidae (*see also* individual taxa) 111, 113
     croaking, *Trichopsis vittatus* 111
     dwarf, *Colisa lalia* 113
     kissing, *Helostoma temmincki* 113
     pearl, *Trichogaster leeri* 113
     thick-lip, *Colisa labiosa* 113
     three-spot, *Trichogaster trichopterus* 113
governor's plum, *Flacourtia indica* 27
grackle, common, *Quisculus quiscula* 145
Gracula religiosa, hill myna 141, 154
gramma, royal, *Gramma loreto* 116
Gramma loreto, royal gramma 116
Grand Cayman 94
Graphognathus spp., white-fringed beetles 83
Grapsidae (*see also* individual taxa) 192
grass(es), Poaceae 32
     Bahia, *Paspalum notatum* 28, 96, 214, 269, 310
     cane, *see Neyraudia reynaudiana*
     cogon, *see Imperata cylindrica*
     cord, *Spartina* spp. 12
     crowfoot, *Dactyloctenium aegyptium* 290
     cutthroat, *Panicum abscisum* 217
     Johnson, *Sorghum halepense* 290, 333
     molasses, *Melinis minutiflora* 28
     muhly, *Muhlenbergia* sp. 48, 280–281
     napier, *Pennisetum purpureum* 29
     para, *Brachiaria mutica* 269
     red root, *Lachnanthes caroliniana* 217
     Rhodes, *Chloris gayana* 90
     saw, *Cladium jamaicense* 48, 191, 279–281
     smut, *Sporobolus jacquemontii* 290
     torpedo, *Panicum repens* 25, 41, 50, 56, 269,

     290, 322–323, 341, 351
     vassey, *Paspalum urvillei* 290
grasshoppers, Orthoptera (*see also* individual taxa) 79
grazing 73
Great Lakes 105, 336–337
Great Smoky Mountains National Park 11
Great White Heron National Wildlife Refuge 164, 268, 271
Greater Antilles 97, 133, 136
greenbrier, *Smilax* 71
greenhouses 88
Green Swamp West property 324
gribbles 187, 191, 195–196
Grimshawe Hammock 65
Gromphadorhina sp., giant Madagascan hissing cockroach 7, 81, 98
grosbeak, blue, *Guiraca caerulea* 147
grouper, panther, *see* barramundi cod, *Cromileptes altivelis*
grub, white sugarcane 247
Grus
     americana, whooping crane 309
     canadensis floridanus, Florida sandhill crane 309
Gryllotalpa major 364
Guam 137, 304, 314, 333
guapote, jaguar, *Cichlasoma managuense* 111, 115
guava(s) 269
     Psidium cattleianum 29
     Psidium guajava 29, 214, 269
     Psidium littorale, strawberry guava 29
guenons 316
guinea fowl, helmeted, *Numida meleagris* 306
guinea pigs 316
Guiraca caerulea, blue grosbeak 147
Gulf Breeze, Florida 153
Gulf County 174
Gulf Stream, Florida 228
guppy, *Poecilia reticulata* 111
Guyana 119–120
Gymnocorymbus ternetzi, black tetra 112

habitat
     creation 40, 53
     destruction 147, 178–179, 363
     modification 11, 39–40, 55, 123, 137, 165, 178, 354, 361
Haematobia irritans, horn fly 87, 92
Haematopinus spp., hog lice and short-nose cattle lice 81
hairstreak
     Bartram's, *Strymon acis* 78
     maesites, *Chlorostrymon maesites* 78
Haiti 94
half flower, *Scaevola taccada* var. *sericea* 25, 269
Haliplanella
     lineata 189–190
     luciae 189–190

Hamilton County 169, 171
hammock communities 44, 49, 172–173, 292
    coastal 48, 281
    mahogany 285
    mesic 178
    subtropical 167
    tropical 41, 49, 53, 65–74, 143, 215, 275,
        279–281, 323, 340
hamsters 316
Harmonia dimidiata 83
Harpullia arborea 67–68
Harris, Lake 54
Hawaii
    biological control in 247, 254, 256, 262
    invertebrates of 83, 97, 194, 223
    plants of 11, 33–34, 332
    vertebrates of 11, 119, 126–127, 146, 152,
        154, 178, 217, 311
hawthorne 256
heartworm, *Dirofilaria immitis* 175
hedgehogs 158
    Erinaceus algirus 301
helminths 178
Heloderma suspectum, Gila monster 303
Helostoma temmincki, kissing gourami 113
Hemichromis letourneauxi, African jewelfish 111
Hemidactylus, see gecko(s)
hemiepiphytes 68
Hemiptera *(see also* bugs and individual taxa) 79,
    88
heptachlor 221
Heraclides aristodemus, Schaus' swallowtail 78
herbicide(s), use of 210, 215, 226, 260–261, 323
    on melaleuca 241–242, 280, 283, 320
    on other species 226–227, 278, 281, 285, 323
    on water hyacinth 230, 259–260, 309,
        324–325
Hernando County 170, 216, 291
Heros severus, banded cichlid 113
Herpailurus yagouaroundi, see jaguarundi, *Felis
    yaguouaroundi*
herpes B virus 183, 302
Herpestes, see mongoose(s)
herpetofauna *(see also* individual taxa) 44, 58,
    123–138, 312
Hesperioidea, butterflies *(see also* individual taxa)
    79, 94
Heterandria formosa, least killifish 117
Heteroperryia hubrichi, Brazilian pepper sawfly
    262, 322
Heteropneustidae, air-sac catfishes 300
Hialeah, Florida 148, 164
Hibiscus tiliaceus, mahoe 27
High Springs, Florida 171
Highland Beach, Florida 276, 278
Highlands County 58, 147, 168, 170–174, 176,
    183, 185, 259
Highlands Hammock State Park 185
Hillsborough Community College 210
Hillsborough County 168–171, 209–211, 232,
    308

Hillsborough River State Park 173
Hippobosca longipennis 86
Hippoboscidae, louse flies 86
Hippodamia variegata 83
hippopotamuses 316
Hirundo rustica, barn swallow 147
Hispaniola 83, 133, 136
histoplasmosis 144, 149
Hobe Sound 58
Hobe Sound National Wildlife Refuge 268,
    270–271
hogs, wild, *see* pig(s)
Hole-in-the-Doughnut 132, 207–209, 276,
    281–284, 286
holly, dahoon, *Ilex cassine* 71
Hollywood, Florida 58, 224
Holmes County 169
Homestead, Florida 47, 153, 218, 276
Homestead Air Force Base, Florida 150
Homoptera *(see also* individual taxa) 79, 88–91
honeysuckle, Japanese, *Lonicera japonica* 22, 25,
    30, 269
Hong Kong 91, 154, 194
hookworm, *Ancylostoma pluridentatum* 181
Hoplias malabaricus, trahira 113
Hoplopsyllus glacialis affinis 181
horses 87, 183
house sparrow, *Passer domesticus* 5–6, 141,
    153–154
Humane Society of the United States 301
hunting preserves 158, 298, 306
hurricane(s)
    Andrew 41–42, 49, 54, 63–74, 94, 115, 150,
        158, 167, 215, 294, 323, 340, 361
    Betsy 277, 284, 340
    Donna 277, 284, 340
    effects on animals 66, 71–72, 94, 115, 143,
        150, 158, 166–167, 169, 340
    effects on plants 41–42, 49, 54, 215, 275,
        277–278, 281, 284, 294, 323, 340
    Kate 361
Hyalella azteca 50
hybridization 136, 142, 147–148, 150, 176,
    181
Hydrellia 261
    balciunasi, Australian hydrilla fly 102, 261
    pakistanae, Asian hydrilla fly 261
hydrilla, *Hydrilla verticillata* 228
    biological control of 84, 97, 102, 107, 248,
        254, 260–261
    disturbance and 51, 53–54
    effects of 14, 53–56, 239, 253
    introduction and extent of in Florida 41, 46
    management of 227, 230, 237–239, 242–243,
        259, 290, 318–319, 366
    spread of 15, 59, 178
Hydrilla verticillata, see hydrilla
Hydrochaeris hydrochaeris, capybara 159,
    169–170
Hydrocyninae, African tigerfishes 300
Hydroides elegans 191

hydrological regime
  alteration of 51–52, 73, 288
  effects on invasibility 54–55, 73, 206
  effects of on plants 48, 55, 59, 208, 214, 275,
    280–281, 284, 322
Hydrozoa (*see also* individual taxa) 190
hyenas 316
hygro, green, *see* hygrophila
hygrophila, *Hygrophila polysperma* 27, 319
Hyla 126
  cinerea 134–135
  squirella 134–135
Hymenoptera (*see also* individual taxa) 79, 90–93
Hypera postica, alfalfa weevil 83–84
Hypericum perforatum, St. John's-wort, Klamath
    weed 248, 257
Hypoderma
  bovis, northern cattle grub 87
  lineatum, common cattle grub 87
Hypophthalmichthys
  molitrix, silver carp 300
  nobilis, *see* carp, bighead
Hypostomus sp., armored catfish 111
hyptage, *Hyptage benghalensis* 27
Hyptage benghalensis, hyptage 27

ibex, *Capra ibex* 305
ice plant, African, *Mesembryanthemum
    crystallinum* 10–11
Icerya purchasi, cottony-cushion scale 83, 90,
    247
Ichneumonidae (*see also* individual taxa) 79
Ictaluridae, freshwater catfishes 111
Icterus pectoralis, spot-breasted oriole 141, 143,
    152
Idaho 127
idris 316
iguanas 7, 124, 131–132
Ilex cassine, dahoon holly 71
Imperata brasiliensis, Brazilian satintail 27, 33,
    45
Imperata cylindrica, cogon grass
  effects of 45, 341
  extent and spread of in Florida 27, 33, 45, 51,
    56, 213–214, 351
  management of 289–20, 310, 322–323
implantation, delayed 177
inbreeding depression 6, 309
inch plant, *Callisia fragrans* 26
India 91, 146, 159, 247
Indian River, Florida 191–192
Indian River County 41, 158, 168–169, 268, 290
Indian River Lagoon 45, 189, 211–212
indigo, *Indigofera* spp. 210
Indigofera spp., indigo 210
influenza, swine 183
Ingraham Highway, Old 276, 278
inkberry, *Scaevola plumieri* 293
insects (*see also* individual taxa) 77–79
  arrival and effects of in Florida 75–76, 80, 84,
    97, 145, 188, 362
  and biological control 153–154, 236, 252, 256

eradication campaigns against 221–223
interactions with nonindigenous species
    42–43, 45, 48, 217
number of species of in Florida 75–79
parasitoids of 10, 12, 82, 85, 87–88
intertidal assemblages 189
introgression 11–12
invasibility, factors affecting 14–17, 51–53, 63,
    73–74, 114, 121–122, 126, 130, 132,
    157–158, 189, 219, 361
invasiveness 21, 32–35, 57–58, 60, 91, 102, 196,
    57–58, 363
invertebrates (*see also* individual taxa) 189, 366
  effect of on native ecosystems 362–363
  effects of nonindigenous species on 50, 57,
    197–198
  number and status of in Florida 101–108,
    187–201
Ipomoea 72–73, 81–82, 362
  aquatica, water spinach 24, 269, 332
  microdactyla, morning glory 82
  tenuissima, morning glory 82
Iridomyrmex humilis, Argentine ant 92
Island Bay National Wildlife Refuge 268
Isopoda (*see also* individual taxa) 189–191,
    193–195
Isoptera, termites 79, 93
Israel 154
Istokpoga, Lake 176, 179, 259
Italy 247

J. N. "Ding" Darling National Wildlife Refuge
    268, 270–271
J. W. Corbett Wildlife Management Area 185
Jack Dempsey, *Cichlasoma octofasciatum* 111
jackals 316
jackrabbit, black-tailed, *Lepus californicus*
    158–159, 166, 177, 181
Jackson County 147, 172
Jacksonville, Florida 150, 167, 191, 318
Jacquemontia reclinata, beach clustervine 293
Jacquinia keyensis, joeweed 293
jaguars 158, 316
jaguarundi, *Felis yagouaroundi* 157, 159,
    172–173, 179, 270
Jamaica 133, 152, 255
jambolan, *Syzygium cumini* 29, 68
Janet Butterfield Brooks preserve 216
Japan 90
jasmine(s) 33, 65–66, 73, 144, 215–216
  Arabian, *Jasminum sambac* 28
  day, *Cestrum diurnum* 26
  Gold Coast, *Jasminum dichotomum* 24, 67–68
  Jasminum fluminense 28, 67–68
  orange, *Murraya paniculata* 28
Jasminum 33, 65–66
  dichotomum, Gold Coast jasmine 24, 67–68
  fluminense, jasmine 28, 67–68
  sambac, Arabian jasmine 28
Java 146
Java plum, *Syzygium cumini* 29, 68
jay, Florida scrub, *Aphelocoma coerulescens* 147

Jefferson County 53, 165, 169, 268
jewelfish, African, *Hemichromis letourneauxi* 111
joeweed, *Jacquinia keyensis* 293
John U. Lloyd Beach State Recreation Area 288
Jonathan Dickinson State Park 185, 216
Jordanella floridae, flagfish 117
jungle fowl, red, *Gallus gallus* 306
Jupiter, Florida 336
Jupiter Island, Florida 211–212
jurisdictional problems 346–352
Justicia cooleyi, Cooley's water willow 292

Kansas 127
karibaweed, *Salvinia molesta* 248
ked, sheep, *Melophagus ovinus* 86
Kendall, Florida 152
kestrel, American, *Falco sparverius* 180
Key Biscayne 128
Key Largo 171, 180, 218
Key Lois 163
Key West, Florida 152, 192, 222, 318
Key West National Wildlife Refuge 267–268, 271
killifish(es) 117
    bluefin, *Lucania goodei* 117
    goldspotted, *Floridichthys carpio* 117
    least, *Heterandria formosa* 117
    pike, *Belonesox belizanus* 111, 120, 270
    Seminole, *Fundulus seminolis* 117
kinkajou 158
Kinosternon 126
kiskadee, great, *Pitanga sulphuratus* 10
Kissimmee River 174, 207, 214–215
Kissimmee Valley 50
kite, Everglade, *Rostrhamus sociabilis* 104, 107, 270
Klamath weed, *Hypericum perforatum* 248, 257
kopsia, *Ochrosia parviflora* 28
kudzu, *Pueraria montana* 22, 25, 46, 53–55, 290, 323, 333–334

La Belle, Florida 230
Labeo chrysophekadion, black sharkminnow 112
Labeotropheus sp., mbuna 113
Labidesthes sicculus, brook silverside 117
lacewings, Neuroptera 79
Lacey Act 297, 330, 335
Lachnanthes caroliniana, red root grass 217
LaCrosse, Wisconsin 336
Laemophloeidae, flat bark beetles 79, 83
Lagomorpha, rabbits (*see also* individual taxa) 159, 165–166
Laguncularia racemosa, white mangrove 162
Lake City, Florida 147, 153
Lake County 54, 86, 162
Lake Okeechobee Aquatic Weed Harvesting Demonstration Project 214–215
Lake Placid, Florida 174
Lake Wales, Florida 161, 171, 217
Lake Woodruff National Wildlife Refuge 268
Lake Worth, Florida 136
lampreys, Petromyzonidae 300
Lampropeltis 136–137

Lampyridae, fireflies 79, 84
langurs 316
lantana, *Lantana camara* 12, 24, 213
Lantana camara, lantana 12, 24, 213
Larra
    analis 92
    bicolor 92
Lates, nile perches 300
lather leaf, *Colubrina asiatica* 24, 48, 51, 212, 269, 271, 277, 285
lavender, sea, *Argusia gnaphalodes* 293
lead tree, *Leucaena leucocephala* 28, 269, 271
leaf miner, citrus, *Phyllocnistis citrella* 93, 97
leaf tiers, *Episimus* 262
leafhopper, sugarcane, *Perkinsiella saccharicida* 89, 247
leafwing, Florida, *Anaea troglodyta* 78
Lee County 84, 136, 165, 173, 222, 232, 240, 268
Leeward Islands 5, 255–256
Leiocephalus
    carinatus, curly-tailed lizard 124, 128–129, 131–132 170
    personatus 124–125, 131
    schreibersi 124, 131
Lemuridae, lemurs 301
lemurs, Lemuridae 301
Leon County 141, 147, 152, 165, 292
leopards 316
Lepidoptera, moths and butterflies (*see also* individual taxa) 75, 78–79, 93–95, 102, 256
Lepidosaphes
    beckii, purple scale 89
    laterochitinosa 89
Lepisosteus platyrhincus, Florida gar 114
Lepomis
    cyanellus, green sunfish 111, 118, 300
    humilis, orange-spotted sunfish 111, 118
    macrochirus, bluegill sunfish 46
    microlophus, redear sunfish 46
leporinus, banded, *Leporinus fasciatus* 112
Leporinus fasciatus, banded leporinus 112
leprosy 183–184
Leptinotarsa decemlineata, Colorado potato beetle 9, 81
Leptodictya tabida, sugarcane lace bug 88
Leptosermoideae (*see also* individual taxa) 251–252
leptospires 183
Lepus californicus, black-tailed jackrabbit 158–159, 166, 177, 181
Lesser Antilles 142
Leucaena leucocephala, lead tree 28, 269, 271
Levy County 169, 172, 268, 292
Liberia 194
lice (*see also* individual taxa) 81
Ligia exotica 189–190
Ligustrum 69
    sinense, privet 28
Liguus fasciatus, Florida tree snail 179
lime berry, *Triphasia trifoliata* 29

Limnoria 191
  pfefferi 190
  saseboensis 190
limpets 189
Linepithema humile, Argentine ant 10
Linnaeus 247
Linognathus 81
lion, *Panthera leo* 158, 302, 316
lionfish, *Pterois volitans* 112, 115
Liothrips ichini, Brazilian pepper thrips 262, 322
Liposarcus
  disjunctivus, vermiculated sailfin catfish 111
  multiradiatus, sailfin catfish 111
Litoria caerulea 123–126, 134
Little Crane Key 163
Little St. George Island 176
Littorina littorea, European periwinkle 11
livebearers, Poeciliidae (*see also* individual taxa)
  111–112, 117
Lixophaga diatraeae 87
lizard(s) (*see also* individual taxa) 125, 179,
  302–303
  curly-tailed, *Leiocephalus carinatus* 124,
    128–129, 131–132 170
  scrub, *Sceloporus woodi* 179
loach(e)s, Cobitidae (*see also* individual taxa) 111
  kuhli, *Acanthophthalmus kuhli* 299
Lochloosa, Lake 230, 239
locust, *Nomadacris septemfasciata* 154
Loggerhead Key, *see* Key Lois 163
Long Lake, Washington 227
Long Pine Key 132
Lonicera 32
  japonica, Japanese honeysuckle 22, 25, 30,
    269
Lophyrotoma zonalis 261
loquat, *Eriobotrya japonica* 144
Lord Howe Island 156
Loricariidae, suckermouth catfishes 111
Louisiana 127, 142, 149, 177, 183, 217, 225,
  261, 314
louse
  human, *Pediculus humanus* 81
  little blue cattle, *Solenoptes capillatus* 81
love bug, *Plecia nearctica* 85, 97, 184
Lower Sugarloaf Key 166
Lower Suwanee National Wildlife Refuge 268,
  272
Loxahatchee National Wildlife Refuge 268,
  270–273, 330
Lucania goodei, bluefin killifish 117
Lulu, Florida 170
lungworms, swine 183
Luxilus chrysocephalus isolepis, southern striped
  shiner 111
Lygaeidae, seed bugs (*see also* individual taxa) 79
Lygodium 41, 46, 310, 321
  japonicum, Japanese climbing fern 28, 263,
    269, 271
  microphyllum, Old World climbing fern 25,
    40–41, 51, 53, 56, 216, 323
Lymantria dispar, European gypsy moth 7–8, 10,

  94, 247, 253, 333–334, 362
Lymantriidae, tussock moths (*see also* individual
  taxa) 94
lynx, European and Canadian 316
Lynx rufus, bobcat 180–182, 316
Lyrodus
  bipartitus 192, 196
  massa 1192, 196
  medilobatus (= *L. mediolobatus*) 192

Macaca
  mulatta, rhesus monkey 159, 162–164, 178,
    183–184
  nemestrina, pig-tailed macaque 163
macaque(s) 316
  "bob-tailed" 163
  crab-eating 7
  pig-tailed, *Macaca nemestrina* 163
macaw, chestnut-fronted, *Ara severa* 141, 154
Macfadyena unguis-cati, cat's claw 28
Maclay State Gardens 292
Macropodus opercularis, paradisefish 113
Madagascar 154
magpie-robin, Seychelles, *Copsychus sechellarum*
  156
mahoe
  Hibiscus tiliaceus 27
  seaside, *Thespesia populnea* 29, 269
Maine 153
maintenance control 221, 229–243, 324, 366
maize 95
Malapteruridae, African electric catfishes 300
malaria parasite, avian, *Plasmodium relictum*
  capistranoae 9–10, 13, 146
malathion 225
mallard, *Anas platyrhynchos* 11–12, 147–148
mambas 7
mammals (*see also* individual taxa) 42–43, 48–49,
  56, 59, 157, 186, 263, 361–362
mandrills 316
mangabeys 316
Mangalitza color pattern 184
mangrove(s)
  black, *Avicennia germinans* 162, 164
  red, *Rhizophora mangle* 11, 162, 193–195,
    363
  white, *Laguncularia racemosa* 162
mangrove systems
  nonindigenous animals in 161, 163, 173, 178,
    194–195
  nonindigenous plants in 48, 278, 280–281
  status and ecological restoration of in Florida
    143, 210, 212, 360
Manilkara zapota, sapodilla 28
mantis, Chinese, *Tenodera aridifolia* 75–76, 98
Marathon, Florida 160
Margarodidae 90
Mariana Islands 64, 68–69
mariculture (*see also* aquaculture) 188, 199–200
Marion County 161–162, 168, 170–171
marisa, golden-horn, *Marisa cornuaurietus* 104,
  107, 270

Marisa cornuaurietus, golden-horn marisa 104, 107, 270
marlberry, *Ardisia escallonioides* 69–70
Marquesas 168
marshes 48, 188–189, 191, 210, 270, 275, 278, 280–281
Martesia
  americana 191, 196
  funisicola 191, 196
  striata 191, 196
martin, purple, Progne subis 147
Martin County 45–46, 58, 211–212, 216, 268, 291
Martinique 247
Mascarene Islands 64, 69, 154
Massachusetts 7, 261
mast wood, *Calophyllum calaba* 26
Masterpiece Gardens 161
Matheson Hammock 65–74, 215
Matheson Hammock County Park 170
Matlacha Pass National Wildlife Refuge 268
Mauritius 64, 152, 154
mayapple, *Podophyllum peltatum* 292
mayflies, Ephemeroptera 79, 101
mbuna, *Labeotropheus* sp. 113
mealworm(s) 76
  giant, *Zophoba* sp. 98
mealybug(s),Pseudococcidae 90
  citrus, *Planococcus citri* 83
  Rhodes-grass, *Antonina graminis* 90
  Pseudococcidae 90
measuring worms, Geometridae 79
medfly, *see* fly, Mediterranean fruit
Megastigmus transvaalensis 93, 97
melaleuca, *Melaleuca quinquenervia*
  damage caused by 9, 252–253, 268–271, 291–292, 334, 336, 339
  extent and effects of in Florida 40–41, 48–49, 53–55, 66, 268–271, 273, 280, 321, 351
  interaction of with fire regime and habitat change 11, 51, 53, 55–56, 66
  interaction of with other species 56–58, 281
  introduction, invasiveness, and spread of 14, 25, 30, 52, 54–55, 351
  management of 229–230, 239–243, 251–252, 260–263, 273, 277, 282–283, 285–286, 310, 320–322, 330, 353
Melaleuca quinquenervia, see melaleuca
Melanaphis sacchari, an aphid 89
melania(s) 107–108
  faune, *Melanoides turricula* 104, 107
  melania, quilted, *Tarebia granifera* 104, 107
  red-rimmed, *Melanoides tuberculata* 104, 107
Melanoides
  tuberculata, red-rimmed melania 104, 107
  turricula, faune melania 104, 107
Meleagris gallopavo, turkey 182, 293
Melia azedarach, chinaberry 25
Melinis minutiflora, molasses grass 28
Melophagus ovinus, sheep ked 86
Melopsittacus undulatus, budgerigar, parakeet 141, 143, 150

meningitis, cryptococcal 149
Mercierella enigmatica, see Ficopomatus enigmaticus
Merremia tuberosa, wood rose 28, 65, 68, 215–216, 323
Merritt Island National Wildlife Refuge 268, 270–272
Metamasius
  callizona 77, 84
  hemipterus, cane weevil 83
Metarhizium anisopliae, green muscardine fungus 247
Metynnis sp., silver dollar 112
Mexico 83, 97, 152, 159, 183, 192, 259, 301
Miami, Florida 318
  birds in 141, 150–152, 156
  invertebrates in 188, 191, 223–225
  mammals in 161, 164, 166, 172, 186
  species introductions through 139, 188, 191, 298, 331
Miami Beach, Florida 150
Miami International Airport 22, 77, 166–167, 299, 301
Miami Rock Ridge forest preserves 64
Miami Springs, Florida 141, 150
Michigan 142
microalgae 198
microarthropods (*see also* individual taxa) 178
Micropterus
  n. sp., shoal bass 116
  salmoides floridanus, largemouth bass 46, 114, 117, 119,
Micrurus 126
midge(s) (*see also* individual taxa) 261
  Ceratopogonidae (*see also* individual taxa) 101
  Chaoboridae (*see also* individual taxa) 101
  Chironomidae (*see also* individual taxa) 86, 101, 105
milfoil, Eurasian water, *Myriophyllum aquaticum* 227, 260–261, 269
milkweed vine, *Morrenia odorata* 260, 262
mimosa, cat-claw, *Mimosa pigra* 25, 33, 58, 332–333, 351
Mimosa pigra, cat-claw mimosa 25, 33, 58, 332–333, 351
minimum viable population size 6
minnow(s)
  Cyprinidae 111–112
  fathead, *Pimephales promelas* 112
mirex 221
Miridae, plant bugs (*see also* individual taxa) 79
Misgurnus anguillicaudatus, oriental weatherfish 111
Mississippi 149, 165, 183
Mississippi River system 105
mite(s)
  Asian honeybee, *Varroa jacobsoni* 80, 91
  broad 80
  citrus rust, *Phyllocoptruta oleivora* 80
  honeybee tracheal, *Acarapis woodi* 80, 91
  spider 80
mitigation 293, 285

Mojave desert 135
moles 179, 316
Mollusca (*see also* individual taxa) 103, 191–192
Molossus molossus tropidorhynchus, velvety free-
    tailed bat 159–161
Molothrus
    ater, brown-headed cowbird 145
    bonariensis, shiny cowbird 270
Monaco 198
mongoose(s) 158, 185–186, 253, 255
    Herpestes auropunctatus 185–186, 254
    Herpestes javanicus 247
monitors, Asian water 7
monkey(s)
    green African savanna 7
    guereza 316
    howler 7, 316
    proboscis 316
    rhesus, *Macaca mulatta* 159, 162–164, 178,
        183–185, 294, 302, 315
    squirrel, *Saimiri sciureus* 159, 161
    vervet, *Cercopithecus aethiops* 159, 161–162,
        183
Monomorium pharaonis, pharaoh ant 92
monophagy 245
Monroe, Lake 86
Monroe County 268
    birds in 141, 149, 150, 152,
    mammals in 158, 166, 171, 180
    other species in 48, 78, 82, 84, 107, 120, 131,
Monroe Station 48
Montserrat 94
morning glories 362
    Ipomoea microdactyla 82
    Ipomoea tenuissima 82
Morone spp., *see* bass
Morrenia odorata, milkweed vine 260, 262
mosquito(es), Culicidae 79, 86, 120, 160, 232,
    249–250
    Asian tiger, *Aedes albopictus* 86, 337
    forest day, *see* mosquito, Asian tiger
    yellow-fever, *Aedes aegypti* 86
mosquitofish 249–250
    Gambusia holbrooki 120
moth(s) 93
    banana, *Opogona sacchari* 95
    borer, sugarcane, *Diatraea saccharalis* 87
    cactus, *Cactoblastis cactorum* 5, 77, 94–95,
        218, 247–248, 255–257, 362, 364
    clearwing, Sesiidae 79
    clothes, Tineidae 95
    diamondback, *Plutella xylostella* 95
    ermine, Yponomeutidae 95
    European gypsy, *Lymantria dispar* 7–8, 10,
        94, 247, 253, 333–334, 362
    gelechiid, Gelechiidae 93–94
    leaf-roller, Tortricidae 79
    Parapoynx diminutalis 102
    pyralid, Pyralidae (*see also* individual taxa)
        94–95, 102
    snow-white linden, *Ennomos subsignarius* 153

tussock, Lymantriidae (*see also* individual taxa)
    94
water-hyacinth, *Sameodes albiguttalis* 260
water-lettuce, *Spodoptera pectinicornis* 261
winter, *Operophtera brumata* 249
mouse (mice) (*see also* individual taxa) 316
    Anastasia beach, *Peromyscus polionotus
        phasma* 293
    beach, *Peromyscus polionotus* 179–180, 182,
        293, 312
    cotton, *Peromyscus gossypinus allapatocola* 271
    house, *Mus musculus* 159, 168, 180, 182–183,
        226
    Key Largo cotton, *Peromyscus polionotus
        telmaphilus* 180
    Perdido Key beach, *Peromyscus polionotus tris-
        syllepsis* 179
Muhlenbergia sp., muhly grass 48, 280–281
Murray, Lake, California 227
Murraya paniculata, orange jasmine 28
Mus musculus, house mouse 159, 168, 180,
    182–183, 226
Musca domestica, house fly 87
Muscidae, muscid flies (*see also* individual taxa)
    87
muskrat, North American, *Ondatra zibethica* 7
mussel(s) 101–102, 104–105, 191
    brown, *Perna perna* 200
    Charru, *Mytella charruana* 191, 201
    zebra, *Dreissena polymorpha* 102–106, 336,
        362
Mustela 10
    frenata, long-tailed weasel 180
    putorius, ferret 158
Myakka River State Park 292, 310
Mycoleptodiscus terrestris 261
Mycteria americana, wood stork 270
Myiopsitta monachus, see parakeet, monk
Mylopharyngodon piceus, snail or black carp 300
myna(s) 316
    common, *Acridotheres tristis* 140–141, 147,
        153–154
    hill, *Gracula religiosa* 141, 154
Myocastor coypus, nutria 158–159, 168–169,
    177–179, 217–218, 225, 314
Myrica cerifera, wax myrtle 71
Myriophyllum
    aquaticum, Eurasian water milfoil 227
    spicatum, Eurasian water milfoil 28
Myrtaceae (*see also* individual taxa) 32, 251–252
myrtle
    downy, *Rhodomyrtus tomentosus* 25, 310
    wax, *Myrica cerifera* 71
Mytella charruana, Charru mussel 191, 201
Myxoma poxvirus 253, 257
Myzus persicae, green peach aphid 83

Nandayus nenday, black-hooded parakeet 141,
    154
Naples, Florida 318, 320
Nassau County 152, 175

Nasua narica, white-nosed coati 159, 170
National (*see also* U.S.)
National Academy of Sciences of the United
   States 263
National Audubon Society 210
National Biological Service 118, 336
National Fisheries Research Centers 336
National Guard 225
National Key Deer National Wildlife Refuge 168,
   270–272
National Marine Fisheries Service 336
National Oceanic and Atmospheric
   Administration 336
National Park Service 117–118, 209, 276–277,
   342, 336
National Plant Germplasm System 334
National Wildlife Refuges (*see also* individual
   refuges) 267–273
Natural Areas Management Division 215
natural enemies (*see also* biological control)
   245–263
Nature Conservancy, The 211–212, 222, 353,
   355, 359
naupaka, beach, *see* scaevola
neem 83
nematodes 259
nemerteans 188
Neochetina
   bruchi, water-hyacinth weevil 260
   eichhorniae, water-hyacinth weevil 260
Neocurtilla hexadactyla, northern mole cricket 92
Neodusmetia sangwani 90
Neohydronomus affinis 260
Neoseps reynoldsi, sand skink 179
Neotoma floridana smalli, Key Largo wood rat
   180, 182, 271, 293
Nephrolepis multiflora, Asian sword fern 28
Netherlands 154
Netta peposaca, rosy-billed pochard 306
Neuroptera, lacewings, ant lions 79
Nevis 94
new association hypothesis 254
New Mexico 127
New York 5–6, 142, 152, 177
New Zealand 10–11, 34, 152, 154, 197, 254
Newcastle disease 146
Neyraudia reynaudiana, Burma reed, cane grass
   25, 44, 56
Nezara viridula, southern green stink bug 88
Nitidulidae, sap beetles 79
Nocomis leptocephalus bellicus, southern bluehead
   chub 111
Nomadacris septemfasciata, locust 154
Nonindigenous Aquatic Nuisance Prevention and
   Control Act 30
North Carolina 149, 363
North Miami, Florida 154
Northwest district of Everglades National Park
   277
Northwest Cape, Florida 276, 281
Notropis baileyi, rough shiner 111

Nova Scotia 11
noxious weed, definition of 341
nudibranchs 191
Numida meleagris, helmeted guinea fowl 306
nutria, *Myocastor coypus* 158–159, 168–169,
   177–179, 217–218, 225, 314
nutrients 40, 45–46, 55, 365

oak, red, *Quercus rubra* 13
oak wilt disease, *Ceratocystis fagacearum* 13
oats, sea, *Uniola paniculata* 212
Ocala, Florida 318
Ocala National Forest 162
ocelot, *Felis pardalis* 158, 185, 316
Ochlockonee River 110
Ochrosia
   elliptica, kopsia 28
   parviflora, kopsia 28
Odocoileus virginianus, white-tailed deer 182,
   271, 306, 310
Odonata, dragonflies, damselflies (*see also*
   individual taxa) 79, 95
Oeceoclades maculata, ground orchid 28
Oestridae, bot and warble flies 87
Oestrus ovis, sheep botfly 87
Office of Technology Assessment 329
Ohio 153
Okaloosa County 141, 169, 176
Okeechobee County 170, 214–215
Okeechobee, Florida 170
Okeechobee, Lake 110, 230
   aerial survey of vegetation near 40, 320–321
   nonindigenous animals of 105–106, 117,
      214–215, 306
   nonindigenous plants of 42, 48, 50, 236,
      240–243, 320, 351, 360
Okeefenokee National Wildlife Refuge 268
Okenia hypogaea, burrowing four-o'clock 293
Oklahoma 303
Oklawaha River 161–162
Old Rhodes Key 167
Oleaceae (*see also* individual taxa) 144
Oligochaeta, segmented worms (*see also*
   individual taxa) 101–102, 188
Oman, Gulf of 116
Ondatra zibethica, North American muskrat 7
Onthophagus, dung beetle 84
Operation Clean Sweep 233
Operophtera brumata, winter moth 249
Ophiosoma ulmi, Dutch elm disease fungus 252,
   364
Opogona sacchari, banana moth 95
Opuntia 94, 247
   corallicola 94, 256
   inermis 255–256
   spinosissima 218
   stricta 94, 255–256
   vulgaris 247
Orange Bowl football stadium 172
Orange County 86, 141, 147, 150, 172
Orange Lake 54, 162, 230, 239

orangutans 316
orchid
    ground, *Oeceoclades maculata* 28
    butterfly, *Encyclia tampensis* 293
orchid tree, *Bauhinia variegata* 26
Oregon 149, 153, 189
Oreochromis
    aureus, blue tilapia, *see* tilapia, blue
    mossambicus, Mozambique tilapia 111, 115,
        300, 308
    niloticus, Nile tilapia 111, 116, 118, 300
oriole, spot-breasted, *Icterus pectoralis* 141, 143,
    152
Orlando 225, 302, 318, 320–321, 331
Ormia depleta 87
Orthoptera, grasshoppers, crickets (*see also* indi-
    vidual taxa) 79, 95–96, 98
Oryctolagus cuniculus, European rabbit 159,
    165–166, 181, 226, 253, 257
oryx, *Oryx gazella* 305
Oryx gazella, oryx 305
Oryza rufipogon, red rice, Asian common wild rice
    28, 226, 352
Oryzaephilus
    acuminatus 83
    mercator, merchant grain beetle 83
    surinamensis, saw-toothed grain beetle 83
Oryzomys argentatus, silver rice rat 178, 182
oscar, *Astronotus ocellatus* 111, 114, 270, 272
Oscar Scherer State Recreation Area 45
Osceola County 50, 173–174, 214–215, 309
oskars 67
Osteoglossidae, bony tongue fishes (*see also* indi-
    vidual taxa) 300
Osteoglossum bicirrhosum 300
Osteopilus septentrionalis, Cuban tree frog 44,
    124, 131–132, 134–135, 270
Ostracoda 101
ostriches 316
Ovatella myosotis 189, 191
Overseas Highway 222
Overstreet, Lake 292
owl
    barn, *Tyto alba* 180
    barried, *Strix varia* 135
oxygen, dissolved 46, 50
Oxyops vitosa 261
oyster, black-lipped pearl, *Pinctada margaritifera*
    189
oyster plant, *Rhoeo spathacea* 29

pacu, *Colossoma* spp. 112, 299
Padda oryzivora, Java sparrow 141, 146
Paederia
    cruddasiana, sewer vine 49, 51, 53–54, 67,
        215–216, 323
    foetida, skunk vine 25, 49, 51, 65, 216, 263,
        292
Pakistan 91, 162
palm(s) 179
    buccaneer, *Pseudophoenix sargentii* 293

thatch, *Thrinax* 179, 285
thatch, *Thrinax morrisii* 179
thatch, *Thrinax radiata* 179
Palm Beach County 268, 311, 330, 336
    invertebrates in 84, 107
    plants in 228, 291
    vertebrates in 114–115, 117–119, 136, 141,
        147, 152, 154, 166, 170, 176
Palmdale, Florida 170
palmetto, saw, *Sabal palmetto* 212
Palominitos 6
Pan troglodytes, chimpanzee 302, 316
Panama City, Florida 167
Panicum
    abscisum, cutthroat grass 217
    repens, torpedo grass 25, 41, 50, 56, 269, 290,
        322–323, 341, 351
panleucopenia virus, feline 181
panther(s) 316
    Florida, *Felis concolor coryi* 180–181,
        271–272, 293, 309, 311
Panthera
    leo, lion 302
    tigris, tiger 302
papaya, *Carica papaya* 64–65, 67–68, 72–73
Papilionoidea, butterflies (*see also* individual taxa)
    79, 94
Papua New Guinea 248
Parachute Key 132
Paradise Key Hammock 66, 70, 285
paradisefish, *Macropodus opercularis* 113
paragonimiasis 107
parakeet(s)
    black-hooded, *Nandayus nenday* 141, 154
    blue-crowned, *Aratinga acuticaudata* 141,
        154
    canary-winged, *Brotogeris versicolurus* 141,
        143, 151
    common pet, *see* budgerigar
    dusky-headed, *Aratinga weddellii* 141, 154
    monk, *Myiopsitta monachus* 7, 141, 143,
        145–146, 150–151, 228, 313, 362
    "Quaker," *see* parakeet, monk
    red-masked, *Aratinga erythrogenys* 141, 154
    rose-ringed, *Psittacula krameri* 141, 154
    shell 316
parallel-evolution model 134
Parapoynx diminutalis, an Asian pyralid moth
    102
Parapristina verticillata 91
parasite(s) 10, 13, 57, 135–136, 180–181,
    183–184, 299
    definition of 245
parasitoid(s) 10, 12, 91–93, 82, 85, 87–88, 99,
    249
    definition of 246
Paratrechina longicornis, crazy ant 92
Parelaphostrongylus tenuis 178
Parlatoria ziziphi, black parlatoria scale 89, 97
Paroaria coronata, red-crested cardinal 141, 154
parrot(s) 146, 155–156, 316

orange-winged, *Amazona amazonica* 141, 154
red-crowned, *Amazona viridigenalis* 141
Pasco county 41, 49, 141, 169, 291, 291
Paspalum
  notatum, Bahia grass 28, 96, 214, 269, 310
  urvillei, vassey grass 290
Passage Key National Wildlife Refuge 268
Passandridae, flat bark beetles 79, 83
Passer domesticus, house sparrow 5–6, 141
Passerina cyanea, indigo bunting 147
Passiflora suberosa, passionflower vine 70
passionflower vine, *Passiflora suberosa* 70
pastureland 171–172, 178, 182, 336
Pavo cristatus, common peafowl 141, 154
pavon, speckled, *Cichla temensis* 113–114, 118,
  120, 307
Payne's Prairie 306
Peace River 184, 213
peacock, speckled, *see* cichlid, peacock, and
  pavon, speckled
peafowl, common, *Pavo cristatus* 141, 154
Pediculus humanus, human louse 81
Pediculoides ventricosus 12
Pediobius foveolatus 82
Pegoscapus jimenezi 70
Pelecypoda, bivalves 101
Pelican Island National Wildlife Refuge 268
Penaeus
  monodon, tiger shrimp 199
  vannamei, white shrimp 199
Pennisetum purpureum, napier grass 29
Pensacola, Florida 152, 318
Pentatomidae, stink bugs 88
pepper, red 252
perch(es)
  climbing, *Anabas testudineus* 113
  nile, *Lates* 300
  Percidae 113, 120
Percidae, perches 113, 120
periwinkle, European, *Littorina littorea* 11
Perkinsiella saccharicida, sugarcane delphacid
  89, 247
Perky Key 165
Perna perna, brown mussel 200
Peromyscus
  gossypinus allapatocola, cotton mouse 271
  polionotus, beach mouse 179–180, 182, 293,
    312
  polionotus phasma, Anastasia beach mouse 293
  polionotus telmaphilus, Key Largo cotton
    mouse 180
  polionotus trissyllepsis, Perdido Key beach
    mouse 179
Persea borbonia, red bay 70
Peru 120
pesticides 221, 223–224
pet industry
  introduction of birds by 139, 142–144, 149,
    155–156
  introduction of other species by 123, 136–137,
    298,–299, 301–304, 356

regulation of 184, 298, 335, 342
Petromyzonidae, lampreys 300
petunia 252
Phaeophyta, brown algae 197
Phasianus colchicus, ring-necked pheasant 144,
  305–306
pheasant, ring-necked, *Phasianus colchicus* 144,
  305–306
Pheidole megacephala, big-headed ant 10
Philephedra tuberculosa 89
Philotrypesis emeryi 93
Phoenicopterus ruber, greater flamingo 147–148
phonotaxis 87
phosphate mining 45, 168, 212–213
Phractocephalus hemioliopterus, red-tail catfish
  112
Phrynosoma cornutum 124–126, 128, 132
Phyllocnistis citrella, citrus leaf miner 93, 97
Phyllocoptruta oleivora, citrus rust mite 80
Phytobius 261
Phytopthora palmivora 262
Piaractus brachypomus, pirapatinga 112
Picayune Strand State Forest 291
Picoides borealis, red-cockaded woodpecker 147,
  291
Pieris 247
  rapae, imported cabbageworm 94
pig(s), *Sus scrofa*
  damage caused by 11, 176, 178, 182, 217,
    271–272, 292, 362
  effects of on other species 10, 13, 17, 59, 156,
    183, 178–180, 181–183
  introduction and status of in Florida 159, 168,
    175–176, 270
  management of 185, 226, 295, 310–311, 314,
    324
  value of 184
pigeon, *see* dove, rock
Pilsbry Hammock 65
Pimelodidae, long-whiskered catfishes 112
Pimephales promelas, fathead minnow 112
Pinctada margaritifera, black-lipped pearl oyster
  189, 191, 201
pine(s) 52
  longleaf, *Pinus palustris* 45, 217, 311
  slash, *Pinus elliottii* var. *densa* 44, 174, 278,
    280–281
pineapple, *Ananas* 84
Pine Island 277
Pine Island National Wildlife Refuge 268
pine rockland 44, 360
pinelands 45, 72, 143, 147, 275, 281
Pinellas County 141–142, 150, 168, 172, 268
Pinellas National Wildlife Refuge 268
Pinus 32–33
  elliottii var. *densa,* slash pine 44, 174, 278,
    280–281
  palustris, longleaf pine 45, 217, 311
pirambeba, *Serrasalmus humeralis* 113, 226,
  311–312
pirambebas, Serrasalminae 112, 299–300

Piranga rubra, summer tanager 147
piranha(s), Serrasalminae 7, 112, 299–300
    red, *Pygocentrus nattereri* 112
    red-bellied, *Serrasalmus nattereri* 301
    redeye, *Serrasalmus rhombeus* 112
pirapatinga, *Piaractus brachypomus* 112
Pistia stratiotes, see water lettuce
Pitanga sulphuratus, great kiskadee 10
Pittman-Robertson Act 121–122
Pittosporum 69
    pentandrum, pittosporum 29, 68
pittosporum, *Pittosporum pentandrum* 29, 68
planning councils, regional 348
Planococcus citri, citrus mealybug 83
plant hoppers, Fulgoroidea (Homoptera) 79, 89
Plantago lanceolata, plantain 256
Plantation Key 150
plants
    introduction and spread of 23–37, 74
    management of 252, 247, 268–271, 317–323
    number of species 76, 263, 361
plantain, *Plantago lanceolata* 256
Plasmodium relictum capistranoae, avian malaria
    parasite 9–10, 13, 146
Platychirograpsus 201
    spectabilis, saber crab 192
    typicus 192
Platydoras costatus, Raphael catfish 112
platyfish
    southern, *Xiphophorus maculatus* 111
    variable, *Xiphophorus variatus* 111
Platygastridae, parasitoid wasps (*see also*
    individual taxa) 91
Platyhelminthes 190
Platypodidae, ambrosia beetles 79, 84
Plecia nearctica, love bug 85, 97, 184
Plutella xylostella, diamondback moth 95
Poaceae, grasses (*see also* individual taxa) 32
pochard, rosy-billed, *Netta peposaca* 306
Podophyllum peltatum, mayapple 292
Poecilia
    hybrids 112
    reticulata, guppy 111
Poeciliidae, livebearers 111–112
Polk County
    animals in 161, 168, 171–172, 176, 184, 217,
    308
    plants in 213, 232, 238–239
Pollution Recovery Trust Fund 349
Polychaeta 188–189, 191
polyphagy 245, 253
Polysiphonia breviarticulata 197, 199
Pomacanthus asfur, Arabian angelfish 116
Pomacanthus xanthometopon, yellow-mask
    angelfish 116
Pomacea
    bridgesi, spike-topped apple snail 103–104,
    106–107, 270
    paludosa, Florida apple snail 106–107
Pomoxis
    annularis, white crappie 112

    nigromaculatus, black crappie 46, 105
Pomponatius typicus 262
popcorn tree, *see* tallow tree, Chinese
poplar trees 247
population growth, human 123, 137
Port St. Lucie, Florida 150
Posidonia oceanica 198
Post, Buckley, Schuh & Jernigan, Inc. (PBS&J)
    211
Potamotrygonidae, freshwater stingrays 300
potato, *Solanum tuberosum* 9, 95, 252
pothos, *see Epipremnum*
poultry farming 85
Pouteria campechiana, canistel 29
poxvirus, *Myxoma* 253, 257
prairie chicken, greater, *Tympanuchus cupido* 305
prairie dog(s) 316
    black-tailed, *Cynomys ludovicianus* 159, 167
Praon palitans 249
predation 134–135, 179–180, 189, 304
Preservation 2000 lands 322, 324
primates (*see also* individual taxa) 158–159,
    161–164, 177, 183, 185, 302–303, 314
Prionitis 197
privet, *Ligustrum sinense* 28
Procyon lotor, raccoon 182
Progne subis, purple martin 147
pronghorn, *Antilocapra americana* 159, 174,
    305–306, 315
Prosopis spp. 33
Pseudaulacaspis
    cockerelli, false oleander scale 89, 97
    pentagona, white peach scale 89, 97
Pseudectroma europaea 90
Pseudemys
    floridana 136
    nelsoni 136
Pseudococcidae, mealybugs 90
Pseudodoras niger, ripsaw catfish 112
Pseudohomalopoda prima 10
Pseudophoenix sargentii, buccaneer palm
    293
pseudorabies 181, 183, 272, 292
Psidium, see guava(s)
psittacosis 149
Psittacula krameri, rose-ringed parakeet 141, 154
Psychotria nervosa, wild coffee 70
Psylloborini, ladybird beetles 82
Pterodoras granulosus, granulated catfish 112
Pterois volitans, lionfish 112, 115
Pteromalidae 91
Pterophyllum scalare, freshwater angelfish 113
Public Health Service 337, 341
Puccinia xanthii 249
Pueraria
    lobata, *see Pueraria montana*
    montana, kudzu 22, 25, 46, 53–55, 290, 323,
    333–334
Puerto Rico 83, 92, 94, 119, 133, 151, 190
Pulex irritans, human flea 96
Puntius conchonius, rosy barb 112

Puntius gelius, dwarf barb 112
Puntius tetrazona, tiger barb 112
purplewing, Florida, *Eunica tatila* 78
Putnam County 168
Pycnonotidae, bulbuls 15, 141, 151–152
Pycnonotus jocosus, red-whiskered bulbul 7, 141–144, 146, 151–152,
Pygocentrus nattereri, red piranha 112
Pylodictus olivaris, flathead catfish 111, 115–116, 118, 121
Pyralidae, moths (*see also* individual taxa) 94–95, 102
pythons 7, 304

quail, button 316
Queensland, Australia 224
quelea, *Quelea quelea* 145
Quelea quelea, quelea 145
Quercus rubra, red oak 13
Quisculus quiscula, common grackle 145

Rabbit Key 166
rabbit(s) 254, 316
  cottontail 181
  European, *Oryctolagus cuniculus* 159, 165–166, 181, 226, 253, 257
  marsh, *Sylvilagus palustris hefneri* 181
rabies 180–181, 183
raccoon, *Procyon lotor* 182
Raccoon Key 163–164, 178
racer, black, *Coluber constrictor* 135
radioactive fallout 184
Ramphotyphlops braminus, Braminy blind snake 124, 131–132, 137
Ramrod Key 165
Rana
  aesopus, gopher frog 291
  sphenocephala 135
ranchland 173
Rankin Key 276, 285
rat(s) 10, 247, 254, 316, 183
  black, *Rattus rattus* 158–159, 168, 179–180, 182–183, 226, 270
  Key Largo wood, *Neotoma floridana smalli* 180, 182, 271, 293
  Norway, *Rattus norvegicus* 158–159, 167–168, 183, 226, 270–271
  silver rice, *Oryzomys argentatus* 178, 182
ratites 316
rattlesnake, Carolina pygmy, *Sistrurus miliaris* 303
Rattus
  norvegicus, Norway rat 158–159, 167–168, 183, 226, 270–271
  rattus, black rat 158–159, 168, 179–180, 182–183, 226, 270
reclamation 212–213
red bay, *Persea borbonia* 70
Red Sea 116
Redd Hammock 65
Remirea maritima, beach star 293

reptile(s) (*see also* individual taxa) 44, 123–138, 263, 272, 316, 360–362
reptile trade 298, 302–303
resistance 257
restoration, ecological 67, 205–219, 288, 294, 335, 352, 365
Rhadinaea 126
Rhagoletis pomonella, apple maggot fly 256
Rhamnaceae (*see also* individual taxa) 24, 212
Rhinecanthus verrucosus, blackpatch triggerfish 116
Rhineura 126
rhinoceroses, Rhinocerotidae 302, 316
Rhizobius lophanthae 82
Rhizophora mangle, red mangrove 11, 162, 193–195, 363
Rhodomyrtus tomentosus, downy myrtle 25, 310
Rhodophyta, red algae (*see also* individual taxa) 197
Rhoeo
  discolor, oyster plant 29
  spathacea, oyster plant 29
rice
  Asian common wild, *Oryza rufipogon* 28, 226, 352
  red, *Oryza rufipogon* 28, 226, 352
Ricinus communis, castor bean 65, 213
rights-of-way, highway 287, 290–291, 350–351
Rivers and Harbors Act 230, 232
robin, American, *Turdus migratorius* 43, 147, 281
rock plowing 207–209, 281, 284
Rodentia, rodents (*see also* individual taxa) 96, 145, 159–160, 166–170, 179, 182, 226, 247, 293
Rodman Dam 163
Rodolia cardinalis, vedalia beetle 83, 90, 247
rosary pea, *Abrus precatorius* 24, 68
rose apple, *Syzygium jambos* 29
Rostrhamus sociabilis, Everglade kite 104, 107, 270
rotenone 267, 312
Rousseau, Lake 230, 234
Royal Palm Hammock 132, 285
Royal Palm Pond 277
rubber tree, Indian, *Ficus elastica* 27
rubber vine, Palay, *Cryptostegia grandiflora* 27
Rubus 69
runoff, agricultural 214, 365
Russia 93, 247
Russian thistle, *Salsola kali* 269
Rynchops nigra, black skimmer 179

Sabal palmetto, saw palmetto 212
Saffir/Simpson Hurricane Scale 64
Sagittaria 53
Saimiri sciureus, squirrel monkey 159, 161
St. George Island, Florida 176
St. Johns County 169
St. Johns National Wildlife Refuge 268
St. Johns River 164–165, 174, 230–233

St. John's-wort, common, *Hypericum perforatum* 248, 257
St. Lucie County 150, 196
St. Lucie River 58
St. Marks National Wildlife Refuge 267–268, 272
St. Marks River 53, 58
St. Petersburg, Florida 141, 150–151, 154, 318
St. Vincent Island 174
St. Vincent National Wildlife Refuge 268, 271–272
sakis, bearded 316
salamanders 178, 302–303
Salminus, dorados 300
Salsola kali, Russian thistle 269
salt cedar, *see* tamarisk
saltern habitat 210
Salvinia
   molesta, karibaweed, giant water fern 248
   rotundifolia, water spangles 269
Sameodes albiguttalis, water-hyacinth moth 260
San Juan, Puerto Rico 151
Sand Key 167
sandalwood, red, *Adenanthera pavonina* 26, 67–68
sandhill community 45, 216, 360
Sanibel Island ix, 136, 165, 180, 222, 230, 240
Sansevieria
   hyacinthoides, bowstring hemp 29
   trifasciata, bowstring hemp 29
Santa Fe River 169–170
Santa Rosa County 147, 169, 172
Sapium sebiferum, see tallow tree, Chinese
sapodilla, *Manilkara zapota* 28
Sarasota, Florida 151–152, 318
Sarasota County 45, 151, 154, 292, 310
Sargassum muticum 196–197, 197, 199
Sarotherodon melanotheron, blackchin tilapia 111, 115
sauger, *Stizostedion canadense* 113
Savannah, Georgia 152
Savannah River 175
Save Our Rivers lands 320, 322, 324
saw grass, *Cladium jamaicense* 48, 53–55, 191, 278–281
sawfly (sawflies) 261
   Brazilian pepper, *Heteroperryia hubrichi* 262, 322
Scaevola
   frutescens, see Scaevola taccada var. sericea
   plumieri, inkberry 293
   sericea, see Scaevola taccada var. sericea
   taccada var. *sericea,* scaevola, half flower 25, 269
scale(s)
   black parlatoria, *Parlatoria ziziphi* 89, 97
   brown soft, *Coccus capparidis* 82, 89
   brown soft, *Coccus hesperidum* 82
   California red, *Aonidiella aurantii* 82, 89
   citrus snow, *Unaspis citri* 89
   coconut, *Aspidiotus destructor* 83, 89
   cottony-cushion, *Icerya purchasi* 83, 90, 247
   euonymus, *Unaspis euonymi* 89

false oleander, *Pseudaulacaspis cockerelli* 89, 97
Florida red, *Chrysomphalus aonidum* 89
   margarodid, Margarodidae 90
   purple, *Lepidosaphes beckii* 89
   soft, Coccidae (Homoptera) 79
   soft, *Philephedra tuberculosa* 89
   tea, *Fiorinia theae* 89
   white peach, *Pseudaulacaspis pentagona* 89, 97
Scapteriscus
   abbreviatus, short-winged mole cricket 95
   borellii, southern mole cricket 87, 95
   vicinus, tawny mole cricket 87, 95
Scarabaeidae, dung beetles 79, 84, 291
Sceloporus woodi, scrub lizard 179
schefflera, *Schefflera actinophylla* 25
Schefflera actinophylla, schefflera 25
Schieffelin, Eugene 152
Schinus terebinthifolius, see Brazilian pepper
Sciurus aureogaster, Mexican gray squirrel 158–159, 166–167, 179
Scolytidae, bark beetles 79, 84
Scolytus multistriatus, European elm bark beetle 252
Scorpaenidae, scorpionfishes 112
scorpionfishes, Scorpaenidae 112
Scotland 154
screwworm, *see* fly, screwworm
scrub habitat 172–174, 179, 360
sea anemones (*see also* individual taxa) 189
sea grasses 197–198
sea lion, California 158
sea squirts 192
sea urchins 198
Sebring, Florida 58
sediment loading 50
seed bank 65, 67, 73, 213, 240, 242, 347
Selenothrips rubrocinctus, redbanded thrips 96
Seminole County 86
Seminole State Forest 292
Serranidae, sea basses 112
Serrasalminae, piranhas and pirambebas 112, 299–300
Serrasalmus
   humeralis, pirambeba 113, 226, 311–312
   nattereri, red-bellied piranha 301
   rhombeus, redeye piranha 112
servals 316
sesban, purple, *Sesbania punicea* 269
Sesbania 33
   punicea, purple sesban 269
   vesicaria, bladderpod 269
Sesiidae, clearwing moths 79
sewer vine, *Paederia cruddasiana* 49, 51, 53–54, 67, 215–216, 323
shade tolerance 67–68, 73
shads, Dorosoma 308
Shark Valley, Everglades 132
sharkminnow, black, Labeo chrysophekadion 112
sheep 86–87, 183
   Barbary, *Ammotragus levuia* 305

shiner(s)
 rough, *Notropis baileyi* 111
 southern striped, *Luxilus chrysocephalus isolepis* 111
shipworms (*see also* individual taxa) 187–188, 190–192, 195–196
shore hoppers, talitrid amphipods 188
shrew(s) 178–179, 316
 Shermans' short-tailed, *Blarina carolinensis shermani* 179
shrimp(s) 191
 tiger, *Penaeus monodon* 199
 white, *Penaeus vannamei* 199
Sialia sialis, eastern bluebird 10, 147
siamangs 316
silkworms 76
 oriental, *Bombyx mori* 98
Silvanidae, flat bark beetles 79, 83
silver dollar, *Metynnis* sp. 112
Silver River 162–163, 302
Silver River State Park 294
Silver Springs, Florida 161, 183–185
silverside, brook, *Labidesthes sicculus* 117
Singapore 153
Sipha flava, yellow sugarcane aphid 83
Siphonaptera, fleas (*see also* individual taxa) 96
Siphonaria pectinata 189, 191
sisal hemp, *Agave sisalana* 26
sissoo, *Dalbergia sissoo* 27
Sistrurus miliaris, Carolina pygmy rattlesnake 303
skimmer, black, *Rynchops nigra* 179
skink, sand, *Neoseps reynoldsi* 179
skunk vine, *Paederia foetida* 25, 49, 51, 65, 216, 263, 292
Smilax, greenbrier 71
snail(s), Gastropoda 101, 103, 189, 191, 253–254
 "albino mystery" 106
 Florida apple, *Pomacea paludosa* 106–107
 Florida tree, *Liguus fasciatus* 179
 giant African, *Achatina fulica* 223–225, 253–254, 366
 spike-topped apple, *Pomacea bridgesi* 103–104, 106–107, 270
snake(s) 125, 179, 302–303
 Braminy blind, *Ramphotyphlops braminus* 124, 131–132, 137
 brown tree, *Boiga irregularis* 5, 137, 304, 314, 333, 336, 362
 eastern indigo, *Drymarchon corais couperi* 180, 272, 291
snakeheads, Channidae 300
Snapper Creek Hammock 215
snowberry, *Chiococca alba* 71
soda apple, tropical, *Solanum viarum* 25, 59, 178, 213, 251–252, 292, 310, 330, 336, 351–352
soil 40, 206, 226, 339
Soil Conservation Service 323
Soil Erosion Service, *see* U.S. Natural Resource Conservation Service

Solanceae 251–252
Solanum 33, 69
 erianthum 68, 72
 torvum, turkey berry 29
 tuberosum, potato 9, 95, 252
 viarum see soda apple, tropical
Solenopsis invicta, see ant, red imported fire
Solenoptes capillatus, little blue cattle louse 81
Sonchus
 asper, prickly sow thistle 269
 oleraceus, common sow thistle 269
Sorghum halepense, Johnson grass 290, 333
South Africa 23, 34, 64, 69, 152, 154
South Carolina 175, 194
South Miami, Florida 150–151
soybeans 227
spangles, water, *Salvinia rotundifolia* 269
sparrow(s)
 Bachman's, *Aimophila aestivalis* 147
 house 5–6, 141, 153–154
 Java, *Padda oryzivora* 141, 146
Spartina
 alterniflora, North American cord grass 12
 anglica, hybrid cord grass 12
 maritima, a British cord grass 12
specialist, definition of
species richness 101
Sphaerodactylus (*see also* gecko(s)) 126
Sphaeroma 194, 201
 annandalei 194
 destructor, see Sphaeroma terebrans
 peruvianum 195
 quoyanum 194
 terebrans 190, 193–195, 198–200, 363
 vastator, see Sphaeroma terebrans
 walkeri 190, 194
Sphecidae, digger wasps 92
spillover, economic 255
spillover, environmental 255–256
spironema, *Callisia fragrans* 26
Spodoptera pectinicornis, water-lettuce moth 261
sponges 188
Sporobolus jacquemontii, smut grass 290
Spring Hill, Florida 170
squirrel(s) 316
 Mexican gray, *Sciurus aureogaster* 158–159, 166–167, 179
 red-bellied, *see* squirrel, Mexican gray
Staphylinidae, rove beetles 79, 85
Starke Ferry Bridge 163
Starlicide 145
starling, European, *Sturnus vulgaris* 141, 143–145, 152, 154, 270
state government (*see also* individual agencies) 339–356
state-owned lands 287–295
stenophagy 246, 253
Stephen Island, New Zealand 10
Sterna albifrons, least tern 179
stingrays, freshwater, Potamotrygonidae 300
Stizostedion canadense, sauger 113
Stizostedion vitreum, walleye 113

stoats 254
Stock Island 160
Stomoxys calcitrans, stable fly 87
stopper
  Spanish, *Eugenia foetida* 71
  white, *Eugenia axillaris* 71
stork, wood, *Mycteria americana* 270
strangler fig, *Ficus   aurea* fig 70–71
Streptopelia
  "*risoria,*" *see* dove, African collared
  decaocto, Eurasian collared dove 141–142,
    147, 149–150
  roseogrisea, ringed turtledove 141–142,
    149–150, 154, 270
Striga asiatica, witchweed 226, 332
Strix varia, barred owl 135
Strobus spp. 32
strontium–90 184
Strymon acis, Bartram's hairstreak 78
sturgeon, Gulf, *Acipenser oxyrynchus desotoi* 116
Sturnus vulgaris, European starling 141,
    143–145, 152, 154, 270
Styela plicata 192
successional age index 67
sugarcane 88–89, 92, 95, 182
Sugarloaf Key 160
Sumter County 291
Sundanella sibogae 192
sunfish(es), Centrarchidae (*see also* individual
    taxa) 111–112, 117
  bluegill 46
  green, *Lepomis cyanellus* 111, 118, 300
  orange-spotted, *Lepomis humilis* 111, 118
  redear, *Lepomis microlophus* 46
Surface Water Improvement and Management
    Program 210
Suriana maritima, bay cedar 293
Surinam cherry, *Eugenia uniflora* 27, 65
Sus scrofa, see pig(s)
Suwannee River 109–110, 230, 233–235, 352
swallow, barn, *Hirundo rustica* 147
swallowtail, Schaus', *Heraclides aristodemus* 78
sweet potato 95
swine fever, African 183
swordtail, green, *Xiphophorus helleri* 111
Sylvilagus palustris hefneri, Keys marsh rabbit
    181
synergism 13–14, 16–17
Syngonium podophyllum, arrowhead vine 29
Syzygium
  cumini, jambolan, Java plum 29, 68
  jambos, rose apple 29

Tabanidae, horse flies 79
Tachinidae, tachinid flies 87
Tadarida brasiliensis, Brazilian free-tailed bat
    160–162
Taenioplana teredini 190
Tallahassee 47, 165, 318
tallow tree, Chinese, *Sapium sebiferum* 25, 45,
    53, 55, 263, 269, 271, 292, 310
tamarindillo, *Acacia choriophylla* 293

tamarisk, *Tamarix ramosissima* 222
Tamarix spp., Eurasian salt cedar 11
  ramosissima, tamarisk 222
tambaqui, *Colossoma macropomum* 112
Tamiami Trail 136, 276
Tampa, Florida 141, 150–151, 166, 168–169,
    306, 318
Tampa Bay 192, 209–211
Tampa Electric Company 210
tanager
  blue-gray, *Thraupis episcopus* 141
  summer, *Piranga rubra* 147
Tanaidacea 191
Tanganyika, Lake 261
Tantilla 126
Tarebia granifera, quilted melania 104, 107
Tarentola americana 137
taro, *Colocasia esculenta* 27, 269
Taxodium distichum, bald cypress 217–218
Taxodium distichum var. *nutans,* pond cypress
    218
taxon-loop model 133
taxonomy 78–79, 245
Taylor County 165
Tectaria incisa, incised halberd fern 29
Tennessee River 105
Tenodera aridifolia, Chinese mantis 75–76, 98
Tephritidae, fruit flies (*see also* individual taxa)
    79, 87–88
Teredo
  bartschi 192, 196
  clappi 192–193
  furcifera 192, 196
Terminalia catappa, tropical almond 29
termite(s), Isoptera 79, 93
  Asian subterranean, *Coptotermes formosanus*
    93
tern, least, *Sterna albifrons* 179
tetra, black, *Gymnocorymbus ternetzi* 112
Texas 89–90, 127, 142, 146, 160, 164, 183, 197,
    224, 259, 261, 309
Thailand 153, 261
Therioaphis
  maculata, spotted alfalfa aphid 83, 89
  trifolii, yellow clover aphid 249
Thespesia populnea, seaside mahoe 29, 269
thistle, common sow, *Sonchus oleraceus* 269
thistle, prickly sow, *Sonchus asper* 269
Thraupis episcopus, blue-gray tanager 141
Three Lakes Wildlife Management Area 309
Thrinax 285
  morrisii, thatch palm 179
  radiata, thatch palm 179
Thrips palmi, palm thrips 96
thrips, Thysanoptera 96
  alligatorweed, *Amynothrips andersoni* 82, 96,
    248, 260
  Brazilian pepper, *Liothrips ichini* 262, 322
  palm, *Thrips palmi* 96
  redbanded, *Selenothrips rubrocinctus* 96
  western flower, *Frankliniella occidentalis* 96
Thurston County, Washington 227

Thysanoptera, *see* thrips
Tiger Creek preserve 176, 217
tiger(s) 316
    Panthera tigris 302
tigerfish(es)
    African, Hydrocyninae 300
    Erythrinidae (*see also* individual taxa) 300
tilapia(s) 300–301
    banded, *Tilapia sparmanni* 113
    blackchin, *Sarotherodon melanotheron* 111, 115
    blue, *Oreochromis aureus* 7, 111, 115–116, 118, 214, 270, 292, 300, 308, 335, 343
    Mozambique, *Oreochromis mossambicus* 111, 115, 300, 308
    Nile, *Oreochromis niloticus* 111, 116, 118, 300
    spotted, *Tilapia mariae* 111, 114, 119, 270, 307, 313
Tilapia sparmanni, banded tilapia 113
Tillandsia 84
    utriculata 84
Tineidae, clothes moths 95
Tingidae, lace bugs 88
Titusville, Florida 161, 163, 318
toad(s)
    cane 255
    giant marine, *Bufo marinus* 124, 131–132, 247, 270, 272, 313
tobacco 252
tomato 252
Tombigbee River 105
topminnow, golden, *Fundulus chrysotus* 117
tortoise
    desert, *Gopherus agassizi* 135–136
    Galápagos, *Geochelone* spp. 304
    gopher, *Gopherus polyphemus* 44, 136, 179, 272, 281, 291, 293
Tortricidae, leaf-roller moths 79
Torymidae (*see also* individual taxa) 93
toucans 316
toxoplasmosis 149
Toxoptera citricida, brown citrus aphid 89
Toxorhynchites 86
Trachemys scripta 124, 126, 128, 126, 131–132, 136–137
Tradescantia 61
    fluminensis 52–53
Traffic (U.S.A.) 302
Trafford, Lake 54
trahira, *Hoplias malabaricus* 113
trahiras, Erythrinidae 300
Trema micranthum, Florida trema 71
trema, Florida, *Trema micranthum* 71
Trialeurodes vaporariorum, greenhouse whitefly 88
trichinosis 181, 183, 292
Trichogaster, *see* gourami(es)
Trichogramma
    minutum 93
    pretiosum 93
Trichogrammatidae, trichogrammatid wasps 93
Tricholimnas sylvestris, flightless woodhen 156

Trichomycteridae, candiru catfishes 299–300
Trichopsis vittatus, croaking gourami 111
Trichosiphonaphis polygoni 89
Trichospilus diatraeae 97
triggerfish, blackpatch, *Rhinecanthus verrucosus* 116
Trinidad 87
Trioxys utilis 249
Triphasia trifoliata, lime berry 29
tristeza virus 89
tropical almond, *Terminalia catappa* 29
Tropical Wonderland 161, 163
Tropidophis canus 137
Tubificidae, segmented worms (*see also* individual taxa) 102
Tupinambis
    nigropunctatus 131
    teguixin 130, 132
turbidity 46
Turdus migratorius, American robin 43, 147, 281
turkey berry, *Solanum torvum* 29
turkey, *Meleagris gallopavo* 182 293
Turks and Caicos Islands 128
turtle(s) 124–125, 302–303,
    bog, *Clemmys muhlenbergi* 303
    green sea, *Chelonia mydas* 211, 278
    leatherback sea, *Dermochlys coriacea* 211
    loggerhead sea, *Caretta caretta* 211, 278
    sea 42, 179, 271–272, 284, 290
turtledove, ringed, *see* dove, African collared
turtlehead, *Chelone glabra* 256
Two Rivers Ranch 171, 173
Tympanuchus cupido, greater prairie chicken 305
Typhlops
    biminiensis 137
    lumbricalis 137
typhus, murine 183
Tyto alba, barn owl 180

U.S. Agricultural Research Service 334
U.S. Air Force 288
U.S. Army Corps of Engineers 230–233, 318–319, 337, 342, 349
U.S. Coast Guard 337
U.S. Department of Agriculture (*see also* agencies within USDA) 22–23 30, 77, 84, 98, 102, 146, 226–227, 241, 320, 329
U.S. Department of Defense 333, 336–337
U.S. Department of Energy Savannah River Plant 175
U.S. Department of Health and Human Services 337
U.S. Department of Justice 301
U.S. Department of the Interior 333, 335–336
U.S. Environmental Protection Agency 221
U.S. Fish and Wildlife Service 98, 145, 155, 218, 232, 267, 301, 306, 330, 335–336, 341
U.S. Forest Service 333–334
U.S. House of Representatives Committee on Resources 139

U.S. Natural Resource Conservation Service 232,
    334–335
U.S. Soil Conservation Service, *see* Natural
    Resource Conservation Service
U.S. Virgin Islands 190
uakaris 316
Unaspis
    citri, citrus snow scale 89
    euonymi, euonymus scale 89
Undaria pinnatifida 197, 199
Uniola paniculata, sea oats 212
Union County 170
Upper Sugarloaf Key 166
Urocyon cinereoargenteus, gray fox 172, 182
Ursidae, bears 180, 182, 302, 316
Ursus americanus, black bear 180
Uruguay 146

vacant niche hypothesis 305
Vallisneria 53
Vancouver, British Columbia 154
vaquita, la, *Diaprepes abbreviatus* 83
Varroa jacobsoni, Asian honeybee mite 80, 91
veliger larvae 102, 105
Venezuela 191, 307
Venice, Florida 154
Venus's-flytrap, *Dionaea muscipula* 363
Vero Beach 168
Vexar 217
Victorella pavida 192
Victoria, Australia 154
Villa 247
vines (*see also* individual taxa) 64, 215–216
Virginia 127
viruses, plant 89
Vitex trifolia, vitex 269
vitex, *Vitex trifolia* 269
Viverra spp., civets 301
Vogtia malloi, alligatorweed stem borer 82, 95,
    260, 248
Volusia County 86, 165, 173, 268
Vulpes vulpes, red fox 159, 171–172, 177,
    179–180, 182–185
vulture
    black, *Coragyps atratus* 180
    turkey, *Cathartes aura* 180

Wacissa River 53
Wakulla County 53, 165, 268
Wakulla River 53
Wales 154
walleye, *Stizostedion vitreum* 113
Wallop-Breaux Act 121–122
Walton County 169
Warm Springs Regional Fisheries Center 336
Washington County 172
Washington state 227
Wasmannia auropunctata, little fire ant 71–72
wasp(s) (*see also* individual taxa) 259,
    agaonid, Agaonidae 16, 90–91, 363–364
    aphelinid, Aphelinidae 79, 91, 99

digger, Sphecidae 92
encyrtid 97
eulophid 97
fig 16, 90–91, 363–364
ichneumon, Ichneumonidae 79
platygastrid, *Amitus hesperidum* 88, 99
torymid, Torymidae 93, 97
trichogrammatid, Trichogrammatidae 93
water hyacinth, *Eichhornia crassipes*
    biological control of 58, 84, 107, 246,
        259–260
    effects of 11, 54, 57, 239, 253, 271
    introduction, extent, and spread of 24, 41, 50,
        58, 213–214, 269, 332–333, 341, 351
    management of 229–237, 242–243, 259–260,
        309, 317–319, 366
water lettuce, *Pistia stratiotes* 25, 84, 213–214,
    236, 260–261, 269, 318–319, 332
water management district(s) 117–118, 317–325,
    344–345, 352
    Northwest Florida 318, 352
    South Florida 40, 214–215, 241, 317,
        319–320, 322–323
    Southwest Florida 210, 317–319, 322–324
    St. Johns River 317–319, 322–323
    Suwannee River 318
Water Management Lands Trust Fund 322
water milfoil, Eurasian, *Myriophyllum* 28, 227
water spinach, *Ipomoea aquatica* 24, 269, 332
water willow, Cooley's, *Justicia cooleyi* 292
watermelons 183
Watersipora subovoidea 192
weasel(s) 254
    long-tailed, *Mustela frenata* 180
weatherfish, oriental, *Misgurnus anguillicaudatus*
    111
Wedelia trilobata, wedelia 29, 212
wedelia, *Wedelia trilobata* 29, 212
weeds 31, 145, 248
weevil(s) 248, 261, 362
    alfalfa, *Hypera posticata* 83–84
    Anthribidae 79
    Apopka, *Diaprepes abbreviatus* 83–84
    boll, *Anthonomus grandis* 83–84, 92, 277
    Brentidae 79
    cane, *Metamasius hemipterus* 83
    Curculionidae (*see also* individual taxa) 79,
        83–84
    Euhrychiopsis lecontei 261
    hydrilla stem, *Bagous hydrillae* 261
    Indian tuber, *Bagous affinis* 261
    stem-feeding, *Eubrychius* 261
    sweet-potato, *Cyclas formicarius* 83–84
    water-hyacinth, *Neochetina bruchi* 260
    water-hyacinth, *Neochetina eichhorniae* 260
Weohyakapka, Lake 230, 238–239
West Palm Beach 152, 318
West Summerland Key 165
Westlake Park 161
wetlands
    artificial 208–209, 283

ecological restoration of 208–210, 213
extent and management of 275, 352, 319–323
mammals in 177–178, 217, 226, 272
plants in 229, 284, 323
whitefly (whiteflies), Aleyrodidae (*see also*
   individual taxa) 80, 88, 89
   citrus, *Dialeurodes citri* 88
   cloudy-winged, *Dialeurodes citrifolii* 88
   greenhouse, *Trialeurodes vaporariorum* 88
   silverleaf, *Bemisia argentifolii* 80, 88
   sweet-potato, *Bemisia tabaci* 80, 88
whitelist approach 364
Wild Bird Conservation Act 139, 155, 301
wildfires, *see* fire regime
Wisconsin 146, 175
Wisteria sinensis, Chinese wisteria 269
wisteria, Chinese, *Wisteria sinensis* 269
witchweed, *Striga asiatica* 226, 332
Withlacoochee State Forest 291–292, 294
wolf (wolves) 181–182
   red, *Canis rufus* 181, 272, 316
   gray 316
wolverines 316
woman's tongue, *Albizia lebbeck* 26
wood rose, *Merremia tuberosa* 28, 65, 68,
   215–216, 323
woodhen, flightless, *Tricholimnas sylvestris* 156
woodpecker, red-cockaded, *Picoides borealis* 147,
   291
World Heritage Sites 275
worms 101–102, 189, 191

Worth, Lake 115
wren, Stephen Island, *Xenica lyalli* 10

Xanthium strumarium, noogoora burr, common
   cocklebur 249
Xanthomonas campestris pv. *citri,* citrus canker
   bacterium 224
Xenarthra (*see also* armadillo) 159, 164–165
Xenica lyalli, endemic Stephen Island wren 10
Xenopsylla cheopis, oriental rat flea 96
Xeriscape practices 350
Xiphophorus
   helleri, green swordtail 111
   maculatus, southern platyfish 111
   variatus, variable platyfish 111

yellow fever 86
Yponomeutidae, ermine moths 95
Yucatan 97
Yucca aloifolia, Spanish bayonet 269
Yulee, Florida 175

Zellwood, Florida 147
Zenaida
   asiatica, white-winged dove 142, 307
   macroura, mourning dove 147
Zeuxo maledivensis 191
Zoobotryon verticillatum 188, 192
zooplankton 46, 57
zoos 298–299
Zophoba sp., giant mealworm 98